endorsed by edexcel

A Level Mathematics for Edexcel

Core

C3

C4

M R Heylings

OXFORD UNIVERSITY PRESS

Great Clarendon Street, Oxford OX2 6DP

Oxford University Press is a department of the University of Oxford.
It furthers the University's objective of excellence in research, scholarship, and education by publishing worldwide in

Oxford New York

Auckland Cape Town Dar es Salaam Hong Kong Karachi
Kuala Lumpur Madrid Melbourne Mexico City Nairobi
New Delhi Shanghai Taipei Toronto

With offices in

Argentina Austria Brazil Chile Czech Republic France Greece
Guatemala Hungary Italy Japan Poland Portugal Singapore
South Korea Switzerland Thailand Turkey Ukraine Vietnam

First published 2009

British Library Cataloguing in Publication Data

Data available

ISBN: 9780-19-911784 0

5 7 9 10 8 6

Printed in Great Britain by Bell and Bain Ltd, Glasgow.

Series Managing Editor Anna Cox

Acknowledgements

The Publisher would like to thank the following for permission to reproduce photographs:
P38 PA Archive/PA Photos; **p74** Daniel Padavona/Shutterstock; **p90** studiovanpascal/iStockphoto; **p120** Image Source/Corbis; **p138** evirgen/iStockphoto; **p158** servifoto/iStockphoto; **p178** Sylvanie Thomas/Shutterstock; **p192** Dijital Film/iStockphoto; **p206** Jenny Horne/Shutterstock; **p212** Toni Räsänen/Dreamstime; **p260** pixonaut/iStockphoto; **p267** Lawrence Freytag/iStockphoto; **p274** Dmitry Nikolaev/Shutterstock; **p296** Hal Bergman/iStockphoto.

The cover photograph is reproduced courtesy of Frans Lemmens/Photodisc/Getty

The publishers would also like to thank Kathleen Austin, Judy Sadler and Charlie Bond for their expert help in compiling this book.

About this book

Endorsed by Edexcel, this book is designed to help you achieve your best possible grade in Edexcel GCE Mathematics Core 3 and Core 4 units. The material is separated into the two units, C3 and C4. You can use the tabs at the edge of the pages for quick reference.

Each chapter starts with a list of objectives and a 'Before you start' section to check that you are fully prepared. Chapters are structured into manageable sections, and there are certain features to look out for within each section:

Key points are highlighted in a blue panel.

Key words are highlighted in **bold blue** type.

Worked examples demonstrate the key skills and techniques you need to develop. These are shown in boxes and include prompts to guide you through the solutions.

EXAMPLE 3

Divide $3x^2 + 5$ by $x^2 + 1$

Rewrite the numerator to involve a multiple of the denominator:

$$\frac{3x^2+5}{x^2+1} = \frac{3(x^2+1)+2}{x^2+1}$$

Divide each term in the numerator:

$$= 3 + \frac{2}{x^2+1}$$

Derivations and additional information are shown in a panel.

Helpful hints are included as blue margin notes and sometimes as blue type within the main text.

Misconceptions are shown in the right margin to help you avoid making common mistakes.

Investigational hints prompt you to explore a concept further.

Each section includes an exercise with progressive questions, starting with basic practice and developing in difficulty.
Some exercises also include

'stretch and challenge' questions marked with a stretch symbol

and investigations to apply your knowledge in a variety of situations.

INVESTIGATION

10 Use computer software or a graphical calculator to check some of your answers to the questions in this exercise. You may have to use abs (absolute value) button to input a modulus.

At the end of each chapter there is a 'Review' section which includes exam style questions as well as past exam paper questions. There are also two 'Revision' sections per unit which contain questions spanning a range of topics to give you plenty of realistic exam practice.

The final page of each chapter gives a summary of the key points, fully cross-referenced to aid revision. Also, a 'Links' feature provides an engaging insight into how the mathematics you are studying is relevant to real life.

At the end of the book you will find full solutions and an index.
The free CD-ROM contains a key word glossary and a list of essential formulae as well as additional study and revision materials and the student book pages.

Contents

C3

1

Algebra and functions

This chapter will show you how to

- combine algebraic fractions
- understand that some mappings are also functions
- find and use composite functions and inverse functions
- use the modulus function and sketch graphs involving it
- transform graphs of functions using translations, reflections, stretches and combinations of these.

Before you start

You should know how to:

1 Combine numerical fractions.

e.g. Calculate $\frac{3}{4}+\frac{1}{6}=\frac{9}{12}+\frac{2}{12}=\frac{11}{12}$

$\frac{3}{4}\times\frac{1}{6}=\frac{3}{24}=\frac{1}{8}$

2 Find important facts to help you sketch graphs of functions.

e.g. The graph of $f(x)=\frac{1}{x-2}$ passes through the point $\left(0,-\frac{1}{2}\right)$, has a vertical asymptote, $x = 2$, and approaches the value 0 as $x \to \pm\infty$

3 Translate, reflect and stretch the graph of $y = f(x)$

e.g. When the graph of $y = x^2$ is translated +2 units parallel to the x-axis, its equation becomes $y = (x-2)^2$

Check in:

1 Calculate

a $\frac{3}{8}+\frac{2}{3}$

b $\frac{3}{8}\times\frac{2}{3}$

2 Sketch the graphs of

a $y = x(x-2)$

b $y = x^2 + x - 2$

c $y = \frac{1}{x}+2$

3 Write the equation of the image of the graph of $y = f(x)$ when

a $f(x) = x^2$ is reflected in the x-axis

b $f(x) = x^2 - 5x$ is reflected in the y-axis

c $f(x) = x^2 + x$ is translated by $\begin{pmatrix}-5\\0\end{pmatrix}$

d $f(x) = x^2$ is stretched with scale factor 4 parallel to the y-axis.

1.1 Combining algebraic fractions

Addition and subtraction

You can add or subtract algebraic fractions when they have a common denominator.

This method is similar to adding or subtracting numerical fractions.

E.g. $\frac{a}{b} + \frac{x}{y} = \frac{ay}{by} + \frac{bx}{by} = \frac{ay + bx}{by}$

by is the common denominator.

You should find and use the lowest common denominator to keep the calculation as simple as possible.

EXAMPLE 1

Evaluate **a** $\frac{3}{8} + \frac{1}{6}$ **b** $\frac{2}{x^2 + x} + \frac{x - 2}{x^2 - 1}$

---

a $\frac{3}{8} + \frac{1}{6} = \frac{9}{24} + \frac{4}{24} = \frac{13}{24}$

The lowest common denominator is 24.

b Factorise the algebraic expressions first:

$$\frac{2}{x^2 + x} + \frac{x - 2}{x^2 - 1} = \frac{2}{x(x + 1)} + \frac{x - 2}{(x + 1)(x - 1)}$$

$$= \frac{2(x - 1)}{x(x + 1)(x - 1)} + \frac{x(x - 2)}{x(x + 1)(x - 1)}$$

The lowest common denominator is $x(x + 1)(x - 1)$.

$$= \frac{\cancel{2x} - 2 + x^2 - \cancel{2x}}{x(x + 1)(x - 1)}$$

$$= \frac{x^2 - 2}{x(x^2 - 1)}$$

Multiplication and division

You do not need to have a common denominator when you multiply or divide algebraic fractions.

The method is similar to multiplying or dividing numerical fractions.

You can only cancel factors which are common to both the numerator and denominator.

Remember when cancelling brackets to cancel the whole bracket and not just part of it.

EXAMPLE 2

Simplify $\frac{x + 2}{x^2 - x} \times \frac{x(x^2 - 1)}{x^2 - x - 6}$

$$\frac{x + 2}{x^2 - x} \times \frac{x(x^2 - 1)}{x^2 - x - 6} = \frac{\cancel{x + 2}}{x(\cancel{x - 1})} \times \frac{\cancel{x}(\cancel{x - 1})(x + 1)}{(\cancel{x + 2})(x - 3)}$$

$$= \frac{x + 1}{x - 3}$$

Factorise as much as you can before cancelling.

Cancel the fraction down to its simplest form.

C3

Dividing by a fraction is equivalent to multiplying by its reciprocal.

EXAMPLE 3

Simplify $\dfrac{x^2 - 2x}{x^2 - 2x - 3} \div \dfrac{x^2 - 4}{2x - 6}$

$$\frac{x^2 - 2x}{x^2 - 2x - 3} \div \frac{x^2 - 4}{2x - 6} = \frac{x^2 - 2x}{x^2 - 2x - 3} \times \frac{2x - 6}{x^2 - 4}$$

Multiply the first fraction by the reciprocal of the divisor.

$$= \frac{x(x-2)}{(x+1)(x-3)} \times \frac{2(x-3)}{(x-2)(x+2)}$$

$$= \frac{2x}{(x+1)(x+2)}$$

Exercise 1.1

1 Simplify

a $\dfrac{6x^2y}{3xyz}$ b $\dfrac{8ab}{c^2} \times \dfrac{3ac}{4b^2}$ c $\dfrac{3p^2q}{2r} \div \dfrac{pq}{4r^2}$

d $\dfrac{x}{x^2 - 1} \times \dfrac{x+1}{x^2}$ e $\left(\dfrac{a}{b}\right)^2 \times \dfrac{b^2}{a^2 - a}$ f $\dfrac{x+3}{x^2 + x - 2} \times \dfrac{x-1}{x^2 + x - 6}$

g $\dfrac{x}{x^2 - 5x + 6} \div \dfrac{x^2 + 2x}{x^2 - 4}$ h $\dfrac{1}{x} \times \dfrac{x^2 - 9}{x^2 - 3x} \div \dfrac{x+3}{x}$

2 Simplify and express as single fractions.

a $\dfrac{x}{y} + \dfrac{y}{z}$ b $\dfrac{a}{b} + \dfrac{b}{a}$ c $\dfrac{1}{ax} - \dfrac{1}{bx}$

d $\dfrac{1}{xy^2} + \dfrac{1}{x^2y}$ e $\dfrac{1}{x+1} + \dfrac{1}{x}$ f $\dfrac{1}{a-1} - \dfrac{1}{a+1}$

g $\dfrac{1}{(a+1)^2} + \dfrac{1}{a+1}$ h $2 + \dfrac{1}{x}$ i $3 - \dfrac{2}{x+1}$

j $\dfrac{4}{y-2} + \dfrac{2}{3}$ k $\dfrac{2}{x-1} + \dfrac{1}{x^2 - 1}$ l $\dfrac{3}{x} + \dfrac{2}{x^2 + x}$

m $\dfrac{1}{z+1} - \dfrac{2}{z^2 - z - 2}$ n $\dfrac{2}{y+2} - \dfrac{y}{y^2 + 3y + 2}$ o $\dfrac{1}{x^2 + 3x + 2} + \dfrac{2}{x^2 + 4x + 3}$

p $\dfrac{3y}{y^2 - 4} + \dfrac{y+1}{y^2 + y - 2}$ q $\dfrac{2z}{z^2 + 2z - 3} - \dfrac{z}{z^2 - 1}$ r $2 + \dfrac{1}{x^2 + 2x} - \dfrac{1}{x^2 - 4}$

3 Simplify

a $\dfrac{1 - x}{1 - \dfrac{1}{x}}$ b $\dfrac{1 + \dfrac{1}{y}}{1 - \dfrac{1}{y^2}}$ c $\dfrac{x^3 + 1}{x^2 - 1}$ d $\dfrac{x^3 - 1}{x^2 + x} \times \dfrac{x+1}{x-1}$

C3

1.2 Algebraic division

You can use long division to divide a polynomial of degree m by a polynomial of degree n, where $m \geqslant n$.

The method is similar to dividing numbers using long division.

The degree of the resulting polynomial is $m - n$.

EXAMPLE 1

Work out $(x^3 - 2x^2 - x + 2) \div (x + 1)$

Set out as a long division:

$$x + 1 \overline{)\, x^3 - 2x^2 - x + 2}$$

Write each polynomial with the highest power of x on the left.

The lead term of $x + 1$ is x.

Divide the lead term x into x^3 and write x^2 in the answer space:

Multiply this x^2 by $x + 1$.

Write $x^3 + x^2$ below and subtract.

$$\begin{array}{r} x^2 \\ x + 1 \overline{)\, x^3 - 2x^2 - x + 2} \\ \underline{x^3 + x^2} \downarrow \\ -3x^2 - x \end{array}$$

Bring down the next term $-x$.

Write like terms in the same column.

Repeat this cycle until the division is complete.

The next step is to divide the lead term x into $-3x^2$ and write $-3x$ in the answer space:

Compare this method with the numerical long division

$$\begin{array}{r} x^2 - 3x + 2 \\ x + 1 \overline{)\, x^3 - 2x^2 - x + 2} \\ \underline{x^3 + x^2} \downarrow \downarrow \\ -3x^2 - x \downarrow \\ \underline{-3x^2 - 3x} \downarrow \\ 2x + 2 \\ \underline{2x + 2} \\ 0 \end{array}$$

Subtract:

$$\begin{array}{r} 243 \\ 23 \overline{)\,5589} \\ \underline{46}\downarrow\downarrow \\ 98\downarrow \\ \underline{92}\downarrow \\ 69 \\ \underline{69} \\ 0 \end{array}$$

giving $5589 \div 23 = 243$

or $5589 = 243 \times 23$

Since the remainder is 0, 23 must be a factor of 5589.

Since the remainder is 0, $x + 1$ is a factor of $x^3 - 2x^2 - x + 2$.

So $(x^3 - 2x^2 - x + 2) \div (x + 1) = x^2 - 3x + 2$

You could also write this result as $x^3 - 2x^2 - x + 2 = (x + 1) \times (x^2 - 3x + 2)$

You can then expand these brackets to check your answer.

C3

Some divisions result in a remainder.

EXAMPLE 2

Divide **a** 4037 by 16

b $4x^3 + 3x + 7$ by $2x - 1$

a

$$\begin{array}{r|l} & 252 \\ \hline 16 & 4037 \\ & \underline{32}\downarrow\downarrow \\ & \ \ 83\downarrow \\ & \ \ \underline{80}\downarrow \\ & \ \ \ \ 37 \\ & \ \ \ \ \underline{32} \\ & \ \ \ \ \ \ 5 \end{array}$$

The remainder is 5.

So $4037 \div 16 = 252$ remainder 5

$= 252\frac{5}{16}$

or $4037 = (252 \times 16) + 5$

b $4x^3 + 3x + 7$ does not have an x^2-term.
Insert a $0x^2$ to fill the place value for x^2:

The $0x^2$ is similar to the 0 in the number 4037, which acts as place holder in the hundreds column.

$$\begin{array}{r|l} & 2x^2 + x + 2 \\ \hline 2x - 1 & 4x^3 + 0x^2 + 3x + 7 \\ & \underline{4x^3 - 2x^2} \quad \downarrow \quad \downarrow \\ & \qquad 2x^2 + 3x \quad \downarrow \\ & \qquad \underline{2x^2 - x} \quad \downarrow \\ & \qquad\qquad 4x + 7 \\ & \qquad\qquad \underline{4x - 2} \\ & \qquad\qquad\qquad 9 \end{array}$$

The remainder is 9.

So $\frac{4x^3 + 3x + 7}{2x - 1} = 2x^2 + x + 2$ remainder 9

quotient ↓ remainder ↓

$= 2x^2 + x + 2 + \frac{9}{2x - 1}$

↑ divisor

You can also write this result as
$4x^3 + 3x + 7$
$= (2x - 1)(2x^2 + x + 2) + 9$

You can expand these brackets to check your answer.

C3

When the highest power of the numerator is the same as the highest power of the denominator you can use a quicker method.

EXAMPLE 3

Divide $3x^2 + 5$ by $x^2 + 1$

Rewrite the numerator to involve a multiple of the denominator:

$$\frac{3x^2+5}{x^2+1} = \frac{3(x^2+1)+2}{x^2+1}$$

Divide each term in the numerator:

$$= 3 + \frac{2}{x^2+1}$$

Both the numerator and denominator have degree 2.

Compare this example to the numerical division

$$\frac{23}{7} = \frac{(3\times 7)+2}{7} = 3\frac{2}{7}$$

Exercise 1.2

1 In each case, divide the first expression by the second expression.

a $x^3 - 7x^2 + 14x - 8;\quad x - 1$

b $x^3 - x^2 - 26x - 24;\quad x + 1$

c $x^3 - 8x^2 + 5x + 50;\quad x + 2$

d $x^3 - 10x^2 + 31x - 30;\quad x - 2$

e $2x^3 + 5x^2 - 4x - 3;\quad 2x + 1$

f $9x^3 + 5x - 2;\quad 3x - 1$

g $n^3 - 7n + 6;\quad n + 3$

h $2n^3 - 9n^2 + 5n + 6;\quad 2n - 3$

i $9y^3 - 16y - 8;\quad 3y + 2$

j $2a^4 - 5a^3 - 10a + 3;\quad a - 3$

k $8a^4 - 20a^3 + 60a - 18;\quad 2a + 3$

l $4z^3 - 17z^2 + 19z - 5;\quad z^2 - 3z + 1$

m $6x^3 + 7x^2 - 23x + 4;\quad 2x^2 + 5x - 1$

n $x^3 + 1;\quad x + 1$

2 Evaluate these divisions, giving each answer as an integer plus an algebraic fraction.

a $\frac{x+6}{x+2}$

b $\frac{x+1}{x+2}$

c $\frac{x-2}{x+2}$

d $\frac{3x+7}{x+2}$

e $\frac{2x+1}{x-3}$

f $\frac{4x+3}{2x+1}$

g $\frac{x^2-1}{x^2+1}$

h $\frac{3x^2+1}{x^2-1}$

i $\frac{2-x}{1+x}$

j $\frac{x^2+x-1}{x^2-2}$

3 In each case, divide the first expression by the second expression.

a $3x^3 - 5x;\quad x^3$

b $3x^2 + 2;\quad x^2 - 3$

c $4x^2 + 1;\quad 2x^2 - 1$

d $14x^2 + 19;\quad 2x^2 + 3$

4 In each case find the quotient and remainder when the first expression is divided by the second expression.

a $x^3 + 5x^2 - 5x + 1$; $x - 1$

b $x^3 + 7x^2 + 8x - 2$; $x + 2$

c $2x^3 - 13x^2 + 17x + 10$; $x - 3$

d $x^3 - 5x^2 + x + 10$; $x - 2$

e $2n^3 + 7n^2 - 6$; $2n + 3$

f $9n^3 - 22n + 6$; $3n - 4$

g $3n^3 - 11n^2 + 2n + 1$; $3n + 1$

h $x^4 + 2x^3 - 6x - 7$; $x^2 - 3$

i $2x^4 + x^3 + x$; $x^2 + 1$

j $x^3 - 2$; $x^2 - 1$

5 Find the functions f(x) which complete these equations.

a $x^3 - 5x^2 + x + 10 = (x - 2) \times \mathrm{f}(x)$

b $x^3 - 7x^2 - 10x - 2 = (x + 1) \times \mathrm{f}(x)$

c $2x^3 - 11x^2 + 15x - 50 = (x - 5) \times \mathrm{f}(x)$

d $x^4 - 3x^3 + 3x - 1 = (x^2 - 1) \times \mathrm{f}(x)$

6 Find the functions and k values which complete these equations.

a $x^3 - 4x^2 + 4x - 1 = (x - 3) \times \mathrm{f}(x) + k$

b $x^3 + 8x^2 + 6x - 1 = (x - 1) \times \mathrm{f}(x) + k$

c $4x^3 - 3x + 2 = (2x - 1) \times \mathrm{f}(x) + k$

d $x^4 + 3x - 4 = (x^2 - x - 1) \times \mathrm{f}(x) + k$

In part **d** k is not a constant.

C3

7 Divide each polynomial by the given factor and hence factorise it completely.

a $x^3 - 4x^2 + x + 6$ has a factor $x - 2$

b $x^3 - 8x^2 + 19x - 12$ has a factor $x - 4$

c $4x^3 - 13x - 6$ has a factor $2x + 3$

d $x^4 - 13x^2 + 36$ has a factor $x^2 - 4$

e $x^3 + 8$ has a factor $x + 2$

f $x^3 - 8$ has a factor $x - 2$

INVESTIGATION

8 $x^2 - 1$ factorises to give $(x - 1)(x + 1)$ and $x^3 - 1$ factorises to give $(x - 1)(x^2 + x + 1)$.

Factorise $x^4 - 1$ and $x^5 - 1$.

Can you make a deduction about $x^6 - 1$?

Make a statement about $x^n - 1$ and see if you can prove it.

1.3 Mappings and functions

Domain and range

A relationship between two sets of numbers is called a mapping.

You can define a mapping as an equation.

E.g. Consider $y = 3x - 1$
Set X maps onto set Y
under the mapping $y = 3x - 1$
You can also write this
as $x \rightarrow 3x - 1$

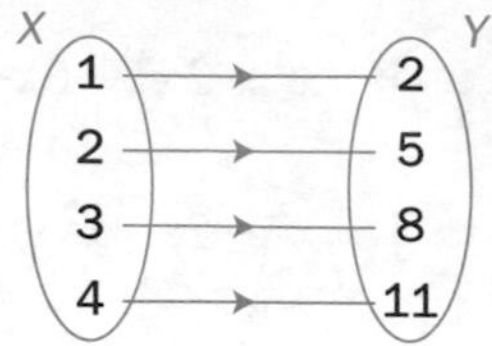

You can also represent a mapping by a set of ordered pairs (x, y) which you can show in a table or as a graph.

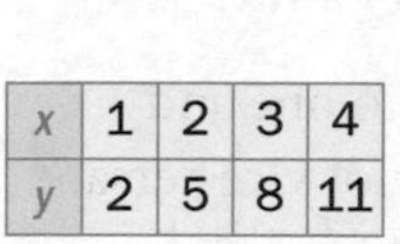

x	1	2	3	4
y	2	5	8	11

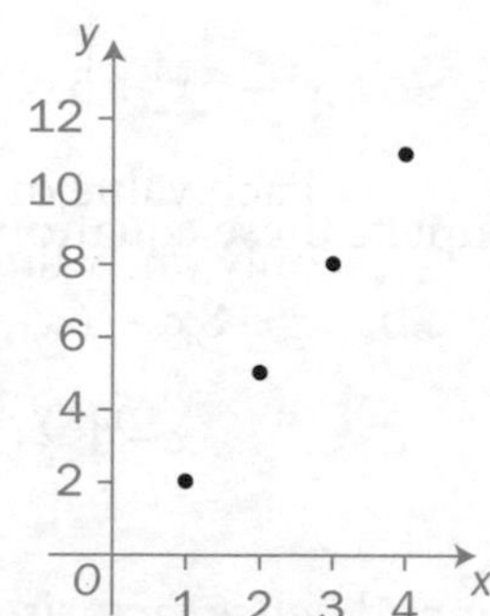

Set X is called the domain of the mapping. Its elements are denoted by x, the independent variable.

Set Y is called the range of the mapping. Its elements are denoted by y, the dependent variable.

You say that 'x maps onto y' or that 'y is the **image** of x'.

A mapping is a function only if each x-value maps onto one and only one y-value.

In the example shown, the mapping is a function.

You can therefore write $y = 3x - 1$
as either $f(x) = 3x - 1$ — You say '$f(x)$ equals $3x - 1$'.
or $f: x \rightarrow 3x - 1$ — You say 'the function f that maps x onto $3x - 1$'.

There are four kinds of mapping.

One-to-one

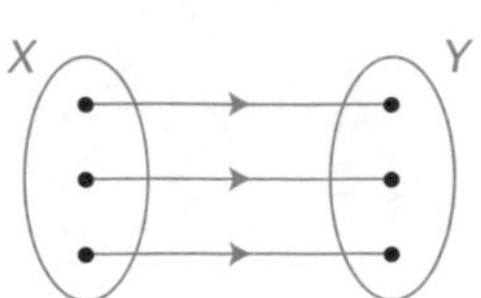

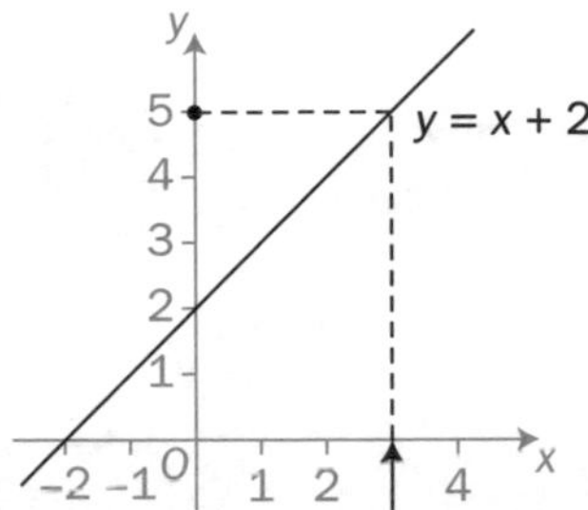

Each value of x maps onto one and only one value of y, and vice versa.

- A one-to-one mapping is a function.

Many-to-one

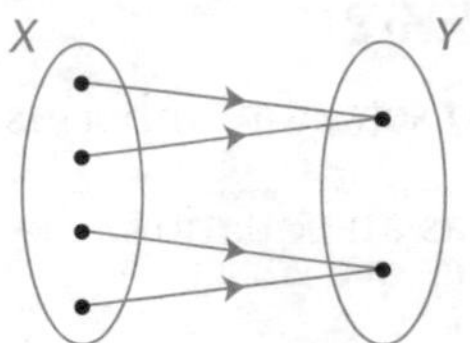

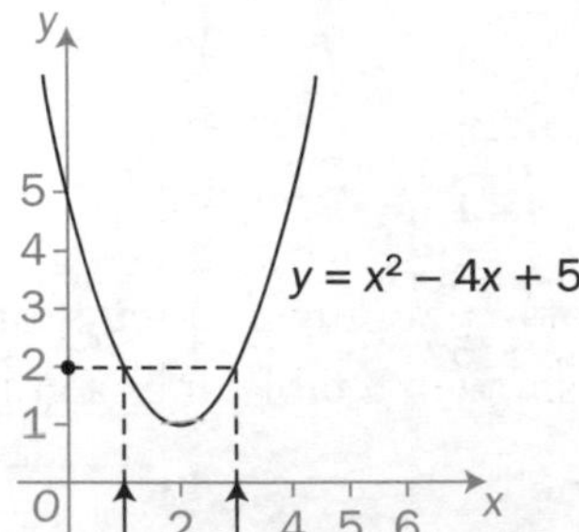

Each value of x maps onto one and only one value of y, but not vice versa.

- A many-to-one mapping is a function.

One-to-many

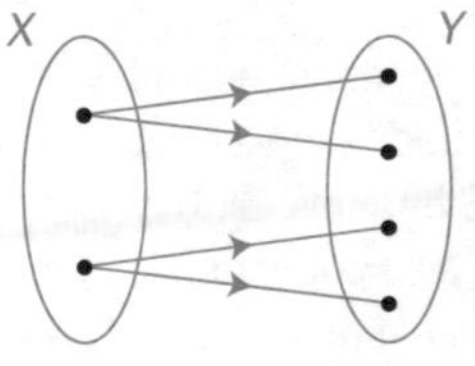

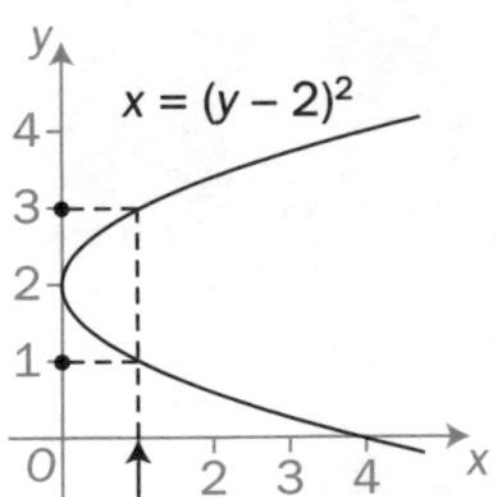

Each x-value maps onto more than one y-value.

- A one-to-many mapping is not a function.

Many-to-many

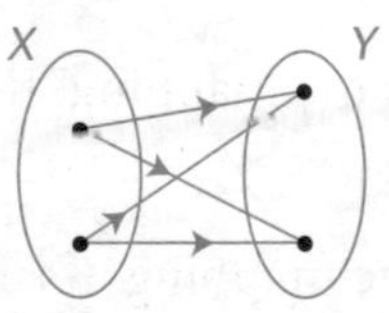

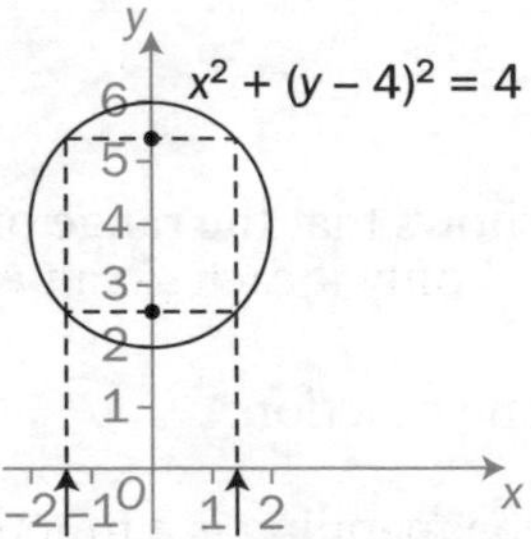

Many x-values map onto many y-values.

- A many-to-many mapping is not a function.

C3

EXAMPLE 1

Find the range of these functions and state the type of the function in each case.

a $f(x) = 2x - 1, x = \{0, 1, 2, 3\}$

b $f(x) = x^2 + 4, x \in \mathbb{R}, -2 < x \leqslant 3$

---

a Sketch the mapping diagram and graph:

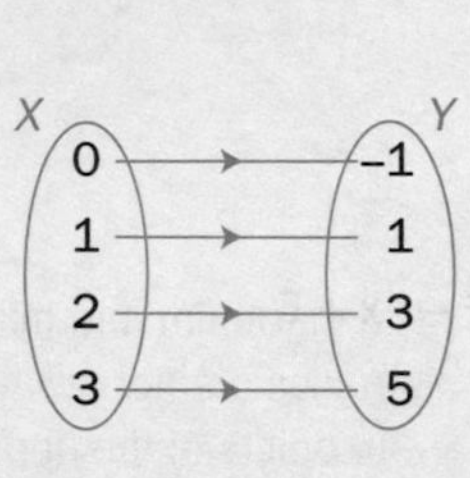

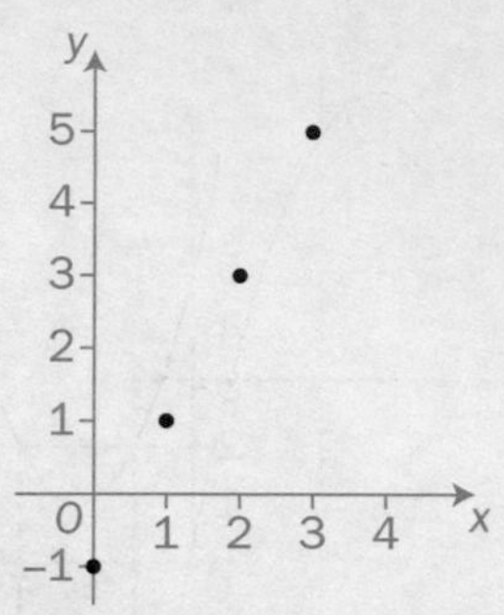

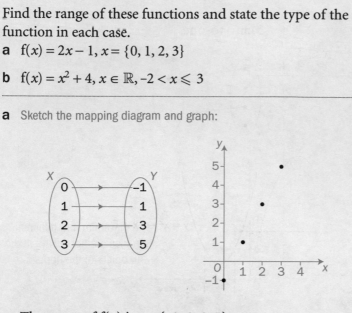

Both the domain and range are discrete as x can take only the four given values.

The range of $f(x)$ is $y = \{-1, 1, 3, 5\}$

$f(x)$ is a one-to-one function.

b Sketch the graph of $f(x)$:

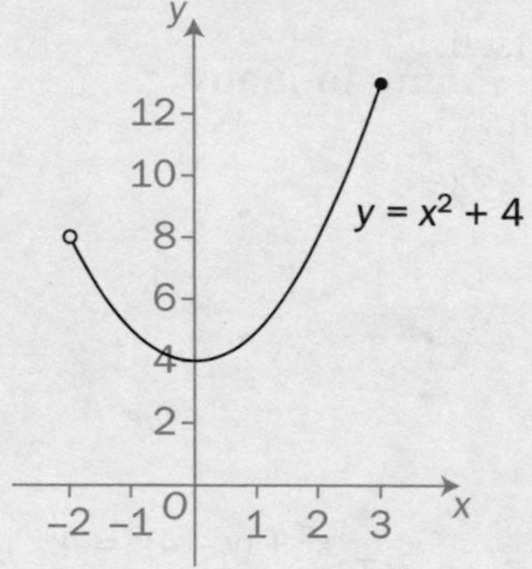

The domain is continuous as it takes all values from −2 to 3 (including 3 but not −2). The range is therefore also continuous. The lowest value of the range is 4 which occurs when $x = 0$.

The graph of $f(x)$ shows that the range of $f(x)$ is $y \in \mathbb{R}, 4 \leqslant y \leqslant 13$

$f(x)$ is a many-to-one function.

C3

EXAMPLE 2

Find the range of the function $f(x) = 4 - x^2$ when the domain is

a $x \in \mathbb{R}$ **b** $-1 < x \leqslant 3$

$x \in \mathbb{R}$ means that the domain contains all real numbers.

Sketch the graph of the function for each of the given domains:

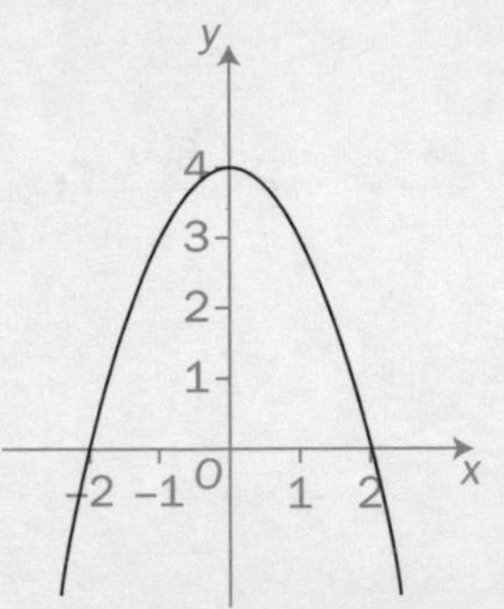

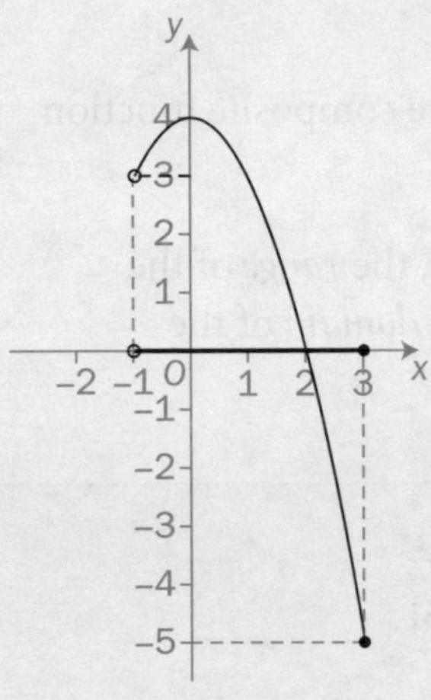

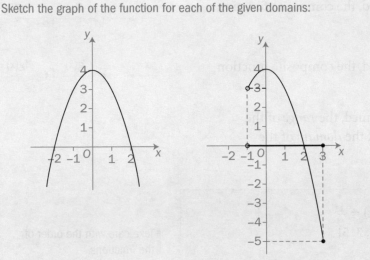

The maximum and minimum y-values are not always at the end points of the domain.

a The points on the graph for $x \in \mathbb{R}$ have y-values from $-\infty$ to 4.

The range of $f(x)$ is
$y \in \mathbb{R}, y \leqslant 4$

b The points on the graph for $-1 < x \leqslant 3$ have y-values from -5 to 4, including -5 itself.

The range of $f(x)$ is
$y \in \mathbb{R}, -5 \leqslant y \leqslant 4$

EXAMPLE 3

a Find the range of $f(x) = \dfrac{8}{(x-2)^2}$ $x \in \mathbb{R}, x \geqslant 1$

b Find the value of k such that $f(k) = 4$

Sketch the graph of $f(x)$.

Refer to C1 for revision of sketching graphs of functions.

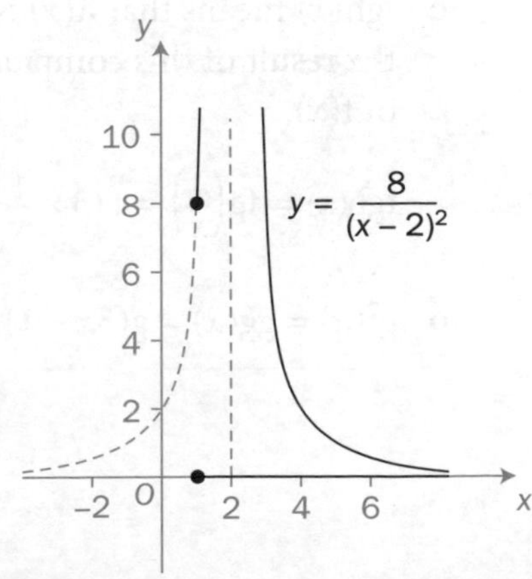

a From the graph, you can see that, for the given domain, the range is
$y \in \mathbb{R}, y > 0$

b When $x = k$, $\dfrac{8}{(k-2)^2} = 4$

$$(k-2)^2 = \frac{8}{4} = 2$$

$$k - 2 = \pm\sqrt{2}$$

giving $k = 2 \pm \sqrt{2}$

The value $x = 2 - \sqrt{2}$ lies outside the domain, so the only possible value of k is $2 + \sqrt{2}$.

Composite functions

A composite function is formed when you combine two (or more) functions. The output from the first function becomes the input to the second function.

When f is applied first and g second, the composite function gf(x) is formed.

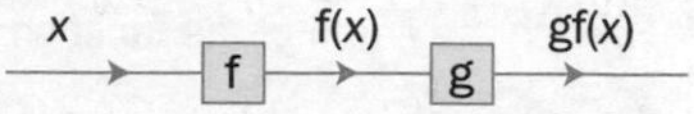

x → [f] → f(x) → [g] → gf(x)

When g is applied first and f second, the composite function fg(x) is formed.

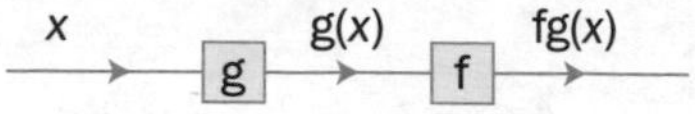

x → [g] → g(x) → [f] → fg(x)

For a composite function to be formed, the *range* of the first function must be all or part of the *domain* of the second function.

C3

EXAMPLE 4

If $f(x) = x^2$, $g(x) = 3x - 1$ and $h(x) = \frac{1}{x}$ find

a fg(x) **b** gf(x)

c fgh(x) **d** $g^2(x)$

Take care with the order of the functions.

---

a fg(x) means that g(x) is the input of f(x).

Draw a flow diagram:

x → [g] → 3x − 1 → [f] → (3x − 1)²

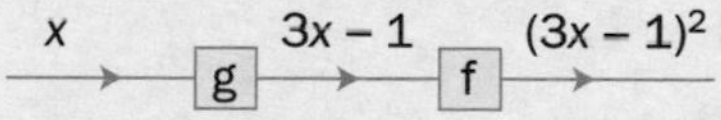

$fg(x) = f(3x - 1) = (3x - 1)^2$

b Draw a flow diagram:

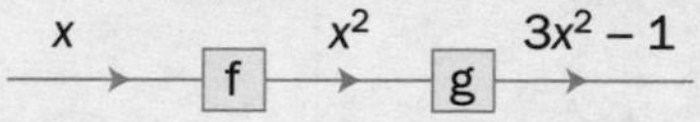

x → [f] → x² → [g] → 3x² − 1

$gf(x) = g(x^2) = 3x^2 - 1$

c fgh(x) means that h(x) is the input of g(x) and then the result of this composite function is the input of f(x).

Draw a flow diagram if you find it helps.

$$fgh(x) = fg\left(\frac{1}{x}\right) = f\left(3 \times \frac{1}{x} - 1\right) = f\left(\frac{3}{x} - 1\right) = \left(\frac{3}{x} - 1\right)^2$$

d $g^2(x) = gg(x) = g(3x - 1) = 3(3x - 1) - 1 = 9x - 4$

$g^2(x)$ means that g(x) is used as the input of g(x).

EXAMPLE 5

Given $f(x) = 2x - 1$ and $g(x) = x^2 + 1$, prove that $gf(x) \neq fg(x)$ except for two values of x.
Find these two values.

$fg(x) = f(x^2 + 1) = 2(x^2 + 1) - 1 = 2x^2 + 1$

$gf(x) = g(2x - 1) = (2x - 1)^2 + 1 = 4x^2 - 4x + 2$

Equate fg(x) and gf(x) and solve to find the values of x:

$$4x^2 - 4x + 2 = 2x^2 + 1$$
$$2x^2 - 4x + 1 = 0$$

giving $\quad x = \dfrac{4 \pm \sqrt{16 - 8}}{4} = 1 \pm \frac{1}{2}\sqrt{2}$

Hence $fg(x) \neq fg(x)$ except when $x = 1 \pm \frac{1}{2}\sqrt{2}$

Exercise 1.3

1 For each function, draw a mapping diagram or a graph, find the range, and state if the function is one-to-one or many-to-one.

a $f(x) = 4x - 3, x = \{0, 2, 4, 6\}$

b $f(x) = 4x - 3, x \in \mathbb{R}, 1 < x < 6$

c $f(x) = 9 - x, x \in \mathbb{R}, 1 < x < 4$

d $f(x) = x^2 + 2, x \in \mathbb{R}$

e $f(x) = +\sqrt{x}, x = \{0, 1, 4, 9\}$

f $f(x) = -\sqrt{x}, x = \{0, 1, 4, 9\}$

g $f(x) = \dfrac{x + 2}{x - 2}, x \in \mathbb{R}, x \neq 2$

h $f(x) = \begin{cases} x^2, & x \in \mathbb{R}, 0 \leqslant x \leqslant 2 \\ 4, & x \in \mathbb{R}, 2 < x \leqslant 6 \end{cases}$

2 **a** Given $g(x) = mx + c$, $g(2) = 8$ and $g(3) = 11$, find m and c.

b Given $h(x) = ax^2 + bx + c$ and $h(0) = 1, h(1) = 0, h(2) = 1$, find a, b, and c.

C3

3 Find the domain of each function.
Say whether the function is one-to-one or many-to-one.

a f: $x \to 3x + 5$ with the range $y = \{5, 20, 35, 50, 65\}$

b f: $x \to x^2$ with the range $y \in \mathbb{R}, 0 \leqslant y \leqslant 9$

c f: $x \to \sqrt{x - 3}$ with the range $y \in \mathbb{R}, 0 \leqslant y \leqslant 3$

d f: $x \to \dfrac{4}{x - 2}$ with the range $y \in \mathbb{R}, 1 \leqslant y \leqslant 4$

4 Sketch each of these functions, given the domain $x \in \mathbb{R}$.
Find the range of each function.

a $f(x) = (x - 2)^2 + 3$

b $f(x) = (x + 4)^2 - 3$

c $f(x) = x^2 - 6x + 10$

d $f(x) = 1 + 2x - x^2$

5 Explain why these mappings are not functions.

a $y = \begin{cases} 3x, & x \in \mathbb{R}, 0 \leqslant x \leqslant 2 \\ 6 - x, & x \in \mathbb{R}, 2 \leqslant x \leqslant 8 \end{cases}$

b $y = \dfrac{6 + x}{2 - x}, x \in \mathbb{R}$

C3

6 Sketch the graph of the function

$$f(x) = \frac{x + 4}{x - 2}, x \in \mathbb{R}, x \neq 2$$

Find the range of $f(x)$ and the values of x which are unchanged by this function.

7 If $f(x) = x^2 + 1$ and $g(x) = 2x - 1$, find

a $f(3)$

b $gf(3)$

c $gf(x)$

d $g(3)$

e $fg(3)$

f $fg(x)$

g $f^2(3)$

h $f^2(x)$

i $g^2(3)$

j $g^2(x)$

8 Find $fg(x)$, $gf(x)$, $f^2(x)$ and $g^2(x)$ when

a $f(x) = x^2 - 2, g(x) = 3x + 4$

b $f(x) = \dfrac{1}{x}, g(x) = 3x^2 + 2, x \neq 0$

c $f(x) = \dfrac{x}{2} + 3, g(x) = 4x - 2$

d $f(x) = \dfrac{1}{x - 1}, g(x) = 2 - x, x \neq 1$

9 When $f: x \to 5 - x$, $g: x \to \sqrt{x}$ and $h: x \to \frac{1}{x}$,
find

a fg **b** fh **c** gh

d fgh **e** hgf **f** f^2

g g^2 **h** h^2

10 a Given that $f(x) = 2x - 1$ and $g(x) = x^2$, show that $fg(x) \neq gf(x)$ for all values of x except $x = \alpha$
Find the value of α.

b Solve the equations

i $fg(x) = 49$
ii $gf(x) = 9$
iii $f^2(x) = 13$

c Find the values of x which map onto themselves under the function fg.

11 Given that $f(x) = 2x + 1$ and $g(x) = 3x + c$,

a find c if $fg(x) = gf(x)$ for all x

b find α if $f^2(\alpha) = \alpha$

12 Given $f(x) = p + qx$, $g(x) = x^2 - 4$ and $h(x) = 3x + 1$,
find p and q such that $hgf(x) = 4(3x^2 + 3x - 2)$

13 a If $f(x) = 2x + 3$,
find i $f^2(x)$
ii $f^3(x)$
iii $f^4(x)$

b If $f(x) = 2x + 3$, find an expression in n for $f^n(x)$.

INVESTIGATION

14 Explain why the composite function $fg(x)$ is not allowed when $f(x) = 2x + 1, x \in \mathbb{R}, -5 \leqslant x \leqslant 5$ and $g(x) = x^2, x \in \mathbb{R}, x \geqslant 0$

How would you change the domains so that the function $fg(x)$ can exist?

C3

1.4 Inverse functions

The **inverse function** $f^{-1}(x)$ of the function $f(x)$ maps the range of $f(x)$ back onto its domain.

E.g. If $f(x) = 3x - 1$ $\quad x \rightarrow \boxed{\times 3} \rightarrow \boxed{-1} \rightarrow 3x - 1$

then $f^{-1}(x) = \frac{x+1}{3}$ $\quad \frac{x+1}{3} \leftarrow \boxed{\div 3} \leftarrow \boxed{+1} \leftarrow x$

The range of $f(x)$ is the domain of $f^{-1}(x)$.
The range of $f^{-1}(x)$ is the domain of $f(x)$.

Both composite functions ff^{-1} and $f^{-1}f$ map x back onto itself,
so $ff^{-1}: x \rightarrow x$ and $f^{-1}f: x \rightarrow x$
or $ff^{-1}(x) = x$ and $f^{-1}f(x) = x$

The notation for the inverse function uses $^{-1}$.
Do not confuse it with the reciprocal.
E.g. $(\sin x)^{-1}$ is the reciprocal $\frac{1}{\sin x}$, whereas $\sin^{-1} x$ is the inverse function of $\sin x$.

The inverse of a one-to-one function is a function.

E.g.

If $\quad f(x) = 3x - 1 \quad$ then $\quad f^{-1}(x) = \frac{x+1}{3}$

and $\quad f(5) = 14 \quad$ so $\quad f^{-1}(14) = 5$

For a one-to-one function, $f(x) = y$ leads to $f^{-1}(y) = x$ with no ambiguity.

The inverse of a many-to-one function is a one-to-many mapping so the inverse is not a function. In such a case, the inverse can be a function only if the domain of $f(x)$ is restricted to certain values to make the inverse a one-to-one mapping.

E.g.

If $\quad f(x) = x^2 \quad$ then $\quad f^{-1}(x) = \pm\sqrt{x}$

and $\quad f(3) = 9 \quad$ so $\quad f^{-1}(9) = \pm 3$

So $f^{-1}(x)$ is not a function.

For $f^{-1}(9)$ to have a unique value, the domain of $f(x)$ must be restricted to non-negative numbers.
In this case, $f^{-1}(x) = +\sqrt{x}$

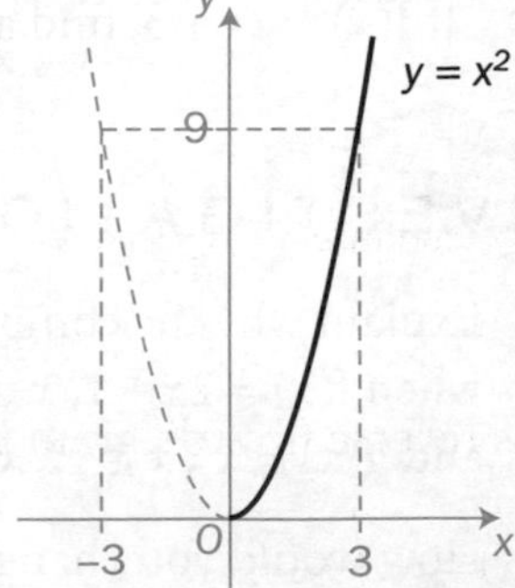

You can find an inverse function

algebraically by

either using a flow diagram in reverse

$x \rightarrow$ [f] $\rightarrow f(x)$

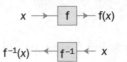

$f^{-1}(x) \leftarrow$ [f^{-1}] $\leftarrow x$

or rearranging the function $y = f(x)$ to give x in terms of y, then interchanging x and y and writing y as $f^{-1}(x)$.

graphically by

reflecting the graph of $y = f(x)$ in the line $y = x$ (which is equivalent to interchanging x and y).

y = x

EXAMPLE 1

Find the inverse function $f^{-1}(x)$ when $f(x) = 5x^2 + 4$, $x \in \mathbb{R}_0^+$

Find the domain and range of the inverse function.

$\mathbb{R}_0^+$ means all positive real numbers and zero.

Sketch a graph to help you to visualise the function and its inverse:

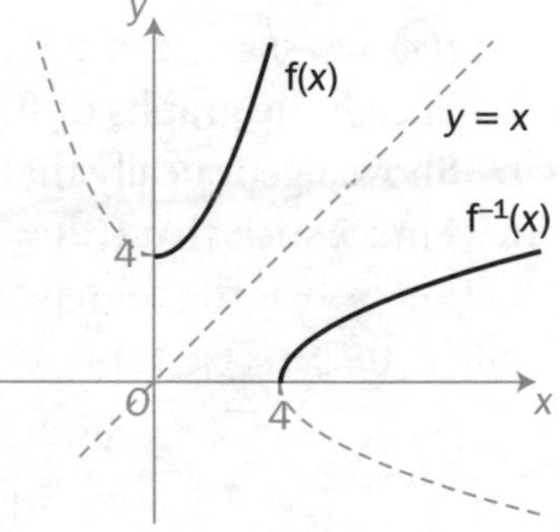

If $x \in \mathbb{R}$, then the inverse of $y = 5x^2 + 4$ would not be a one-to-one mapping and therefore not a function.

By restricting the domain to $\mathbb{R}_0^+$, the inverse is then a one-to-one mapping and a function.

First method

Draw a flow diagram for f(x):

$x \rightarrow$ [square] $\rightarrow$ [$\times 5$] $\rightarrow$ [$+ 4$] $\rightarrow f(x)$

Draw the flow diagram in reverse:

$f^{-1}(x) \leftarrow$ [square root] $\leftarrow$ [$\div 5$] $\leftarrow$ [$- 4$] $\leftarrow x$

This reverse flow diagram gives

$$f^{-1}(x) = +\sqrt{\frac{x-4}{5}}$$

Second method

Write $y = 5x^2 + 4$

Rearrange to make x the subject:

$$x = +\sqrt{\frac{y-4}{5}}$$

Interchange x and y and replace y by $f^{-1}(x)$:

$$y = +\sqrt{\frac{x-4}{5}}$$

$$f^{-1}(x) = +\sqrt{\frac{x-4}{5}}$$

Use the method that you feel most comfortable with. You can use a different method to check your answer.

The domain of $f(x)$ is $\mathbb{R}_0^+$ $(x \geqslant 0)$ and the range of $f(x)$ is $y \geqslant 4$

So, the domain of $f^{-1}(x)$ is $x \geqslant 4$ and the range of $f^{-1}(x)$ is $y \geqslant 0$

C3

EXAMPLE 2

Find the inverse function and its domain and range if
$f(x) = \dfrac{4}{x-3}$ for $x \in \mathbb{R}, x \neq 3$

The domain has $x \neq 3$ because when $x = 3$ the denominator, $x - 3$, is zero and division by zero is not defined.

Make x the subject of $y = \dfrac{4}{x-3}$:

$$y(x-3) = 4$$
$$xy = 3y + 4$$
$$x = \frac{3y+4}{y}$$

Interchange x and y and replace y by $f^{-1}(x)$:

$$y = \frac{3x+4}{x}$$

$$f^{-1}(x) = \frac{3x+4}{x}$$

$f^{-1}(x)$ exists for all real values of x except $x = 0$
So, the domain of $f^{-1}(x)$ is $x \in \mathbb{R}, x \neq 0$
The range of $f^{-1}(x)$ is the domain of $f(x)$ giving $y \in \mathbb{R}, y \neq 3$

C3

EXAMPLE 3

a Find $f^{-1}(x)$ given that
$f(x) = +\sqrt{x+2}, x \in \mathbb{R}, x \geqslant -2$
Sketch the graphs of $f(x)$ and $f^{-1}(x)$.
b Show algebraically that $f^{-1}f(x) = x$
c Find x such that $f(x) = f^{-1}(x)$

$f(x)$ only exists if $x + 2 \geqslant 0$ so the domain of $f(x)$ is limited to $x \geqslant -2$. Also, only the positive value of the square root is allowed, so that $f(x)$ is a function.

a Make x the subject of $y = \sqrt{x+2}$:

$$y^2 = x + 2$$
$$x = y^2 - 2$$

Interchange x and y. Replace y by $f^{-1}(x)$:

$$y = x^2 - 2$$
$$f^{-1}(x) = x^2 - 2$$

The graph of the inverse function is half of the curve $y = x^2$ translated 2 units downwards.
The graph of $f^{-1}(x)$ is a reflection of $f(x)$ in the line $y = x$
From the graph, the domain of $f^{-1}(x)$ is $x \in \mathbb{R}, x \geqslant 0$
and the range of $f^{-1}(x)$ is $y \in \mathbb{R}, y \geqslant -2$

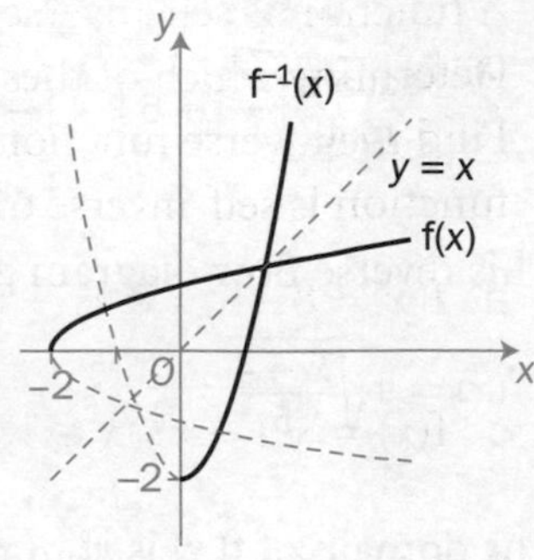

b Substitute $f(x) = +\sqrt{x+2}$ into $f^{-1}f(x)$:

$$f^{-1}f(x) = f^{-1}\left(+\sqrt{x+2}\right) = \left(+\sqrt{x+2}\right)^2 - 2 = x + 2 - 2 = x$$

So, f followed by f^{-1} leaves x unchanged.

EXAMPLE 3 (CONT.)

c $f(x) = f^{-1}(x)$ gives the equation $+\sqrt{x+2} = x^2 - 2$
This is not easy to solve.

However, the graphs of $f(x)$ and $f^{-1}(x)$ intersect on the line $y = x$
So, at this point of intersection, $f(x) = f^{-1}(x) = x$

Solve the two equations $\sqrt{x+2} = x$ and $x^2 - 2 = x$:

Both these equations give $x^2 - x - 2 = 0$
$(x-2)(x+1) = 0$

So the only solution is $x = 2$

$x = -1$ is outside the domain of $f^{-1}(x)$.

Exercise 1.4

1 Find the inverse function $f^{-1}(x)$ of each of these functions, $f(x)$, where the domain of $f(x)$ is $\mathbb{R}$.

a $f(x) = 3x - 2$ **b** $f(x) = 2x + 1$ **c** $f(x) = 2(x + 1)$

d $f(x) = \dfrac{x+3}{2}$ **e** $f(x) = \frac{1}{2}x + 3$ **f** $f(x) = (x+1)^2, \quad x \geqslant -1$

g $f(x) = x^2 + 1, \quad x \geqslant 0$ **h** $f(x) = 6 - x$ **i** $f(x) = \sqrt{x-3}, \quad x \geqslant 3$

2 Find the inverse function $f^{-1}(x)$ of each of these functions, $f(x)$, where the domain of $f(x)$ is $\mathbb{R}$.
In each case, sketch the graphs of $f(x)$ and $f^{-1}(x)$ on the same axes.

a $f(x) = 2x + 1$ **b** $f(x) = 10 - 2x$ **c** $f(x) = \dfrac{x+4}{2}$

d $f(x) = x^2 + 2, x \geqslant 0$ **e** $f(x) = (x-2)^2, x \geqslant 2$ **f** $f(x) = \sqrt{x-4}, x \geqslant 4$

g $f(x) = \sqrt{x+3}, x \geqslant -3$ **h** $f(x) = \dfrac{1}{2-x}, x > 2$ **i** $f(x) = \dfrac{1}{x} + 2, x > 0$

3 A function is self-inverse if the function and its inverse are identical.
Determine which of these functions have an inverse function.
Find the inverse function where one exists and say whether the function is self-inverse or not.

a $f(x) = 8 - x, x \in \mathbb{R}$ **b** $f(x) = \dfrac{12}{x}, x \in \mathbb{R}, x \neq 0$

c $f(x) = \sqrt{4 - x^2}, x \in \mathbb{R}, 0 \leqslant x \leqslant 2$ **d** $f(x) = \sqrt{4 - x^2}, x \in \mathbb{R}, -2 \leqslant x \leqslant 2$

e $f(x) = 8 - x, x \in \mathbb{R}, 0 \leqslant x \leqslant 8$ **f** $f(x) = \dfrac{x}{x-1}, x \in \mathbb{R}, x \neq 1$

C3

4 Find the inverse of $f(x) = 8 - x, x \in \mathbb{R}, x \geqslant 0$ and explain why the function $f(x)$ is not self-inverse.

5 Prove that $g(x) = \frac{x+1}{x-1}, x \in \mathbb{R}, x \neq 1$ is self-inverse.
Find the elements of the domain that map onto themselves under $g(x)$.

6 The function $f(x)$ is defined by $f(x) = x^2 - 6x + 14, x \in \mathbb{R}, x \geqslant 3$
By completing the square, draw the graph of $f(x)$ and find its range.
Find the inverse function $f^{-1}(x)$, stating its domain and range.
Sketch its graph on the same axes as $f(x)$.

To complete the square, rewrite in the form $(x - a)^2 + b$
See C1 for revision.

7 For each of these functions, find its range and its inverse function.
State the domain and range of the inverse function.
Sketch the graphs of $y = f(x)$ and $y = f^{-1}(x)$ on the same axes.

You can use the 'completing the square' method.

a $f: x \to x^2 - 4x + 5, x \in \mathbb{R}, x \geqslant 2$

b $f: x \to x^2 - 8x + 21, x \in \mathbb{R}, x \leqslant 4$

c $f: x \to 4x - x^2, x \in \mathbb{R}, x > 2$

d $f: x \to x^2 + 4x + 2, x \in \mathbb{R}, x > -2$

8 For each of these functions, $f(x)$, find $f^{-1}(x)$ and its domain.
Solve the equation $f(x) = f^{-1}(x)$
Find the points where the graphs of $y = f(x)$ and $y = f^{-1}(x)$ intersect.

a $f(x) = \frac{1}{2}x + 4, x \in \mathbb{R}$

b $f(x) = x^2, x \in \mathbb{R}, x \geqslant 0$

c $f(x) = (x - 2)^2, x \in \mathbb{R}, x \geqslant 2$

d $f(x) = x^2 + 8x + 12, x \in \mathbb{R}, x \geqslant -4$

9 **a** Find a, b and c such that $f^{-1}(x) = x^2 + ax + b, x \in \mathbb{R}, x \geqslant c$ is the inverse function of $f(x) = 1 + \sqrt{x+1}, x \in \mathbb{R}, x \geqslant -1$

b Find the values of

i $f(8)$
ii $f^{-1}(8)$
iii x such that $f(x) = f^{-1}(x)$

C3

10 The function g(x) is defined as g: $x \to \frac{2}{x-2}$, $x \in \mathbb{R}$, $x \neq 2$
Find the inverse function $g^{-1}(x)$ and its domain.
Solve the equation $g(x) = g^{-1}(x)$

11 Find the inverse $g^{-1}(x)$ of the function $g(x) = \frac{3x+1}{x-3}$, $x \in \mathbb{R}$, $x \neq 3$. And show that g(x) is self-inverse.
Sketch the graphs of g(x) and $g^{-1}(x)$.
Find the values of x which map onto themselves.

Find the horizontal asymptote by letting $x \to \infty$. Also find the vertical asymptote.

12 The function $h(x) = \frac{cx+1}{x-2}$, $x \in \mathbb{R}$, $x \neq 2$ is self-inverse.
Find the value of c.
Find the values of x which map onto themselves.

13 Prove that $f(x) = \frac{\alpha x + \beta}{x - \alpha}$, $x \neq \alpha$ is self-inverse for all α and β.
Find the values of x which map onto themselves.

14 Sketch the graph of the function $f(x) = \frac{x^2+1}{x^2-1}$, $x \in \mathbb{R}_0^+$, $x \neq 1$
and its inverse $f^{-1}(x)$ on the same axes.
State the domain and range of the inverse function.
Explain why the solution of the equation $f(x) = f^{-1}(x)$ is also a solution of $x^3 - x^2 - x - 1 = 0$
Show that an approximate solution is $x = 1.84$

INVESTIGATION

15 Which one of these four functions has an inverse function?
Explain why the other three functions do not have inverse functions.

$f_1(x) = x^2 - 5$, $x \in \mathbb{R}$

$f_2(x) = \frac{3}{x^2}$, $x \in \mathbb{R}$

$f_3(x) = 5 - 2x$, $x \in \mathbb{R}^+$

$f_4(x) = \sin x$, $x \in \mathbb{R}$, $0 \leqslant x \leqslant \pi$

C3

1.5 The modulus function

The **modulus** of a number $|x|$ is its magnitude or absolute value.

In general, when $x \geqslant 0$, $|x| = x$
when $x < 0$, $|x| = -x$

On your calculator, $|x|$ may be written as ABS(x).

E.g.

$|2| = 2$ and $|-2| = 2$

$|x| < 2$ can also be written as $-2 < x < 2$

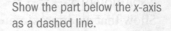

An empty circle shows that the end value is not included.

$|x| \geqslant 2$ can also be written as $x \geqslant 2$ or $x \leqslant -2$

A filled circle on a number line shows that the end value is included.

The graph $y = |f(x)|$

$|f(x)|$ is always positive, so the graph $y = |f(x)|$ always lies above the x-axis.

EXAMPLE 1

Sketch the graph of $y = |2x - 3|$

Sketch the graph of $y = 2x - 3$

Where $y < 0$, reflect the graph in the x-axis:

Show the part below the x-axis as a dashed line.

The reflected line has the equation $y = -2x + 3$

The function $f(x) = |2x - 3|$ can be written as

$$f(x) = \begin{cases} 2x - 3, x \geqslant 1\frac{1}{2} \\ -2x + 3, x < 1\frac{1}{2} \end{cases}$$

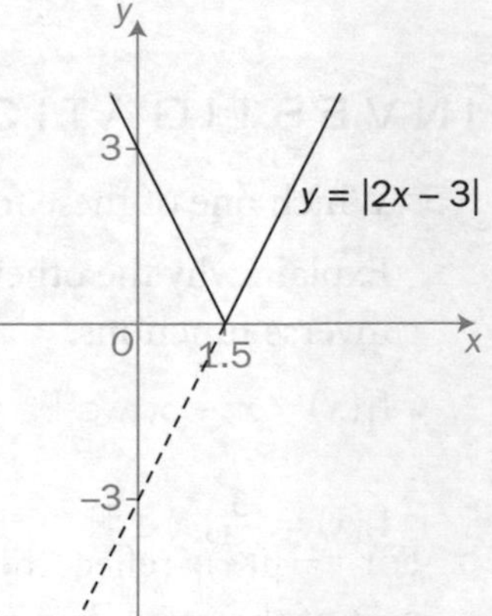

The graph of $y = f(|x|)$

$|x|$ is always positive. So, the two points on the graph with $x = k$ and $x = -k$ have the same value of $f(|x|)$. The graph $y = f(|x|)$ is therefore symmetrical about the y-axis.

EXAMPLE 2

Sketch the graph of $y = 2|x| - 3$

Sketch the graph of $y = 2x - 3$ for $x \geqslant 0$ and reflect it in the y-axis:

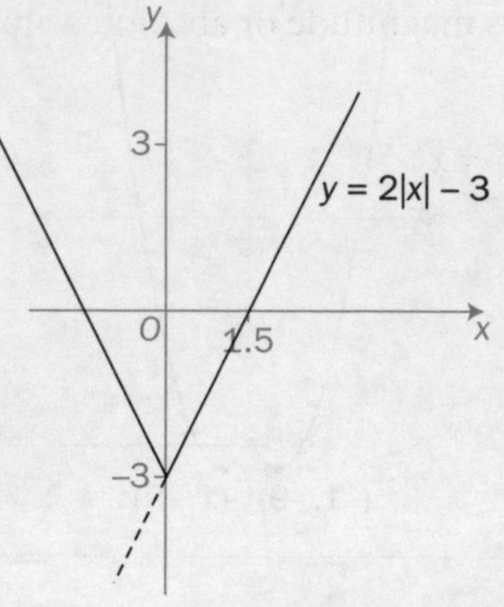

The reflection has the equation $y = -2x - 3$

The function $f(x) = 2|x| - 3$ can be written as

$$f(x) = \begin{cases} 2x - 3, & x \geqslant 0 \\ -2x - 3, & x < 0 \end{cases}$$

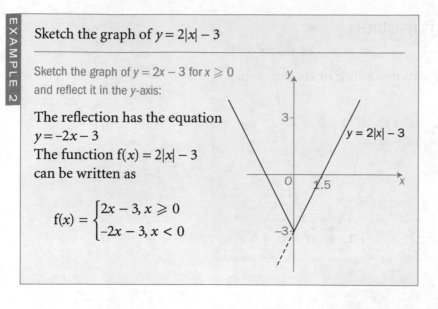

EXAMPLE 3

Sketch the graphs of

a $y = f(x)$ **b** $y = |f(x)|$ **c** $y = f(|x|)$

when $f(x) = x^2 - 2x - 8$

a $y = x^2 - 2x - 8$
$= (x - 4)(x + 2)$
When $x = 0$, $y = -8$
When $y = 0$, $x = 4$ or -2

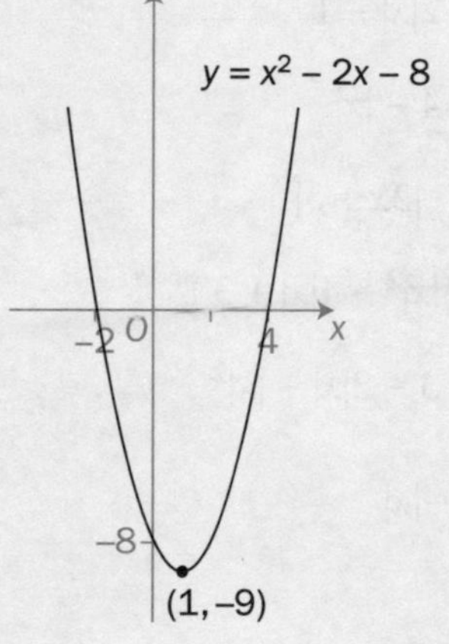

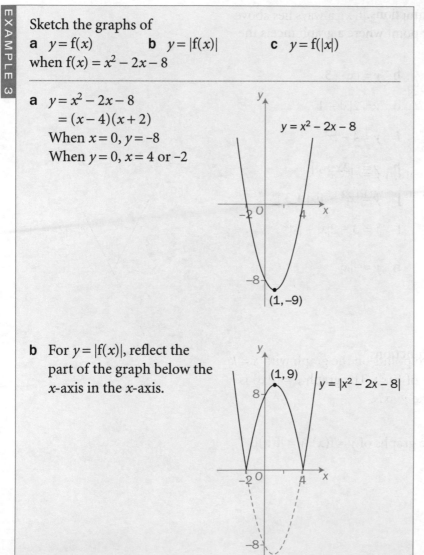

Differentiating or completing the square shows that there is a minimum value at (1, –9).

You could use computer software or a graphical calculator to check these results.

b For $y = |f(x)|$, reflect the part of the graph below the x-axis in the x-axis.

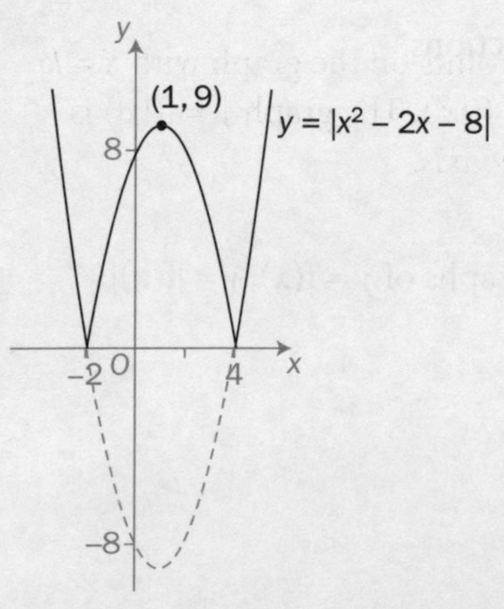

The solution to part **c** is shown on the next page.

C3

EXAMPLE 3 (CONT.)

c For $y = f(|x|)$, reflect the part of the graph to the right of the y-axis in the y-axis:

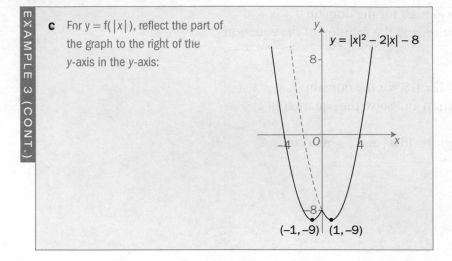

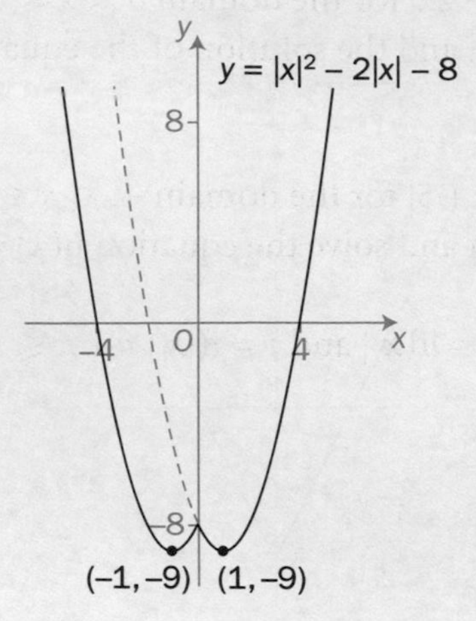

Exercise 1.5

1 Sketch the graphs of these functions.
Give the coordinates of any point where a graph meets the x-axis or y-axis.

a $y = |x - 3|$ **b** $y = |x| - 3$

c $y = |2x - 1|$ **d** $y = 2|x| - 1$

e $y = |4 - x|$ **f** $y = 4 - |x|$

g $y = |2x + 3|$ **h** $y = -|2x + 3|$

i $y = |x^2 - 4x + 3|$ **j** $y = |x|^2 - 4|x| + 3$

k $y = |3 - 2x + x^2|$ **l** $y = 3 - 2|x| + |x|^2$

m $y = \left|\frac{6}{x}\right|$ **n** $y = -|x|$

o $y = |\log_{10} x|$

2 Sketch the graphs of these functions.

a $y = |x + 4| + |x - 1|$

b $y = |x - 2| + |x - 1|$

3 On separate axes, sketch the graphs of $y = f(x)$, $y = |f(x)|$ and $y = f(|x|)$ where

a $f(x) = x^2 - 3x - 4$

b $f(x) = 4x - x^2$

C3

4 Sketch the graph of $g(x) = |6 - 2x|$ for the domain $0 \leqslant x \leqslant 8$
Find the range of the function and the solution of the equation $g(x) = 4$

5 Sketch the graph of $h(x) = |2x + 5|$ for the domain $-4 \leqslant x \leqslant 1$
Find the range of the function and solve the equation $h(x) = 3$

6 Sketch the graph of $y = f(x)$, $y = |f(x)|$ and $y = f(|x|)$ for $\pi \leqslant x \leqslant \pi$ where

a $y = |\sin x|$

b $y = \sin |x|$

c $y = |\cos x|$

d $y = \cos |x|$

7 Find the point of intersection of the graphs of $y = |5 - x|$ and $y = |x + 1|$

8 Find all points of intersection of the graphs of $y = |x + 2|$ and $y = 2|x| - 4$

9 How many solutions are there to the simultaneous equations $y = (|x| - 2)^2$ and $y = |2x - 4|$?
Find them.

INVESTIGATION

10 Use computer software or a graphical calculator to check some of your answers to the questions in this exercise. You may have to use **abs** (absolute value) button to input a modulus.

1.6 Solving modulus equations and inequalities

You can use a graphical or an algebraic method to find solutions to equations and inequalities which involve modulus signs.

EXAMPLE 1

Solve $|2x + 1| = 5$

Algebraic method

$$(2x + 1)^2 = 5^2$$
$$4x^2 + 4x + 1 = 25$$
$$4x^2 + 4x - 24 = 0$$
$$x^2 + x - 6 = 0$$
$$(x + 3)(x - 2) = 0$$
$$x = -3 \text{ or } 2$$

The solutions are $x = -3$ and $x = 2$

Graphical method

Find the points of intersection of $y = |2x + 1|$ and $y = 5$:

At P, $2x + 1 = 5$ so $x = 2$

The reflection of $y = 2x + 1$ in the x-axis is $y = -2x - 1$:

At Q, $-2x - 1 = 5$

$-2x = 6$ so $x = -3$

The solutions are

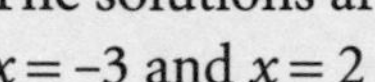

$x = -3$ and $x = 2$

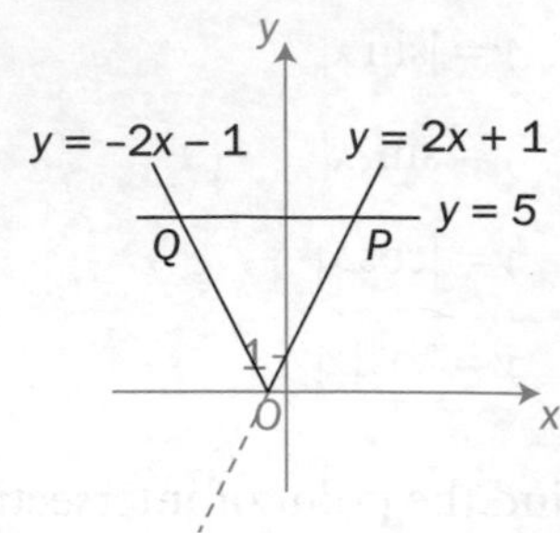

EXAMPLE 2

Solve **a** $|x| + 4 = 3x$ **b** $|x| + 4 < 3x$

Sketch the graphs of $y = |x| + 4$ and $y = 3x$:

a The two graphs have only one point of intersection, P.

At P, $x + 4 = 3x$

$4 = 2x$

The solution is $x = 2$

b For $|x| + 4 < 3x$, you need the graph of $y = |x| + 4$ to be below the graph of $y = 3x$

This occurs to the right of P.

The solution is $x > 2$

The graph of $y = |x| + 4$ involves a reflection in the y-axis.

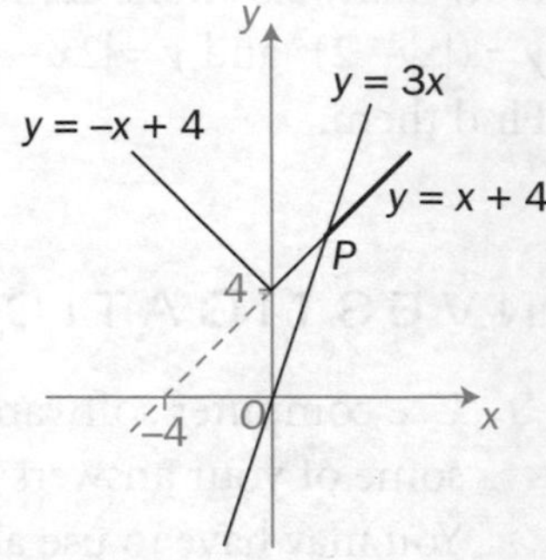

EXAMPLE 3

How many solutions has the equation $|x^2 - 3x| = |x - 2|$?
Find the solution nearest to $x = 3$

Sketch the graphs of $y = |x^2 - 3x|$ and $y = |x - 2|$:

The two graphs intersect four times.
So, there are four solutions to the equation $|x^2 - 3x| = |x - 2|$

The solution nearest to $x = 3$ is given by the point P.

At P, the graphs of $y = x - 2$ and $y = -(x^2 - 3x)$ intersect.

So $x - 2 = -(x^2 - 3x)$

$x^2 - 2x - 2 = 0$ giving $x = 1 \pm \sqrt{3}$

Reject the invalid value $1 - \sqrt{3}$ which is negative.
The required solution is $x = 1 + \sqrt{3} = 2.73$ (to 2 d.p.)

Both graphs involve a reflection in the x-axis.

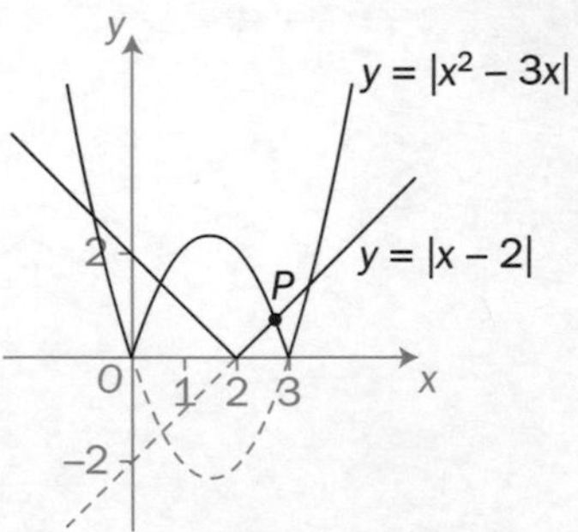

Reflecting $y = x^2 - 3x$ in the x-axis gives $y = -(x^2 - 3x) = -x^2 + 3x$

Use the quadratic formula.
See 01 for revision.

Exercise 1.6

1 Use these two diagrams to help you solve

a

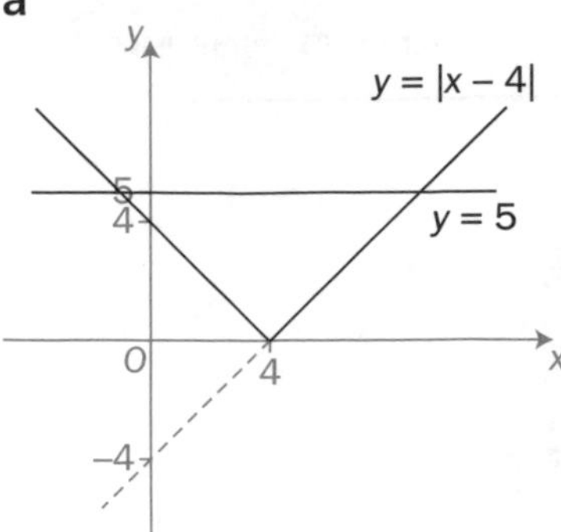

b

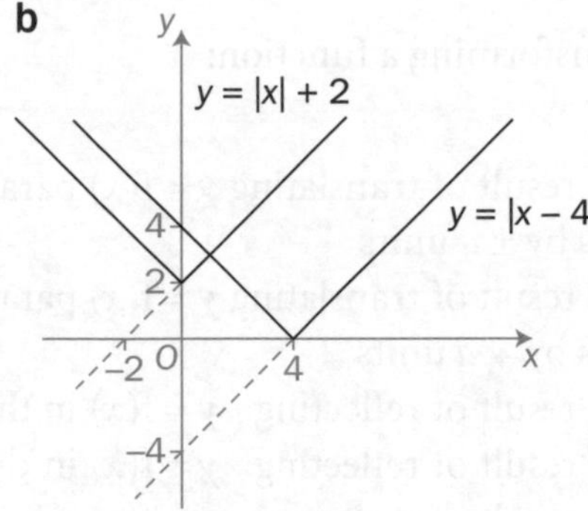

i $|x-4|=5$ **ii** $|x-4|<5$ **i** $|x-4|=|x|+2$ **ii** $|x-4|>|x|+2$

2 Solve these equations and inequalities.

a **i** $|x+2|=5$ **ii** $|x+2|<5$

b **i** $|2x+3|=7$ **ii** $|2x+3|>7$

c **i** $|3x-1|=8$ **ii** $|3x-1|\geqslant 8$

d **i** $|6-2x|=2$ **ii** $|6-2x|\leqslant 2$

e **i** $|x-2|=|x|$ **ii** $|x-2|<|x|$

f **i** $|3x-2|=|x+1|$ **ii** $|3x-2|<|x+1|$

3 Solve these equations.

a $|x+1|=x+4$ **b** $|x|+1=x+4$ **c** $|2x+3|=3x-2$

d $|2x+3|=6-x$ **e** $2|x|+3=4x$ **f** $|2x-4|=|x|+2$

g $|x^2-4x|=3x-6$ **h** $|x|^2-4|x|=x$

4 Solve these inequalities.

a $|2x-1|>x+1$ **b** $2|x|-1<x+1$ **c** $|3x-6|<|x|$

d $|2x-2|\geqslant|x|-3$ **e** $|x^2-3x|\leqslant 2|x|-4$ **f** $|x^2-4|<|x|+2$

5 Solve **a** $|x^2-x-6|=|x|+2$ **b** $\left|\dfrac{x-2}{x+3}\right|<4$

INVESTIGATION

6 If $|x^2+bx+c|=x^2+bx+c$ for all x, find a condition which b and c must satisfy.

C3

1.7 Transformations of graphs of functions

Recall the rules of transforming a function:

Refer to 01 for revision.

- $y = \mathrm{f}(x) \pm a$ is the result of translating $y = \mathrm{f}(x)$ parallel to the y-axis by $\pm a$ units
- $y = \mathrm{f}(x \pm a)$ is the result of translating $y = \mathrm{f}(x)$ parallel to the x-axis by $\mp a$ units
- $y = -\mathrm{f}(x)$ is the result of reflecting $y = \mathrm{f}(x)$ in the x-axis
- $y = \mathrm{f}(-x)$ is the result of reflecting $y = \mathrm{f}(x)$ in the y-axis
- $y = a\mathrm{f}(x)$ is the result of stretching $y = \mathrm{f}(x)$ parallel to the y-axis by a scale factor of a
- $y = \mathrm{f}(ax)$ is the result of stretching $y = \mathrm{f}(x)$ parallel to the x-axis by a scale factor of $\frac{1}{a}$

You can combine these transformations to give new functions.

E.g. When $y = \mathrm{f}(x)$ is transformed by:

- translations of $\begin{pmatrix} a \\ 0 \end{pmatrix}$ and then $\begin{pmatrix} 0 \\ b \end{pmatrix}$, the result is $y = \mathrm{f}(x - a) + b$
- a stretch of scale factor 2 parallel to the x-axis and then a reflection in the x-axis, the result is $y = -\mathrm{f}\left(\frac{x}{2}\right)$

The order in which you do the transformations can sometimes affect the final function.

EXAMPLE 1

The quadratic function $\mathrm{f}(x) = x^2 - 3x$ is reflected in the y-axis and then translated by the vector $\begin{pmatrix} 2 \\ 0 \end{pmatrix}$.

Find the equation of the final function.

After reflection in the y-axis, the function becomes

$$\mathrm{f}(x) = (-x)^2 - 3(-x) = x^2 + 3x$$

The reflection in the y-axis changes $\mathrm{f}(x)$ to $\mathrm{f}(-x)$.

After the translation, the function now becomes

$$\mathrm{f}(x) = (x-2)^2 + 3(x-2)$$
$$= x^2 - 4x + 4 + 3x - 6$$
$$= x^2 - x - 2$$

The translation changes the new $\mathrm{f}(x)$ to $\mathrm{f}(x - 2)$.

The equation of the final function is $\mathrm{f}(x) = x^2 - x - 2$

You can check the answer using a graph-plotter.

EXAMPLE 2

Describe the two transformations needed to transform the graph of $y = x^2$ into the graph of $y = 4 - x^2$

Work out the result of a reflection in the x-axis:

If $y = f(x) = x^2$, then after reflection in the x-axis
$y = -f(x) = -x^2$

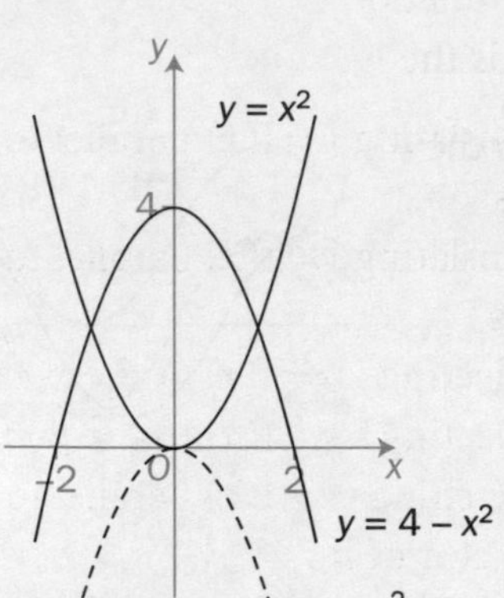

Now work out the result of translating the new function, $y = -x^2$, parallel to the y-axis 4 units upwards:

After the translation, $y = f(x) = -x^2$ is transformed into $y = f(x) + 4$
$= 4 - x^2$

These are the two required transformations.

You can use a sketch to check your work.

In Example 2, the order matters: the reflection must be first and the translation must be second. If you did the translation first and the reflection second, then the final graph would have the equation $y = -(x^2 + 4) = -x^2 - 4$

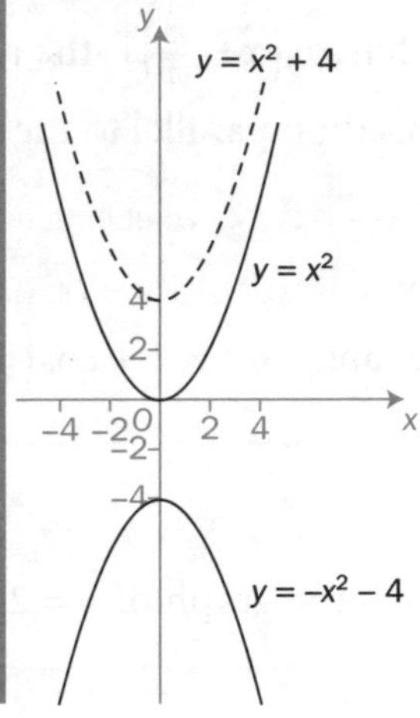

C3

EXAMPLE 3

Sketch the graph of the function $y = 3\sin 2x$, $x \in \mathbb{R}$ and find its range.

If $y = \sin x$, then $y = \sin 2x$ is the result of a stretch parallel to the x-axis with scale factor $\frac{1}{2}$.

The 'stretch' is really a 'squash'.

If $y = \sin 2x$, then $y = 3\sin 2x$ is the result of a stretch parallel to the y-axis with scale factor 3.

The range of $y = 3\sin 2x$ is $-3 \leqslant x \leqslant 3$ (or $|x| \leqslant 3$).

Each point on $y = \sin 2x$ maps onto a point three times further away from the x-axis.

The stretch parallel to the x-axis has changed the period from 2π to π (360° to 180°).

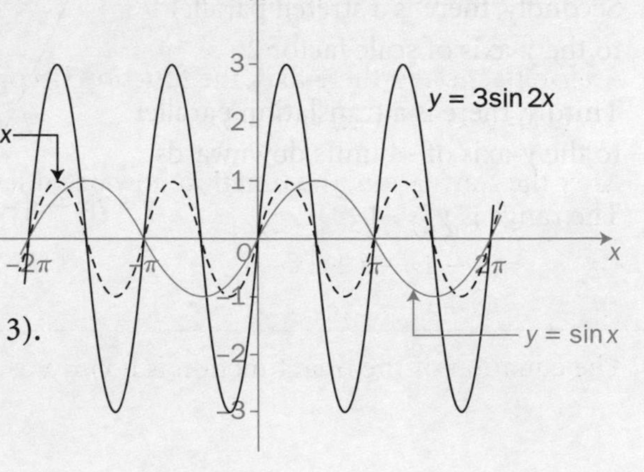

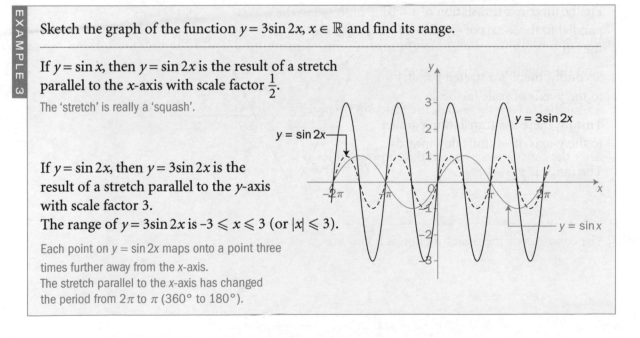

EXAMPLE 4

Sketch the graph of $y = 1 + \cos\left(x - \frac{\pi}{4}\right)$, $x \in \mathbb{R}$ and find its range.

If $y = \cos x$, then $y = \cos\left(x - \frac{\pi}{4}\right)$ is the result of a translation parallel to the x-axis of $+\frac{\pi}{4}$ units to the right.

$y = f(x - a)$ is the result of translating $y = f(x)$ parallel to the x-axis by $+a$ units.

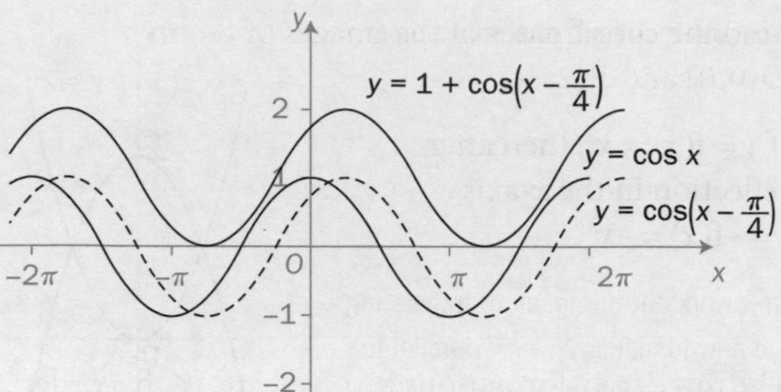

If $y = \cos\left(x - \frac{\pi}{4}\right)$, then $y = 1 + \cos\left(x - \frac{\pi}{4}\right)$ is the result of a translation parallel to the y-axis of +1 unit upwards.

$y = f(x) + a$ is the result of translating $y = f(x)$ parallel to the y-axis by $+a$ units.

The range of $y = 1 + \cos\left(x - \frac{\pi}{4}\right)$ is $0 \leqslant x \leqslant 2$.

By not involving any stretches in this problem, you have not changed the period of 2π.

In this problem, the order in which you do the two translations does not matter.

C3

EXAMPLE 5

Sketch the graph of $y = 2|x + 3| - 4$, $x \in \mathbb{R}$ and find its range.

Three transformations of $y = |x|$ are involved.

Firstly, there is a translation of $y = |x|$ parallel to the x-axis of –3 units to the left.

Secondly, there is a stretch parallel to the y-axis of scale factor 2.

Thirdly, there is a translation parallel to the y-axis of –4 units downwards.

The range is $y \geqslant -4$

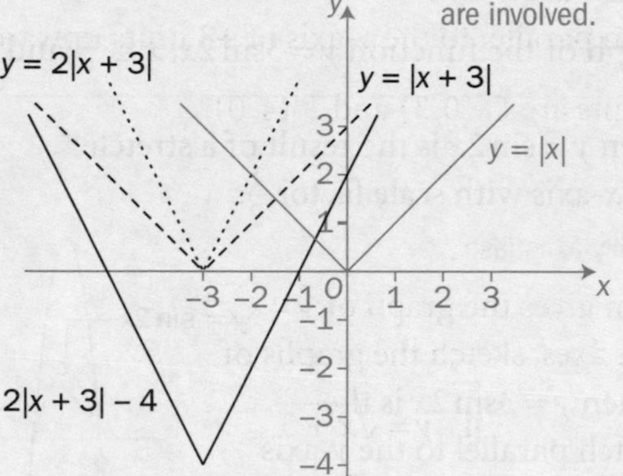

EXAMPLE 6

This diagram shows the graph of $y = f(x)$

On the same axes, sketch the graph of $y = 3 - f\left(\frac{1}{2}x\right)$

Find the coordinates of the images of points $O(0,0)$ and $P(2,3)$.

This function is defined by its graph rather than by an equation.

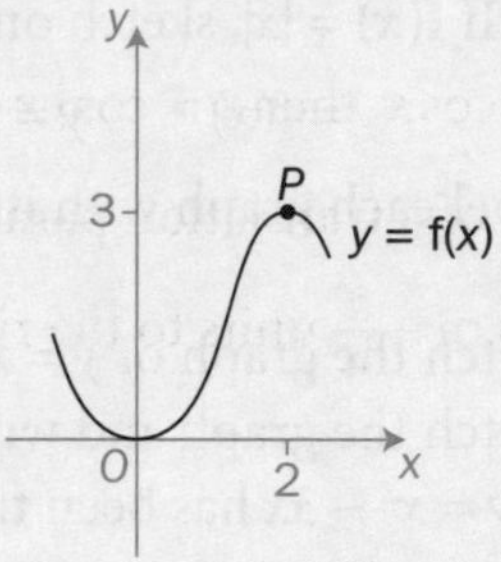

The three transformations involved are, in this order:

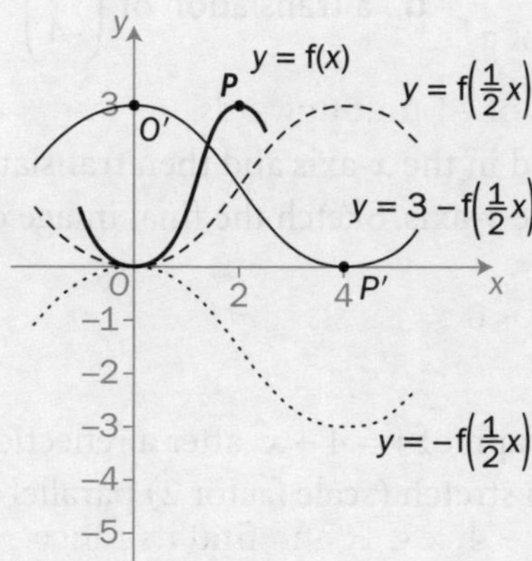

i a stretch parallel to the x-axis of scale factor 2

ii a reflection in the x-axis

iii a translation parallel to the y-axis of +3 units upwards.

The image points are $O'(0,3)$ and $P'(4,0)$.

Exercise 1.7

1 This diagram gives the graph of $y = \sqrt{x}$

On the same axes, sketch the graphs of

i $y = \sqrt{x}$ **ii** $y = \sqrt{x+3}$

iii $y = \sqrt{x-3}$ **iv** $y = \sqrt{x} + 3$

v $y = \sqrt{x} - 3$

Label each sketch with its equation.

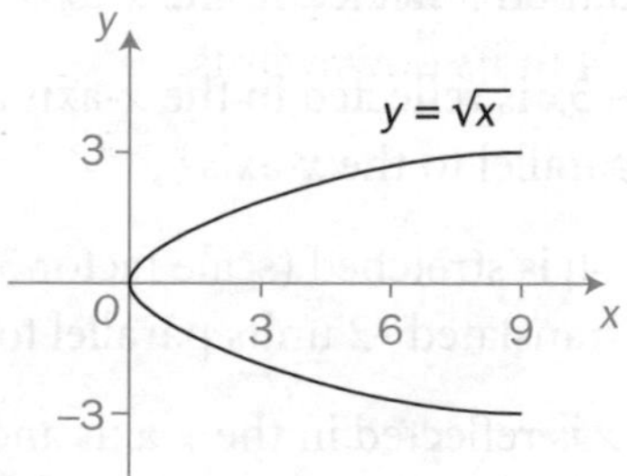

2 On the same axes, sketch the graphs of

$y = x^2$ $y = x^2 + 3$ $y = x^2 - 3$ $y = (x+3)^2$ $y = (x-3)^2$

Label each sketch with its equation.

C3

3 **a** If $f(x) = |x|$, sketch on the same diagram the graphs of

$$y = f(x), \qquad y = f(2x), \qquad y = 2f(x)$$

b If $f(x) = |x|$, sketch on the same diagram the graphs of

$$y = f(x) \qquad y = f(x+2) \qquad y = f(x-2)$$

Label each graph with its equation.

4 Sketch the graph of $y = x^2 - 3x$
Sketch the graph and write the equation after the graph of $y = x^2 - 3x$ has been transformed by

a a reflection in the x-axis

b a reflection in the y-axis

c a translation of $\begin{pmatrix} 2 \\ 0 \end{pmatrix}$

d a translation of $\begin{pmatrix} 0 \\ -4 \end{pmatrix}$.

5 The graph of $y = x^2$ is reflected in the x-axis and then translated 5 units upwards parallel to the y-axis. Sketch the final image of $y = x^2$ and write its equation.

6 Sketch the graph of $y = 4 - x^2$

a Sketch the image of the graph of $y = 4 - x^2$ after a reflection in the x-axis followed by a stretch (scale factor 2) parallel to the x-axis.

b Sketch the image of the graph of $y = 4 - x^2$ after a reflection in the y-axis followed by a stretch (scale factor 2) parallel to the x-axis.

Write the equations for the resulting graphs.

7 Sketch the graph of $y = f(x)$ and its image when

a $f(x) = x^2 - 4x + 3$ is stretched (scale factor 2) parallel to the y-axis and then reflected in the x-axis

b $f(x) = x^2 - 3x$ is reflected in the x-axis and then translated +1 units parallel to the x-axis

c $f(x) = |x - 4|$ is stretched (scale factor 3) parallel to the x-axis and then translated −2 units parallel to the x-axis

d $f(x) = \sin x$ is reflected in the y-axis and translated +1 unit parallel to the y-axis

e $f(x) = \sin x$ is reflected in the x-axis and translated by the vector $\begin{pmatrix} \frac{\pi}{2} \\ 3 \end{pmatrix}$.

In each case, write the equation of the image.

8 Describe the transformations of the graph of $y = \cos x$ which result in a graph with the equation

a $y = 1 + 2\cos x$

b $y = 3 + \cos\left(x + \frac{\pi}{4}\right)$

c $y = 3 - \cos 2x$

d $y = 4\cos\left(x - \frac{\pi}{2}\right)$

e $y = 2\cos 3x$

f $y = \frac{1}{2}\cos(-x)$

9 On the same axes, sketch the graphs of $y = \tan x$ and $y = 1 + \tan\left(x - \frac{\pi}{2}\right)$ for $0 \leqslant x \leqslant \pi$

10 The graph shows the function $y = f(x)$

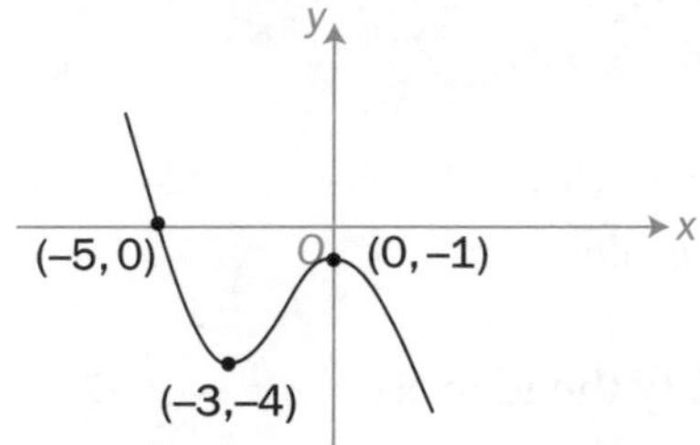

a The graph of the function $y = f(x)$ is transformed into the graph of $y = f\left(\frac{1}{2}x\right) + 1$

i Describe the transformations which have taken place.

ii Sketch the graph of $y = f(x)$ and its image on the same diagram.

b Repeat when $f(x)$ is transformed into $y = 2 - 3f(x - 1)$

INVESTIGATION

11 Which two transformations result in the graph of $y = f(x)$ being

a enlarged with scale factor k and centre $(0, 0)$

b rotated through 180° about the origin $(0, 0)$?

Write the equation of the image of $y = f(x)$ in each case.

C3

Review 1

1 Express each of these expressions as a single fraction in its simplest form.

a $\dfrac{1}{x-1}+\dfrac{2}{2x+3}$ **b** $\dfrac{2}{x+1}-\dfrac{2x}{x+2}$

c $\dfrac{3}{(x+1)(x-4)}+\dfrac{3}{(x-1)(x-4)}$ **d** $\dfrac{2a^3(a+b)}{3b}\times\dfrac{6b^2}{a^2-b^2}$

e $\dfrac{3m(m-2)}{m-1}\div\dfrac{m^2-m-2}{m^2}$ **f** $\dfrac{4x^2}{(x-1)^2}\times\dfrac{x^2-3x+2}{8x^3(x-2)}$

2 a Express $\dfrac{x}{(x+1)(x+3)}+\dfrac{x+12}{x^2-9}$ as a single fraction in its simplest form. [(c) Edexcel Limited 2001]

b Hence, find the solution of $\dfrac{x}{(x+1)(x+3)}+\dfrac{x+12}{x^2-9}=-1\dfrac{1}{2}$

3 a Find the quotient and remainder when

i $4x^3+4x^2-5x-3$ is divided by $2x+1$

ii x^3-7x+8 is divided by $x-2$

b Find the function $f(x)$ and the constant k which satisfy the identity $2x^3-3x^2+4\equiv(x-1)\,f(x)+k$

c Divide $f(x)=x^3-2x^2-9x+18$ by $x-2$ and so factorise $f(x)$ completely.

4 The functions f and g are defined by $f(x)=x^2-1, x\in\mathbb{R}, x\geqslant 0$ and $g(x)=2x-1, x\in\mathbb{R}$

a Find the values of

i $f(3)$ ii $g(3)$ iii $fg(3)$

iv $gf(3)$ v $f^{-1}(3)$ vi $g^{-1}(3)$

b Find algebraic expressions for

i $fg(x)$ ii $gf(x)$

iii $f^{-1}(x)$ iv $gf^{-1}(x)$

c Prove that there are no solutions to the equation $fg(x)=gf(x)$

C3

5 The function f is defined over the domain $0 \leqslant x \leqslant 4$ by

$$f(x) = x \qquad 0 \leqslant x < 2$$

$$f(x) = 4 - x \qquad 2 \leqslant x \leqslant 4$$

a Sketch the graph of f over its domain.

b Find all the values of x for which $f(x) = \frac{x+4}{4}$ [(c) Edexcel Limited 2002]

6 The function g is defined by g: $x \rightarrow 4x^2 + 9, x \in \mathbb{R}, x \geqslant 0$

a Find the inverse of g.

b Find the value of $g^{-1}(18)$.

7 Prove that the function $f(x) = \frac{x+2}{x-1}, x \in \mathbb{R}, x \neq 1$ is self-inverse.
Also prove that the only elements of the domain which map onto themselves are $x = 1 \pm \sqrt{3}$

Self-inverse means that the function and its inverse are identical.

8 Given that $f(x) = x^2 - 5x + 6$, sketch the graphs of

a $y = |f(x)|$

b $y = f(|x|)$

C3

9 This diagram shows the graph of the function $y = f(x), -1 \leqslant x \leqslant 6$

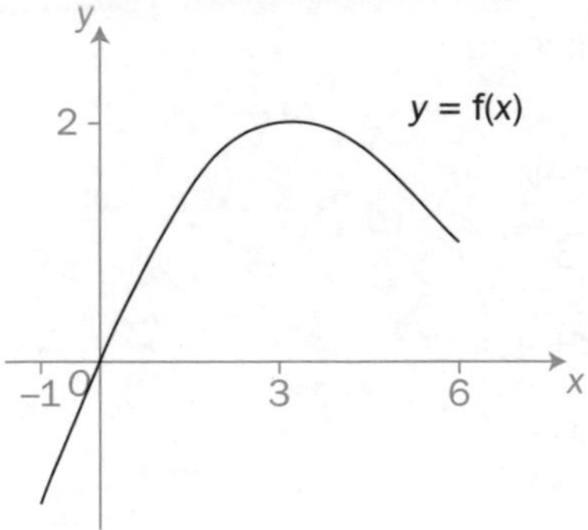

Sketch, on separate diagrams, the graphs of

a $y = f(x) - 2$

b $y = |f(x)|$

c $y = f(|x|)$

Give the coordinates of any turning points on your three diagrams.

10 The graphs of $y = x - 2$ and $y = 1$ are shown on this diagram.

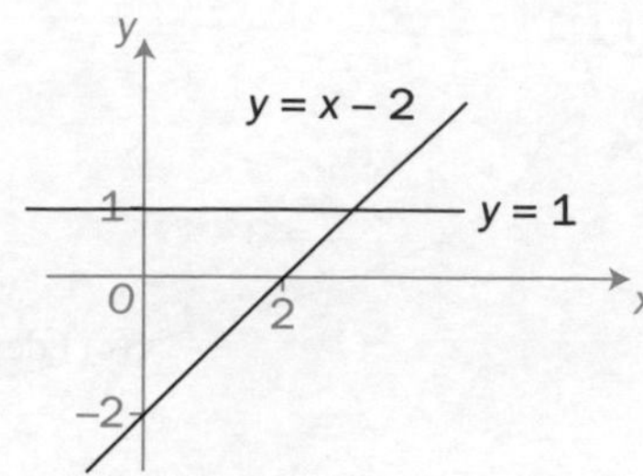

Sketch the graphs of $y = |x - 2|$ and $y = 1$ on the same diagram and find their points of intersection.

11 Sketch the graphs of

a $y = |2x - 3|$ **b** $y = 2|x| - 3$

c $y = |4 - x|$ **d** $y = 4 - |x|$

e $y = |x^2 - 5x + 4|$ **f** $y = |x|^2 - 5|x| + 4$

g $y = |2x - x^2|$ **h** $y = 2|x| - |x|^2$

C3

12 Solve the equations

a $|3x - 2| = x + 2$ **b** $3|x| - 2 = x + 2$

c $|2x - 3| = |4 - x|$ **d** $2|x| - 3 = |4 - x|$

e $|6 - 2x| = x + 3$ **f** $|6 - 2x| = |x| + 3$

13 Solve the inequalities

a $|2x - 3| > |4 - x|$ **b** $2|x| - 3 < |4 - x|$

c $|2x - 4| < x$ **d** $|6 - 2x| > x + 3$

e $6 - 2|x| \geqslant 1 - x$ **f** $(|x| - 2)^2 \leqslant x + 4$

14 Solve

a $|x^2 - 4| = x + 2$

b $|x^2 - 4| \geqslant x + 2$

c $|x^2 - 4| \geqslant |x| + 2$

15 The graph of the function $f(x) = 3 - \sqrt{x}, x \geqslant 0$ is shown on this diagram.

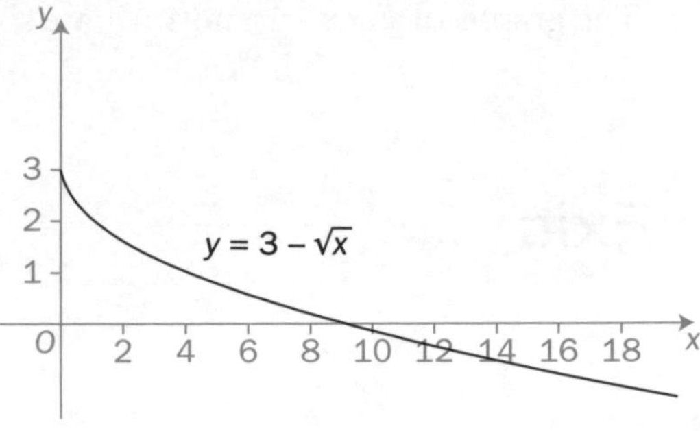

a Give the range of f(x).

b Find the values of
i $f^{-1}f(6)$ ii $f^{-1}(-1)$

c Sketch the graph of $y = |f(x)|$

Find the values of the constant c if the equation $|f(x)| = c$ has exactly two roots.

16 This diagram shows the graph of $y = f(x)$

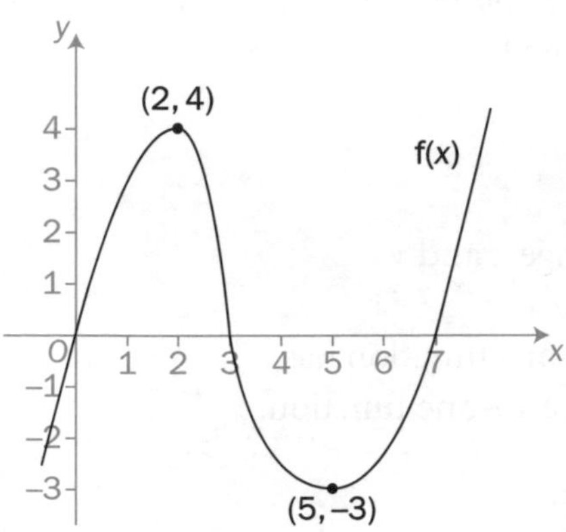

Sketch on separate diagrams the graphs of

a $y = \frac{1}{2}f(x+1)$ **b** $y = f\left(\frac{1}{2}x\right) + 1$

17 The graph of $y = x^2$ is transformed onto each of these graphs by successive transformations.
Describe the transformations which have taken place in each case.

a $y = 3x^2 - 5$ **b** $y = \frac{1}{2}(x-2)^2$

c $y = 1 - 2x^2$

18 $f(x) = \frac{2x+5}{x+3} - \frac{1}{(x+3)(x+2)}, x > -2$

a Express f(x) as a single fraction in its simplest form.

b Hence show that $f(x) = 2 - \frac{1}{x+2}, x > -2$

c The curve $y = \frac{1}{x}, x > 0$, is mapped onto the curve $y = f(x)$, using three successive transformations T_1, T_2 and T_3, where T_1 and T_3 are translations.
Describe fully T_1, T_2 and T_3.

C3

Exit

Summary

Refer to

- Algebraic division follows the same procedure as numerical long division. 1.2
- A **function** maps each element of the domain onto *one and only one* element of the range. 1.3
- The **composite function** gf(x) means that the function f is applied first and the function g is applied second. The function gf(x) only exists if the range of f is part of the domain of g. 1.3
- To find the **inverse function** $f^{-1}(x)$ of f(x)

  either use a flow diagram in reverse

  or make x the subject of $y = f(x)$ and then interchange x and y

  or reflect the graph of $y = f(x)$ in the line $y = x$.

  A one-to-one function always has an inverse. A many-to-one function has an inverse only if the domain is restricted to make it a one-to-one function. 1.4
- If $|x| < a$, then $-a < x < a$. If $|x| > a$, then $x > a$ or $x < -a$.

  To sketch the graph of $y = |f(x)|$, sketch $y = f(x)$ and reflect any part *below* the x-axis in the x-axis.

  To sketch the graph of $y = f(|x|)$, sketch $y = f(x)$ for $x > 0$ only and reflect it in the y-axis. 1.5, 1.6
- You can apply a combination of transformations to the graph of $y = f(x)$. The order of the transformations may affect the result. 1.7

Links

Many situations in real life involving several variables and uncertainty can be modelled using mathematics.

Computation using complex numerical methods is used in forecasting weather, and mathematics is essential in understanding and predicting climate change.

Even though, in reality, the variables involved are complicated, they can be simplified using a mathematical model which can then be used to understand the system being studied and to make predictions.

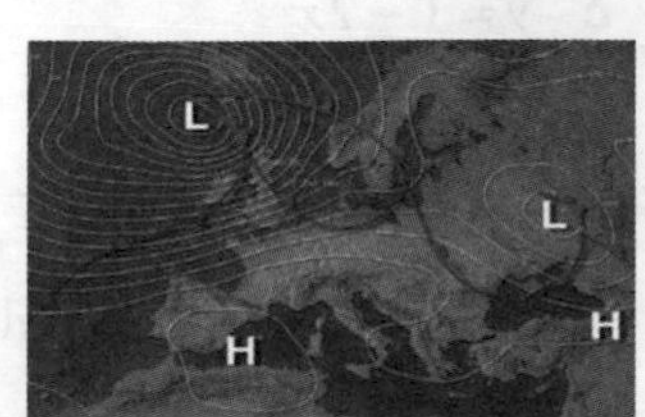

2 Trigonometry

This chapter will show you how to
- use reciprocal and inverse trigonometric functions
- solve trigonometric equations and prove identities
- use the compound angle formulae
- use the double angle and half angle formulae
- find equivalent forms for $a\cos\theta + b\sin\theta$.

Before you start

You should know how to:

1 Use the special triangles.

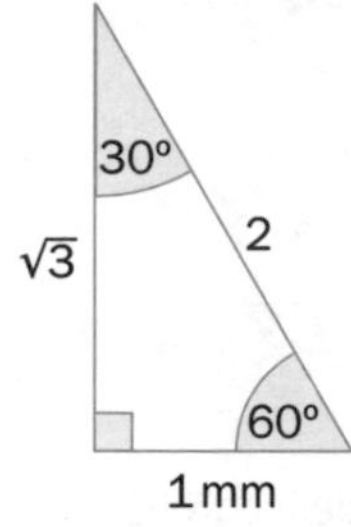

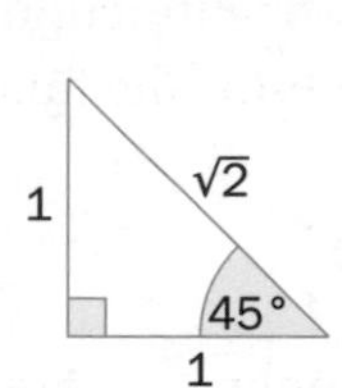

e.g. $\sin 30° + \tan 45° = \frac{1}{2} + 1 = 1.5$

2 Find $\cos\theta$ when θ is acute and $\tan\theta$ is known.

e.g. If $\tan\theta = \frac{7}{24}$,

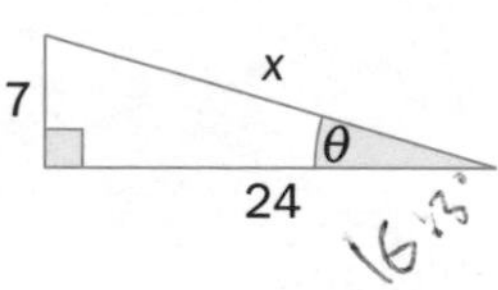

then $x = \sqrt{7^2 + 24^2} = 25$

and $\cos\theta = \frac{24}{25}$

3 Find angles in all four quadrants.

e.g. If $\cos\theta = -\frac{1}{2}$, then θ is in the second and third quadrants.

Possible values of $\theta = 180° \pm 60° = 120°$ or $240°$

4 Use identities involving $\sin\theta$, $\cos\theta$ and $\tan\theta$.

e.g. If $\sin\theta = 0.6$ and θ is acute, then $\sin^2\theta + \cos^2\theta = 1$ gives $\cos\theta = \sqrt{1 - (0.6)^2} = \sqrt{0.64} = 0.8$

Check in:

1 Find the values of

a $\sin 60° + \cos 30°$

b $\tan 60° + \tan 45°$

c $\sin^2 30° + \sin^2 60°$

d $\sin^2 45° - \cos^2 45°$

2 a If θ is acute and $\sin\theta = \frac{8}{17}$, find the values of $\cos\theta$ and $\tan\theta$.

b If θ is obtuse and $\sin\theta = \frac{4}{5}$, find the values of $\cos\theta$ and $\tan\theta$.

3 Find θ such that $0° \leqslant \theta \leqslant 360°$ when

a $\sin\theta = 0.3$

b $\tan\theta = -2$

c $\cos\theta = \frac{1}{4}$

4 a If θ is obtuse and $\sin\theta = 0.4$, find $\cos\theta$.

b Prove the identity $1 + \frac{1}{\tan^2\theta} \equiv \frac{1}{\sin^2\theta}$

2.1 Reciprocal trigonometric functions

The reciprocal trigonometric functions, secant, cosecant and cotangent, are defined as

$$\sec\theta = \frac{1}{\cos\theta} \qquad \operatorname{cosec}\theta = \frac{1}{\sin\theta} \qquad \cot\theta = \frac{1}{\tan\theta} = \frac{\cos\theta}{\sin\theta}$$

Be careful not to confuse these. You would expect cosec to be related to cos and sec to sin.

EXAMPLE 1

Find, to 3 decimal places, the values of

a $\sec 40°$ **b** $\cot 162°$ **c** $\operatorname{cosec}\frac{2\pi}{5}$

---

a $\sec 40° = \frac{1}{\cos 40°} = \frac{1}{0.76604} = 1.305$

b $\cot 162° = \frac{1}{\tan 162°} = \frac{1}{-\tan 18°} = \frac{1}{-0.32492} = -3.077$

c $\operatorname{cosec}\frac{2\pi}{5} = \frac{1}{\sin\frac{2\pi}{5}} = \frac{1}{0.95106} = 1.051$

162° is in the second quadrant where tan is negative.

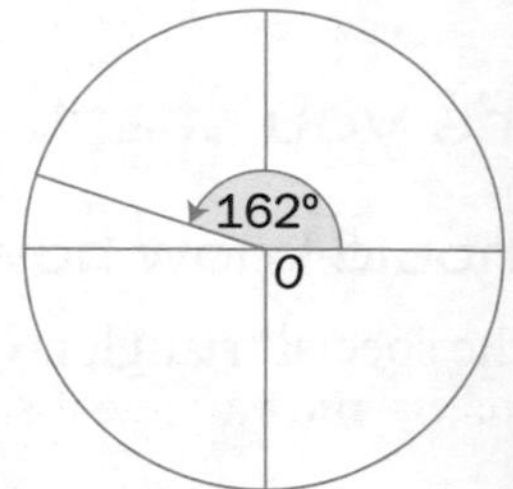

$\frac{2\pi}{5}$ radians $= \frac{360°}{5} = 72°$

The graphs of sec θ, cosec θ and cot θ

Starting with the graph of $y = \cos\theta$, you can sketch the graph of its reciprocal $y = \sec\theta$

sec θ is not defined when $\cos\theta = 0$

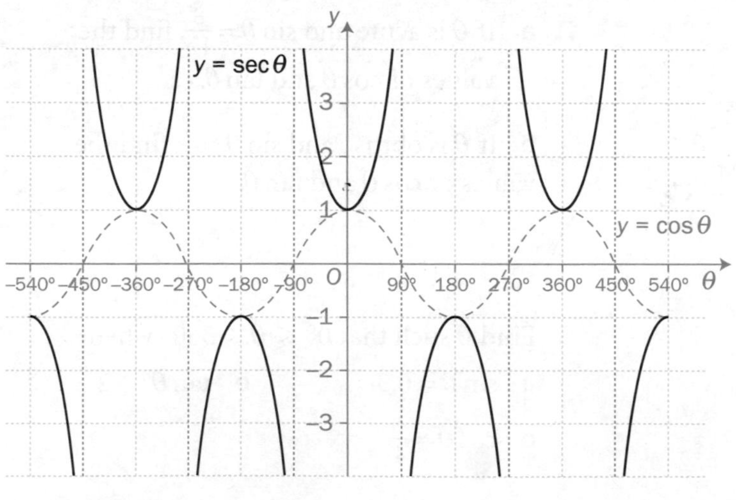

$y = \cos\theta$ and $y = \sec\theta$ both have a period of 360° (2π radians).

When $\cos\theta = \pm 1$, $\sec\theta = \pm 1$
When $\cos\theta = 0$, $\sec\theta = \infty$ and its graph has vertical asymptotes.

The domains and ranges of $\cos\theta$ and $\sec\theta$ are:

	Domain	Range
$y = \cos\theta$	$\theta \in \mathbb{R}$	$y \in \mathbb{R}, -1 \leqslant y \leqslant 1$
$y = \sec\theta$	$\theta \in \mathbb{R}, \theta \neq \pm 90°, \pm 270°, \ldots$	$y \in \mathbb{R}, y \geqslant 1, y \leqslant -1$

When $\cos\theta$ is positive, $\sec\theta$ is also positive. When $\cos\theta$ is negative, $\sec\theta$ is also negative.

C3

Similarly, you can sketch the graph of $y = \operatorname{cosec}\theta$

cosec θ is not defined when $\sin\theta = 0$

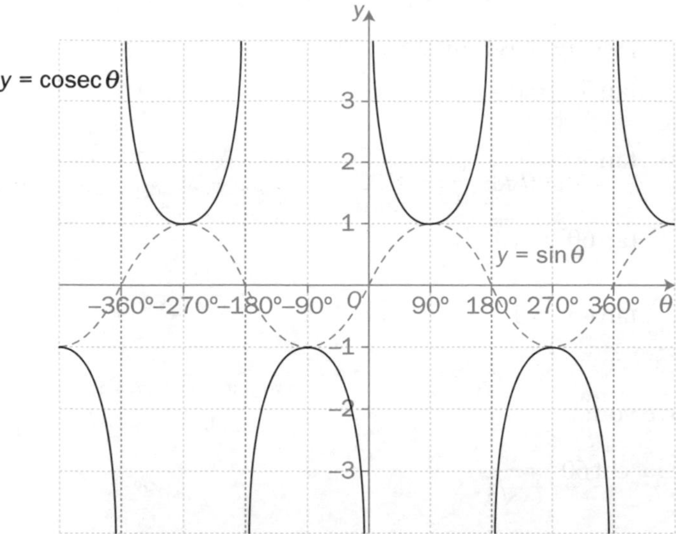

$y = \sin\theta$ and $y = \operatorname{cosec}\theta$ have periods of 360° (2π radians).

When $\sin\theta = \pm1$, $\operatorname{cosec}\theta = \pm1$
When $\sin\theta = 0$, the graph of $\operatorname{cosec}\theta$ has a vertical asymptote.

The domains and ranges of $\sin\theta$ and $\operatorname{cosec}\theta$ are:

When $\sin\theta$ is positive, $\operatorname{cosec}\theta$ is also positive. When $\sin\theta$ is negative, $\operatorname{cosec}\theta$ is also negative.

	Domain	Range
$y = \sin\theta$	$\theta \in \mathbb{R}$	$y \in \mathbb{R}, -1 \leqslant y \leqslant 1$
$y = \operatorname{cosec}\theta$	$\theta \in \mathbb{R}, \theta \neq 0°, \pm180°, \pm360°, \ldots$	$y \in \mathbb{R}, y \geqslant 1, y \leqslant -1$

The graph of $\cot\theta$ is shown here.

cot θ is not defined when $\sin\theta = 0$

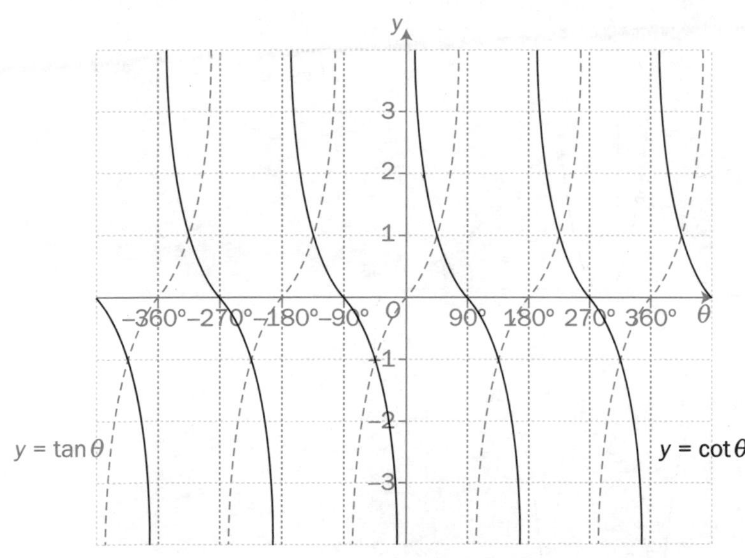

$y = \tan\theta$ and $y = \cot\theta$ have periods of 180° (π radians).

When $\tan\theta = \pm1$, $\cot\theta = \pm1$
When $\tan\theta = 0$, the graph of $\cot\theta$ has vertical asymptotes.

The domains and ranges of $\tan\theta$ and $\cot\theta$ are:

	Domain	Range
$y = \tan\theta$	$\theta \in \mathbb{R}, \theta \neq \pm90°, \pm270°, \ldots$	$y \in \mathbb{R}$
$y = \cot\theta$	$\theta \in \mathbb{R}, \theta \neq 0°, \pm180°, \ldots$	$y \in \mathbb{R}$

You should be familiar with all of these graphs when θ is in radians as well as in degrees.

C3

Standard trigonometric results

The standard trigonometric results from these two special triangles are:

See C2 Chapter 12 for revision.

$\sin 0° = 0 \qquad \cos 0° = 1 \qquad \tan 0° = 0$

$\sin 30° = \frac{1}{2} \qquad \cos 30° = \frac{\sqrt{3}}{2} \qquad \tan 30° = \frac{1}{\sqrt{3}}$

$\sin 60° = \frac{\sqrt{3}}{2} \qquad \cos 60° = \frac{1}{2} \qquad \tan 60° = \sqrt{3}$

$\sin 45° = \frac{1}{\sqrt{2}} \qquad \cos 45° = \frac{1}{\sqrt{2}} \qquad \tan 45° = 1$

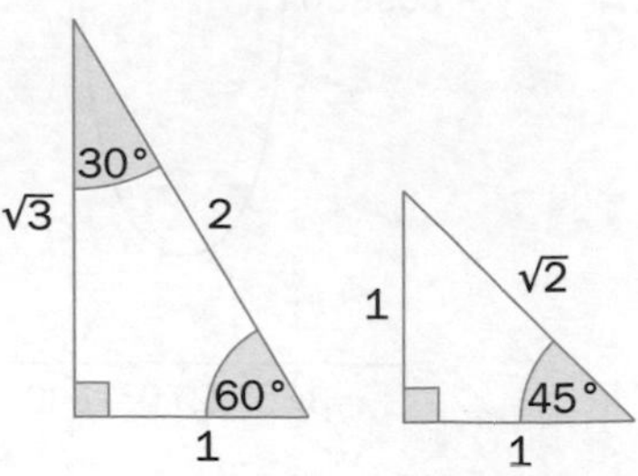

You should also know these results in radians, where 180° = π radians.

The same two special triangles also give you:

$\operatorname{cosec} 60° = \frac{2}{\sqrt{3}} \qquad \sec 60° = 2 \qquad \cot 60° = \frac{1}{\sqrt{3}}$

$\operatorname{cosec} 30° = 2 \qquad \sec 30° = \frac{2}{\sqrt{3}} \qquad \cot 30° = \sqrt{3}$

$\operatorname{cosec} 45° = \sqrt{2} \qquad \sec 45° = \sqrt{2} \qquad \cot 45° = 1$

and from the trigonometric graphs you get:

$\sin 90° = 1 \qquad \cos 90° = 0 \qquad \tan 90° = \infty$

$\sin 180° = 0 \qquad \cos 180° = -1 \qquad \tan 180° = 0$

The triangle shown gives

$$\tan\theta = \frac{a}{b} = \frac{\frac{a}{c}}{\frac{b}{c}} = \frac{\sin\theta}{\cos\theta}$$

By Pythagoras' Theorem, $a^2 + b^2 = c^2$

Divide by c^2: $\sin^2\theta + \cos^2\theta = 1$

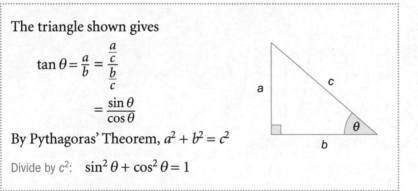

See C2 for revision.

You have the following results:

$\tan\theta = \frac{\sin\theta}{\cos\theta} \qquad \sin^2\theta + \cos^2\theta = 1$

$\sin\theta = \cos(90° - \theta) \qquad \cos\theta = \sin(90° - \theta)$

Consider $\cot\theta = \dfrac{1}{\tan\theta} = \dfrac{1}{\frac{\sin\theta}{\cos\theta}} = \dfrac{\cos\theta}{\sin\theta}$

Use Pythagoras' Theorem.

Divide $a^2 + b^2 = c^2$ by b^2: $\left(\frac{a}{b}\right)^2 + 1 = \left(\frac{c}{b}\right)^2$ so $\tan^2\theta + 1 = \sec^2\theta$

Divide $a^2 + b^2 = c^2$ by a^2: $1 + \left(\frac{b}{a}\right)^2 = \left(\frac{c}{a}\right)^2$ so $1 + \cot^2\theta = \operatorname{cosec}^2\theta$

You have

$\cot\theta = \dfrac{\cos\theta}{\sin\theta}$

$\tan\theta = \cot(90° - \theta)$

$\sec^2\theta = 1 + \tan^2\theta$

$\operatorname{cosec}^2\theta = 1 + \cot^2\theta$

You need to be able to recall all of these trigonometric values and formulae. They are **not** provided in the formulae booklet in the examination.

EXAMPLE 2

Find, giving each answer as a surd, the exact values of

a $\sec 330°$ **b** $\operatorname{cosec} 225°$ **c** $\cot\dfrac{7\pi}{6}$

a 330° is in the fourth quadrant where cosine is positive.

$$\sec 330° = \frac{1}{\cos 330°} = \frac{1}{\cos 30°} = \frac{1}{\frac{\sqrt{3}}{2}} = \frac{2}{\sqrt{3}} = \frac{2\sqrt{3}}{3}$$

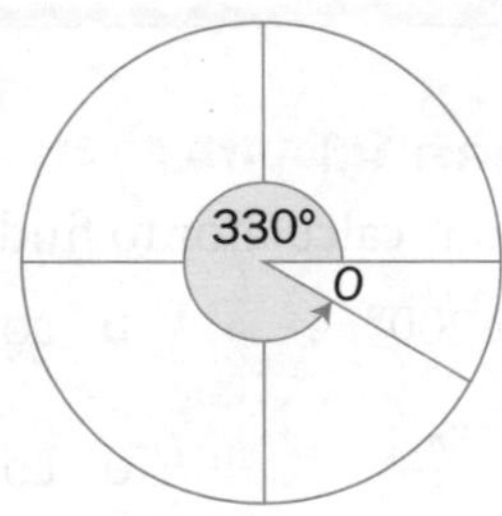

b 225° is in the third quadrant where sine is negative.

$$\operatorname{cosec} 225° = \frac{1}{-\sin 45°} = \frac{1}{-\frac{1}{\sqrt{2}}} = -\sqrt{2}$$

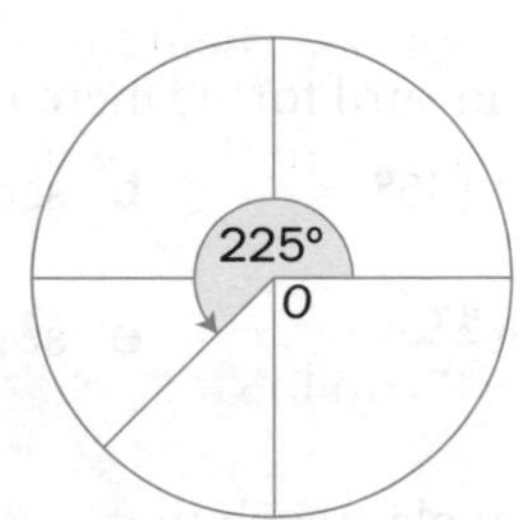

c $\dfrac{7\pi}{6}$ radians $= \dfrac{7 \times 180°}{6} = 210°$ is in the third quadrant where tangent is positive.

$$\cot\frac{7\pi}{6} = \cot 210° = \frac{1}{\tan 210°} = \frac{1}{\tan 30°} = \frac{1}{\frac{1}{\sqrt{3}}} = \sqrt{3}$$

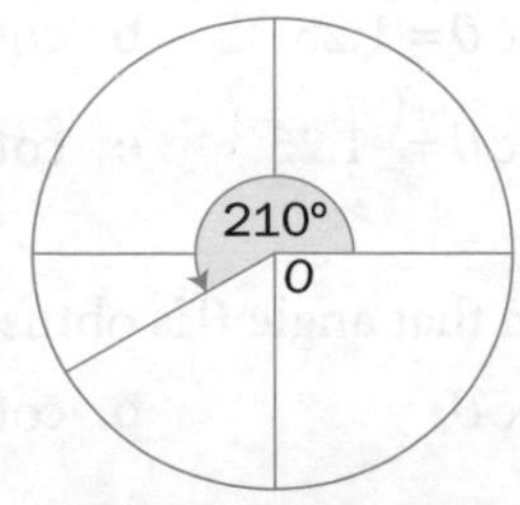

C3

EXAMPLE 3

If $\cos\theta = -\frac{5}{13}$ and θ is an obtuse angle, find the exact value of $\cot\theta$.

Method 1

Use $1 + \tan^2\theta = \sec^2\theta$ with $\sec\theta = -\frac{13}{5}$:

$$\tan^2\theta = \left(-\frac{13}{5}\right)^2 - 1$$
$$= \frac{169}{25} - 1$$
$$= \frac{144}{25}$$
$$\tan\theta = \pm\frac{12}{5}$$

But θ is obtuse, so $\tan\theta$ is negative.

So $\tan\theta = -\frac{12}{5}$

and $\cot\theta = \frac{1}{\tan\theta}$

$$= -\frac{5}{12}$$

Method 2

Draw a right-angled triangle for a first quadrant angle, ϕ, where $\cos\phi = \frac{5}{13}$:

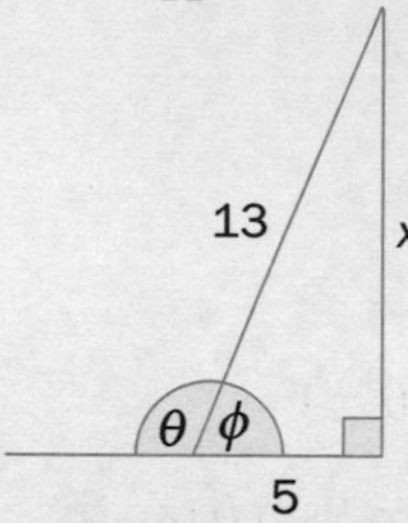

Use Pythagoras' theorem:

$$x = \sqrt{169 - 25} = 12 \text{ and}$$
$$\cot\phi = \frac{5}{12}$$

Angle θ is obtuse, so $\tan\theta$ and $\cot\theta$ are negative.

$$\cot\theta = -\cot\phi = -\frac{5}{12}$$

Exercise 2.1

1 Use your calculator to find, to 3 significant figures, the values of

a $\sec 200°$ **b** $\cot 130°$ **c** $\operatorname{cosec} 340°$

d $\sec\frac{3\pi}{5}$ **e** $\cot\frac{5\pi}{6}$ **f** $\operatorname{cosec}\frac{2\pi}{9}$

2 Find, in surd form where needed, the values of

a $\cot 135°$ **b** $\sec 120°$ **c** $\operatorname{cosec} 210°$

d $\cot\frac{4\pi}{3}$ **e** $\sec\frac{7\pi}{4}$ **f** $\operatorname{cosec}\frac{3\pi}{2}$

3 Find angle θ such that $-180° < \theta < 180°$ when

Use the graphs of these functions to help you.

a $\sec\theta = 1.25$ **b** $\cot\theta = 2.5$ **c** $\operatorname{cosec}\theta = 3.0$

d $\sec\theta = -1.25$ **e** $\cot\theta = -3.5$ **f** $\operatorname{cosec}\theta = -2.0$

4 Given that angle θ is obtuse and $\tan\theta = -\frac{8}{15}$, find the value of

a $\sec\theta$ **b** $\cot\theta$

5 Angle α is a reflex angle and $\cos\alpha = \frac{9}{41}$
Find the value of

a $\operatorname{cosec}\alpha$ b $\cot\alpha$

6 Write each of these expressions as a power of $\sec\beta$, $\operatorname{cosec}\beta$ or $\cot\beta$.

a $\frac{1}{\tan^2\beta}$ b $\frac{\sec\beta}{\cos^2\beta}$

c $\frac{1-\cos^2\beta}{\sin^3\beta}$ d $\frac{\cot^2\beta\sec^2\beta}{\sin^3\beta}$

7 Find the value of $\cot\theta$ when

a $2\sin\theta = 3\cos\theta$ b $4\tan\theta = 1$

c $\cos\theta\sin\theta = \sin^2\theta$ d $\cos\theta = 9\sin\theta\tan\theta$

e $\sin\theta = 3\tan^2\theta\cos\theta$ f $\operatorname{cosec}\theta = 2$

8 Simplify these expressions.

a $\cot x\tan x$ b $\cot x\sin x\tan x$

c $\cot^2 x\sec x\sin x$ d $\operatorname{cosec} x\sec x\sin^2 x$

e $\sin x(\cot x\cos x + \sin x)$ f $\frac{1}{\cos^2\alpha} - \frac{\sec^2\alpha}{\operatorname{cosec}^2\alpha}$

g $\frac{1}{\operatorname{cosec}^2\alpha} + \cot^2\alpha\sin^2\alpha$ h $\frac{1}{\sin^2\alpha} - \frac{\cot\alpha}{\tan\alpha}$

i $\sec\alpha\left(\frac{1}{\cos\alpha} - \frac{\tan\alpha}{\operatorname{cosec}\alpha}\right)$

INVESTIGATION

9 What transformation maps the graph of $y = \sec x$ onto $y = \sec(x - 90°)$?

What two transformations are needed to map the graph of $y = \cot x$ onto $y = \cot(90° - x)$?

Use computer software to compare the graphs of $y = \sec(x - 90°)$ and $y = \operatorname{cosec} x$ and to compare the graphs of $y = \cot(90° - x)$ and $y = \tan x$.

What can you deduce?

Is this investigation sufficient as proof of your results?

C3

2.2 Trigonometric equations and identities

An equation is true for some, but *not all*, values of the variables involved.

An identity is true for *all* values of the variables involved.

There are two strategies for proving identities.
You prove:

either that the LHS of the identity is equal to the RHS (or vice versa).

or that both sides of the identity are equal to a common third expression.

It is usual to start with the more complicated side of the identity.

EXAMPLE 1

Prove that $\tan x \sin x + \cos x \equiv \sec x$

Manipulate the LHS of the identity:

Choose one side of the identity to work on.

$$\text{LHS} = \tan x \sin x + \cos x = \frac{\sin x}{\cos x} \times \sin x + \cos x$$

$$= \frac{\sin^2 x + \cos^2 x}{\cos x}$$

$$= \frac{1}{\cos x} = \sec x = \text{RHS}$$

So the identity is proved.

EXAMPLE 2

Prove that $\dfrac{\cos x \tan x}{\operatorname{cosec}^2 x} \equiv \dfrac{\cos^3 x}{\cot^3 x}$

Show that both sides are equal to a common expression:

$$\text{LHS} = \cos x \times \frac{\sin x}{\cos x} \times \sin^2 x = \sin^3 x$$

Remember $\operatorname{cosec}^2 x = \dfrac{1}{\sin^2 x}$

$$\text{RHS} = \cos^3 x \times \frac{\sin^3 x}{\cos^3 x} = \sin^3 x$$

$\cot^3 x = \dfrac{1}{\tan^3 x} = \dfrac{\cos^3 x}{\sin^3 x}$

Hence, LHS ≡ RHS and the identity is proved.

EXAMPLE 3

Solve the equation $1 + \tan x = \sec^2 x$ for $0 \leqslant x \leqslant 2\pi$

Substitute $1 + \tan^2 x$ for $\sec^2 x$:

$$1 + \tan x = 1 + \tan^2 x$$
$$0 = \tan^2 x - \tan x$$
$$0 = \tan x(\tan x - 1)$$
$$\text{so} \quad \tan x = 0 \text{ or } \tan x = 1$$

$\tan x = 0$ gives $x = 0$ or 2π

or $\tan x = 1$ gives $x = \frac{\pi}{4}$ or $\frac{5\pi}{4}$

The solutions are $x = 0, \frac{\pi}{4}, \frac{5\pi}{4}, 2\pi$

tan is positive in the first and third quadrants.

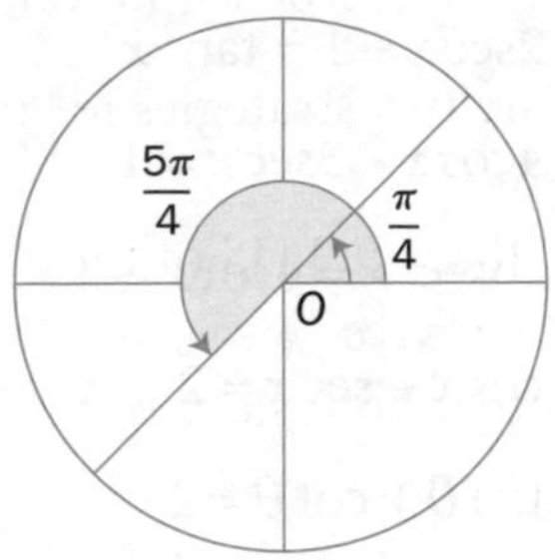

Exercise 2.2

1 Solve these equations for $-180° \leqslant x \leqslant 180°$

See C2 for revision of translations of trignometric functions.

a $\sec(x - 10°) = 3$

b $\operatorname{cosec}(x + 20°) = -4$

c $\cot(x + 30°) = -2$

d $\cot 2x = \frac{1}{2}$

e $\sec(2x + 40°) = -2$

f $\operatorname{cosec}\left(\frac{1}{2}x - 10°\right) = -\frac{3}{2}$

g $3\cos x = \sec x$

h $4\cot x = 3\tan x$

i $3\cos x - \cot x = 0$

j $4\sin x = 3\tan x$

k $2\cot x = \operatorname{cosec} x$

l $\cot x = \tan x$

2 Solve these equations for $-\pi \leqslant \theta \leqslant \pi$

a $\sec^2\theta = 2$

b $\cot^2\theta = 3$

c $4\sin\theta = 3\operatorname{cosec}\theta$

d $\tan\theta = 4\sin\theta\cos\theta$

C3

3 Solve these equations for $0 \leqslant x \leqslant 360°$

a $2 + \sec^2 x = 4\tan x$

b $2\cot x = \tan x + 1$

c $3\sin x - 2\operatorname{cosec} x = 1$

d $2\sec x - 1 = \tan^2 x$

e $4\cos x - 3\sec x = 1$

f $\operatorname{cosec}^2 x = 4\cot x - 3$

g $\cos x + \sec x = 2$

h $\tan\theta + \cot\theta = 2$

i $\tan\theta + 3\cot\theta = 5\sec\theta$

4 Rewrite each pair of equations as one equation in terms of x and y.

a $x = 4\sec\alpha,\ y = 2\tan\alpha$

b $x = 3\operatorname{cosec}\alpha,\ y = 2\cot\alpha$

c $x = 4\cos\alpha,\ y = 3\tan\alpha$

d $x = 1 - \sin\alpha,\ y = 1 + \cos\alpha$

e $x = 3\cos\alpha,\ y = 4 + \tan\alpha$

f $x = a\sin\alpha,\ y = b\sec\alpha$

5 Describe the transformations which map the graph of the first equation onto the graph of the second equation.

a $y = \sec x,\ y = \sec\left(\frac{1}{2}x\right)$

b $y = \sec x,\ y = \frac{1}{2}\sec 2x$

c $y = \operatorname{cosec} x,\ y = 2\operatorname{cosec}(x + 90°)$

d $y = \cot x,\ y = -\cot(-x)$

C3

6 Find the points of intersection of these pairs of curves for $-360° \leqslant x \leqslant 360°$, giving answers to 2 significant figures where necessary.

  a $y = 1 + \cos x$, $y = 2\sin^2 x$

  b $y = 1 + 2\tan x$, $y = 1 + \sec x$

  c $y = \sec x$, $y = 1 + \cos x$

  d $y = \tan x - \sin^2 x$, $y = (\cos x - \sec x)^2$

7 Prove these identities.

  a $\tan\theta\sin\theta + \cos\theta \equiv \sec\theta$

  b $\tan\theta + \cot\theta \equiv \sec\theta\operatorname{cosec}\theta$

  c $\sec\theta - \tan\theta \equiv \dfrac{1}{\sec\theta + \tan\theta}$

  d $\dfrac{1 - \tan^2\theta}{1 + \tan^2\theta} \equiv 1 - 2\sin^2\theta$

  e $\dfrac{\sin\theta}{1 + \tan\theta} \equiv \dfrac{\cos\theta}{1 + \cot\theta}$

  f $(1 + \sec\theta)(1 - \cos\theta) \equiv \tan\theta\sin\theta$

  g $\dfrac{\sin\theta}{1 + \cos\theta} + \dfrac{1 + \cos\theta}{\sin\theta} \equiv \dfrac{2}{\sin\theta}$

  h $\dfrac{1}{\cot\theta + \operatorname{cosec}\theta} \equiv \dfrac{1 - \cos\theta}{\sin\theta}$

  i $\dfrac{\tan^2\theta + \cos^2\theta}{\sin\theta + \sec\theta} \equiv \sec\theta - \sin\theta$

INVESTIGATION

8 Use a graphical package on a computer to check your answers to all the questions in this exercise.

For example, you can check question **1** part **a** by plotting the graph of $y = \sec(x - 10°)$ and finding the points where $y = 3$.
For question **3** part **a**, you can plot the graphs of $y = 2 + \sec^2 x$ and $y = 4\tan x$ and then find their points of intersection.

C3

2.3 Inverse trigonometric functions

If $f(x) = \sin x$, then the **inverse function** is $f^{-1}(x) = \arcsin x$ or $\sin^{-1} x$.

Do not confuse $\sin^{-1} x$ with $(\sin x)^{-1}$.

E.g. $\sin \frac{\pi}{6} = 0.5$, so $\arcsin 0.5 = \frac{\pi}{6}$
arcsin 0.5 is the angle whose sine is 0.5.

$f(x) = \sin x$ is a many-to-one function.
Its domain must be restricted to $-\frac{\pi}{2} \leqslant x \leqslant \frac{\pi}{2}$ for it to have an inverse.

See Chapter 1 for revision of inverse functions.

The graph of $y = \arcsin x$ is a reflection of $y = \sin x$ in the line $y = x$

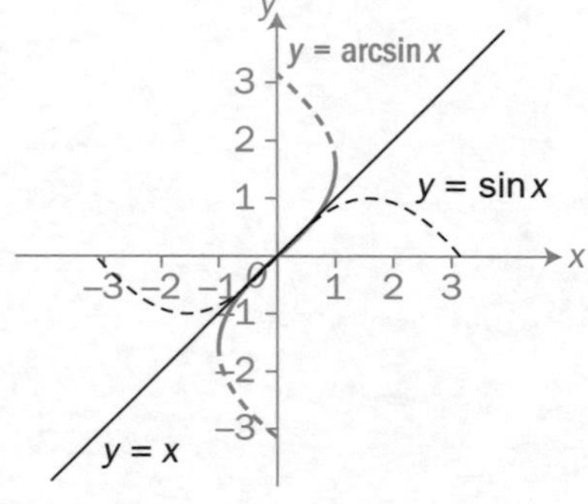

For the reflection to work, the scales on the two axes must be the same, with angles measured in radians.

The graphs show their domains and ranges as:

	Domain	Range
$f(x) = \sin x$	$-\frac{\pi}{2} \leqslant x \leqslant \frac{\pi}{2}$	$-1 \leqslant y \leqslant 1$
$f^{-1}(x) = \arcsin x$	$-1 \leqslant x \leqslant 1$	$-\frac{\pi}{2} \leqslant y \leqslant \frac{\pi}{2}$

The **principal value** of $\arcsin x$ is the unique value of $\arcsin x$ within the allowed range $-\frac{\pi}{2} \leqslant y \leqslant \frac{\pi}{2}$

This is shown as the continuous blue line on the graph.

Similarly, provided that domains are restricted, $y = \cos x$ and $y = \tan x$ have inverse functions $y = \arccos x$ and $y = \arctan x$

The alternative notations are $\cos^{-1} x$ and $\tan^{-1} x$.

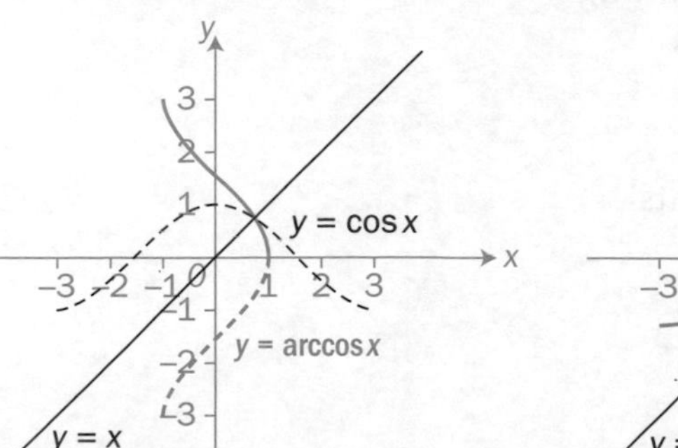

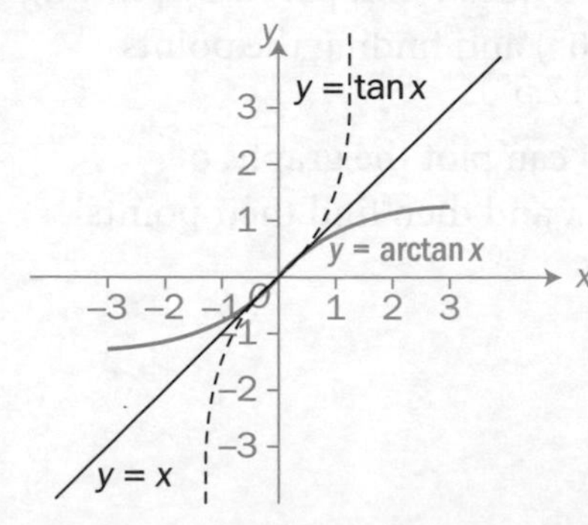

The graphs show their domains and ranges as:

	Domain	Range
$f(x) = \cos x$	$0 \leqslant x \leqslant \pi$	$-1 \leqslant y \leqslant 1$
$f^{-1}(x) = \arccos x$	$-1 \leqslant x \leqslant 1$	$0 \leqslant y \leqslant \pi$

The principal values of arccos x are in the range $0 \leqslant y \leqslant \pi$

	Domain	Range
$f(x) = \tan x$	$-\frac{\pi}{2} \leqslant x \leqslant \frac{\pi}{2}$	$y \in \mathbb{R}$
$f^{-1}(x) = \arctan x$	$x \in \mathbb{R}$	$-\frac{\pi}{2} \leqslant y \leqslant \frac{\pi}{2}$

The principal values of arctan x are in the range $-\frac{\pi}{2} \leqslant y \leqslant \frac{\pi}{2}$

EXAMPLE 1

Find the values of

a $\arccos \frac{\sqrt{3}}{2}$

b $\arctan(-1)$

c $\arcsin(-0.3)$

For $\arccos\left(\frac{\sqrt{3}}{2}\right)$, you need to find the angle whose cosine is $\frac{\sqrt{3}}{2}$.

Use the special triangles:

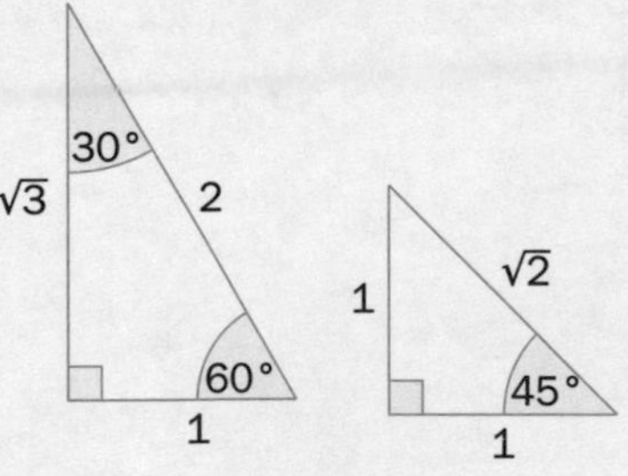

a $\arccos \frac{\sqrt{3}}{2} = \frac{\pi}{6}$ (or 30°)

b $\arctan(-1) = -\frac{\pi}{4}$ (or –45°)

c Use a calculator:

$\arcsin(-0.3) = -0.305$ (or –17.5°)

The principal values of arctan are in the first and fourth quadrants. So, an angle with a negative tangent (in this case, –1) is in the fourth quadrant.

A negative principal value of arcsin is a fourth quadrant angle.

C3

EXAMPLE 2

Find $\tan\left(\arcsin\frac{3}{4}\right)$.

You need to find $\tan\theta$ given that $\theta = \arcsin\frac{3}{4}$;

that is, find $\tan\theta$ given that $\sin\theta = \frac{3}{4}$

Take angle θ to be an acute angle.

Sine is positive, so angle θ is in the first or second quadrant.

Method 1

$\tan\theta = \dfrac{\sin\theta}{\cos\theta}$

You know $\sin\theta = \frac{3}{4}$ and so you need $\cos\theta$.

Use $\sin^2\theta + \cos^2\theta = 1$:

$$\cos\theta = \sqrt{1 - \left(\frac{3}{4}\right)^2}$$
$$= \frac{\sqrt{7}}{4}$$

Take +ve square root for cosine in the first quadrant.

$$\text{Hence, } \tan\theta = \frac{\frac{3}{4}}{\frac{\sqrt{7}}{4}}$$
$$= \frac{3}{\sqrt{7}}$$

Method 2

Draw a triangle to show $\sin\theta = \frac{3}{4}$:

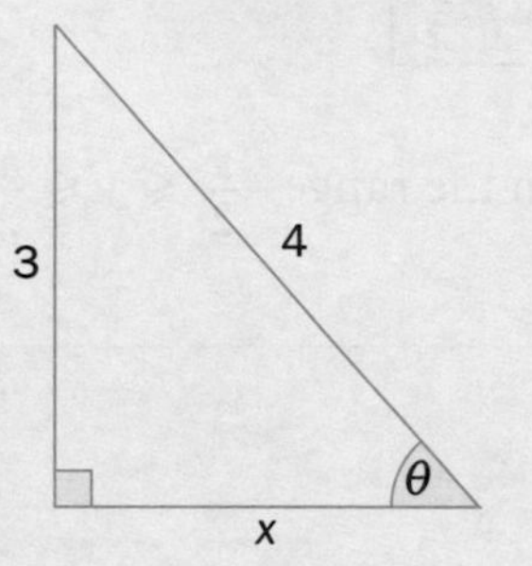

Use Pythagoras' Theorem:

$x^2 + 3^2 = 4^2$

$x = \sqrt{16 - 9} = \sqrt{7}$

So $\tan\theta = \dfrac{3}{\sqrt{7}}$

and $\tan\left(\arcsin\frac{3}{4}\right) = \dfrac{3}{\sqrt{7}}$

Exercise 2.3

1 Giving answers in terms of π, find

a $\arcsin\left(\frac{1}{\sqrt{2}}\right)$ **b** $\arctan\left(\sqrt{3}\right)$

c $\arccos 1$ **d** $\arctan(-1)$

e $\arccos\left(\frac{\sqrt{3}}{2}\right)$ **f** $\arcsin 0$

g $\arccos\left(-\frac{1}{\sqrt{2}}\right)$ **h** $\arctan\left(-\frac{1}{\sqrt{3}}\right)$

2 Find, as a surd where necessary, for angles between 0° and 180°

a $\sin x$, given that $x = \arccos \frac{1}{3}$

b $\tan\left(\arccos \frac{3}{4}\right)$

c $\cos(\arcsin 1)$

d $\sin\left(\arcsin \frac{5}{8}\right)$

e $\cos[\arctan(-1)]$

f $\sin[\arccos(-0.5)]$

g $\tan\left[\arccos\left(-\frac{2}{3}\right)\right]$

3 Find the values of

a $\arccos \frac{\sqrt{3}}{2} + \arcsin\left(-\frac{1}{2}\right)$

b $\arctan 1 - \arctan(-1)$

4 **a** If $\theta = \arcsin x$, find in terms of x

i $\cos\theta$ ii $\tan\theta$

b Express $\sec(\arccos x)$ as an algebraic expression in x.

5 **a** Given that $\alpha = \arctan x$, express $\sin\alpha + \cos\alpha$ in terms of x.

b Given that $x = \tan\theta$, find $\operatorname{arccot} x$ in terms of θ.

6 Prove these identities.

a $\arctan x \equiv \frac{\pi}{2} - \operatorname{arccot} x$

b $\arcsin x + \arccos x \equiv \frac{\pi}{2}$

c $\arctan x + \arctan\left(\frac{1}{x}\right) \equiv \frac{\pi}{2}$

INVESTIGATION

7 Explore how to input the inverse trigonometric functions using a computer's graphical package.

C3

2.4 Compound angle formulae

You can find an expression for $\sin(A + B)$ in terms of the trigonometric ratios of angle A and angle B.

It is **not** true that $\sin(A + B) = \sin A + \sin B$
You can prove this statement using a counter-example:

Let $A = B = 45°$

$$\sin(A + B) = \sin(45° + 45°) = \sin 90° = 1$$

and $\sin A + \sin B = \frac{1}{\sqrt{2}} + \frac{1}{\sqrt{2}} = \frac{2}{\sqrt{2}} = \sqrt{2} \neq 1$

So, $\sin(A + B) \neq \sin A + \sin B$

In fact, $\sin(A + B) = \sin A\cos B + \cos A\sin B$

Consider angles A and B in the right-angled triangles OPQ and OQR:

Triangles OMN and RMQ are similar so $\angle MRQ = \angle MON =$ angle A

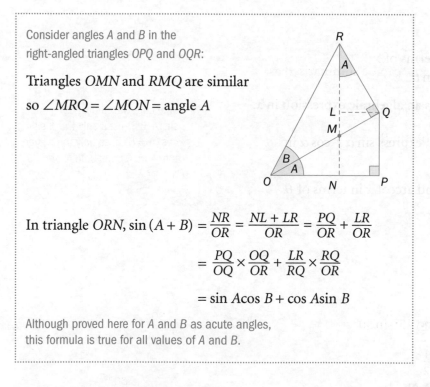

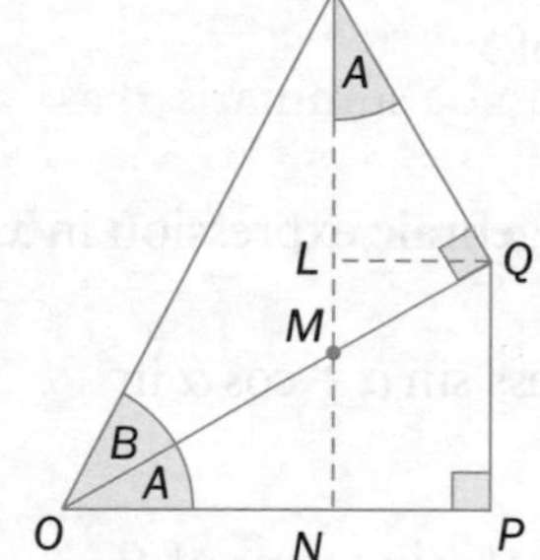

In triangle ORN, $\sin(A + B) = \frac{NR}{OR} = \frac{NL + LR}{OR} = \frac{PQ}{OR} + \frac{LR}{OR}$

$$= \frac{PQ}{OQ} \times \frac{OQ}{OR} + \frac{LR}{RQ} \times \frac{RQ}{OR}$$

$$= \sin A\cos B + \cos A\sin B$$

NL = PQ

Although proved here for A and B as acute angles, this formula is true for all values of A and B.

Replacing B by $-B$ and using
$\sin(-B) = -\sin B$ and $\cos(-B) = \cos B$
gives

$$\sin(A - B) = \sin A\cos(-B) + \cos A\sin(-B)$$

or $\sin(A - B) = \sin A\cos B - \cos A\sin B$

Angle $-B$ is in the fourth quadrant where sine is negative and cosine is positive.

C3

Similarly, you can derive formulae for $\cos(A \pm B)$:

$$\cos(A+B) = \cos A\cos B - \sin A\sin B$$
and $$\cos(A-B) = \cos A\cos B + \sin A\sin B$$

Try this for yourself.

You can derive expressions for $\tan(A+B)$ and $\tan(A-B)$ using $\tan\theta = \dfrac{\sin\theta}{\cos\theta}$:

$$\tan(A+B) = \frac{\sin(A+B)}{\cos(A+B)} = \frac{\sin A\cos B + \cos A\sin B}{\cos A\cos B - \sin A\sin B}$$

Divide all terms by $\cos A\cos B$.

$$= \frac{\dfrac{\sin A\cos B}{\cos A\cos B} + \dfrac{\cos A\sin B}{\cos A\cos B}}{\dfrac{\cos A\cos B}{\cos A\cos B} - \dfrac{\sin A\sin B}{\cos A\sin B}} = \frac{\tan A + \tan B}{1 - \tan A\tan B}$$

Cancel as shown and use $\tan\theta = \dfrac{\sin\theta}{\cos\theta}$

By replacing B by $-B$, you can derive the formula

$$\tan(A-B) = \frac{\tan A - \tan B}{1 + \tan A\tan B}$$

These **compound angle formulae** can be summarised as:

$$\sin(A \pm B) = \sin A\cos B \pm \cos A\sin B$$
$$\cos(A \pm B) = \cos A\cos B \mp \sin A\sin B$$
$$\tan(A \pm B) = \frac{\tan A \pm \tan B}{1 \mp \tan A\tan B}$$

Take care with the signs:
$\sin(A \pm B)$ uses only the $\pm$ sign
$\cos(A \pm B)$ uses only the $\mp$ sign
$\tan(A \pm B)$ uses both $\pm$ and $\mp$.

EXAMPLE 1

Find $\cos 15°$ as a surd.

Use the formula $\cos(A-B) = \cos A\cos B + \sin A\sin B$:

$$\cos 15° = \cos(45° - 30°)$$
$$= \cos 45°\cos 30° + \sin 45°\sin 30°$$

Use the special triangles.

$$= \frac{1}{\sqrt{2}} \times \frac{\sqrt{3}}{2} + \frac{1}{\sqrt{2}} \times \frac{1}{2}$$
$$= \frac{\sqrt{3}+1}{2\sqrt{2}}$$
$$= \tfrac{1}{4}\sqrt{2}(\sqrt{3}+1)$$

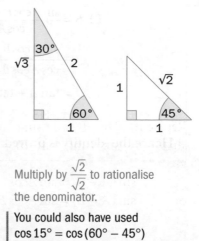

Multiply by $\dfrac{\sqrt{2}}{\sqrt{2}}$ to rationalise the denominator.

You could also have used $\cos 15° = \cos(60° - 45°)$

C3

EXAMPLE 2

Find the value of $\cos(\alpha+\beta)$ when α is acute, β is obtuse, $\cos\alpha = \frac{3}{5}$ and $\sin\beta = \frac{5}{13}$

From the triangles, you have

$$\sin\alpha = \frac{4}{5} \text{ and } \cos\beta = -\frac{12}{13}$$

so $\cos(\alpha+\beta) = \cos\alpha\cos\beta - \sin\alpha\sin\beta$

$$= \frac{3}{5}\times\left(-\frac{12}{13}\right) - \frac{4}{5}\times\frac{5}{13}$$

$$= \frac{-36-20}{5\times 13} = -\frac{56}{65}$$

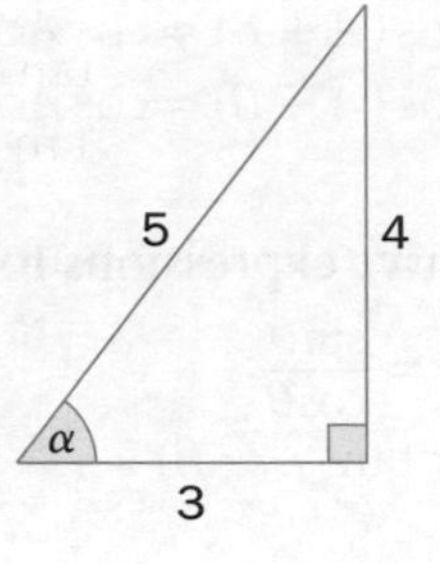

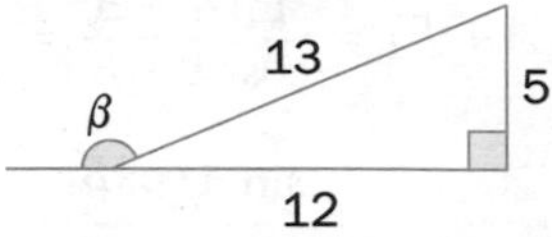

EXAMPLE 3

Solve the equation $\sin(\theta - 30°) = 3\cos\theta$ for $0° < \theta < 360°$

Expand the brackets: $\sin\theta\cos 30° - \cos\theta\sin 30° = 3\cos\theta$

$$\sin\theta\times\frac{\sqrt{3}}{2} - \cos\theta\times\frac{1}{2} = 3\cos\theta$$

$$\frac{\sqrt{3}}{2}\sin\theta = \frac{7}{2}\cos\theta$$

$$\tan\theta = \frac{7}{\sqrt{3}}$$

$\tan\theta$ is positive, so θ is in the first and third quadrants.

For $0° < \theta < 360°$, $\theta = 76.1°$ or $256.1°$

EXAMPLE 4

Prove the identity $\dfrac{\sin(A+B)}{\cos A\cos B} \equiv \tan A + \tan B$

Show that the LHS is equivalent to the RHS:

$$\text{LHS} = \frac{\sin A\cos B + \cos A\sin B}{\cos A\cos B}$$

$$= \frac{\sin A\cancel{\cos B}}{\cos A\cancel{\cos B}} + \frac{\cancel{\cos A}\sin B}{\cancel{\cos A}\cos B}$$

$$= \tan A + \tan B$$

$$= \text{RHS}$$

Using $\tan\theta = \dfrac{\sin\theta}{\cos\theta}$

Hence the identity is proved.

C3

EXAMPLE 5

Solve the equation $\arctan(1+x) + \arctan(1-x) = \arctan 2$

Let $\alpha = \arctan(1+x)$: $\quad \tan\alpha = 1 + x$

Let $\beta = \arctan(1-x)$: $\quad \tan\beta = 1 - x$

The equation to solve is now $\alpha + \beta = \arctan 2$

or $\tan(\alpha+\beta) = 2$

Consider $\tan(\alpha+\beta) = \dfrac{\tan\alpha + \tan\beta}{1 - \tan\alpha\tan\beta} = \dfrac{(1+x)+(1-x)}{1-(1+x)(1-x)}$

$$= \frac{2}{1-(1-x^2)} = \frac{2}{x^2}$$

Hence $\dfrac{2}{x^2} = 2$

The solution is $x = \pm 1$

Exercise 2.4

1 Write as a single trigonometric ratio and so find the exact value of

a $\sin 35°\cos 10° + \cos 35°\sin 10°$ b $\sin 70°\cos 10° - \cos 70°\sin 10°$

c $\cos 40°\cos 10° + \sin 40°\sin 10°$ d $\cos 80°\cos 40° - \sin 80°\sin 40°$

e $\dfrac{\tan 70° - \tan 45°}{1 + \tan 70°\tan 45°}$ f $\dfrac{\tan 100° + \tan 35°}{1 - \tan 100°\tan 35°}$

2 Simplify

a $\sin 2A\cos A + \cos 2A\sin A$ b $\cos 3\alpha\cos 2\alpha - \sin 3\alpha\sin 2\alpha$

c $\dfrac{\tan 2x + \tan x}{1 - \tan 2x\tan x}$ d $\dfrac{1 + \tan 3x\tan x}{\tan 3x - \tan x}$

3 Write each expression as a single trigonometric ratio.

a $\dfrac{1}{\sqrt{2}}\cos x - \dfrac{1}{\sqrt{2}}\sin x$ b $\dfrac{\sqrt{3}}{2}\cos x + \dfrac{1}{2}\sin x$

c $\dfrac{\sqrt{3} + \tan x}{1 - \sqrt{3}\tan x}$ d $\dfrac{1 + \tan x}{1 - \tan x}$

e $\sin(90° - x)\cos x + \cos(90° - x)\sin x$ f $\cos^2 x - \sin^2 x$

4 By expanding $\sin(A - B)$, show that

a $\sin(90° - A) = \cos A$ b $\sin(180° - A) = \sin A$

5 Find the exact value of

a $\sin 15°$ b $\cos 75°$ c $\tan 75°$

d $\tan 15°$ e $\tan 105°$ f $\sec 75°$

C3

6 Given acute angles α and β such that $\sin\alpha = \frac{12}{13}$ and $\tan\beta = \frac{3}{4}$, find

a $\sin(\alpha+\beta)$ **b** $\tan(\alpha+\beta)$ **c** $\sec(\alpha+\beta)$

7 If $\sin\theta = \frac{4}{5}$ and $\sin\phi = \frac{8}{17}$ where θ is acute and ϕ is obtuse, find

a $\cos(\theta-\phi)$ **b** $\sin(\theta-\phi)$ **c** $\cot(\theta-\phi)$

8 Angles A and B are obtuse and acute respectively, such that $\tan A = -2$ and $\tan B = \sqrt{5}$. Find the value of

a $\cot(A+B)$ **b** $\sin(A-B)$

9 **a** If $\tan(\alpha+\beta) = 4$ and $\tan\alpha = 3$, find $\tan\beta$.

b If $\sin(\alpha+\beta) = \cos\beta$ and $\sin\alpha = \frac{3}{5}$, find $\tan\beta$.

c If $\tan(\alpha-\beta) = 5$, find $\tan\alpha$ in terms of $\tan\beta$.

10 If $\sin(\theta+\phi) = \cos\phi$, show that $\tan\theta + \tan\phi = \sec\theta$

11 Solve these equations for $0 < \theta < 360°$

a $3\sin\theta = \sin(\theta+45°)$ **b** $2\cos\theta = \cos(\theta+30°)$

c $2\sin\theta + \sin(\theta+60°) = 0$ **d** $\tan(\theta-45°) = 3\cot\theta$

e $\sin(\theta-60°) = 3\cos(\theta-30°)$ **f** $\sin(\theta+90°) = \tan\theta$

g $\tan(60°-\theta) = \tan(\theta-45°)$ **h** $\sin\theta + \cos\theta = \frac{1}{2}$

12 Prove these identities.

a $\sin(\alpha+30°) + \sin(\alpha-30°) \equiv \sqrt{3}\sin\alpha$

b $(\sin\alpha + \cos\alpha)(\sin\beta + \cos\beta) \equiv \sin(\alpha+\beta) + \cos(\alpha-\beta)$

c $\dfrac{\sin(\alpha+\beta)}{\cos\alpha\cos\beta} \equiv \tan\alpha + \tan\beta$

d $\sin\left(\frac{\pi}{4}+\alpha\right) + \sin\left(\frac{\pi}{4}-\alpha\right) \equiv \sqrt{2}\cos\alpha$

e $\cos(\alpha-\beta) - \cos(\alpha+\beta) \equiv 2\sin\alpha\sin\beta$

f $\dfrac{\cos(\alpha+\beta)}{\sin\alpha\sin\beta} \equiv \cot\alpha\cot\beta - 1$

g $\cos(\alpha+\beta)\cos(\alpha-\beta) \equiv \cos^2\alpha - \sin^2\beta$

h $\sin^2\left(\theta+\frac{\pi}{4}\right) + \cos^2\left(\theta-\frac{\pi}{4}\right) \equiv 1 + 2\sin\theta\cos\theta$

i $\cot(\alpha+\beta) \equiv \dfrac{\cot\alpha\cot\beta - 1}{\cot\alpha + \cot\beta}$

13 Find

i the greatest value **ii** the least value

that each of these expressions can have, given that θ varies with $0° < \theta < 360°$

a $\sin\theta\cos 40° + \cos\theta\sin 40°$

b $\cos\theta\cos 20° - \sin\theta\sin 20°$

Give the values of θ at which the greatest and least values occur in each case.

14 If $\tan\left(\theta + \frac{\pi}{3}\right) = \frac{1}{3}$, show that $\tan\theta = 2 - \frac{5}{3}\sqrt{3}$

15 Prove that $\arcsin\left(\frac{1}{2}\right) + \arcsin\left(\frac{1}{3}\right) \equiv \arcsin\left(\frac{2\sqrt{2} + \sqrt{3}}{6}\right)$

16 Solve these equations.

a $\arctan x = \arctan 7 - \arctan 2$

b $\arcsin x + \arccos\left(\frac{12}{13}\right) = \arcsin\left(\frac{4}{5}\right)$

c $x = \arcsin k + \arcsin\left(\sqrt{1 - k^2}\right)$

17 Prove these identities.

a $\arctan x + \arctan\left(\frac{1}{x}\right) \equiv \frac{\pi}{2}$

b $\arcsin x + \arccos x \equiv \frac{\pi}{2}$

INVESTIGATION

18 Use graphical software on a computer to check your answers to any equations in this exercise and to confirm any identities that you have proved.
For example, for question **11 a**, draw the graphs of $y = 3\sin\theta$ and $y = \sin(\theta + 45°)$

C3

Double angle and half angle formulae

Double angle formulae

Let $A = B$ in the expansions of $\sin(A + B)$, $\cos(A + B)$ and $\tan(A + B)$:

$A + A = 2A$ is called a double angle.

$$\sin 2A = 2\sin A\cos A$$

$$\cos 2A = \cos^2 A - \sin^2 A$$

$$\tan 2A = \frac{2\tan A}{1 - \tan^2 A}$$

These are the **double angle formulae.**

There are two other forms of the formula for $\cos 2A$:

Use $\cos^2 A + \sin^2 A = 1$

Substitute $\cos^2 A = 1 - \sin^2 A$ into the formula for $\cos 2A$:

$$\begin{aligned}\cos 2A &= \cos^2 A - \sin^2 A \\ &= (1 - \sin^2 A) - \sin^2 A \\ &= 1 - 2\sin^2 A\end{aligned}$$

Similarly, substituting $\sin^2 A = 1 - \cos^2 A$
gives $\cos 2A = 2\cos^2 A - 1$

Work through this proof on your own.

The three versions of the formula for $\cos 2A$ are

$$\cos 2A = \begin{cases}\cos^2 A - \sin^2 A \\ 2\cos^2 A - 1 \\ 1 - 2\sin^2 A\end{cases}$$

You can rearrange the identities for $\cos 2A$ to give expressions for $\cos^2 A$ and $\sin^2 A$:

Rearrange $\cos 2A = 2\cos^2 A - 1$:

$$2\cos^2 A = 1 + \cos 2A$$

Divide through by 2:

$$\cos^2 A = \tfrac{1}{2}(1 + \cos 2A)$$

Similarly, $\sin^2 A = \frac{1}{2}(1 - \cos 2A)$

Work through this proof on your own.

C3

$$\cos^2 A = \frac{1}{2}(1 + \cos 2A)$$

$$\sin^2 A = \frac{1}{2}(1 - \cos 2A)$$

Learn these formulae. They are **not** in the formulae booklet.

EXAMPLE 1

Find the solutions of $\cos 2\theta + 3\sin\theta = 2$ for $0 < \theta < \pi$

Choose an identity which gives the equation in terms of $\sin\theta$ or $\cos\theta$ only:

$$\cos 2\theta + 3\sin\theta = 2$$

Substitute $\cos 2\theta = 1 - 2\sin^2\theta$:

$$1 - 2\sin^2\theta + 3\sin\theta = 2$$

Rearrange to equate to 0:

$$2\sin^2\theta - 3\sin\theta + 1 = 0$$

This is a quadratic equation in $\sin\theta$.

Factorise:

$$(2\sin\theta - 1)(\sin\theta - 1) = 0$$

Solve for $\sin\theta$: $\sin\theta = +\frac{1}{2}$ or $+1$

$\sin\theta = \frac{1}{2}$ gives a first or second quadrant angle.

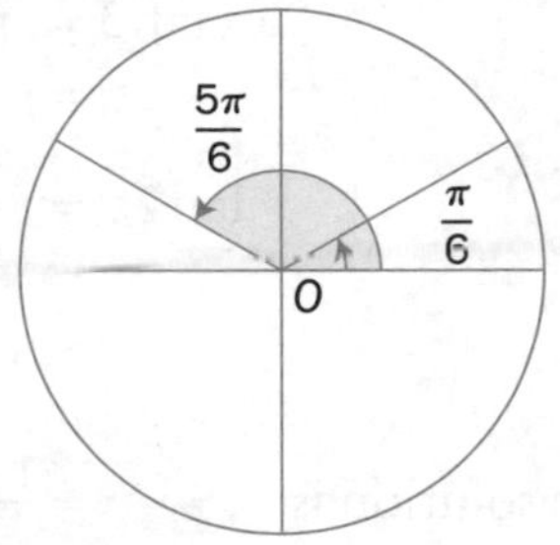

For $0 < \theta < \pi$,

$\sin\theta = +\frac{1}{2}$ gives $\theta = \frac{\pi}{6}$ or $\pi - \frac{\pi}{6} = \frac{5\pi}{6}$

and $\sin\theta = +1$ gives $\theta = \frac{\pi}{2}$

The solutions are $\theta = \frac{\pi}{6}, \frac{\pi}{2}$ and $\frac{5\pi}{6}$.

EXAMPLE 2

Prove the identity $\cot\alpha - \tan\alpha \equiv 2\cot 2\alpha$

Show that the LHS of the identity is equivalent to the RHS:

Choose one side of the identity to work with.

$$\begin{aligned} \text{LHS} &= \frac{\cos\alpha}{\sin\alpha} - \frac{\sin\alpha}{\cos\alpha} \\ &= \frac{\cos^2\alpha - \sin^2\alpha}{\sin\alpha\cos\alpha} \\ &= \frac{\cos 2\alpha}{\sin\alpha\cos\alpha} \\ &= \frac{2\cos 2\alpha}{2\sin\alpha\cos\alpha} \\ &= \frac{2\cos 2\alpha}{\sin 2\alpha} \\ &= 2\cot 2\alpha \\ &= \text{RHS} \end{aligned}$$

The identity is proved.

C3

EXAMPLE 3

Find y in terms of x given that $x = \sec\theta$ and $y = \cos 2\theta$

Express x in terms of $\cos\theta$:

$$x = \sec\theta = \frac{1}{\cos\theta}$$

Rearrange to make $\cos\theta$ the subject:

$$\cos\theta = \frac{1}{x}$$

Use the double angle formula to express y in terms of $\cos\theta$:

$$y = \cos 2\theta = 2\cos^2\theta - 1 = 2\left(\frac{1}{x^2}\right) - 1 = \frac{2}{x^2} - 1$$

Choose the double angle formula which converts $\cos 2\theta$ to $\cos\theta$ only.

So, $y = \dfrac{2}{x^2} - 1$

EXAMPLE 4

Given that $2\arctan 3 = \text{arccot}\, x$, find x.

Let $\arctan 3 = \alpha$ so that $\tan\alpha = 3$
Let $\text{arccot}\, x = \beta$ so that $\cot\beta = x$ and $\tan\beta = \dfrac{1}{x}$

Substitute α and β into the equation:

$$2\arctan 3 = \text{arccot}\, x$$

$$2\alpha = \beta$$

So $$\tan 2\alpha = \frac{2\tan\alpha}{1 - \tan^2\alpha} = \tan\beta$$

$$\frac{2\times 3}{1 - 3^2} = \frac{1}{x}$$

The solution is $$x = \frac{1-9}{6} = -1\frac{1}{3}$$

Half angle formulae

You can also change between single angles and half angles.

You can find the **half angle formulae** by replacing $2A$ by A and replacing A by $\frac{1}{2}A$ in the double angle formulae.

$$\sin A = 2\sin\frac{A}{2}\cos\frac{A}{2}$$

$$\cos A = \begin{cases} \cos^2\frac{A}{2} - \sin^2\frac{A}{2} \\ 2\cos^2\frac{A}{2} - 1 \\ 1 - 2\sin^2\frac{A}{2} \end{cases}$$

$$\tan A = \frac{2\tan\frac{A}{2}}{1 - \tan^2\frac{A}{2}}$$

You can also find expressions for $\cos^2\frac{A}{2}$ and $\sin^2\frac{A}{2}$.

$$\cos^2\frac{A}{2} = \frac{1}{2}(1 + \cos A)$$
$$\sin^2\frac{A}{2} = \frac{1}{2}(1 - \cos A)$$

Try to prove these results yourself.

EXAMPLE 5

Given that angle θ is acute and $\sin\theta = 0.6$ find the value of $\cos\frac{\theta}{2}$.

Firstly find the value of $\cos\theta$:

$$\cos^2\theta = 1 - \sin^2\theta$$

Using $\sin^2\theta + \cos^2\theta = 1$

Substitute $\sin\theta = 0.6$:

$$\cos^2\theta = 1 - (0.6)^2$$
$$= 0.64$$

Hence $\cos\theta = \pm\sqrt{0.64} = \pm 0.8 = 0.8$ as θ is acute.

Now use $\cos^2\frac{\theta}{2} = \frac{1}{2}(1 + \cos\theta)$ and substitute $\cos\theta = 0.8$:

$$\cos^2\frac{\theta}{2} = \frac{1}{2}(1 + 0.8) = 0.9$$

Hence $\cos\frac{\theta}{2} = \pm 0.949$ to 3 s.f.

θ is acute, so $\cos\frac{\theta}{2} = 0.949$ to 3 s.f.

EXAMPLE 6

Solve the equation $\sin\alpha = \cos\frac{\alpha}{2}$ for $0 \leqslant \alpha \leqslant \pi$

Change LHS into half angles:

$$2\sin\frac{\alpha}{2}\cos\frac{\alpha}{2} = \cos\frac{\alpha}{2}$$
$$2\sin\frac{\alpha}{2}\cos\frac{\alpha}{2} - \cos\frac{\alpha}{2} = 0$$

Factorise by taking out $\cos\frac{\alpha}{2}$ as a common factor:

$$\cos\frac{\alpha}{2}\left(2\sin\frac{\alpha}{2} - 1\right) = 0$$

either $\cos\frac{\alpha}{2} = 0$ *or* $\sin\frac{\alpha}{2} = \frac{1}{2}$

$\sin\frac{\alpha}{2}$ is positive, so $\frac{\alpha}{2}$ is in the first or second quadrant.

$\frac{\alpha}{2} = \frac{\pi}{2}, \frac{3\pi}{2}, \ldots$ $\qquad$ $\frac{\alpha}{2} = \frac{\pi}{6}, \frac{5\pi}{6}, \ldots$

$\alpha = \pi, 3\pi, \ldots$ $\qquad$ $\alpha = \frac{\pi}{3}, \frac{5\pi}{3}, \ldots$

For $0 \leqslant \alpha \leqslant \pi$, the solutions are $\alpha = \frac{\pi}{3}$ or π

Exercise 2.5

1 Write each expression as a single trigonometric ratio.

a $2\sin 23°\cos 23°$ **b** $\cos^2 42° - \sin^2 42°$ **c** $\dfrac{2\tan 70°}{1-\tan^2 70°}$

d $2\cos^2 50° - 1$ **e** $2\sin 3\theta\cos 3\theta$ **f** $1-2\sin^2 4\theta$

g $\frac{1}{2}(1+\cos 40°)$ **h** $1+\cos 2\theta$ **i** $\sin\theta\cos\theta$

j $\dfrac{2\tan 3\theta}{1-\tan^2 3\theta}$ **k** $\cos^2\dfrac{\pi}{5} - \sin^2\dfrac{\pi}{5}$ **l** $\dfrac{1-\tan^2 4\theta}{2\tan 4\theta}$

m $1+\cos\theta$ **n** $\sec\theta\operatorname{cosec}\theta$ **o** $\cot\theta - \tan\theta$

2 Find the exact value of each expression.
Give each answer as a surd where necessary.

a $2\sin\dfrac{\pi}{12}\cos\dfrac{\pi}{12}$ **b** $2\cos^2\dfrac{\pi}{8} - 1$ **c** $1-2\sin^2\dfrac{\pi}{8}$

d $\dfrac{1-\tan^2 22\frac{1}{2}°}{\tan 22\frac{1}{2}°}$ **e** $1-\sin^2 75°$ **f** $\dfrac{\sin\frac{\pi}{8}}{\sec\frac{\pi}{8}}$

C3

3 Find the values of $\sin 2\alpha$, $\cos 2\alpha$ and $\tan 2\alpha$ when

a $\cos\alpha = \frac{3}{5}$ **b** $\sin\alpha = -\frac{1}{3}$ **c** $\tan\alpha = -\frac{5}{12}$

4 Find the values of $\cos\theta$ and $\sin\theta$ when

a $\cos\frac{\theta}{2} = \frac{3}{4}$ **b** $\cos\frac{\theta}{2} = \frac{1}{3}$

5 Find the values of $\sin x$, $\cos x$ and $\tan x$ when x is acute and

a $\cos 2x = \frac{17}{25}$ **b** $\sin 2x = \frac{4}{9}\sqrt{5}$ **c** $\tan 2x = \frac{3}{4}$

6 Find the values of $\sin\frac{\alpha}{2}$, $\cos\frac{\alpha}{2}$ and $\tan\frac{\alpha}{2}$ when α is acute and

a $\cos\alpha = \frac{1}{9}$ **b** $\sin\alpha = \frac{3}{5}$ **c** $\tan\alpha = \frac{4}{3}$

7 Find y in terms of x given that

a $x = \sin\alpha$, $y = \cos 2\alpha$

b $x = 3\tan\alpha$, $y = \tan 2\alpha$

c $x = 3\sec\alpha$, $y = \cos 2\alpha$

8 Solve the equations for $0 \leqslant \theta \leqslant 360°$

a $2\sin\theta\cos\theta = \frac{1}{\sqrt{2}}$

b $\cos^2\theta - \sin^2\theta = \frac{\sqrt{3}}{2}$

c $\sin\frac{\theta}{2}\cos\frac{\theta}{2} = \frac{1}{\sqrt{8}}$

d $\cos 2\theta = \sin\theta$

e $\cos 2\theta + 3\sin\theta + 1 = 0$

f $\sin 2\theta = \cos\theta$

g $\cos 2\theta - \cos\theta + 1 = 0$

h $\tan 2\theta = 3\tan\theta$

i $\cos\theta - 2 = 3\cos 2\theta$

j $\sin 2\theta - 1 = \cos 2\theta$

k $\sin 2\theta + \sin\theta = \tan\theta$

l $4\sin\theta = \sin\frac{\theta}{2}$

m $\tan\theta = 6\tan\frac{\theta}{2}$

n $3\cos\frac{\theta}{2} = 2 + \cos\theta$

o $\sin\theta = \cot\frac{\theta}{2}$

p $\cos\theta = 5\sin\frac{\theta}{2} + 3$

q $\cos 2\theta = \tan 2\theta$

r $\sin\frac{\theta}{2} - 2 = 3\cos\theta$

s $\tan\theta\tan 2\theta = 2$

t $4\sin^2\theta + 5\sin 2\theta\cos\theta = 4$

u $\sin\theta + 2\cos\theta = 1$

9 Prove these identities.

a $\frac{1 - \cos 2A}{\sin 2A} \equiv \tan A$

b $\sin 2A \equiv \frac{2\tan A}{1 + \tan^2 A}$

c $\sec 2A + \tan 2A \equiv \frac{\cos A + \sin A}{\cos A - \sin A}$

d $\cot A - \tan A \equiv 2\cot 2A$

e $\tan A + \cot A \equiv 2\operatorname{cosec} 2A$

f $\operatorname{cosec} 2A + \cot 2A \equiv \cot A$

g $2\operatorname{cosec} 2A \equiv \sec A \operatorname{cosec} A$

h $\operatorname{cosec} A - \cot A \equiv \tan\frac{A}{2}$

i $\tan A \sec\frac{A}{2} \equiv 2\sin\frac{A}{2}\sec A$

10 Prove that $2\arctan 2 + \arctan 3 = \operatorname{arccot} 3$

11 Use $3A = 2A + A$ to prove that
$\sin 3A = 3\sin A - 4\sin^3 A$
and $\cos 3A = 4\cos^3 A - 3\cos A$
Find an expression for $\tan 3A$ in terms of $\tan A$.

C3

2.6 The equivalent forms for $a\cos\theta + b\sin\theta$

Consider $y = 3\cos\theta + 4\sin\theta$

Compare the graphs of $y = 3\cos\theta$, $y = 4\sin\theta$ and $y = 3\cos\theta + 4\sin\theta$

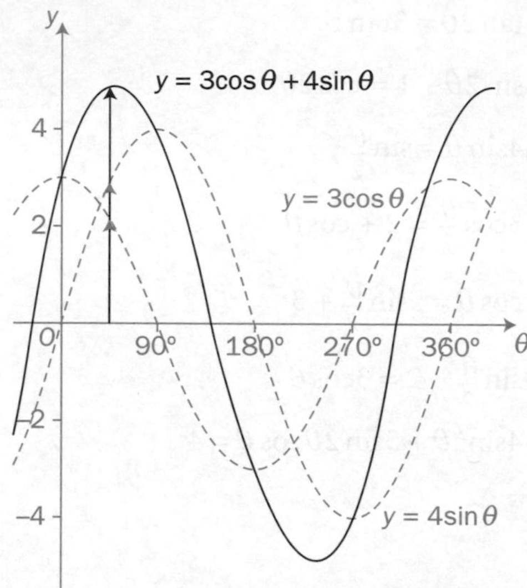

For each value of x, you can add the two y-values on the two blue curves to give the y-value on the black curve as the arrows show.

The graph of $y = 3\cos\theta + 4\sin\theta$ is a transformation of the basic sine curve $y = \sin\theta$ under a stretch (scale factor 5) parallel to the y-axis and a translation of about 37° to the left.

This diagram shows $y = \sin\theta$ transformed into $y = 5\sin(\theta + 37°)$

The 37° is only approximate.

You will find a more accurate value later.

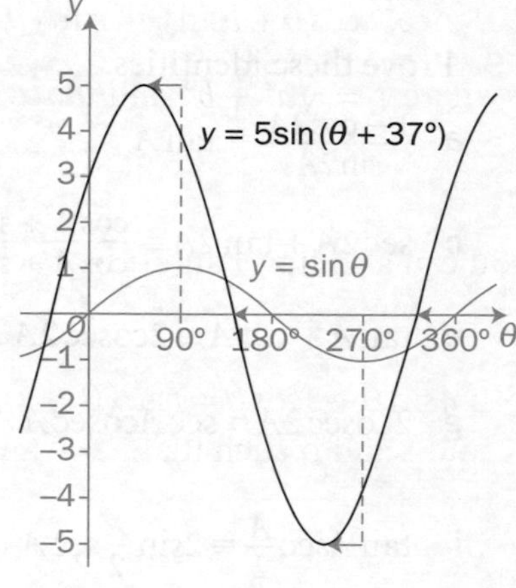

Alternatively, the graph of $y = 3\cos\theta + 4\sin\theta$ is a transformation of the basic cosine curve $y = \cos\theta$ under a stretch (scale factor 5) parallel to the y-axis and a translation of about 53° to the right.

This diagram shows $y = \cos\theta$ transformed into $y = 5\cos(\theta - 53°)$

The 53° is only approximate.

You will find a more accurate value later.

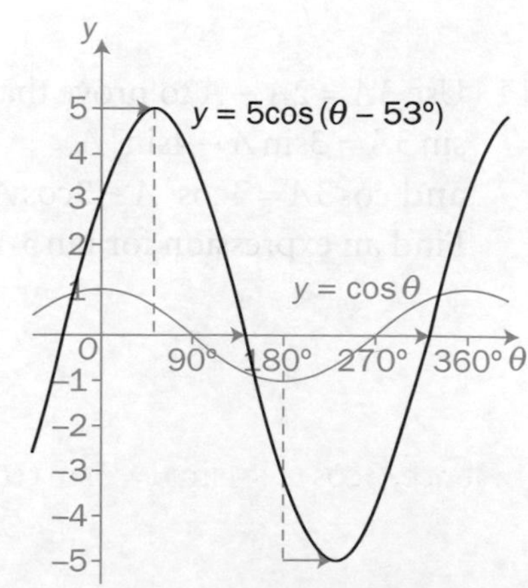

In general, $a\cos\theta \pm b\sin\theta$ is equivalent to $r\sin(\theta \pm \alpha)$ or $r\cos(\theta \pm \alpha)$, where r is positive and α is an acute angle.

Let $a\cos\theta + b\sin\theta = r\sin(\theta + \alpha)$

$$= r\sin\theta\cos\alpha + r\cos\theta\sin\alpha$$

Equate the coefficients of $\sin\theta$ and $\cos\theta$:

$$a = r\sin\alpha$$
$$b = r\cos\alpha$$

Divide these two equations:

$$\frac{a}{b} = \frac{r\sin\alpha}{r\cos\alpha} = \tan\alpha$$

Cancel through by r.

$$\alpha = \arctan\left(\frac{a}{b}\right)$$

Square a and b and add:

$$a^2 + b^2 = r^2(\sin^2\alpha + \cos^2\alpha) = r^2$$

$\sin^2\alpha + \cos^2\alpha = 1$

So $$r = \sqrt{a^2 + b^2}$$

r is positive.

Hence, $a\cos\theta + b\sin\theta = r\sin(\theta + \alpha)$

where $r = \sqrt{a^2 + b^2}$ and $\tan\alpha = \frac{a}{b}$

You can also find r and α for each of $r\sin(\theta \pm \alpha)$ and $r\cos(\theta \pm \alpha)$.

Try this yourself.

EXAMPLE 1

If $4\sin\theta + 3\cos\theta = r\sin(\theta + \alpha)$, find r and α such that $r > 0$ and α is acute.

Expand $r\sin(\theta + \alpha)$: $4\sin\theta + 3\cos\theta = r\sin\theta\cos\alpha + r\cos\theta\sin\alpha$

Both sides contain + signs.

Compare coefficients of $\sin\theta$ and $\cos\theta$:

$$4 = r\cos\alpha \qquad [1]$$
$$3 = r\sin\alpha \qquad [2]$$

Divide equation [2] by equation [1]: $\frac{3}{4} = \frac{r\sin\alpha}{r\cos\alpha} = \tan\alpha$

α is acute, so $\sin\alpha$, $\cos\alpha$ and $\tan\alpha$ are all positive.

$$\alpha = \arctan\frac{3}{4} = 36.9°$$

Square equations [1] and [2] and add: $3^2 + 4^2 = r^2(\sin^2\alpha + \cos^2\alpha)$

$$= r^2$$
$$r = \sqrt{25} = 5$$

r is positive so ignore $r = -5$

Hence, $3\cos\theta + 4\sin\theta = 5\sin(\theta + 36.9°)$

You can check your answer by using a graphical package on a computer.

C3

Exercise 2.6

1 In each equation find the values of r and α, where $r > 0$ and α is acute. Give r as a surd where appropriate and α to the nearest 0.1°.

a $4\cos\theta + 3\sin\theta = r\cos(\theta - \alpha)$ **b** $5\sin\theta + 12\cos\theta = r\sin(\theta + \alpha)$

c $\cos\theta - 2\sin\theta = r\cos(\theta + \alpha)$ **d** $4\sin\theta - 2\cos\theta = r\sin(\theta - \alpha)$

e $3\sin\theta - 4\cos\theta = r\sin(\theta + \alpha)$ **f** $8\cos\theta + 15\sin\theta = r\cos(\theta + \alpha)$

2 Solve these equations, given that $0° \leqslant \theta \leqslant 360°$

a $5\cos\theta + 12\sin\theta = 6$ **b** $2\cos\theta - 3\sin\theta = 1$

c $8\sin\theta + 15\cos\theta = 10$ **d** $3\sin\theta - 5\cos\theta = 4$

3 Prove that

a $\cos\theta + \sin\theta = \sqrt{2}\sin\left(\theta + \frac{\pi}{4}\right)$ **b** $\cos\theta - \sin\theta = \sqrt{2}\cos\left(\theta + \frac{\pi}{4}\right)$

4 Prove that

a $\sqrt{3}\cos\theta + \sin\theta = 2\sin\left(\theta + \frac{\pi}{3}\right)$ **b** $\cos\theta - \sqrt{3}\sin\theta = 2\cos\left(\theta + \frac{\pi}{3}\right)$

5 Solve these equations, given that $-\pi \leqslant \theta \leqslant \pi$

a $\cos\theta + \sqrt{3}\sin\theta = 2$ **b** $\cos\theta + \sin\theta = \frac{1}{\sqrt{2}}$

c $3\cos\theta + \sin\theta = 1$ **d** $\cos\theta + 2\sin\theta = 2$

6 Solve these equations, given that $-180° \leqslant \theta \leqslant 180°$

a $\cos 2\theta + 2\sin 2\theta = 1$ **b** $2\cos 3\theta - 6\sin 3\theta = 5$

c $6\cos\frac{\theta}{2} + 8\sin\frac{\theta}{2} = 3$ **d** $\sin\frac{\theta}{2} - 4\cos\frac{\theta}{2} = 1$

7 **a** Show that $\sqrt{2}\cos\theta + \sqrt{3}\sin\theta$ can be written in the form $r\cos(\theta - \alpha)$, where $r > 0$ and α is acute. Find the values of r and α.

b Hence, find the maximum value of $\sqrt{2}\cos\theta + \sqrt{3}\sin\theta$ and the smallest positive value of θ at which a maximum occurs.

c Find the minimum value of $\frac{1}{\sqrt{2}\cos\theta + \sqrt{3}\sin\theta}$ and the smallest positive value of θ at which a minimum occurs.

8 **a** Express $8\cos 2\theta - 6\sin 2\theta$ in the form $r\cos(2\theta + \alpha)$ where $r > 0$ and α is acute. State the values of r and α.

b Find the minimum value of $8\cos 2\theta - 6\sin 2\theta$ and the smallest positive value of θ at which a minimum value occurs.

c Find the maximum value of $8\cos 2\theta - 6\sin 2\theta$ and the smallest positive value of θ at which a maximum value occurs.

9 **a** Show that $5\sin\theta - 12\cos\theta \equiv r\sin(\theta - \alpha)$ for $r > 0$ and α acute. Give the values of r and α.

b Find the maximum and minimum values of $5\sin\theta - 12\cos\theta$ and the smallest positive values of α at which they occur.

Find the required stationary values of these expressions and, in each case, give the smallest positive value of θ at which they occur.

c maximum of $5\sin\theta - 12\cos\theta + 20$

d maximum of $20 - (6\cos\theta + 8\sin\theta)$

e minimum of $\dfrac{20}{6\cos\theta + 8\sin\theta}$

f minimum of $\dfrac{15}{5\sin\theta - 12\cos\theta + 2}$

10 **a** Express $3\sin\theta - 2\cos\theta$ in the form $r\sin(\theta - \alpha)$ such that $r > 0$ and α is an acute angle.

b Sketch the graph of $y = 3\sin\theta - 2\cos\theta$ for $-360° < \theta < 360°$, labelling all points at which the graph crosses the axes within this interval.

c Describe the transformations which the graph of $y = \sin\theta$ undergoes to become the graph of $y = 3\sin\theta - 2\cos\theta$

11 Two alternating electrical currents are combined so that the resultant current I is given by $I = 2\cos\omega t - 4\sin\omega t$ where the constant $\omega = 4$ and t is the time ($t > 0$). Find the maximum value of I and the smallest value of t at which it occurs.

INVESTIGATION

12 Let $4\cos\theta - 3\sin\theta \equiv r_1\cos(\theta + \alpha_1) \equiv r_2\cos(\theta - \alpha_2)$

$$\equiv r_3\sin(\theta - \alpha_3) \equiv r_4\sin(\theta + \alpha_4)$$

Find the values of r_1, r_2, r_3, r_4 and α_1, α_2, α_3, α_4.

Use computer software to draw five sinusoidal graphs to confirm your results.

1 a If θ is acute and $\tan\theta = \sqrt{2}$, find in surd form

i $\cot\theta$ ii $\sec\theta$ iii $\operatorname{cosec}\theta$

b If θ is obtuse and $\sin\theta = \frac{5}{13}$, find as fractions

i $\operatorname{cosec}\theta$ ii $\cot\theta$ iii $\sec\theta$

2 Solve these equations where $0° \leqslant \theta \leqslant 180°$

a $\operatorname{cosec}\theta = 4$

b $\cot(\theta + 20°) = 4$

c $\cot(2\theta - 30°) = 3$

d $\sec(3\theta - 80°) = 3$

e $\sin^2 2\theta = \frac{1}{2}$

f $\sec^2\left(\frac{1}{2}\theta\right) = 4$

3 Prove these identities.

a $(\sin^2\theta - 2\cos^2\theta)\sec^2\theta \equiv \sec^2\theta - 3$

b $\dfrac{\tan\theta + 1}{\sin\theta} \equiv \dfrac{\cot\theta + 1}{\cos\theta}$

c $\dfrac{\operatorname{cosec}\theta + \sec\theta}{\cot\theta + \tan\theta} \equiv \sin\theta + \cos\theta$

d $(\cot\theta + \operatorname{cosec}\theta)^2 \equiv \dfrac{1 + \cos\theta}{1 - \cos\theta}$

4 Solve these equations where $-180° \leqslant \theta \leqslant 180°$

a $2\cot\theta + \tan\theta = 3$

b $6\sin\theta = 1 + \operatorname{cosec}\theta$

c $\cot\theta - 3\cos\theta = 0$

d $\tan^2\theta - 7 = 2\sec\theta$

e $3\cos^2\theta = 2\sin\theta\cos\theta$

f $\tan\theta = 3 - 2\cot\theta$

5 a Describe the successive transformations which map the graph of $y = \sin\theta$ onto

i $y = 2\sin\left(\theta + \frac{\pi}{2}\right)$ ii $y = 3 - \sin 2\theta$ iii $y = 1 + 2\sin\left(\theta - \frac{\pi}{4}\right)$

b Sketch, on one diagram, the graphs of $y = \cos\theta$ and $y = 3 - \cos 2\theta$, for $-180° \leqslant x \leqslant 180°$, giving all points of intersection with the coordinate axes.

6 a Find, in terms of π, the principal value of

i $\arcsin\left(\frac{1}{2}\right)$ ii $\arctan(\sqrt{3})$ iii $\arccos(-1)$

b Find, as a surd, the value of

i $\cos\left(\arcsin\frac{\sqrt{3}}{2}\right)$ ii $\sin(\arctan 1)$ iii $\tan\left(\arccos\frac{1}{2}\right)$

C3

7 a Given that $\sin^2\theta + \cos^2\theta \equiv 1$, show that $1 + \tan^2\theta \equiv \sec^2\theta$

b Solve, for $0 \leqslant \theta < 360°$, the equation $2\tan^2\theta + \sec\theta = 1$ giving your answers to 1 decimal place. [(c) Edexcel Limited 2005]

8 a Sketch, on the same diagram, the graphs of $y = \cos x$ and $y = \sec x$ for $-270° \leqslant x \leqslant 270°$

b On the same diagram, also sketch the graph of $y = \sec(x - 90°)$, stating the transformation which maps $\sec x$ onto $\sec(x - 90°)$.

9 Given that $\sin A = \frac{15}{17}$ and $\sin B = \frac{3}{5}$ where A and B are both acute angles, find

a $\sin(A + B)$

b $\cos(A + B)$

c $\tan(A - B)$

10 Use the expansions of $\sin(A \pm B)$ and $\cos(A \pm B)$ to find

a $\sin 15°$ as a surd by substituting $A = 60°$ and $B = 45°$

b $\sin(A + B)$ where A is obtuse, $\sin A = \frac{5}{13}$ and $B = 15°$

11 Solve these equations for $0° \leqslant \theta \leqslant 360°$

a $\cos(\theta + 30°) = 2\sin\theta$

b $\cos 2\theta + \cos\theta + 1 = 0$

c $\cos 2\theta = \sin\theta$

d $\tan^2\theta = 2\sec\theta - 1$

e $1 + \sin\frac{\theta}{2} = 3\cos\theta$

f $\tan 2\theta = 3\tan\theta$

g $4\tan\theta + 3\tan\frac{\theta}{2} = 0$

h $3\sin\theta = 2 + \operatorname{cosec}\theta$

i $\sin\theta + \cos\theta = 1$

12 Prove these identities.

a $\frac{\cos 2\theta}{\cos\theta + \sin\theta} \equiv \cos\theta - \sin\theta$

b $\cot(\alpha + \beta) \equiv \frac{\cot\alpha\cot\beta - 1}{\cot\alpha + \cot\beta}$

c $\sin 3\theta + \sin\theta \equiv 2\sin 2\theta\cos\theta$

d $\tan\alpha + \cot\alpha \equiv 2\operatorname{cosec} 2\alpha$

C3

13 Solve these equations for $-180° \leqslant \theta \leqslant 180°$

a $\cos\theta + \cos 2\theta = 2$

b $\sin 2\theta = \frac{1}{2}\sin^2\theta$

c $4\tan 2\theta \tan\theta = 1$

14 Rewrite each pair of equations as an equation in terms of x and y.

a $x = \sin\alpha,\ y = \cos 2\alpha$

b $x = \frac{1}{2}\sec\alpha,\ y = \cos 2\alpha$

c $x = \tan 2\alpha,\ y = \tan\alpha$

15 a Given that $\sin x = \frac{3}{5}$, use an appropriate double angle formula to find the exact value of $\sec 2x$.

b Prove that
$\cot 2x + \operatorname{cosec} 2x \equiv \cot x \quad \left(x \neq \frac{n\pi}{2}, n \in \mathbb{Z}\right)$ [(c) Edexcel Limited 2004]

16 Find the values of

a $\arcsin\left(\frac{\sqrt{3}}{2}\right)$

b $\arccos\left(-\frac{1}{2}\right)$

c $\sin\left(\arctan\left(\frac{2}{3}\right)\right)$

d $\tan\left(\arcsin\left(\frac{2}{3}\right)\right)$

17 If $\alpha = \arcsin\left(\frac{1}{3}\right)$ and $\beta = \arccos\left(\frac{3}{5}\right)$, find the values of $\sin(\alpha + \beta)$.

18 Given that $\alpha + \beta = \arctan(5\sqrt{3} + 8)$ and $\beta = \arctan\left(\frac{1}{2}\right)$, use the expansion of $\tan(\alpha + \beta)$ to find the acute angle α.

19 a Prove that, for all values of x,
$\cos x - \cos(x + 60°) \equiv \cos(x - 60°)$

b Use the fact that $36° = 120° - 84°$ to find the exact value of α given that $\sin\alpha = \sin 84° - \sin 36°$

c For $0 \leqslant x \leqslant 360°$, solve the equation
$\sin(60° + 2x) - 4\sin 2x = 1 + \sin(60° - 2x)$
giving your answer in degrees correct to 1 decimal place.

20 a If $\sin(x + 30) = 2\sin(x - 30°)$, prove that $\tan x = \frac{2}{3}\sqrt{3}$

b Solve the equation $2 - 2\cos 2\theta = \sin 2\theta$,
for $0 \leqslant \theta \leqslant 360°$, giving answers correct to 0.1° where necessary.

21 a Prove the identity $\frac{1-\tan^2 x}{1+\tan^2 x} \equiv \cos 2x$

b Hence, prove that $\tan \frac{\pi}{12} = \sqrt{7-4\sqrt{3}}$

22 Find the values of r and α, where $r > 0$ and angle α is acute, when

a $12\sin\theta + 5\cos\theta \equiv r\sin(\theta+\alpha)$

b $8\sin\theta - 15\cos\theta \equiv r\sin(\theta-\alpha)$

c $2\cos\theta + \sin\theta \equiv r\cos(\theta-\alpha)$

d $\cos\theta - \sin\theta \equiv r\cos(\theta+\alpha)$

23 a Find the values of r and α such that $3\sin\theta + 4\cos\theta \equiv r\sin(\theta+\alpha)$, where $r > 0$ and α is an acute angle.

b Write down the maximum value of $3\sin\theta + 4\cos\theta$.

c Solve the equation $3\sin\theta + 4\cos\theta = 2$ for $0 < \theta < 360°$

24 a Find the maximum and minimum values of each of these expressions and the smallest positive values of θ at which they occur.
i $3\sin\theta + \cos\theta$
ii $\cos\theta - 2\sin\theta$

b Solve these equations for $0° \leqslant \theta < 360°$
i $3\sin\theta + \cos\theta = 2$
ii $\cos\theta - 2\sin\theta = 1$

25 $f(x) = 12\cos x - 4\sin x$

a Given that $f(x) = R\cos(x+\alpha)$, where $R \geqslant 0$ and $0 \leqslant \alpha \leqslant 90°$, find the value of R and the value of α.

b Hence, solve the equation
$12\cos x - 4\sin x = 7$
for $0 \leqslant x \leqslant 360°$, giving your answers to one decimal place.

c i Write down the minimum value of $12\cos x - 4\sin x$.
ii Find, to 2 decimal places, the smallest positive value of x for which this minimum value occurs.

C3

2 Exit

Summary

Refer to

- $\sec\theta = \dfrac{1}{\cos\theta}$ $\quad$ $\operatorname{cosec}\theta = \dfrac{1}{\sin\theta}$ $\quad$ $\cot\theta = \dfrac{1}{\tan\theta}$

  $\tan\theta = \cot(90° - \theta)$ $\quad$ $\sec^2\theta = 1 + \tan^2\theta$ $\quad$ $\operatorname{cosec}^2\theta = 1 + \cot^2\theta$ 2.1, 2.2
- The **inverse trigonometric functions** are arcsin x, arccos x and arctan x.
  Their **principal values** are unique values within the allowed range.
  $\theta = \arcsin x$ exists within the allowed range $-\frac{\pi}{2} \leqslant \theta \leqslant \frac{\pi}{2}$
  $\theta = \arccos x$ exists within the allowed range $0 \leqslant \theta \leqslant \pi$
  $\theta = \arctan x$ exists within the allowed range $-\frac{\pi}{2} \leqslant \theta \leqslant \frac{\pi}{2}$ 2.3
- The **compound angle formulae** are
  $\sin(A \pm B) = \sin A\cos B \pm \cos A\sin B$ $\quad$ $\tan(A \pm B) = \dfrac{\tan A \pm \tan B}{1 \mp \tan A\tan B}$
  $\cos(A \pm B) = \cos A\cos B \mp \sin A\sin B$ 2.4
- The **double angle formulae** are

$$\sin 2A = 2\sin A\cos A \qquad \cos 2A = \begin{cases} \cos^2 A - \sin^2 A \\ 2\cos^2 A - 1 \\ 1 - 2\sin^2 A \end{cases} \qquad \tan 2A = \frac{2\tan A}{1 - \tan^2 A}$$ 2.5

- The **half angle formulae** are

$$\sin A = 2\sin\frac{A}{2}\cos\frac{A}{2}$$

$$\tan A = \frac{2\tan\frac{A}{2}}{1 - \tan^2\frac{A}{2}} \qquad \cos A = \begin{cases} \cos^2\frac{A}{2} - \sin^2\frac{A}{2} \\ 2\cos^2\frac{A}{2} - 1 \\ 1 - 2\sin^2\frac{A}{2} \end{cases} \qquad \cos^2\frac{A}{2} = \frac{1}{2}(1 + \cos A) \qquad \sin^2\frac{A}{2} = \frac{1}{2}(1 - \cos A)$$ 2.5

- $a\cos\theta \pm b\sin\theta$ can take the equivalent forms $r\cos(\theta \pm \alpha)$ or $r\sin(\theta \pm \alpha)$,
  where r is positive and α is an angle. 2.6

Links

Trigonometry is behind the technology of modern Satellite Navigation (Sat Nav) systems.

Sat Nav uses the Global Positioning System (GPS) which relies on a collection of satellites, orbiting the earth and transmitting data. Information about how the satellites orbit, and their position at a particular time, allows a GPS receiver to calculate its position on the surface of the Earth using basic trigonometry.

Combined with some maps and planning software, this is the basis of in-car Sat Nav technology.

Revision 1

1 Simplify $\frac{3}{1+x}-\frac{3}{2+x}-\frac{2}{(2+x)^2}$ and express your answer as a single fraction.

2 Simplify as far as possible

a $\frac{x^3+x^2-2x}{x^2-1}$

b $\frac{x}{x+1}+\frac{2}{x+2}$

3 Find the quotient and remainder when $x^4+4x^3+2x^2+x-5$ is divided by x^2+x+1.

4 Given that $x^4-3x^3+7x^2-8x+5 \equiv (x^2-2x+1)\times \mathrm{Q}(x)+\mathrm{R}(x)$ find the two function $\mathrm{Q}(x)$ and $\mathrm{R}(x)$.

5 a Find the quotient and remainder when x^2-3x+2 is divided into $2x^3-x^2-9x+6$.

b Hence, or otherwise, find the values of the constants λ and μ so that there is no remainder when $2x^3-x^2+\lambda x+\mu$ is divided by x^2-3x+2.

6 a Express $4\sin\theta+3\cos\theta$ in the form $R\sin(\theta+\alpha)$, where $R>0$ and α is an acute angle.

b Hence, solve the equation $4\sin\theta+3\cos\theta=\frac{5}{2}$ for $0° < \theta < 360°$

7 a Express $\cos\theta+4\sin\theta$ in the form $R\cos(\theta-\alpha)$ where α is an acute angle. Give the exact value of R and the value of α correct to the nearest degree.

b Hence, or otherwise solve the equation $\cos\theta+4\sin\theta=3$ for $0<\theta<360°$

8 By writing $5\sin\theta-12\cos\theta$ in the form $R\sin(\theta-\alpha)$ where $R>0$ and $0°<\alpha<90°$, find

a the greatest possible value of $5\sin\theta-12\cos\theta$

b the smallest possible value of θ for which the greatest value occurs.

9 Given that $\mathrm{f}(x)=9-(x+2)^2, \quad x\in\mathbb{R}$

a find the range of $\mathrm{f}(x)$

b state whether $\mathrm{f}^{-1}(x)$ exists or not

c find the value of $\mathrm{ff}(-4)$.

10 a Express $x^2 - 4x + 1$ in the form $(x - a)^2 + b$.

b Given that the function f is defined by f: $x \to x^2 - 4x + 1, x \in \mathbb{R}, x \geqslant 2$,
find **i** the range of f **ii** the inverse function f^{-1}.

c Sketch the graphs of the functions f and f^{-1} on the same axes.

11 a Describe the transformations which are needed to transform the graph of $y = x^3$ into the graph of $y = 1 - 2x^3$
Indicate the order in which the transformations occur.

b The graph of $y = x^2 - 2x + 5$ is reflected in the y-axis and then translated by $\begin{pmatrix} -1 \\ 0 \end{pmatrix}$. Find the equation of the final image in its simplest form.

12 The functions f and g are defined by

f: $x \to x^2 - 5x + 4, \quad x \in \mathbb{R}, \quad 2 \leqslant x \leqslant 5$
g: $x \to kx - 2, \quad x \in \mathbb{R}$, where k is a constant.

a Find the range of the function f.

b If $gf(5) = 2$, find the value of k.

13 a The function f is defined by f: $x \to 5x, \quad x \in \mathbb{R}$.
Write down $f^{-1}(x)$ and state the domain of f^{-1}.

b The function g is defined by g: $x \to 3x^2 - 2, \quad x \in \mathbb{R}$.
Find $gf^{-1}(x)$ and state the range of gf^{-1}.

14 This sketch shows the curve with equation $y = f(x), x \in \mathbb{R}, 0 \leqslant x \leqslant a$
The curve meets the coordinate axes at the points $(a, 0)$ and $(0, b)$.

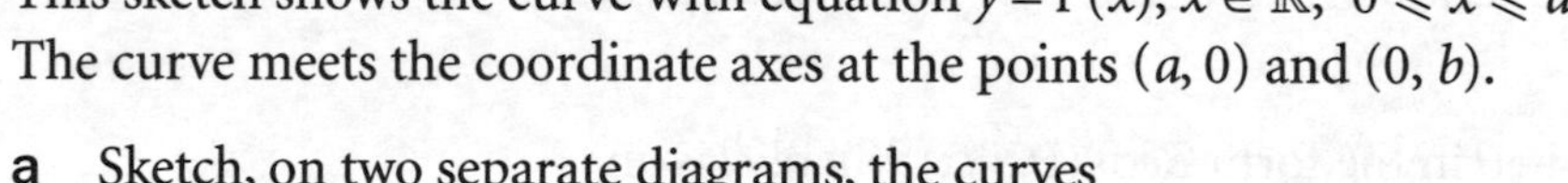

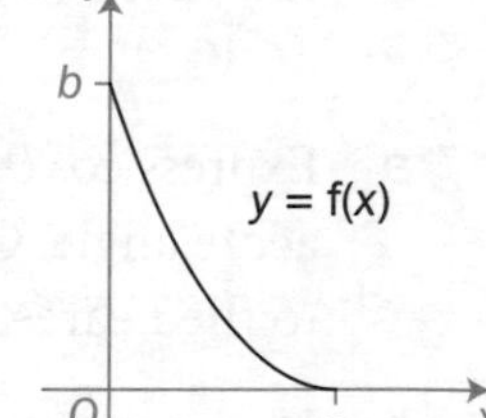

a Sketch, on two separate diagrams, the curves
i $y = f^{-1}(x)$
ii $y = 4f\left(\frac{x}{2}\right)$
marking the coordinates of all points where these two curves meet the coordinates axes.

b If f defined by $f(x) = (x - 3)^2, \quad x \in \mathbb{R}, \ 0 \leqslant x \leqslant 3$, find the value of a and b. State the range of f.

c The function g is defined by $g(x) = 1 + \sqrt{x}, \quad x \in \mathbb{R}, \ x > 0$
Find $gf(x)$, giving your answer in its simplest form.

15 a Sketch the graphs of $y = |2x + 1|$ and $y = |x - 1|$ on the same axes.

b Solve the equation $|2x + 1| = |x - 1|$

16 **a** Solve these equations.

i $|3x - 1| = 8$ ii $|x| + 3 = 2x$ iii $|x + 3| = 2x$

b Solve these inequalities.

i $|3x - 1| > 8$ ii $|x| + 2 \geqslant \frac{1}{2}x$ iii $|x - 1| < 2x + 1$

17 This sketch shows the curve $y = \mathrm{f}(x)$, $x \in \mathbb{R}$. Point $(1, 3)$ is a turning point on the curve. The x-axis and the line $x = 3$ are both asymptotes to the curve.

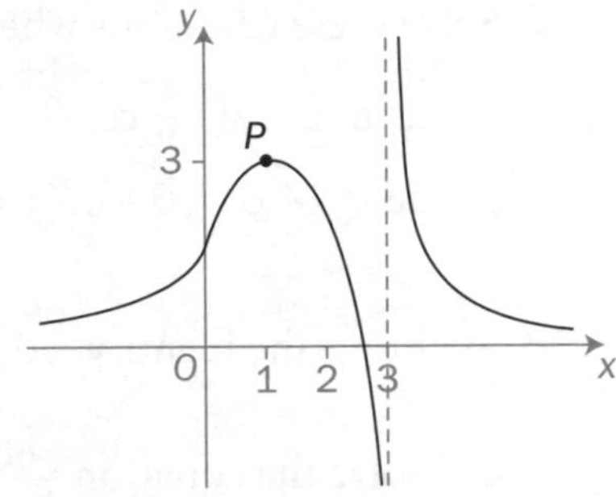

Sketch, on three separate diagrams, the graphs of

a $y = |\mathrm{f}(x)|$ **b** $y = \mathrm{f}(|x|)$ **c** $y = \mathrm{f}(x - 3)$

showing any asymptotes and the coordinates of any maximum or minimum points.

18 **a** Sketch the graphs of $y = |3x - 2|$ and $y = \frac{1}{x}$ on the same diagram.

b Use your graphs to say why there is only one solution of the equation $x|3x - 2| - 1 = 0$

c Use algebra to find the solution of the equation $x|3x - 2| - 1 = 0$

19 This diagram shows a sketch of the curve $y = \mathrm{f}(x)$, $x \geqslant -1$. The curve passes through the origin O, has a maximum value at the point $P(3, 2)$ and has the x-axis as an asymptote.

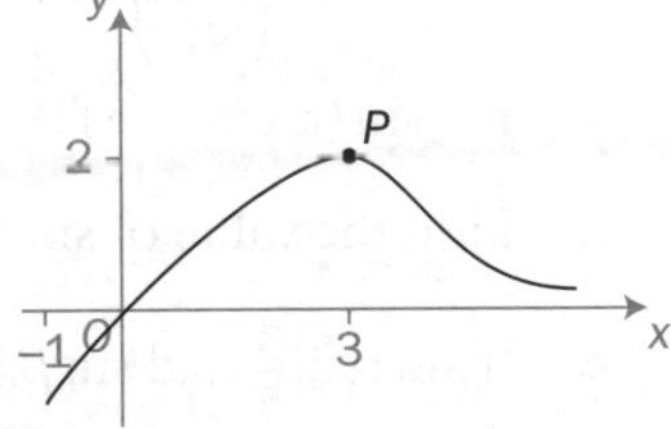

On separate diagrams, draw sketches of the curves with these equations

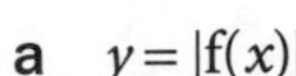
a $y = |\mathrm{f}(x)|$ **b** $y = \mathrm{f}(|x|)$ **c** $y = \mathrm{f}(x + 3)$

On each sketch, indicate the coordinates of points at which the curves have turning points and the coordinates of points where the curves meet the x-axis.

20 This figure shows part of the graph of $y = \mathrm{f}(x)$, $x \in \mathbb{R}$. The graph consists of two line segments that meet at the point $(1, a)$, $a < 0$. One line meets the x-axis at $(3, 0)$. The other line meets the x-axis at $(-1, 0)$ and the y-axis at $(0, b)$, $b < 0$.

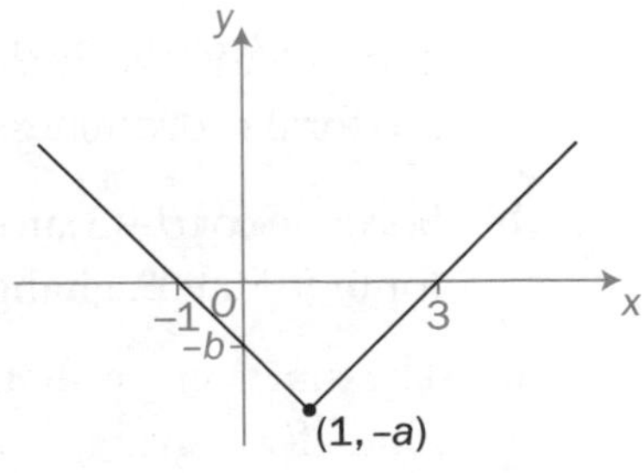

On separate diagrams, sketch the graphs with equations

a $y = \mathrm{f}(x + 1)$ **b** $y = \mathrm{f}(|x|)$

Indicate clearly on each sketch the coordinates of any points of intersection with the axes.

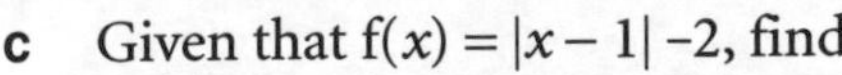
c Given that $\mathrm{f}(x) = |x - 1| - 2$, find

i the value of a and the value of b

ii the value of x for which $\mathrm{f}(x) = 5x$

[(c) Edexcel Limited 2005]

21 Prove these identities.

a $\dfrac{\sin 2\theta}{1-\cos 2\theta} \equiv \cot\theta$ **b** $\cot\theta - \tan\theta \equiv 2\cot 2\theta$

c $\dfrac{\cos\theta}{\sec\theta} + \dfrac{1}{1+\cot^2\theta} \equiv 1$ **d** $\dfrac{\tan^2\theta+1}{1-\tan^2\theta} \equiv \sec 2\theta$

22 Solve these equations where $0° \leqslant \theta \leqslant 180°$

a $2\sin^2\theta = 3(1-\cos\theta)$ **b** $\sec^2\theta = 2(2\tan\theta - 1)$

c $2\cos\theta + \cos(\theta + 60°) = 0$ **d** $\cos 2\theta + \cos\theta + 1 = 0$

23 a Prove the identity $6 - 3\sec^2\theta \equiv \dfrac{6\cos 2\theta}{1+\cos 2\theta}$

b Solve the equation $\dfrac{6\cos 2\theta}{1+\cos 2\theta} = 13 - 11\tan\theta$ for $0° < \theta < 360°$

24 a Prove the $\cos 3\alpha \equiv 4\cos^3\alpha - 3\cos\alpha$ by substituting $(2\alpha + \alpha)$ for 3α.

b Solve the equation $\sec 2\alpha \cos 6\alpha + 1 = 0$ for $0° < \alpha < 90°$

25 a Find, in radians, the values of

i $\arcsin\left(\dfrac{1}{\sqrt{2}}\right)$

ii $\sin(\arctan\sqrt{3})$

b Find the value of $\sin[\arctan(-1)]$

c If $\sin\alpha = \dfrac{3}{5}$ and $\sin\beta = \dfrac{8}{17}$ where α is acute and β is obtuse, find the value of $\cos(\alpha - \beta)$.

26 a i Express $(12\cos\theta - 5\sin\theta)$ in the form $R\cos(\theta + \alpha)$, where $R > 0$ and $0 < \alpha < 90°$

ii Hence solve the equation

$$12\cos\theta - 5\sin\theta = 4$$

for $0 < \theta < 90°$, giving your answer to 1 decimal place.

b Solve $8\cot\theta - 3\tan\theta = 2$ for $0 < \theta < 90°$, giving your answer to 1 decimal place. [(c) Edexcel Limited 2004]

C3

3 Exponentials and logarithms

This chapter will show you how to
- discover the value of the irrational number e
- use natural (or Napierian) logarithms
- use the exponential function $y = e^x$ and its inverse function $y = \ln x$
- draw graphs of functions which involve e^x and $\ln x$
- solve equations which involve e^x and $\ln x$
- use exponential and logarithmic functions to solve real-life problems.

Before you start

You should know how to:

1 Calculate a^x and $\log_a x$ for different values of a and x.

e.g. If $a = \frac{1}{2}$ and $x = -3$,

then $a^x = \left(\frac{1}{2}\right)^{-3} = 2^3 = 8$

2 Use the laws of logarithms.

e.g. If $\log_{10} y = 3$, then $y = 10^3 = 1000$

3 Find the inverse function of $f(x)$.

e.g. If $f(x) = 3x^2 + 1$, $x \in \mathbb{R}$, then undoing the operations in reverse order gives

$f^{-1}(x) = +\sqrt{\frac{x-1}{3}}$, $x \in \mathbb{R}$, $x \geqslant 1$

4 Reflect, stretch or translate a graph and find its new equation.

e.g. When the graph of $y = x^2 - x - 1$ is translated by $\begin{pmatrix}3\\0\end{pmatrix}$, the equation of the new graph is $y = (x-3)^2 - (x-3) - 1$ giving $y = x^2 - 7x + 11$

Check in:

1 Calculate the value of

a a^x for $a = -3$, $x = -2$

b a^{2x+1} for $a = \frac{1}{9}$, $x = -\frac{1}{4}$

c $\log_a(x^2 + 2x + 1)$ for $a = 10$, $x = 9$

2 Find x, y and z given that $\log_3 x = 2$, $\log_{10}(y+1) = \frac{1}{2}$ and $z = \log_2 16$

3 Find the inverse function, stating its domain and range, when

a $f(x) = 3x + 5$

b $f(x) = \frac{x-3}{2}$

c $f(x) = +\sqrt{2x-1}$

4 Find the equation of the resulting curve when $y = x^2 + 2$ is

a stretched (scale factor 3) parallel to the y-axis and then reflected in the y-axis

b translated by $\begin{pmatrix}0\\-5\end{pmatrix}$ and then stretched (scale factor 2) parallel to the x-axis.

The exponential function, e^x

a^x is an exponential function for all values of a.

You can use a table to draw the graph of a particular exponential function, for example $y = 2^x$:

x	-2	-1	0	1	2
y	$2^{-2} = \frac{1}{4}$	$2^{-1} = \frac{1}{2}$	$2^0 = 1$	$2^1 = 2$	$2^2 = 4$

You can also draw graphs of other members of the family of curves $y = a^x$

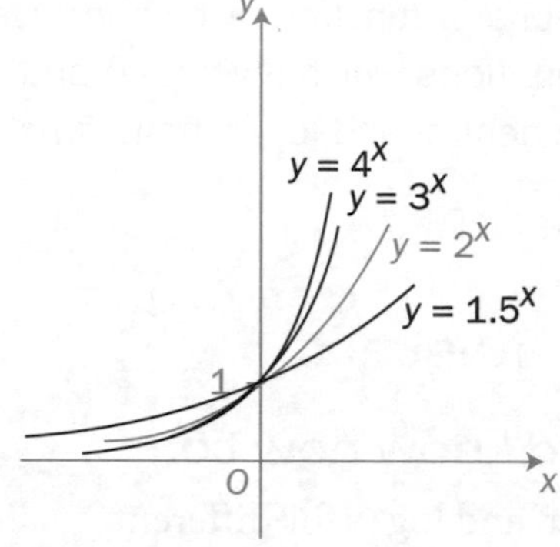

All the curves pass through the point $(0, 1)$ because $y = a^0 = 1$ for all values of a.

To investigate the gradient of the curves at the point $P(0, 1)$ from first principles, you calculate the gradient of the chord PQ for small values of δx.

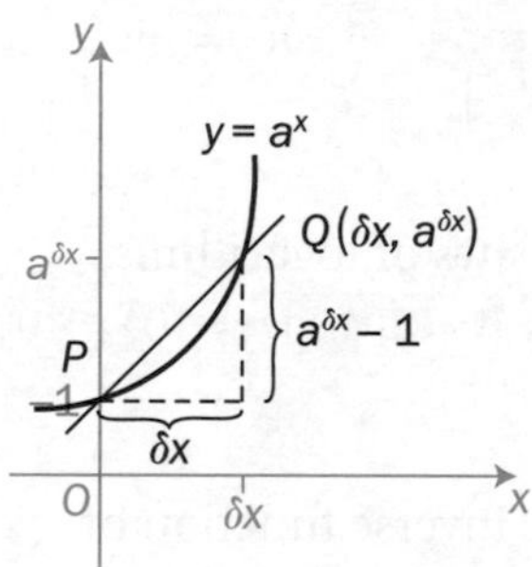

Gradient of chord $PQ = \frac{\delta y}{\delta x} = \frac{a^{\delta x} - 1}{\delta x}$

In the limit, as $\delta x \to 0$,
the gradient of the chord $PQ \to$ the gradient of the tangent at P.

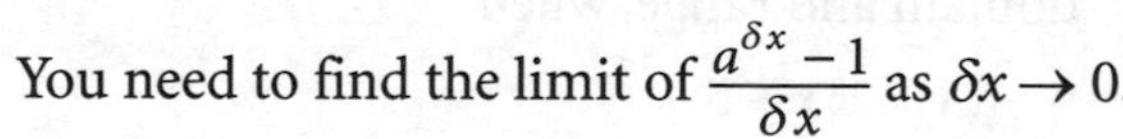

You need to find the limit of $\frac{a^{\delta x} - 1}{\delta x}$ as $\delta x \to 0$.

Let δx take a very small value, say $\delta x = 0.0001$

This spreadsheet shows the values of $\frac{a^{\delta x} - 1}{\delta x}$ for graphs with different values of a:

a	$\frac{(a^{\delta x} - 1)}{\delta x}$
2.0	0.6932
2.2	0.7885
2.4	0.8755
2.6	**0.9556**
2.8	**1.0297**
3.0	1.0987

Between $a = 2$ and $a = 3$, there is a point on the curve $y = a^x$ where the gradient at P is exactly 1. This value is between $a = 2.6$ and $a = 2.8$

You can get a more accurate approximation of this value by looking at the values between 2.6 and 2.8.

Altering the values in the spreadsheet gives:

a	$\frac{(a^{\delta x}-1)}{\delta x}$
2.60	0.9556
2.65	0.9746
2.70	**0.9933**
2.75	**1.0117**
2.80	1.0297

The value you want is between $a = 2.70$ and $a = 2.75$

Altering the values in the spreadsheet again gives:

a	$\frac{(a^{\delta x}-1)}{\delta x}$
2.70	0.9933
2.71	**0.9970**
2.72	**1.0007**
2.73	1.0044
2.74	1.0080

The value you want is slightly less than 2.72.

This value of a is known as **the exponential function, e,** where e = 2.718 28 to 5 decimal places.

e is an irrational number (like π).

The graph of $y = e^x$ has a gradient of 1 at the point (0, 1).

Sometimes e^x is written as $\exp(x)$.

EXAMPLE 1

Sketch the graphs, for $x \in \mathbb{R}$, of

a $y = e^{-x}$ **b** $y = 2e^x - 3$ **c** $y = e^{2x+1}$

State the range of each function.

a The graph of $y = e^{-x}$ is the reflection of $y = e^x$ in the y-axis. Its range is $y \in \mathbb{R}$, $y > 0$

b The graph of $y = 2e^x - 3$ is the result of a stretch (scale factor 2) of $y = e^x$ parallel to the y-axis followed by a translation of –3 downwards parallel to the y-axis.
Its range is $y \in \mathbb{R}$, $y > -3$

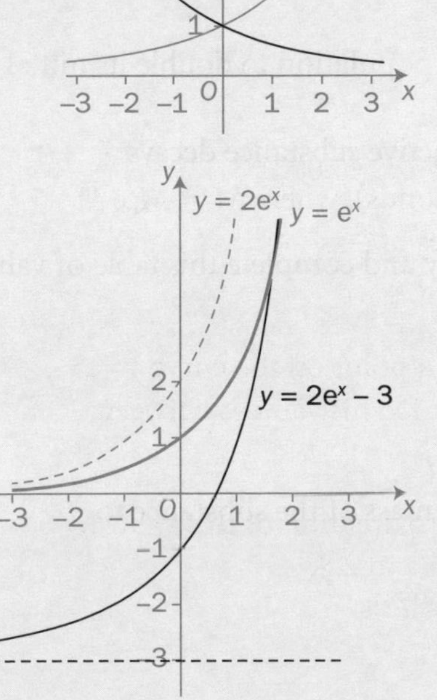

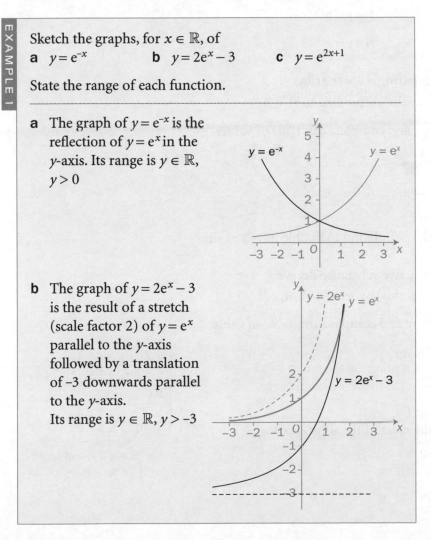

The solution to part **c** is shown on the next page.

C3

EXAMPLE 1 (CONT.)

c $y = e^{2x+1} = e^{2x} \times e^1 = e \times e^{2x}$

The graph of $y = e^{2x+1}$ is the result of a stretch $\left(\text{scale factor } \frac{1}{2}\right)$ of $y = e^x$ parallel to the x-axis

followed by a stretch (scale factor e) parallel to the y-axis.

Its range is $y \in \mathbb{R}, y > 0$

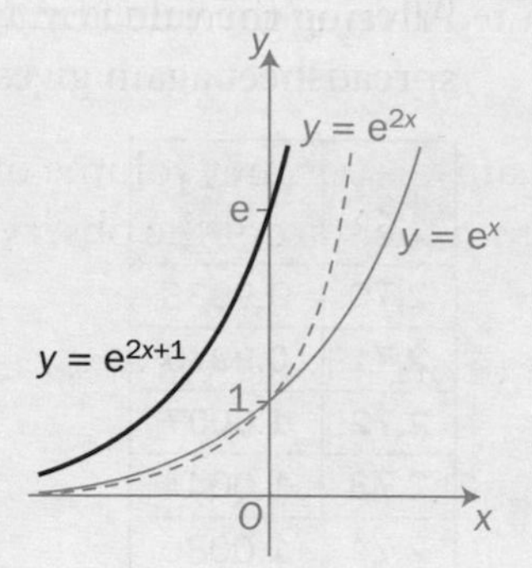

Exercise 3.1

1 Sketch the graphs of these functions for the domain $x \in \mathbb{R}$.
State the transformations of $y = e^x$ which are involved.

a $y = 1 + e^{-x}$ **b** $y = 1 - e^{-x}$

c $y = 3e^x + 2$ **d** $y = 2 - 3e^x$

e $y = 3e^{2x}$ **f** $y = e^{x+1}$

g $y = e^{x-2}$ **h** $y = e^{-x+2}$

2 The population, P, of rats infesting a sewer grows exponentially over time, t (weeks), according to $P = Ae^{\frac{t}{20}}$

Find the value of A and copy and complete this table of values.

t	0	5	10	15	20
P	100				

Draw the graph of P against t.
How long does it take for the population to double its initial size?

3 A mass of M units of a radioactive substance decays exponentially over time t (seconds), where $M = M_0e^{-\frac{t}{10}}$

Find the value of M_0 and copy and complete this table of values.

t	0	5	10	15	20
M	6				

Draw the graph of M against t.
How long does it take for the mass of the substance to reduce to 3 units?

This time period is known as the *half-life* of the substance.

C3

4 The growth of algae in a polluted river is governed by the equation

$$N = N_0 e^{\alpha t},$$

where N is the number of organisms per unit volume of river water, t is the time in weeks from the start of the observation, and N_0 and α are constants.

a After 4 weeks, the number of organisms N is observed to be double the initial number.
Find the value of α to 4 significant figures.

b If $N_0 = 20$, what is the value of N after 10 weeks of observation?

c How many weeks does it take for N to treble its initial value?

d Give a reason why this model of pollution is unrealistic.

5 A radio-active substance decays such that its mass M at time t (hours) is given by $M = M_0 e^{-kt}$, where M_0 and k are both constants.

a If $k = 0.006\,93$ show that the half-life of this substance is 100 hours.

b How long does the substance take to decay to a tenth of its original mass?

6 The number of cells N which are infected with a virus were observed to change with time t (hours) as given by

$$N = 200 - 50e^{-2t}$$

a Construct a table of values of N for $0 \leqslant t \leqslant 15$ and draw the graph of N against t.

b How many infected cells were there initially?

c What is the limiting value of N as time increases?

d What series of transformations of $N = e^{-2t}$ result in the given relationship?

7 State the transformation of the graph of $y = e^x$ which result in

a $y = e^{2x+3} + 4$

b $y = e^{2x-1} - 4$

c $y = e^{2-x} - 3$

INVESTIGATION

8 Find other examples of exponential growth and decay which are governed by the equations

$$y = Ae^{kt} \text{ and } y = Ae^{-kt}$$

E.g. What is Newton's law of cooling?

C3

3.2 The logarithmic function, ln x

The inverse of the exponential function $y = e^x$ is **the logarithmic function** $y = \log_e x$

You can show this by taking logarithms in any base n and finding x in terms of y.

Logarithms with base e are called natural logarithms.

They are usually written as $\ln x$ rather than $\log_e x$.

The function $y = e^x$ has the inverse function $y = \ln x$

Their graphs are reflections of each other in the line $y = x$

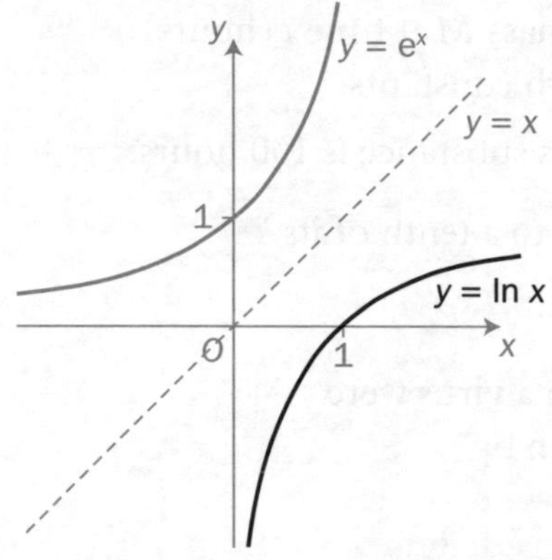

- $y = e^x$ has the x-axis as an asymptote
- $y = \ln x$ has the y-axis as an asymptote
- for $y = e^x$, the domain is $x \in \mathbb{R}$ and the range is $y \in \mathbb{R}, y > 0$
- for $y = \ln x$, the domain is $x \in \mathbb{R}, x > 0$ and the range is $y \in \mathbb{R}$
- e^x is positive for all values of x,
- $\ln x$ does not exist for negative values of x.

EXAMPLE 1

Find the inverse function $f^{-1}(x)$ for the function $f(x) = 1 + e^{-x}$
Sketch the graphs of $f(x)$ and $f^{-1}(x)$ on the same diagram, labelling any intersections with the coordinate axes.
Give the domain and range of $f^{-1}(x)$.

To find the inverse, let $y = 1 + e^{-x}$:

$$y - 1 = e^{-x}$$

Take logarithms:

$$\ln(y-1) = \ln(e^{-x})$$
$$= -x \ln e \quad [\ln e = 1]$$
$$= -x$$

Interchange x and y:

$$\ln(x-1) = -y$$
$$y = -\ln(x-1)$$

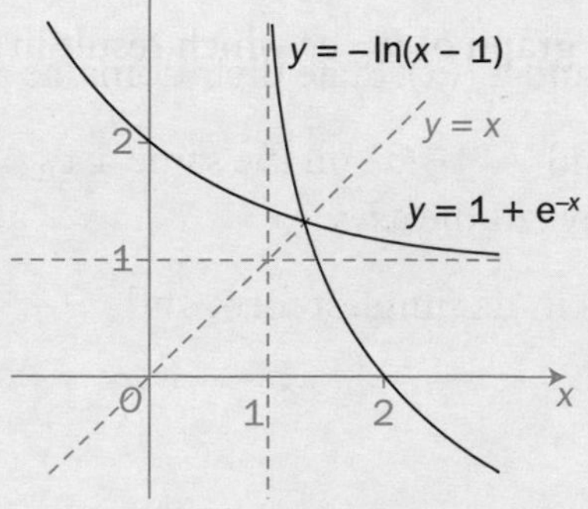

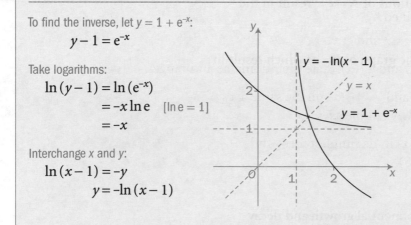

The graph of $y = 1 + e^{-x}$ is the graph of $y = e^x$ reflected in the y-axis and then translated upwards by 1 unit.
The graph of $y = -\ln(x-1)$ is the graph of $y = \ln x$ translated to the right by 1 unit and then reflected in the y-axis.
These transformations are the inverses of each other.

Hence the inverse function is $f^{-1}(x) = -\ln(x-1)$

The domain of $f^{-1}(x)$ is $x \in \mathbb{R}, x > 1$ and its range is $y \in \mathbb{R}$.

Exercise 3.2

1 **a** Find the values (to 3 significant figures) of

i $e^{1.5}$ ii $e^{-1.5}$ iii $\frac{1}{e^{1.5}}$

iv $\ln 1.5$ v $\ln\left(\frac{1}{1.5}\right)$ vi $\frac{1}{\ln 1.5}$

b Simplify

i $\ln(e^5)$ ii $\ln(e^x)$ iii $e^{\ln 5}$

iv $e^{\ln x}$ v $\ln(2e^5)$ vi $e^{2\ln 5}$

2 Describe the transformations of $y = \ln x$ which result in each of these functions. Sketch the graphs of these functions, labelling any points where the graphs intersect the coordinate axes. State the domain and range of each function.

a $y = 2 + \ln x$ **b** $y = 3 - \ln x$ **c** $y = \ln(x + 2)$ **d** $y = \ln(x - 2)$

e $y = 1 + 3\ln x$ **f** $y = 1 + \ln(2x)$ **g** $y = 1 - \ln x$ **h** $y = 2\ln(1 - x)$

3 **a** Describe the successive transformations of the graph of $y = \ln x$ which produce the graph of $y = 2\ln(x + 3) + 1$

b Find the points at which this graph cuts the x-axis and y-axis.

C3

4 Find the inverse function $f^{-1}(x)$ for each function $f(x)$. Sketch the graphs of $f(x)$ and $f^{-1}(x)$ on the same diagram, labelling any intersections with the coordinate axes. In each case, give the domain and range of $f^{-1}(x)$.

a $y = 1 + 2\ln x, \quad x > 0$ **b** $y = \ln(x + 4), \quad x > -4$

c $y = 3 + e^{-x}, \quad x \in \mathbb{R}$ **d** $y = 2 - e^{\frac{1}{2}x}, \quad x \in \mathbb{R}$

5 The functions f and g are defined by

$f: x \to \ln(6 - 2x), x \in \mathbb{R}, x < 3$ and $g: x \to e^{\frac{1}{2}x}, x \in \mathbb{R}$.

a Find expressions for $f^{-1}(x)$ and $g^{-1}(x)$. State their domains and ranges.

b Sketch the curves $y = f(x)$ and $y = f^{-1}(x)$ on the same axes, stating the points where they cut the axes.

c Find an expression for $gf(x)$ in its simplest form and calculate the value of $gf(-5)$.

INVESTIGATION

6 What is semi-log and log-log graph paper? Explore why and how scientists and engineers use it to draw graphs of $y = Ae^{kx}$ and other exponential curves.

3.3 Equations involving e^x and $\ln x$

You can use the laws of logarithms and the techniques from unit Core 2 to solve equations involving e^x and $\ln x$.

EXAMPLE 1

Solve the equation $e^{3x+2} = 25$

Take logarithms to base e of both sides:

$$\ln(e^{3x+2}) = \ln 25$$

$$(3x+2)\ln e = \ln 25$$

$\log_a(x^n) = n\log_a x$

$$x = \frac{\ln 25 - 2}{3} = 0.406 \text{ to 3 s.f.}$$

$\log_a a = 1$ so $\ln e = 1$

EXAMPLE 2

Solve the equations **a** $\ln(3x+1) = \frac{1}{2}$ **b** $e^{2x} = 2e^x + 8$

a Use $\log_a x = y \Leftrightarrow x = a^y$:

$$3x + 1 = e^{\frac{1}{2}}$$

$3x + 1 = \sqrt{e}$

$$x = \frac{\sqrt{e} - 1}{3} = 0.216 \text{ to 3 s.f.}$$

b Rearrange: $e^{2x} - 2e^x - 8 = 0$

Let $y = e^x$: $y^2 - 2y - 8 = 0$

$y = e^x$ is a useful substitution. Remember it.

$$(y-4)(y+2) = 0$$

$$y = 4 \text{ or } -2$$

Hence $e^x = 4$ or -2

$e^x > 0$ for all x, so the solution is $x = \ln 4 = 1.39$ to 3 s.f.

Exercise 3.3

1 Solve these equations. Give your answers to 3 s.f.

a $e^x = 9$ **b** $e^x = 7.39$ **c** $4e^x + 2 = 36$ **d** $e^{4x} + 2 = 36$

e $4e^{-x} + 2 = 36$ **f** $e^{-x} = 7.39$ **g** $7 - 2e^{-x} = 4$ **h** $e^{x+2} = 20$

i $2e^{3x-1} = 10$ **j** $e^{4x-1} = \frac{1}{2}$ **k** $e^{3-x} = 1$ **l** $e^{2x-3} \times e^x = 0.1$

2 Solve these equations.

a $\ln x = 6$ **b** $\ln(x+1) = 2$ **c** $\ln(2x-1) = 0$

d $\ln(2x+1) = 1$ **e** $\ln(3-x) = 2$ **f** $\ln(5-2x) = 2.78$

3 Solve these equations.

a $e^{2x} - 3e^x + 2 = 0$ **b** $e^{2x} + 12 = 7e^x$ **c** $e^{2x} = e^x + 2$

Use the substitution $y = e^x$

d $2e^{2x} = e^x + 10$ **e** $4e^{2x} - 8e^x + 3 = 0$ **f** $e^{2x} - 16 = 0$

4 Find the three points on the graph of $y = 2e^{3x+1}$ where

a $y = 20$ **b** $y = 2$ **c** $y = 1$

5 Find the point of intersection of the curves $y = 2e^{2x} - 1$ and $y = 3e^{x} - 2$

6 The graph of the function $f(x) = \ln(2x - 5), x \in \mathbb{R}, x > k$ has a vertical asymptote $x = k$

a Find the value of k. **b** Find x such that $f(x) = -1$

7 **a** Find the equation of the asymptote to the curve $y = 3e^{2x} + 5$

b Find the points of intersection with the curve $y = 8e^{x}$

8 Find the point where the curve $y = e^{x} + 1$ intersects the curve $y = 6e^{-x}$

9 The population, P millions, of a country is growing exponentially as given by $P = 15e^{kt}$ where t is the time in years.
Given that $P = 20$ when $t = 5$, find the value of k.
Calculate the population when

a $t = 0$ **b** $t = 10$

10 The mass, m, of a radioactive material at a time t is given by $m = m_0e^{-kt}$ where k and m_0 are constants.
If $m = \frac{9}{10}m_0$ when $t = 10$, find the value of k.
Also find the half-life of the material.

The half-life is the time taken for the material to decay to half of its original mass.

C3

11 The isotope strontium-90 is present in radioactive fallout and has a half-life of 29 years. Its mass m at a time t (years) is given by $m = m_0e^{-kt}$ where k and m_0 are constants.
Find how many years elapse before any of this isotope in the atmosphere decays to a hundredth of its original mass.

12 A hot liquid cools in such a way that the temperature difference, θ, between its actual temperature and the room temperature is given by $\theta = ke^{-\alpha t}$ where t is the time and k and α are constants. When the room is at a constant temperature 20 °C and the initial temperature of the liquid is 70 °C, the temperature falls to 60 °C after two minutes.
Find the temperature of the liquid after 10 minutes.

INVESTIGATION

13 The graph of $y = A\sin\omega t$ can represent an oscillation of amplitude A over a time t, where A and ω are constants.
If $A = A_0e^{kt}$, where A_0 and k are constants, then the amplitude A changes with time. Use computer software to draw the graphs of $y = A_0\sin\omega t$ and $y = A_0e^{kt}\sin\omega t$ for $t > 0$ for different values of A_0 and k.

The variable y can represent, for example, the noise of a siren or an electrical current.

Start your exploration with $A_0 = 2$, $k = -0.2$ and $\omega = 1$

Review 3

1 Solve these equations for x, giving answers to 2 significant figures.

a $3e^x = 10$ **b** $e^{1-2x} = 0.5$

c $\ln(2x - 1) = 3$ **d** $\ln(1 - x) = -3$

2 Sketch the graph of $y = e^x$
Sketch the graphs of these functions, stating the transformations of $y = e^x$ which occur.

a $y = e^{-x}$ **b** $y = 2e^x + 3$

c $y = 3 - e^{2x}$ **d** $y = e^{|x|}$

3 Sketch the graph of $y = \ln x$
Sketch the graphs of these functions, stating the transformations of $y = \ln x$ which occur.

a $y = 2 - \ln x$ **b** $y = 2\ln x + 1$

c $1 - 2\ln x$ **d** $y = |\ln x|$

4 Solve the equations

a $e^{2x} + e^x = 6$ **b** $e^x + e^{-x} = 2.5$

5 Find the exact solutions of

a $e^{2x+3} = 6$ **b** $\ln(3x + 2) = 4$ [(c) Edexcel Limited 2003]

6 Given that $f(x) = e^x$ and $g(x) = \frac{1}{2}e^x - 1$, find

a the values of $f(2)$, $f^{-1}(2)$ and $g^{-1}(2)$, to 2 decimal places

b the functions $f^{-1}(x)$ and $g^{-1}(x)$.

7 Find the inverse function $f^{-1}(x)$ for each function $f(x)$.
Sketch the graphs of $f(x)$ and $f^{-1}(x)$ on the same diagram, labelling any intersections with the coordinate axes.
In each case give the domain and range of $f^{-1}(x)$.

a $f(x) = 3 + \ln x$ **b** $f(x) = \ln(x - 2)$

c $f(x) = 2 + e^{-x}$ **d** $f(x) = 1 - e^{2x}$

8 The population, P, of a certain organism grows exponentially over time t (days) according to $P = Ae^{\frac{t}{20}}$
Find the value of A and complete this table of values.
Draw the graph of P against t.

t	0	5	10	15	20
P	5				

Calculate how long it takes for the population to double its initial size.

9 The temperature T °C in a boiler rises exponentially over 24 hours so that, after a time t hours, $T = T_0e^{\frac{t}{20}}$

a Given that the temperature is 165 °C after 10 hours, find the value of T_0.

b What is the temperature after 24 hours?

10 Trees in a certain location are infected by a disease.
The number of unhealthy trees, N, was observed to change over time t (in years) as given by $N = 200 - Ae^{-\frac{t}{20}}$

a If there are 91 unhealthy trees after 10 years, find the value of A.

b How many unhealthy trees were there initially?

c What is the limiting value of N as time increases?

11 Find, giving your answer to 3 significant figures where appropriate, the value of x for which

a $3^x = 5$

b $\log_2(2x + 1) - \log_2 x = 2$

c $\ln \sin x = -\ln \sec x$, in the interval $0 < x < 90°$ [(c) Edexcel Limited 2005]

3

Exit

Summary

Refer to

- $f(x) = a^x$ is an exponential function for all values of a. $f(x) = e^x$ is **the exponential function**, where the irrational number e has a value of 2.71 828 to 5 decimal places. 3.1
- $f(x) = e^x$ has the inverse function $f^{-1}(x) = \ln x$, where **the natural logarithm** $\ln x$ denotes $\log_e x$. The graph of $y = \ln x$ is the reflection of the graph of $y = e^x$ in the line $y = x$ 3.2
- You can solve the equation $e^{ax+b} = q$ by taking natural logarithms of both sides. You can solve the equation $\ln(ax + b) = q$ by rewriting it as $ax + b = e^q$ 3.3
- Exponential growth occurs when a variable y changes with time t and $y = Ae^{kt}$ where $k > 0$. Exponential decay occurs when a variable y changes with time t and $y = Ae^{-kt}$ where $k > 0$ 3.3

Links

Exponentials and logarithms are used to understand and quantify many natural phenomena.

The decibel scale, used to measure sound, is based on logarithms.
The equation $\mathbf{dB} = 10\log(\mathbf{I}/\mathbf{I_0})$ is used to compute the intensity of a sound, where **dB** is a unit of sound in decibels, **I** is the intensity of the sound, and $\mathbf{I_0}$ is the softest sound that a human ear can detect.

This equation can be used to calculate certain values, such as the threshold for noise pollution.

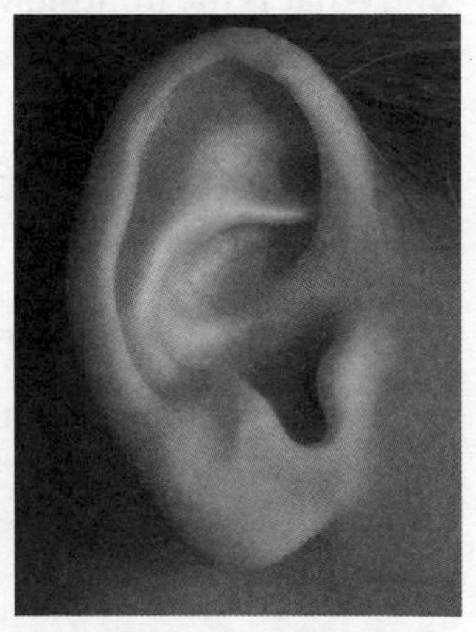

4

Differentiation

This chapter will show you how to differentiate

- the three basic trigonometric functions ($\sin x$, $\cos x$, $\tan x$) and the exponential and logarithmic functions (e^x and $\ln x$)
- the sums, differences, products and quotients of these functions
- composite functions formed by having functions within functions
- functions of the type $x = f(y)$

Before you start

You should know how to:

1 Find the gradient of the tangent at a point P on a curve using the method of small increments.

See C2 Section 7.1.

2 Find the gradient of a tangent to a curve.

e.g. If $y = 3x^2 - 4x + 1$, then $\frac{dy}{dx} = 6x - 4$

The gradient of the tangent at (1,0) is $6 \times 1 - 4 = 2$

3 Find the equation of a straight line if you know its gradient and a point on the line.

e.g. The point (1,2) is on a straight line, gradient 3.

Substitute in $y = mx + c$: $2 = 3 \times 1 + c$ so $c = -1$

The equation of the line is $y = 3x - 1$

4 Find a stationary value of a function and decide if it is a maximum or minimum.

e.g. If $y = x^2 - 6x + 1$, then $\frac{dy}{dx} = 2x - 6$

$\frac{dy}{dx} = 0$ when $x = 3$ and $y = 3^2 - 18 + 1 = -8$

$\frac{d^2y}{dx^2} = 2 > 0$, so -8 is a minimum point.

5 Use the laws of logarithms.

e.g. $\log_{10} 500 = \log_{10}(5 \times 10^2)$

$= \log_{10} 5 + \log_{10} 10^2$

$= \log_{10} 5 + 2\log_{10} 100$

$= \log_{10} 5 + 2$

Check in:

1 Draw the graph of $y = x^2$ accurately. The point $P(1,1)$ is fixed and point Q moves from (3,9) to (1,1). Draw the chord PQ in several positions and find its gradient each time. Compare these values with the gradient of the tangent at P.

2 Find the gradient of the tangent to these curves at the point (1,2).

a $y = x^2 + 2x - 1$ **b** $y = x^3 - x + 2$

c $y = (2 - x)(3 - x)$

3 Find the equation of the straight line which

a has a gradient of 4 and passes through the point (3,2)

b passes through the points (3,1) and (5,5).

4 Find the turning points on the curve $y = x^3 - 3x^2 - 9x + 1$ and determine whether they are maximums or minimums.

5 Prove that $\log_3 \sqrt{18} = 1 + \frac{1}{2}\log_3 2$

4.1 Trigonometric functions

You can find the value of the derivative of the function $y = \sin x$ graphically from the gradient of the tangent at a given point.

See C1 and C2 for revision of differentiation.

When differentiating trigonometric functions you must work in radians.

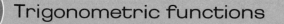

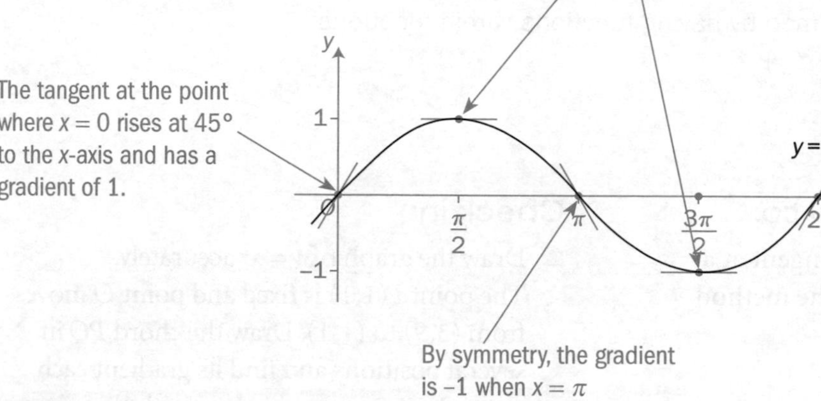
The tangent is parallel to the x-axis when $x = \frac{\pi}{2}$ and $\frac{3\pi}{2}$ and the gradient is 0.
The tangent at the point where $x = 0$ rises at 45° to the x-axis and has a gradient of 1.
The gradient is also 1 when $x = 2\pi$
$y = \sin x$
By symmetry, the gradient is −1 when $x = \pi$

For $y = \sin x$

x	0	$\frac{\pi}{2}$	π	$\frac{3\pi}{2}$	2π
$\frac{dy}{dx}$	1	0	−1	0	1

You can find other values using computer software to draw the curve and its tangents.

which, as a graph, gives these points:

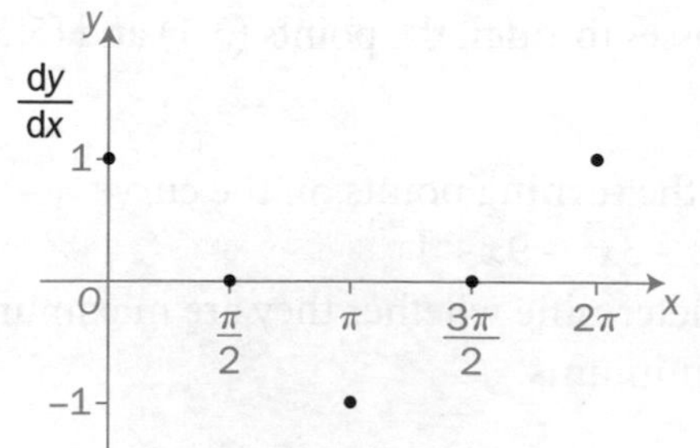

The values in the table and on the graph are the same as those for the function $y = \cos x$

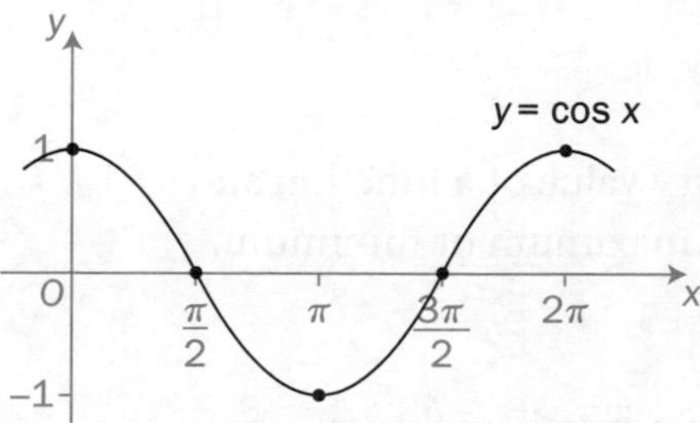
$y = \cos x$

This suggests that the derivative of $y = \sin x$ is closely related to the function $y = \cos x$

C3

The derivatives of sin x, cos x and tan x.

Consider the function $y = \sin x$, where x is in radians.

Let $P(x, y)$ and $Q(x + \delta x, y + \delta y)$ be two points on the graph of $y = \sin x$

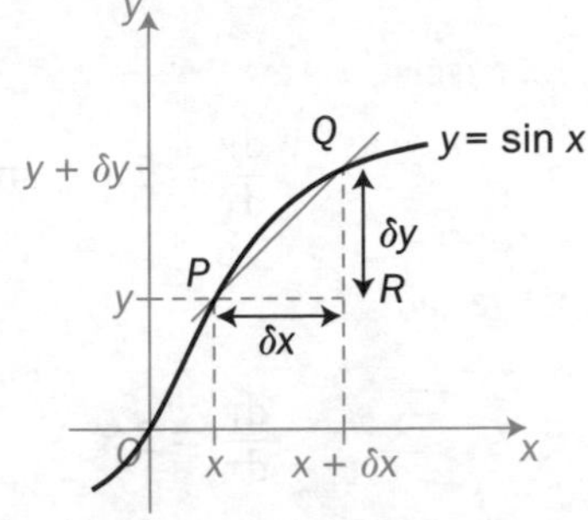

For P, $y = \sin x$
For Q, $y + \delta y = \sin(x + \delta x)$

The gradient of the chord $PQ = \dfrac{QR}{PR}$

$$= \frac{\sin(x + \delta x) - \sin x}{(x + \delta x) - x}$$

$$= \frac{\sin x \cos \delta x + \cos x \sin \delta x - \sin x}{\delta x}$$

Use the expansion of $\sin(A + B)$ from Chapter 2.

If P is fixed and Q approaches P, then $\delta x \to 0$
$\cos \delta x \to 1$
and $\sin \delta x \to \delta x$

You can test these statements by finding the sine and cosine of small angles (in radians) on your calculator.

and, in the limit, the chord PQ becomes the tangent at P.

The phrase 'in the limit' indicates that δx has reduced to zero and $\dfrac{\delta y}{\delta x}$ has become an exact differential $\dfrac{dy}{dx}$.

So, as $\delta x \to 0$, the gradient of $PQ \to \dfrac{\sin x \times 1 + \cos x \times \delta x - \sin x}{\delta x}$

$$\to \frac{\cos x \times \delta x}{\delta x}$$

$$\to \cos x$$

In the limit, when Q reaches P, the gradient of the tangent at P is $\cos x$.

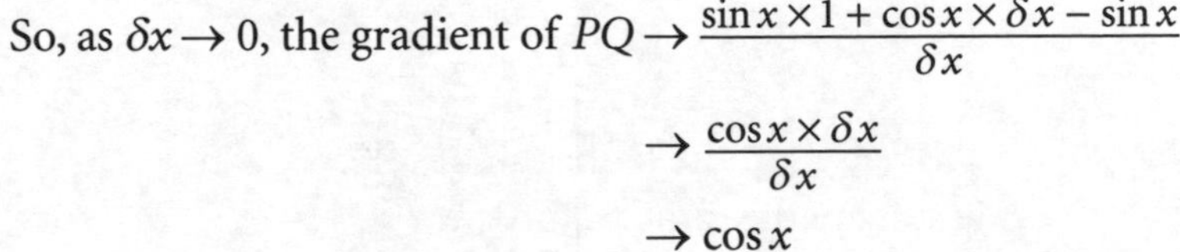

$$\frac{d(\sin x)}{dx} = \cos x$$

x is in radians in all of these results.

In a similar way, you can show that

$$\frac{d(\cos x)}{dx} = -\sin x$$

The derivative of $\cos x$ has a negative sign.

$$\frac{d(\tan x)}{dx} = \sec^2 x$$

Derive the results for $\cos x$ and $\tan x$ for yourself.

C3

EXAMPLE 1

Find the gradient of the curve $y = x^2 + \cos x$ at the point where $x = \frac{\pi}{2}$

You have $y = x^2 + \cos x$

Differentiate wrt x:

wrt means 'with respect to'.

$$\frac{dy}{dx} = 2x - \sin x$$

Substitute $x = \frac{\pi}{2}$:

$$\frac{dy}{dx} = 2 \times \frac{\pi}{2} - \sin\frac{\pi}{2}$$

$$= \pi - 1$$

The required gradient is $\pi - 1$.

EXAMPLE 2

Find the equation of the normal to the curve $y = 1 + 2\tan x$ at the point $(0, 1)$.

You have $y = 1 + 2\tan x$

Differentiate wrt x:

$$\frac{dy}{dx} = 2\sec^2 x$$

Substitute $x = 0$:

$$= 2\sec^2 0$$

$\sec x = \frac{1}{\cos x}$

$$= \frac{2}{\cos^2 0}$$

$$= \frac{2}{1} = 2$$

The gradient of the tangent at the point $(0, 1)$ is 2.

The gradient of the normal at $(0, 1)$ is $-\frac{1}{2}$.

If the gradients of the tangent and normal are m and m', then $mm' = -1$ or $m' = -\frac{1}{m}$

The y-intercept of the normal is 1.

See C1 Chapter 2 for revision of equations of straight lines.

So, the equation of the normal is $y = -\frac{1}{2}x + 1$

EXAMPLE 3

Find the values of θ for which $y = 3 + \sin\theta$ has a maximum value.

You have $y = 3 + \sin\theta$

Differentiate wrt θ: $\frac{dy}{d\theta} = \cos\theta = 0$

for stationary values when $\theta = \frac{\pi}{2}, \frac{3\pi}{2}, \frac{5\pi}{2}, \ldots$

Differentiate again: $\frac{d^2y}{d\theta^2} = -\sin\theta$

At $\theta = \frac{\pi}{2}$ $\frac{d^2y}{d\theta^2} = -\sin\frac{\pi}{2} = -1 < 0$

which indicates a maximum value of y — $\frac{d^2y}{d\theta^2} < 0$ indicates a maximum

At $\theta = \frac{3\pi}{2}$ $\frac{d^2y}{d\theta^2} = -\sin\frac{3\pi}{2} = +1 > 0$

which indicates a minimum value of y — $\frac{d^2y}{d\theta^2} > 0$ indicates a minimum

At $\theta = \frac{5\pi}{2}$ $\frac{d^2y}{d\theta^2} = -\sin = \frac{5\pi}{2} = -1 < 0$

which indicates a maximum value of y

and so on.

Maximum values of y occur when

$$\theta = \frac{\pi}{2}, \frac{5\pi}{2}, \frac{9\pi}{2}, \ldots$$

$$= \frac{\pi}{2} + 2n\pi \text{ where } n = 0, \pm1, \pm2, \ldots$$

Exercise 4.1

1 Differentiate these functions with respect to x.

a $x^2 + \sin x$

b $6x - \cos x$

c $\sin x + \tan x$

d $\sin x + x + 1$

e $3\cos x - 4\tan x$

f $3x^2 + 2x - 1 - \frac{1}{2}\cos x$

C3

2 Find the gradient of the tangent to each of these curves at the point with the given value of x.

a $y = 2x + \tan x$ when $x = \frac{\pi}{4}$

b $y = x^2 - \sin x$ when $x = \frac{\pi}{2}$

c $y = 6x + 2\cos x$ when $x = \pi$

d $y = \cos x + \sin x$ when $x = \frac{\pi}{2}$

e $y = 3\tan x - 2\cos x$ when $x = 0$

3 Find the equation of the tangent to the curve $y = \cos x$ at the point where $x = \frac{\pi}{4}$

4 Find the equation of the tangent to the curve $y = \sin x$ at the point where $x = \frac{\pi}{4}$. At which point does this tangent cut the x-axis?

5 Find the equation of the normal to the curve $y = \sin x + \cos x$ at the point $\left(\frac{\pi}{2}, 1\right)$. At which point does this normal intersect the x-axis?

C3

6 Find the smallest positive value of θ for which

a the gradient of the curve $y = \theta + \sin\theta$ has the value $\frac{1}{2}$

b the gradient of the curve $y = \sin\theta - \sqrt{3}\cos\theta$ has the value 0.

7 Find the smallest positive value of θ for which these functions have maximum values.

a $y = \sin\theta - 2$ b $y = \sin\theta + 2\cos\theta$

8 Find the smallest positive value of θ for which these functions have minimum values.

a $y = 5 + \cos\theta$ b $y = \cos\theta - \sin\theta$

9 Find the maximum and minimum values of the function $f(x) = 3\sin x + 4\cos x$ for $-\pi \leqslant x \leqslant \pi$

Show that, if α and β are the smallest positive values of x at the maximum and minimum values respectively, then $\alpha - \beta = \pi$

10 An object moves in a straight line such that its distance y from a fixed point at a time t is given by $y = A\sin t$, where A is a positive constant.

a Find the velocity of the object $\frac{dy}{dt}$ and prove that $\frac{dy}{dt} = \sqrt{A^2 - y^2}$

b Prove that $\frac{d^2y}{dt^2} + y = 0$

11 A pendulum bob swings such that the angle θ that the pendulum makes with the vertical is given by $\theta = \theta_0(\sin t + \cos t)$, where θ_0 is a constant.

a Find the value of θ when

i $t = 0$ ii $t = \frac{\pi}{2}$

b Find the value of its angular velocity $\frac{d\theta}{dt}$ when

i $t = 0$ ii $t = \frac{\pi}{2}$

c Explain why the time taken for one oscillation of the bob is 2π.

12 A particle moves in a straight line so that its distance, y metres, from a fixed point O is given by

$y = 5\cos t + 3\sin t$

where t is the time in seconds after it has begun to move.

a How far is the particle from O at the start of the motion?

b Find the first two times that the particle is at rest.

c What is the particle's acceleration after one second of motion and when does the particle next have this acceleration?

C3

INVESTIGATION

13 Construct a table of values with these headings for $0 \leqslant x \leqslant 0.5$ in radians.

x (radians)	x (degrees)	sin x	cos x	tan x

a When x is in radians, for what range of values of x is

i $\sin x \approx x$ ii $\tan x \approx x$

so that they differ by no more than 10% of the value of x?

b The diagram shows the graph of $y = 1 - kx^2$

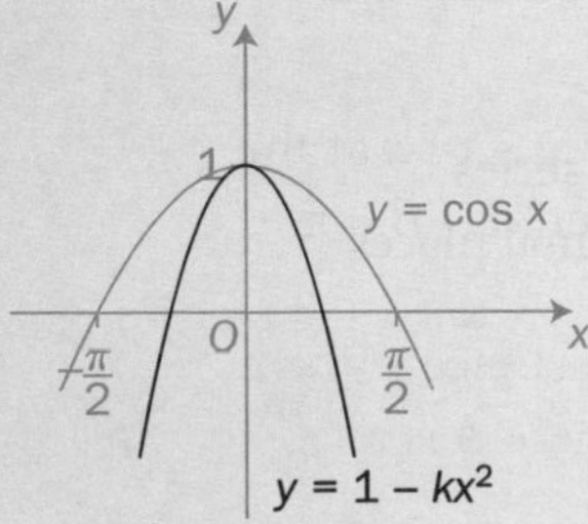

Find a value of k such that

$\cos x \approx 1 - kx^2$ for small values of x (in radians).

4.2 The exponential function, e^x

Suppose you want to find a function that stays unchanged when it is differentiated.

Start the search with the simple function $y = 1$

For $y = 1$, $\frac{dy}{dx} = 0$

Now work backwards.

In each step, make $\frac{dy}{dx}$ the same as y from the previous step and integrate to give a new y:

$$\frac{dy}{dx} = 0 \quad \Rightarrow \quad y = 1$$

$$\frac{dy}{dx} = 1 \quad \Rightarrow \quad y = 1 + x$$

$$\frac{dy}{dx} = 1 + x \quad \Rightarrow \quad y = 1 + x + \frac{x^2}{2}$$

$$\frac{dy}{dx} = 1 + x + \frac{x^2}{2} \quad \Rightarrow \quad y = 1 + x + \frac{x^2}{2} + \frac{x^3}{2 \times 3}$$

Continue this process to produce a series with an infinite number of terms:

$$y = 1 + x + \frac{x^2}{1\times 2} + \frac{x^3}{1\times 2\times 3} + \frac{x^4}{1\times 2\times 3\times 4} + \dots + \frac{x^n}{n!} + \dots$$

This series differentiates to give the same series.

By substituting different values of x, you can calculate the sum of the series and record the results in a table.

You need to also show that this series converges no matter what value is given to x.

For $x = 0 \quad y = 1$

For $x = 1 \quad y = 1 + 1 + \frac{1}{2} + \frac{1}{6} + \frac{1}{24} + \dots = 2.718282\dots$

For $x = 2 \quad y = 1 + 2 + \frac{4}{2} + \frac{8}{6} + \frac{16}{24} + \dots = 7.389056\dots$ and so on.

x	0	1	2	3	4
y	1	2.718282...	7.389056...	20.085537...	54.598150...
		$= 2.718282^1$	$= 2.718282^2$	$= 2.718282^3$	$= 2.718282^4$

Recall from Chapter 3 that the number e = 2.718282...

Like $\sqrt{2}$ and π, the number e is irrational.
It cannot be given an exact numerical value, but this series allows you to calculate the value of e to as many decimal places as you wish.

You can also use negative values of x in the series.

For $x = -1$, $\quad y = 1 - 1 + \frac{1}{2} - \frac{1}{6} + \frac{1}{24} - \dots = \frac{1}{2.718282} = e^{-1}$

x	-1	0	1	2	3	4	...	n	...
y	e^{-1}	$e^0 = 1$	e	e^2	e^3	e^4	...	e^n	...

Try substituting other negative values of x into the series.

C3

You can use these values to draw the graph of $y = e^x$:

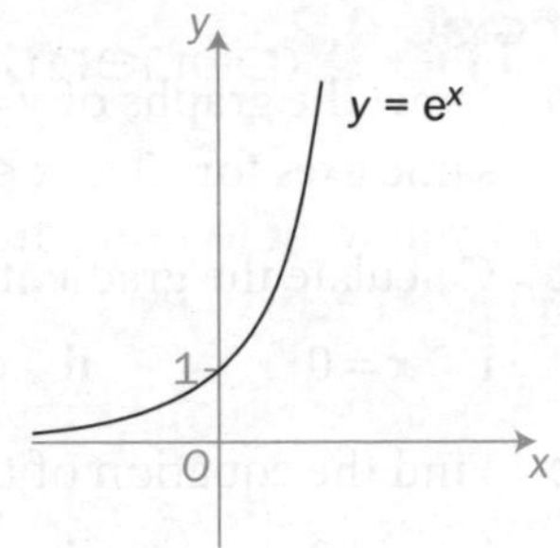

It is the exponential function and is one member of the family of exponential functions $y = a^x$ which you first met in unit C2.

In summary,

if $y = e^x$, then $\frac{dy}{dx} = e^x$ or $\frac{d(e^x)}{dx} = e^x$

EXAMPLE 1

Find the gradient of the tangent to the curve $y = (e^x + 1)^2 - e^{2x}$ at the point where $x = 0$

You have

$$\begin{aligned} y &= (e^x + 1)^2 - e^{2x} \\ &= e^{2x} + 2e^x + 1 - e^{2x} \\ &= 2e^x + 1 \end{aligned}$$

Differentiate wrt x: $\frac{dy}{dx} = 2e^x$

At $x = 0$

$$\begin{aligned} \frac{dy}{dx} &= 2e^0 \\ &= 2 \times 1 = 2 \end{aligned}$$

The gradient of the tangent is 2 at the point where $x = 0$

EXAMPLE 2

Find the point on the curve $y = e^x - x$ at which y has a stationary value. Determine the nature of the stationary value.

Differentiate $y = e^x - x$ wrt x: $\frac{dy}{dx} = e^x - 1$

At a stationary point, $\frac{dy}{dx} = 0$

$$\begin{aligned} e^x - 1 &= 0 \\ e^x &= 1 \\ x &= 0 \end{aligned}$$

and $y = e^0 - 0 = 1$

There is a stationary value at the point $(0, 1)$.

Differentiate again: $\frac{d^2y}{dx^2} = e^x$

At the point $(0, 1)$ $\frac{d^2y}{dx^2} = e^0 = 1 > 0$

At a minimum point, $\frac{d^2y}{dx^2}$ is positive.

Hence, the stationary value at the point $(0, 1)$ is a minimum.

C3

Exercise 4.2

1 **a** Draw the graphs of $y = 2^x$, $y = e^x$ and $y = 3^x$ accurately on the same axes for $-3 \leqslant x \leqslant 3$

b Calculate the gradient of $y = e^x$ when

i $x = 0$ ii $x = 1$

c Find the equation of the tangent to the curve $y = e^x$ when

i $x = 0$ ii $x = 1$

2 Differentiate these functions with respect to x.

a $3x^2 + e^x$ **b** $4e^x + 6x - 1$

c $5\sin x + 2e^x$ **d** $3\tan x - \frac{1}{2}e^x$

e $(e^x + 1)(x - 1) - xe^x$ **f** $(e^x + 1)(e^{-x} - 1) - e^{-x}$

3 Find the gradient of the tangent to each of these curves at the point with the given value of x.
Give each answer **i** in terms of e **ii** as a decimal to 3 significant figures.

a $y = e^x - x$ when $x = 2$

b $y = 3x^2 - 4e^x$ when $x = 1$

c $y = e^x - \cos x$ when $x = \frac{\pi}{2}$

d $y = x^2 + 2x + 1 - e^x$ when $x = -1$

e $y = \tan x + 2e^x$ when $x = 0$

4 Find the equation of the tangent to each curve at the point where $x = 0$

a $y = 3x + 2e^x$

b $y = 3e^x - \sin x + 5$

5 Find the equation of the normal to the curve $y = \frac{1}{2}e^x + 2x^2$ at the point $\left(0, \frac{1}{2}\right)$.

6 Prove that the normal to the curve $y = 1 - x + e^x$ at the point $(1, e)$ passes through the point $(e^2, -1)$.

7 Line L_1 is the tangent to the curve $y = 2e^x - x$ at the point $(0, 2)$.
Line L_2 is the tangent to the curve $y = \sin x - x^2$ at the origin.
Prove that the point of intersection of L_1 and L_2 is $(2, 4)$.

C3

8 Find the values of x for which these functions have stationary values.
Determine the nature of the stationary values.

a $f(x) = 1 - x + e^x$

b $f(x) = 1 + x - e^x$

c $f(x) = 1 + 4x - 2e^x$

9 Sketch the graph of the function $y = x - e^x$
Indicate any stationary values.

10 Consider the function $y = e^{2x} - 3e^x + 2$

a By letting $t = e^x$, find the points where the graph of $y = e^{2x} - 3e^x + 2$ cuts the x-axis.

This substitution gives a quadratic equation in t.

b Find the value of x at which y has a stationary value.
Find whether the stationary value is a maximum or a minimum.

11 Sketch the curve $y = x^2 + e^x$

INVESTIGATION

12 Use a spreadsheet with these headings, where $f(x) = a^x$, $f(x + \delta x) = a^{x+\delta x}$

and the gradient $g(x) = \dfrac{f(x + \delta x) - f(x)}{\delta x}$

a	x	δx	f(x)	f(x + δx)	g(x)

The function $g(x)$ gives the gradient of the chord PQ which, for small values of δx such as 0.0001, is approximately equal to the gradient of the tangent at P.

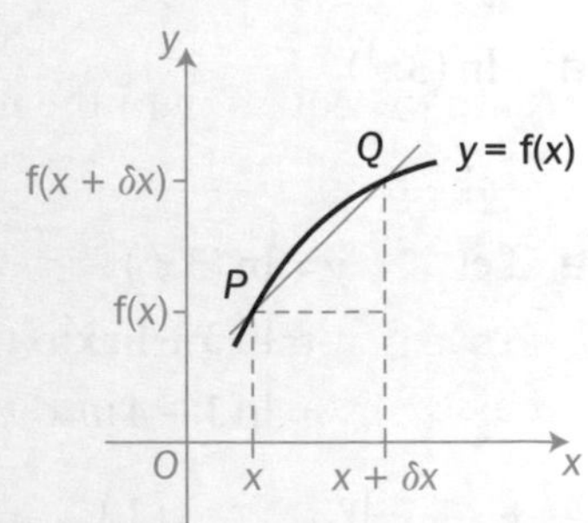

Use the spreadsheet for $a \neq e$ (say, $a = 4$) and confirm that $f(x) \neq g(x)$ for any value of x.
Repeat for other values of $a \neq e$

Now let $a = e$
Confirm the result of Chapter 3 that the gradient at the point $(0, 1)$ is e.

Show that, for other values of x, $f(x)$ and $g(x)$ have (almost) the same value. You have now illustrated (but not proved) that $f(x) = e^x$ stays unchanged when you differentiate it.

C3

The logarithmic function, ln x

$y = \ln x$ is the inverse of the function $y = e^x$

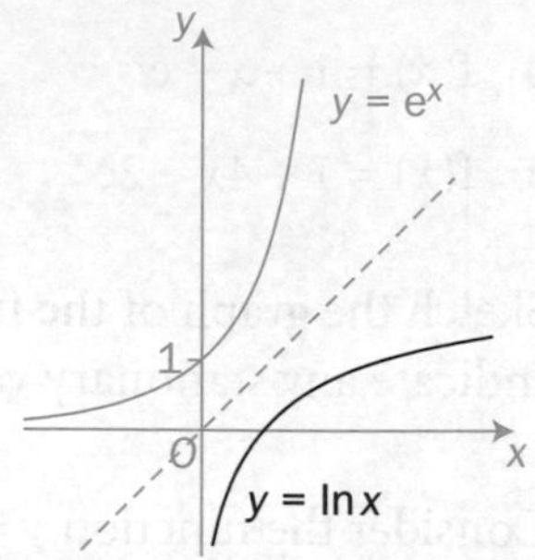

The graph of $y = \ln x$ is the reflection of the graph of $y = e^x$ in the line $y = x$

Consider the derivative of $y = \ln x$

Rewrite the relationship: $x = e^y$

$\log_a b = c$ implies $b = a^c$

Differentiate with respect to y: $\frac{dx}{dy} = e^y$

You differentiate x wrt y.

The derivative of e^y wrt y is e^y.

So $\frac{dy}{dx} = \frac{1}{\frac{dx}{dy}}$

$= \frac{1}{e^y}$

$= \frac{1}{x}$

The result $\frac{dy}{dx} = \frac{1}{\frac{dx}{dy}}$ is shown later in this chapter.

If $y = \ln x$, then $\frac{dy}{dx} = \frac{1}{x}$ or $\frac{d(\ln x)}{dx} = \frac{1}{x}$

EXAMPLE 1

Find the differential of

a $\ln(3x^4)$ **b** $\ln\left(\frac{3}{\sqrt{x}}\right)$

a Let $y = \ln(3x^4)$

$= \ln 3 + \ln x^4$

$= \ln 3 + 4\ln x$

So $\frac{dy}{dx} = 0 + 4\left(\frac{1}{x}\right)$

$= \frac{4}{x}$

b Let $y = \ln\left(\frac{3}{\sqrt{x}}\right)$

$= \ln 3 - \ln x^{\frac{1}{2}}$

$= \ln 3 - \frac{1}{2}\ln x$

So $\frac{dy}{dx} = 0 - \frac{1}{2}\left(\frac{1}{x}\right)$

$= -\frac{1}{2x}$

Where possible, simplify the logarithmic function to make it easier to differentiate.

$\sqrt{x} = x^{\frac{1}{2}}$

Exercise 4.3

1 Differentiate these functions with respect to x.

a $2\ln x$ b $\ln x^2$ c $\ln(4x^3)$

d $\ln\left(\frac{1}{2}x\right)$ e $\ln\left(\frac{1}{2x}\right)$ f $\ln\sqrt{5x^3}$

g $\ln\left(\frac{1}{\sqrt{x}}\right)$ h $\ln(e^{2x})$ i $\ln(xe^x)$

j $\ln(x^2 e^{2x})$

2 Find the gradient of the tangent to each of these curves at the point with the given value of x.

a $y = 2x + \ln x$ when $x = 2$

b $y = x^3 - \ln x^2$ when $x = 1$

c $y = \ln\sqrt{x} + e^x$ when $x = 1$

d $y = \ln\left(\frac{1}{3x}\right) + \frac{1}{3x}$ when $x = -1$

e $y = \ln x^{\pi} + \sin x$ when $x = \frac{\pi}{2}$

3 Find the points at which the graphs of these functions have zero gradient. Determine whether the functions have a maximum or a minimum at these points.

a $f(x) = x - \ln x$ b $f(x) = x^2 - \ln(x^2)$

4 Find the equation of the tangent to the curve $y = x - \ln x$ at the point where $x = 2$. At which point does the tangent cut the x-axis?

5 Find the equation of the normal to the curve $y = 2\ln x^3$ at the point where $x = 1$. Find the distance between the points at which the normal cuts the x- and y-axes.

6 Show that the point $(1, e)$ lies on the curve $y = e^x + \ln\sqrt{x}$
Find the area of the triangle bounded by the two coordinate axes and the tangent to the curve at this point.

INVESTIGATION

7 Differentiate $y = \log_{10} x$ by first changing the base of the logarithm to base e.
You can then differentiate $y = \ln x$ to find $\frac{d(\log_{10} x)}{dx}$.

Try other bases. What do you notice?

C3

4.4 The product rule

Consider two functions $u = \mathrm{f}(x)$ and $v = \mathrm{g}(x)$ which are multiplied so that $y = uv$

$y = \mathrm{f}(x) \times \mathrm{g}(x)$

Let δx be a small increase in x which produces small increases of δu, δv and δy in u, v and y respectively.

You have $$y + \delta y = (u + \delta u)(v + \delta v)$$
$$= uv + u\delta v + v\delta u + \delta u\delta v$$

Subtract $y = uv$: $$\delta y = u\delta v + v\delta u + \delta u\delta v$$

Divide each term by δx: $$\frac{\delta y}{\delta x} = \frac{u\delta v}{\delta x} + \frac{v\delta u}{\delta x} + \frac{\delta u\delta v}{\delta x} \qquad (1)$$

As $\delta x \to 0$ $\quad \frac{\delta y}{\delta x} \to \frac{\mathrm{d}y}{\mathrm{d}x}, \frac{\delta v}{\delta x} \to \frac{\mathrm{d}v}{\mathrm{d}x}, \frac{\delta u}{\delta x} \to \frac{\mathrm{d}u}{\mathrm{d}x}$

and $\quad \delta u \to 0, \delta v \to 0$

As $x \to 0$, $\frac{\delta u \delta v}{\delta x}$ tends to zero.

In the limit from (1), $\frac{\mathrm{d}y}{\mathrm{d}x} = u\frac{\mathrm{d}v}{\mathrm{d}x} + v\frac{\mathrm{d}u}{\mathrm{d}x}$

C3

In the limit, if $y = uv$ then
$$\frac{\mathrm{d}y}{\mathrm{d}x} = u\frac{\mathrm{d}v}{\mathrm{d}x} + v\frac{\mathrm{d}u}{\mathrm{d}x}$$
or $$\frac{\mathrm{d}}{\mathrm{d}x}[\mathrm{f}(x) \times \mathrm{g}(x)] = \mathrm{f}(x) \times \mathrm{g}'(x) + \mathrm{g}(x) \times \mathrm{f}'(x)$$

Think of the **product rule** as:
(1st function × derivative of the 2nd) + (2nd function × derivative of the 1st)

EXAMPLE 1

Differentiate **a** $x^2 \ln x$ **b** $(x^3 + 2)\sin x$ **c** $x^2 \tan x$

a Let $y = x^2 \ln x$

$u = x^2 \qquad v = \ln x$

$\frac{\mathrm{d}u}{\mathrm{d}x} = 2x \qquad \frac{\mathrm{d}v}{\mathrm{d}x} = \frac{1}{x}$

Differentiating both u and v wrt x.

So $\frac{\mathrm{d}y}{\mathrm{d}x} = u\frac{\mathrm{d}v}{\mathrm{d}x} + v\frac{\mathrm{d}u}{\mathrm{d}x}$

$= x^2 \times \frac{1}{x} + \ln x \times 2x$

$= x + 2x\ln x$

EXAMPLE 1 (CONT.)

b Let $y = (x^3 + 2)\sin x$

$$u = x^3 + 2 \qquad v = \sin x$$

$$\frac{du}{dx} = 3x^2 \qquad \frac{dv}{dx} = \cos x$$

So $\dfrac{dy}{dx} = u\dfrac{dv}{dx} + v\dfrac{du}{dx} = (x^3 + 2)\cos x + 3x^2 \sin x$

c When $y = x^2 \tan x$, $\dfrac{dy}{dx} = x^2 \times \sec^2 x + \tan x \times 2x$

$$= x^2 \sec^2 x + 2x\tan x$$

$\dfrac{d(\tan x)}{dx} = \sec^2 x$

Exercise 4.4

1 Differentiate these functions with respect to x.

a $x\ln x$ **b** $x^2\sin x$ **c** $e^x\tan x$

d $(x^2 - 3x + 1)e^x$ **e** $e^x\ln x$ **f** $\sin x\ln x$

g $x^{-2}\cos x$ **h** $\dfrac{1}{x}e^x$ **i** $(x^3 - 1)\ln x$

j $\sqrt{x}\tan x$ **k** $(x^2 - 1)(x^3 - x^2 + 1)$ **l** $(x^3 + 1)(x^2 - 2x + 4)$

2 Find the gradient of the tangent to the curve $y = x^2\sin x$ at the point where $x = \dfrac{\pi}{2}$

3 Find the equation of the tangent to the curve $y = x^3 e^x$ at the point where

a $x = 0$ **b** $x = -3$

4 Find the equation of the normal to the curve $y = (x^2 - 2x + 1)(x^3 - 4x^2 + 1)$ at the point $(1,0)$.

5 a Prove that the equation of the normal to the curve $y = \sin x \times \ln x$ at the point where $x = 1$ is given by $y = 1.19(1 - x)$

b Find the area of the triangle enclosed by the normal and the x- and y-axes.

6 Find the turning points on the curve with the equation $y = x^2(x + 1)^8$
Determine the nature of each turning point and sketch the curve.

7 Find the stationary value of the function $f(x) = x^3\ln x$ and show that it is a minimum.

8 Given that $y = \sin x\cos x$, prove that $\dfrac{dy}{dx} = 2\cos^2 x - 1$

Find the stationary points on the graph of $y = \sin x\cos x$ for $0 \leqslant x \leqslant \pi$ and determine whether they are maxima or minima.

4.5 The quotient rule

Consider two functions $u = \mathrm{f}(x)$ and $v = \mathrm{g}(x)$ which are divided to give $y = \frac{u}{v}$

$y = \frac{\mathrm{f}(x)}{\mathrm{g}(x)}$

Let δx be a small increase in x which produces small increases of δu, δv and δy in u, v and y respectively.

You have
$$y + \delta y = \frac{u + \delta u}{v + \delta v}$$

Subtract $y = \frac{u}{v}$:
$$\delta y = \frac{u + \delta u}{v + \delta v} - \frac{u}{v}$$

Combine the two fractions on the RHS:
$$= \frac{v(u + \delta u) - u(v + \delta v)}{v(v + \delta v)}$$

Expand and simplify:
$$= \frac{v\delta u - u\delta v}{v^2 + v\delta v}$$

Divide by δx:
$$\frac{\delta y}{\delta x} = \frac{v \times \frac{\delta u}{\delta x} - u \times \frac{\delta v}{\delta x}}{v^2 + v\delta v} \quad (1)$$

As $\delta x \to 0$ $\quad \frac{\delta y}{\delta x} \to \frac{\mathrm{d}y}{\mathrm{d}x}, \frac{\delta u}{\delta x} \to \frac{\mathrm{d}u}{\mathrm{d}x}, \frac{\delta v}{\delta x} \to \frac{\mathrm{d}v}{\mathrm{d}x}$

and $\delta u \to 0, \delta v \to 0$

In the limit from (1), $\frac{\mathrm{d}y}{\mathrm{d}x} = \frac{v\frac{\mathrm{d}u}{\mathrm{d}x} - u\frac{\mathrm{d}v}{\mathrm{d}x}}{v^2}$

In the limit, if $y = \frac{u}{v}$ then
$$\frac{\mathrm{d}y}{\mathrm{d}x} = \frac{v \times \frac{\mathrm{d}u}{\mathrm{d}x} - u \times \frac{\mathrm{d}v}{\mathrm{d}x}}{v^2}$$

or
$$\frac{\mathrm{d}}{\mathrm{d}x}\left(\frac{\mathrm{f}(x)}{\mathrm{g}(x)}\right) = \frac{\mathrm{g}(x) \times \mathrm{f}'(x) - \mathrm{f}(x) \times \mathrm{g}'(x)}{(\mathrm{g}(x))^2}$$

Think of the **quotient rule** as:
(bottom × derivative of top) – (top × derivative of bottom), all divided by $(\text{bottom})^2$

EXAMPLE 1

Differentiate $\frac{\sin x}{\ln x}$

You have $y = \frac{\sin x}{\ln x}$

where $u = \sin x$ and $v = \ln x$

Take care not to mix up the functions u and v.

$$\frac{du}{dx} = \cos x \quad \text{and} \quad \frac{dv}{dx} = \frac{1}{x}$$

$$\text{So} \quad \frac{dy}{dx} = \frac{\ln x \times \cos x - \sin x \times \frac{1}{x}}{(\ln x)^2}$$

$$= \frac{x \ln x \cos x - \sin x}{x(\ln x)^2}$$

Multiply both numerator and denominator by x.

EXAMPLE 2

Differentiate $\frac{x^2 + 4x + 1}{\cos x}$

You have $y = \frac{x^2 + 4x + 1}{\cos x}$

$$\text{So} \quad \frac{dy}{dx} = \frac{\cos x \times (2x + 4) - (x^2 + 4x + 1) \times (-\sin x)}{(\cos x)^2}$$

$\frac{d(\cos x)}{dx} = -\sin x$

$$= \frac{(2x + 4)\cos x + (x^2 + 4x + 1)\sin x}{\cos^2 x}$$

$$= \frac{(2x + 4) + (x^2 + 4x + 1)\tan x}{\cos x}$$

EXAMPLE 3

Find the derivatives of **a** $\tan x$ **b** $\cot x$ **c** $\operatorname{cosec} x$ **d** $\sec x$

These results are given on the formulae list. You do not need to memorise them.

a Let $y = \tan x = \frac{\sin x}{\cos x}$

$$\frac{dy}{dx} = \frac{\cos x \times \cos x - \sin x \times (-\sin x)}{\cos^2 x}$$

$$= \frac{\cos^2 x + \sin^2 x}{\cos^2 x}$$

$$= \frac{1}{\cos^2 x}$$

$$= \sec^2 x$$

The solutions to parts **b**, **c** and **d** are on the next page.

C3

EXAMPLE 3 (CONT.)

b Let $y = \cot x = \dfrac{\cos x}{\sin x}$

$$\frac{dy}{dx} = \frac{\sin x \times (-\sin x) - \cos x \times \cos x}{\sin^2 x}$$

$$= \frac{-(\sin^2 x + \cos^2 x)}{\sin^2 x}$$

$$= \frac{-1}{\sin^2 x} = -\text{cosec}^2 x$$

c Let $y = \text{cosec}\, x = \dfrac{1}{\sin x}$

$$\frac{dy}{dx} = \frac{\sin x \times 0 - 1 \times \cos x}{\sin^2 x}$$

$$= \frac{-\cos x}{\sin^2 x}$$

$$= -\frac{1}{\sin x} \times \frac{\cos x}{\sin x}$$

$$= -\text{cosec}\, x \cot x$$

d Let $y = \sec x = \dfrac{1}{\cos x}$

$$\frac{dy}{dx} = \frac{\cos x \times 0 - 1 \times (-\sin x)}{\cos^2 x}$$

$$= \frac{\sin x}{\cos^2 x}$$

$$= \frac{1}{\cos x} \times \frac{\sin x}{\cos x} = \sec x \tan x$$

For some functions, you need to use both the product rule and the quotient rule.

EXAMPLE 4

Differentiate $\dfrac{x^2 e^x}{\sin x}$

Firstly, consider the function $f(x) = x^2 e^x$, which is the product of x^2 and e^x.

Differentiate f(x) using the product rule: $f'(x) = x^2 e^x + e^x \times 2x$

$$= x^2 e^x + 2xe^x$$

Let $y = \dfrac{x^2 e^x}{\sin x}$ and use the quotient rule:

$$\frac{dy}{dx} = \frac{\sin x (x^2 e^x + 2xe^x) - x^2 e^x \cos x}{\sin^2 x}$$

$$= \frac{xe^x (x \sin x + 2 \sin x - x \cos x)}{\sin^2 x}$$

C3

Exercise 4.5

1 Differentiate these functions with respect to x.

a $\dfrac{x}{\sin x}$ b $\dfrac{x^2}{\tan x}$ c $\dfrac{x^2+1}{x}$ d $\dfrac{\ln x}{x}$

e $\dfrac{e^x}{\sin x}$ f $\dfrac{x+1}{e^x}$ g $\dfrac{\sqrt{x}}{\sin x}$ h $\dfrac{\cos x}{x^2}$

i $\dfrac{1-x}{1+x}$ j $\dfrac{x^2+1}{x^2-1}$ k $\dfrac{2x^3-1}{3x^2-1}$ l $\dfrac{x^2}{\ln x}$

2 Find the gradient of the tangent at the point where $x=0$ for the curve

a $y=\dfrac{3x-2}{2x+1}$ b $y=\dfrac{\tan x}{x+1}$

3 The tangent to the curve $y=\dfrac{x}{x^3+1}$ at the point $\left(1, \dfrac{1}{2}\right)$ cuts the x-axis at the point P. Find the coordinates of P.

4 Find the equation of the normal at the point (0, 0) on the curve $y=\dfrac{2x}{e^x}$

5 Find $\dfrac{d^2y}{dx^2}$ if $y=\dfrac{1+x}{x}$

6 Prove that the maximum value of $f(x)=\dfrac{\ln x}{x^3}$ is $\dfrac{1}{3e}$

7 Use both the product rule and the quotient rule to differentiate these functions.

a $\dfrac{\sin x\cos x}{x}$ b $\dfrac{x\sin x}{\cos x}$ c $\dfrac{xe^x}{\ln x}$

d $\dfrac{e^x\ln x}{x}$ e $\dfrac{x^2\ln x}{e^x}$ f $\dfrac{x\sin x}{\ln x}$

INVESTIGATION

8 You can differentiate $y=\dfrac{e^x\sin x}{x^2}$

- either by first using the product rule on $e^x\sin x$ followed by the quotient rule on the whole expression
- or by first using the quotient rule on $\dfrac{e^x}{x^2}$ followed by the product rule on the whole expression.

Show that these two methods give the same answer.

Is one method easier than the other?

C3

4.6 The chain rule

A **composite function** is a 'function of a function'.

You can differentiate 'functions of functions' using the **chain rule**.

Consider $y = g(u)$ and $u = f(x)$, so that $y = g(f(x))$

Examples of composite functions:
$\ln(x^3 - 1)$ $\quad e^{\sin x}$
$(x^2 - 2x - 1)^4$ $\quad \sqrt{2 + x^3}$

Let δx be a small increase in x which produces small increases of δu and δy in u and y respectively.

Then $\dfrac{\delta y}{\delta x} = \dfrac{\delta y}{\delta u} \times \dfrac{\delta u}{\delta x}$

As $\delta x \to 0$, $\quad \dfrac{\delta y}{\delta x} \to \dfrac{dy}{dx}, \quad \dfrac{\delta y}{\delta u} \to \dfrac{dy}{du}$ and $\dfrac{\delta u}{\delta x} \to \dfrac{du}{dx}$ (1)

In the limit from (1), $\dfrac{dy}{dx} = \dfrac{dy}{du} \times \dfrac{du}{dx}$

In the limit, if $y = g(f(x))$, then $\dfrac{dy}{dx} = \dfrac{dy}{du} \times \dfrac{du}{dx}$

Think of the chain rule as:
derivative of 'outer' function wrt u × derivative of 'inner' function wrt x

EXAMPLE 1

Differentiate **a** $\sin(x^2 + 4x + 3)$ **b** $\ln(\sin x)$

a Let $y = \sin(x^2 + 4x + 3)$ $\quad u = x^2 + 4x + 3 \quad y = \sin u$

$$\frac{du}{dx} = 2x + 4 \qquad \frac{dy}{du} = \cos u$$

Differentiating u wrt x and y wrt u.

Apply the chain rule: $\dfrac{dy}{dx} = \dfrac{dy}{du} \times \dfrac{du}{dx}$

$= \cos u \times (2x + 4)$

Substitute $u = x^2 + 4x + 3$: $= (2x + 4)\cos(x^2 + 4x + 3)$

b Let $y = \ln(\sin x)$ $\quad u = \sin x \quad y = \ln u$

$$\frac{du}{dx} = \cos x \qquad \frac{dy}{du} = \frac{1}{u}$$

Apply the chain rule: $\dfrac{dy}{dx} = \dfrac{1}{u} \times \cos x$

Substitute $u = \sin x$: $= \dfrac{1}{\sin x} \times \cos x$

$= \dfrac{\cos x}{\sin x} = \cot x$

$\dfrac{\cos x}{\sin x} = \dfrac{1}{\tan x} = \cot x$

C3

EXAMPLE 2

Find $\frac{dy}{dx}$ when **a** $y = \cos^3 x$ **b** $y = \sqrt{x^3 - 1}$

a $y = \cos^3 x$, so if $u = \cos x$ then $y = u^3$

$$\frac{dy}{du} = 3u^2 = 3\cos^2 x$$

$$\frac{du}{dx} = -\sin x$$

so $\frac{dy}{dx} = 3\cos^2 x \times (-\sin x)$

$= -3\cos^2 x \sin x$

b $y = \sqrt{x^3 - 1}$, so if $u = x^3 - 1$ then $y = \sqrt{u} = u^{\frac{1}{2}}$

$$\frac{dy}{du} = \frac{1}{2}u^{-\frac{1}{2}} = \frac{1}{2}(x^3 - 1)^{-\frac{1}{2}}$$

$$\frac{du}{dx} = 3x^2$$

so $\frac{dy}{dx} = \frac{1}{2}(x^3 - 1)^{-\frac{1}{2}} \times 3x^2$

$= \frac{3x^2}{2\sqrt{x^3 - 1}}$

EXAMPLE 3

a Differentiate $2\ln\left(x\sqrt{x^2 + 1}\right)$

b Find the equation of the tangent to the curve $y = 2\ln\left(x\sqrt{x^2 + 1}\right)$ at the point where $x = 1$

a Let $y = 2\ln\left(x\sqrt{x^2 + 1}\right)$

Where possible, simplify functions before you differentiate.

$= 2\ln x + 2\ln(x^2 + 1)^{\frac{1}{2}}$

$= 2\ln x + \ln(x^2 + 1)$

$\ln(a \times b) = \ln a + \ln b$

Differentiate, using the chain rule for $\ln(x^2 + 1)$:

$y = \ln u$ and $u = x^2 + 1$

so $\frac{dy}{du} = \frac{1}{u}$ and $\frac{du}{dx} = 2x$

$$\frac{dy}{dx} = 2\left(\frac{1}{x}\right) + 2x \times \left(\frac{1}{x^2 + 1}\right)$$

$$= 2\left[\frac{x^2 + 1 + x^2}{x(x^2 + 1)}\right]$$

$$= \frac{2(2x^2 + 1)}{x(x^2 + 1)}$$

b When $x = 1$ $y = 2\ln\left(1 \times \sqrt{2}\right) = 2\ln 2^{\frac{1}{2}} = \ln 2$

$$\frac{dy}{dx} = \frac{2 \times 3}{1 \times 2} = 3$$

The equation of the tangent is $\frac{y - \ln 2}{x - 1} = 3$

Using $\frac{y - y_1}{x - x_1} = m$

$y - \ln 2 = 3(x - 1)$

$y = 3x + \ln 2 - 3$

C3

Exercise 4.6

1 Differentiate with respect to x

a $\sin(x^2+1)$ b $\cos(x^3-1)$ c $\tan(x^2)$

d $\ln(x^2+2x+3)$ e $\ln(x^3+1)$ f $\ln(x^3)$

Simpify part **f** before you differentiate.

g $\ln(\sin x)$ h $\sin(\ln x)$ i $\ln(\ln x)$

j $e^{\sin x}$ k e^{x+4} l e^{x^2}

m $(x^2+1)^5$ n $(2x+6)^7$ o $(x^3-1)^{-2}$

p $\sqrt{x^2-1}$ q $\sqrt{x^3+1}$ r $\dfrac{1}{x-1}$

s $\dfrac{2}{x^2+3x-1}$ t $\dfrac{3}{\sqrt{2x+1}}$ u $\dfrac{1}{\sqrt{x^2+1}}$

Write part **r** as $(x-1)^{-1}$ first. You could use the quotient rule instead of the chain rule but the working is longer.

v $\dfrac{1}{\ln x}$ w $\dfrac{2}{e^x+1}$ x $\sqrt[3]{2x-1}$

2 Find $f'(\theta)$ when

a $f(\theta)=\sin(\theta^2)$ b $f(\theta)=\sin^2\theta$

c $f(\theta)=\sqrt{\sin\theta}$ d $f(\theta)=\tan(\theta^3)$

e $f(\theta)=\tan^3\theta$ f $f(\theta)=\dfrac{1}{\sqrt{\tan\theta}}$

3 Find $\dfrac{dy}{dx}$ when

a $y=\dfrac{1}{\sin x}$ b $y=e^{e^x}$

c $y=\dfrac{4}{x^2+4x-1}$ d $y=4e^{\frac{1}{x}}$

e $y=\ln\sqrt{x-1}$ f $y=\ln\left(\dfrac{\sin x}{1-\cos x}\right)$

Simplify the function in parts **e** and **f** before you differentiate as in question **1** part **f**.

4 Differentiate with respect to x.

a $y=\sin kx$ b $y=\cos kx$

c $y=\tan kx$ d $y=\ln kx$

e $y=e^{kx}$

The differential of $\tan kx$ is in the formulae list provided in the exam. None of the other results here are provided.

5 Find the equation of the tangent to the curve $y=(2x-4)^3$ at the point where $x=3$

6 Find the point on the curve $f(x)=(2x-1)^4$ where the gradient is 1.

When the gradient is 1, $f'(x)=1$

C3

7 Find the gradient of the curve $y = \ln\sqrt{1+\sin x}$ at the point where $x = \frac{\pi}{4}$

Simplify the logarithm before you differentiate.

8 Find the equation of the tangent to the curve $y = \sqrt{x^2+16}$ at the point (3,5).

9 Differentiate with respect to x, where a and b are constants.

a $\sqrt{a^2+b^2x^2}$ **b** $\frac{1}{\sqrt{ax^2+b}}$ **c** $(a^2x^2-b^2)^{\frac{2}{3}}$

10 Given that a and b are constants, find the only value of x for which $\sqrt{\frac{a}{ax^4+b}}$ has a stationary value.

11 **a** Prove that the curve $y = \ln\left(\frac{1+e^x}{e^x}\right)$ has no stationary values.

b Show that the tangent to this curve at the point where $x = 0$ has the equation $x + 2y = \ln\alpha$.
Find the value of α.

12 **a** Prove that the curve $y = \frac{8}{x^2} + \frac{1}{(1-x)^2}$ has only one turning point.

b Show that the coordinates of the turning point are $\left(\frac{2}{3}, 27\right)$.

c i Find an expression for $\frac{d^2y}{dx^2}$.

ii Hence, prove that the turning point is a minimum.

13 Find the position and nature of the stationary points of $f(x) = \sin^2 x$

INVESTIGATION

14 Show that the function $y = \ln(x^2+1)$ has only one stationary value. Determine its nature and position.

Sketch the graph of the function.

Check your answers by using computer software to draw the graph.

Investigate the properties of $\ln(x^2+k)$ for different values of k.

C3

4.7 Further applications

You can combine the product, quotient and chain rules to differentiate more complicated functions.

EXAMPLE 1

Differentiate $y = (\sqrt{x^2+1})\sin x$

The function $y = (\sqrt{x^2+1})\sin x$ is the product of $\sqrt{x^2+1}$ and $\sin x$.

$\sqrt{x^2+1}$ is a function of a function.

You have $u = \sqrt{x^2+1} = (x^2+1)^{\frac{1}{2}}$ and $v = \sin x$

Use the chain rule to differentiate u:

$$\frac{du}{dx} = \frac{1}{2}(x^2+1)^{-\frac{1}{2}} \times 2x$$

Use the product rule:

$$\frac{dy}{dx} = \sqrt{x^2+1} \times \cos x + \sin x \times \frac{1}{2}(x^2+1)^{-\frac{1}{2}} \times 2x$$

$$= (\sqrt{x^2+1})\cos x + \frac{x\sin x}{\sqrt{x^2+1}}$$

$$= \frac{(x^2+1)\cos x + x\sin x}{\sqrt{x^2+1}}$$

EXAMPLE 2

Differentiate $y = \ln\sin(x^2+x+1)$

You have $y = \ln u$ and $u = \sin v$ and $v = x^2 + x + 1$

Use the chain rule: $\frac{dy}{dx} = \frac{dy}{du} \times \frac{du}{dv} \times \frac{dv}{dx}$

There are three links in the chain rule here.

$$= \frac{1}{u} \times \cos v \times (2x+1)$$

Substitute for u and v: $= \frac{1}{\sin v} \times \cos(x^2+x+1) \times (2x+1)$

$$= \frac{(2x+1)\cos(x^2+x+1)}{\sin(x^2+x+1)}$$

$$= (2x+1)\cot(x^2+x+1)$$

EXAMPLE 3

Differentiate $y = 3^x$

Take the logarithms of both sides to remove the index:

$\ln y = \ln 3^x = x\ln 3$

Now, $\ln y$ is a function of y and y is a function of x.
So, $\ln y$ is a function of a function of x.

From the chain rule you have $\dfrac{d(\ln y)}{dx} = \dfrac{d(\ln y)}{dy} \times \dfrac{dy}{dx} = \dfrac{1}{y} \times \dfrac{dy}{dx}$

$\dfrac{d(\ln y)}{dy} = \dfrac{1}{y}$

Differentiate $\ln y = \ln 3 \times x$ with respect to x:

$$\frac{1}{y} \times \frac{dy}{dx} = \ln 3 \times 1$$

$$\frac{dy}{dx} = y \times \ln 3 = 3^x \ln 3$$

$\ln 3$ is a numerical constant that is multiplying x.

In general, if $y = a^x$ then $\dfrac{dy}{dx} = a^x \ln a$

Now consider a function of the type $x = f(y)$

Examples are $x = \sqrt{y^3 + 1}$ and $x = \ln(y^2 - 3y + 1)$

Let δx be a small increase in x which produces a small increase of δy in y.

Then $\dfrac{\delta y}{\delta x} = \dfrac{1}{\frac{\delta x}{\delta y}}$

As $\delta x \to 0$ and $\delta y \to 0$

then $\dfrac{\delta y}{\delta x} \to \dfrac{dy}{dx}$ and $\dfrac{\delta x}{\delta y} \to \dfrac{dx}{dy}$ giving $\dfrac{dy}{dx} = \dfrac{1}{\frac{dx}{dy}}$

So, for $x = f(y)$, $\dfrac{dy}{dx} = \dfrac{1}{\frac{dx}{dy}}$

EXAMPLE 4

Find $\dfrac{dy}{dx}$ when $x = y^2 + 3$

Differentiate $x = y^2 + 3$ with respect to y: $\dfrac{dx}{dy} = 2y$

So $\dfrac{dy}{dx} = \dfrac{1}{\frac{dx}{dy}} = \dfrac{1}{2y} = \dfrac{1}{2\sqrt{x-3}}$

Find $\dfrac{dy}{dx}$ in terms of y first and then in terms of x.

C3

Exercise 4.7

1 Differentiate with respect to x.

a $(x^2-1)^3\sin x$

b $(x^3+1)^2\tan x$

c $(3x+2)^4\times e^x$

d $(\sqrt{x+1})\times e^x$

e $\dfrac{\sin 2x}{x^2}$

f $(\sqrt{x^2-1})\times\ln x$

g $x\sin(x^2)$

h $x\sin^2 x$

i $x\cos 3x$

j $x^2\tan 2x$

k xe^{2x+1}

l $x^2\ln(x^2-1)$

m $\sin x\cos 2x$

n $\sin 4x\cos x$

o $\sin x\cos^2 x$

p $e^x\sin 3x$

q $e^{-2x}\cos x$

r $\dfrac{e^{3x}}{x}$

s $\dfrac{e^{\sin x}}{x}$

t $e^{-x}\sin 2x$

2 Differentiate with respect to x.

a $\ln\sin(x^2)$

b $\ln\tan(x^2)$

c $\sin^2 3x$

d $\cos^3 4x$

e $\tan^2 2x$

f $\tan^3(x+4)$

g $e^{\sin^2 x}$

h $e^{\ln(x^2+1)}$

Can you see an easier method for part **h** using inverse functions?

3 Find $\dfrac{dy}{dx}$ when

a $y=4^x$

b $y=5^x$

c $y=c^x$ where c is a constant

d $y=2^{2x}$

e $y=5^{2x}$

f $y=x^x$

Can you see a connection between parts a and d?

4 Find the value of x which gives a stationary point on the curve $y=\sqrt{x^2-1}\times e^x$

5 Find the value of $\dfrac{d^2y}{dx^2}$ when $x=0$ and $y=e^x\sin 2x$

6 Find the gradient of the curve $y=\ln\sqrt{1+\sin 2x}$ at the point where $x=\dfrac{\pi}{2}$

C3

7 Find the value of x which gives a stationary point on the curve $y = 5^{2x} - 4 \times 5^x$

8 Find $\frac{dy}{dx}$ in terms of x when

a $x = 3y^2 - 4$

b $x = \sqrt{y^2 + 3}$

c $x = (y + 1)^2$

9 Find $\frac{dy}{dx}$ in terms of y when

a $x = 4y^3 + y$

b $x = 3\sin^4 y$

c $x = 5\tan^3(2y)$

d $x = y \ln y$

e $x = e^{-y} \sin y$

f $x = 3\sin 4y + 4\cos 2y$

g $x = \ln\left(y^2 \times \sqrt{y - 2}\right)$

10 Find the equation of the tangent at the point $(2, 1)$ on the curve $x = 2y^3$

11 Find the equations of the tangents to the curve $x = 4y^2$ at the points where $x = 4$

12 Prove that the curve $x = y^2 e^y$ has no stationary points.

INVESTIGATIONS

13 Find the turning points on the curve $y = e^x \sin 2x$ for $0 \leqslant x \leqslant 2\pi$
Sketch its graph over this range and check your answers using computer software.

14 a Find the turning points of the curve $y = x + \frac{1}{x}$
Sketch its graph.

b Prove that the curve $x = y + \frac{1}{y}$ has no turning points.

Sketch its graph.

c What is the connection between your two sketches?

C3

Review 4

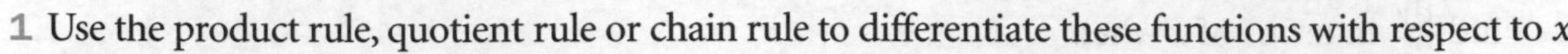

1 Use the product rule, quotient rule or chain rule to differentiate these functions with respect to x.

a $x^3 \tan x$ **b** $\tan^3 x$ **c** $\tan(x^3)$ **d** $\dfrac{\tan x}{x^3}$

e $e^x \sin x$ **f** $e^{\sin x}$ **g** $\dfrac{1}{e^{\sin x}}$ **h** $\cos x \ln x$

i $\sin 3x$ **j** $\tan 6x$ **k** $\ln(2x+3)$ **l** e^{3x-1}

m $\dfrac{\sin x}{e^x}$ **n** $e^x(x^2-3x+4)$ **o** $\dfrac{\sqrt{x}}{x^2+1}$

2 Differentiate these functions with respect to x, given that a and b are constants.

a $y = e^{ax}$ **b** $y = e^{ax+b}$ **c** $y = e^{f(x)}$ **d** $y = \sin(ax)$

e $y = \sin(ax+b)$ **f** $y = \sin[f(x)]$ **g** $y = \tan(ax)$ **h** $y = \tan(ax+b)$

i $y = \tan[f(x)]$ **j** $y = \ln(ax)$ **k** $y = \ln(ax+b)$ **l** $y = \ln[f(x)]$

3 Find the derivatives of these functions with respect to x.

a $e^x \cos 3x$ **b** $e^{3x} \tan x$ **c** $e^{3x} \cos 2x$ **d** $4\tan\left(\frac{1}{2}x\right)$

e $\ln(x^3 \sin x)$ **f** $\ln\left(x\sqrt{x+1}\right)$ **g** $\dfrac{(3x+1)^2}{(2x-1)^3}$ **h** $\ln e^x$

i $\dfrac{\sin 5x}{\cos 5x}$ **j** $x^2 \sin\left(2x + \dfrac{\pi}{2}\right)$ **k** $e^{3\tan x}$ **l** $\ln\left(\dfrac{1-\sin x}{1+\sin x}\right)$

m $x\cot x$ **n** $\ln(\tan x + \sec x)$ **o** $\dfrac{x}{\sec x}$

4 Find the values of x for which these functions have stationary points for $-\pi < x < \pi$. Determine the nature of the stationary points.

a $y = xe^{-x}$ **b** $y = \dfrac{e^x}{\sin x}, x \neq 0$ **c** $y = e^{-x}\cos x$

5 Find the equation of the tangent to the curve $y = \ln\sqrt{1+\sin 2\theta}$ at the point where $\theta = \dfrac{\pi}{2}$

6 The curve C has equation $y = 4x^{\frac{3}{2}} - \ln(5x)$, where $x > 0$.
The tangent at the point on C, where $x = 1$, meets the x-axis at the point A.
Prove that the x-coordinate of A is $\frac{1}{5}\ln(5e)$. [(c) Edexcel Limited 2002]

7 Differentiate with respect to x

a $y = 2^x$ **b** $y = a^x$ **c** $y = \log_{10} x$

8 a Find the equation of the normal to the curve $y = 1 + \sin^2 x$ at the point where $x = \dfrac{\pi}{4}$

b Find the equation of the normal to the curve $y = \tan^2 x$ where $x = \dfrac{\pi}{4}$ and the point where the normal cuts the x-axis.

9 The function $y = \sin(x)$ has the angle x measured in degrees.

a Write the function so that the angle x is in radians.

b Find $\frac{d(\sin x)}{dx}$

10 Find the derivative with respect to x of these functions.

a $f(x) = \ln \tan x$ **b** $f(x) = \ln(\sin x \cos x)$

c $f(x) = (2x - 1)^2(3x + 2)^3$ **d** $f(x) = \ln((2x - 1)^2(3x + 2)^3)$

11 Use the derivatives of cosec x and cot x to prove that

$$\frac{d}{dx}[\ln(\operatorname{cosec} x + \cot x)] = -\operatorname{cosec} x$$

[(c) Edexcel Limited 2004]

12 Given that $y = e^x \sin 2x$, find the value of $\frac{d^2y}{dx^2}$ when $x = 0$

13 If $y = e^{-x}\cos x$, show that $\frac{d^2y}{dx^2} + 2\frac{dy}{dx} + 2y = 0$

14 Find $\frac{d^2y}{dx^2}$ when $y = \frac{1+x}{1-x}$

15 Find $\frac{dy}{dx}$ in terms of x when

a $x = (y + 1)^2$ **b** $x = y^2 + 3$

16 Differentiate with respect to x

a x^3e^{3x} **b** $\frac{2x}{\cos x}$ **c** $\tan^2 x$

d Given that $x = \cos y^2$ find $\frac{dy}{dx}$ in terms of y. [(c) Edexcel Limited 2002]

17 a Differentiate with respect to x

i x^2e^{3x+2} ii $\frac{\cos(2x^3)}{3x}$

b Given that $x = 4\sin(2y + 6)$ find $\frac{dy}{dx}$ in terms of x. [(c) Edexcel Limited 2006]

18 Show that $\frac{dy}{dx}$ for the function $y = \sqrt{2x+3} + \frac{1}{\sqrt{2x+3}}$

is given by $\frac{dy}{dx} = \frac{2(x+1)}{\sqrt{(2x+3)^3}}$

19 Find the turning points on the curve $y = (x^2 - 9)^3$ and determine whether each of them is a maximum or minimum.

C3

4 Exit

Summary

Refer to

- The derivatives of these functions f(x) are: 4.1

$f(x)$	$f'(x)$
$\sin x$	$\cos x$
$\cos x$	$-\sin x$
$\tan x$	$\sec^2 x$
e^x	e^x
$\ln x$	$\frac{1}{x}$

4.2

4.3

- The product rule If $y = uv$, then $\frac{dy}{dx} = u\frac{dv}{dx} + v\frac{du}{dx}$ 4.4
- The quotient rule If $y = \frac{u}{v}$, then $\frac{dy}{dx} = \frac{v\frac{du}{dx} - u\frac{dv}{dx}}{v^2}$ 4.5
- The chain rule If $y = g(u)$ and $u = f(x)$ so that $y = g(f(x))$, 4.6

  then $\frac{dy}{dx} = \frac{dy}{du} \times \frac{du}{dx}$ 4.7

  In particular, $\frac{d(\sin ax)}{dx} = a\cos ax$ and $\frac{d(\cos ax)}{dx} = -a\sin ax$

- If $x = f(y)$, then you can find $\frac{dy}{dx}$ from $\frac{dy}{dx} = \frac{1}{\frac{dx}{dy}}$ 4.7

Links

Mathematics is used extensively in the financial world, which relies on accurate forecasts of the future.

In calculus, a derivative shows how a dependent variable is affected as an independent variable changes.

This can be applied in a financial setting to, for example, predict how the change in price of an underlying asset will affect the related market value. Techniques such as the chain rule are usually used to calculate the mathematical derivatives involved.

C3

5 Numerical methods

This chapter will show you how to
- use graphical methods to solve equations of the type f(x) = g(x)
- use graphical methods to solve equations of the type f(x) = 0
- find non-integer roots of the equation f(x) = 0
- use iterative methods to solve equations
- distinguish between a sequence of convergent and divergent iterations
- represent iterative methods graphically.

Before you start

You should know how to:

1 Rearrange an algebraic expression.

e.g. $x = \frac{5x^2 + 7}{3x + 4}$

becomes $2x^2 - 4x + 7 = 0$

2 Solve equations graphically.

e.g. Plot the graph of $y = x^3 - 4x^2 + 3x$

Hence find that the solutions of

$x^3 - 4x^2 + 3x = 0$ are $x = 0, 1$ or 3.

3 Solve simultaneous equations graphically.

e.g. Plot the graphs of $y = x^2 - x + 1$ and $y = 2x - 1$

Hence find that the solutions of $x^2 - 3x + 2 = 0$ are $x = 1$ and $x = 2$

4 Use a spreadsheet to find the values of f(*x*) for a range of values of *x*.

e.g. For $f(x) = x^3 - x + 2$

x	$x^3 - x + 2$
0	2
1	2
2	8
3	26

Check in:

1 Find f(*x*) such that f(*x*) = 0 when

a $x = \sqrt{\frac{x^3 + 1}{2x}}$ **b** $x = \sqrt[3]{\frac{x^2 - 1}{2x}}$

2 **a** Plot the graph of $y = x^2 - 2x - 4$ for $-2 \leqslant x \leqslant 4$

Hence solve the equation $x^2 - 2x - 4 = 0$

b Plot the graph of $y = \frac{x^2 - 2x + 1}{x - 2}$ for $-2 \leqslant x \leqslant 4$ and find the point where the graph touches the *x*-axis.

3 Plot the graphs of $y = 4 - x^2$ and $y = \frac{1}{x}$ for $|x| \leqslant 3$. Explain how the graphs can be used to solve the equation $x^3 - 4x + 1 = 0$ and find the solutions.

4 Create a spreadsheet to show the values of $f(x) = x^2 + 2x - 3$ and $g(x) = x + 2$ for integer values of *x* where $0 \leqslant x \leqslant 7$. State the value of the integer *a* such that the solution of $x^2 + x - 5 = 0$ lies between *a* and $a + 1$.

5.1 Graphical methods

Many equations, including those derived in real-life scenarios in science, engineering, economics and elsewhere, cannot be solved algebraically.

Instead, you have to use numerical methods to find approximate solutions.

'Solutions' are also called 'roots'.

Intersection of two graphs

One method for finding an approximate solution to an equation of the type $f(x) = g(x)$ is to draw the graphs of $y = f(x)$ and $y = g(x)$ and find the x-values of any points of intersection.

EXAMPLE 1

Find the solution of the equation $\cos x = x$ for $0 < x < \pi$

Sketch the graphs of $y = \cos x$ and $y = x$ on the same axes:

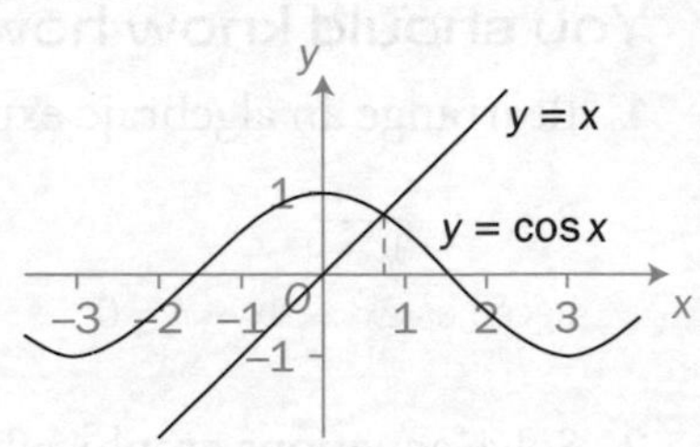

There is only one point of intersection so the equation $\cos x = x$ has only one solution.

The sketch gives a rough approximation of the solution as $x = 0.7$

This is an approximation. It is not an accurate answer.

You can find a more accurate value using

either a graphical calculator or graphical computer software

or an ordinary calculator or computer spreadsheet.

x is in radians.

Tabulate values of x and $\cos x$ with increments in x of 0.01:

For $x \leqslant 0.73$, $x < \cos x$

For $x \geqslant 0.74$, $x > \cos x$

The solution is between 0.73 and 0.74

0.74 and 0.738 are closer together than 0.73 and 0.745, so the solution is between 0.735 and 0.740.

x		$\cos x$
0.70	<	0.765
0.71	<	0.758
0.72	<	0.752
0.73	<	0.745
0.74	>	0.738
0.75	>	0.732
0.76	>	0.725

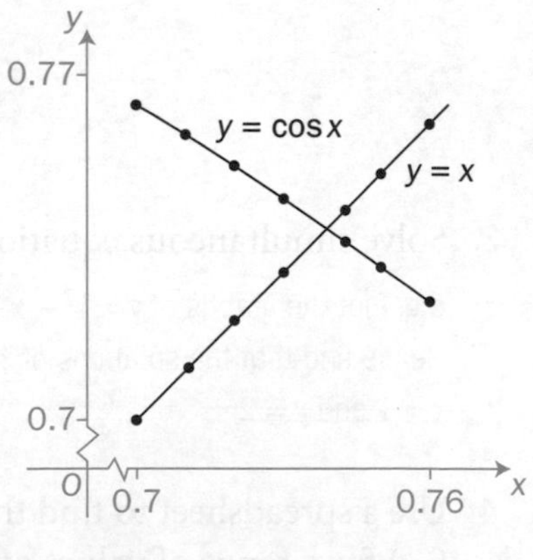

Reduce the increment in x to 0.001:

$x < \cos x$ for $x \leqslant 0.739$

$x > \cos x$ for $x \geqslant 0.740$

The solution is between 0.739 and 0.740 but it is closer to 0.739.

To 3 significant figures, the solution of the equation $\cos x = x$ is $x = 0.739$

x		$\cos x$
0.736	<	0.7412
0.737	<	0.7405
0.738	<	0.7398
0.739	<	0.7391
0.740	>	0.7385

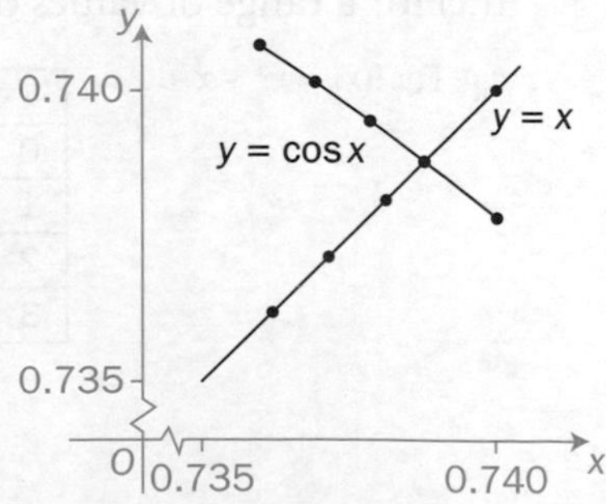

Intersection with the *x*-axis

Another graphical method for solving an equation of the type $f(x) = g(x)$ is to define a new function $h(x) = f(x) - g(x)$ and then find the solution of the equation $h(x) = 0$

You are now searching for the *x*-values of the points at which the graph of $y = h(x)$ crosses (or touches) the *x*-axis.

EXAMPLE 2

Find the solution of the equation $\cos x = x$ for $0 < x < \pi$

This equation is the same as in Example 1.

Let $h(x) = \cos x - x$

Sketch the graph of $h(x) = \cos x - x$:

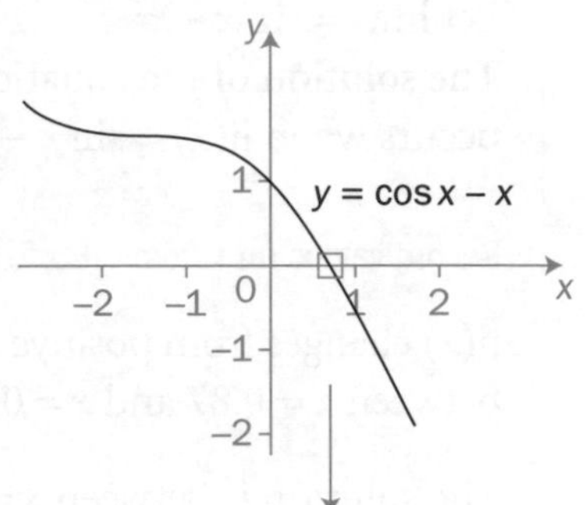

The sketch gives an approximate value of the root as $x = 0.7$

The *x*-value where $y = \cos x - x$ crosses the *x*-axis is the same *x*-value as that at the point of intersection in Example 1.

Use a spreadsheet to record values with increments in *x* of 0.01:

For $x \leqslant 0.73$, $h(x) > 0$

For $x \geqslant 0.74$, $h(x) < 0$

The root is between 0.73 and 0.74.

The root is nearer to 0.74 because −0.002 is closer to 0 than is 0.015.

x	cos *x* − *x*	
0.70	0.065	> 0
0.71	0.048	> 0
0.72	0.032	> 0
0.73	0.015	> 0
0.74	−0.002	< 0
0.75	−0.018	< 0
0.76	−0.035	< 0

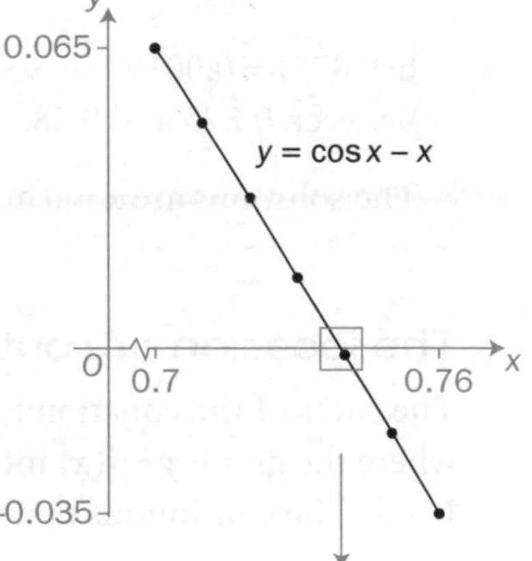

Reduce the increments in *x* to 0.001:

For $x \leqslant 0.739$, $h(x) > 0$

For $x \geqslant 0.740$, $h(x) < 0$

The root is between 0.739 and 0.740

The root is closer to 0.739 than 0.740 because 0.0001 is closer to 0 than is −0.0015.

x	cos *x* − *x*	
0.737	0.0035	> 0
0.738	0.0018	> 0
0.739	0.0001	> 0
0.740	−0.0015	< 0
0.741	−0.0032	< 0

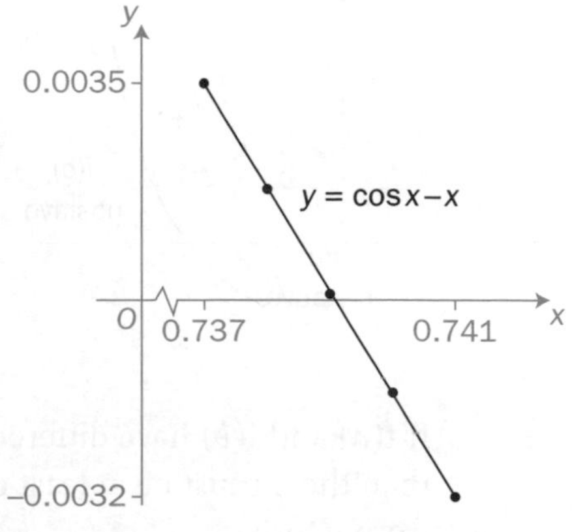

To 3 significant figures, the solution of the equation $\cos x = x$ is $x = 0.739$

Check:
$h(0.7395) = \cos(0.7395) - 0.7395 = -0.0007$ is negative, so the root must lie between $x = 0.739$ and $x = 0.7395$

C3

Sometimes it is best to combine both approaches as shown in Example 3.

EXAMPLE 3

Solve the equation $x^2 = \sin x$ for $-\frac{\pi}{2} < x < \frac{\pi}{2}$

Sketch the graphs of $y = x^2$ and $y = \sin x$:

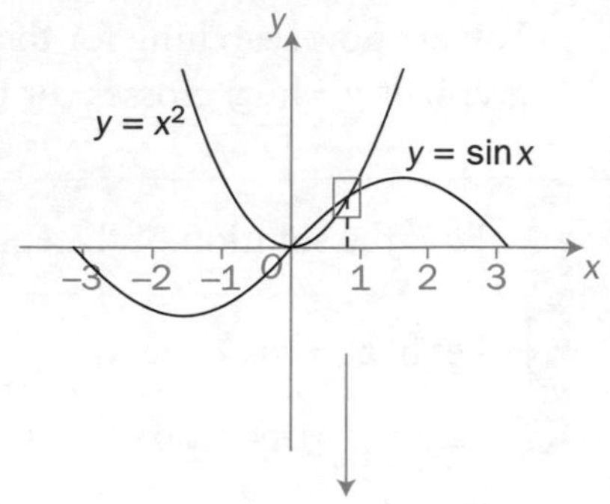

There are two roots.
$x = 0$ is an obvious root.
From the sketch, the other root is approximately $x = 0.8$

Let $h(x) = \sin x - x^2$
The solution of the equation $x^2 = \sin x$ occurs when $h(x) = \sin x - x^2 = 0$

Record values for x from 0.85 to 0.90 in a table:

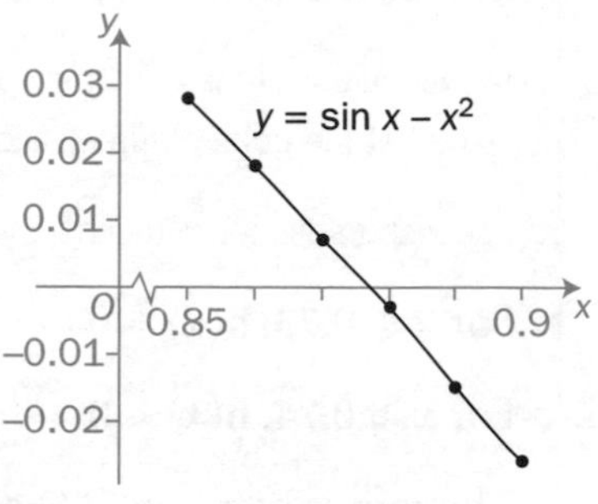

$h(x)$ changes from positive to negative between $x = 0.87$ and $x = 0.88$

x	$\sin x - x^2$
0.85	0.0288 > 0
0.86	0.0182 > 0
0.87	0.0074 > 0
0.88	–0.0037 < 0
0.89	–0.0150 < 0
0.90	–0.0267 < 0

The solution is between $x = 0.87$ and 0.88.

$h(0.875) = 0.0019 > 0$, so the root is between 0.875 and 0.88.

$h(0.875) = \sin(0.875) - 0.875^2$
$= 0.001918...$

The solutions are $x = 0$ and $x = 0.88$ to 2 significant figures.

The location of roots on the x-axis

The roots of the equation $f(x) = 0$ are the x-values of the points where the graph $y = f(x)$ intersects the x-axis.
Let $f(x)$ be continuous between $x = a$ and $x = b$

'Continuous' means that there is no break or discontinuity in the graph.

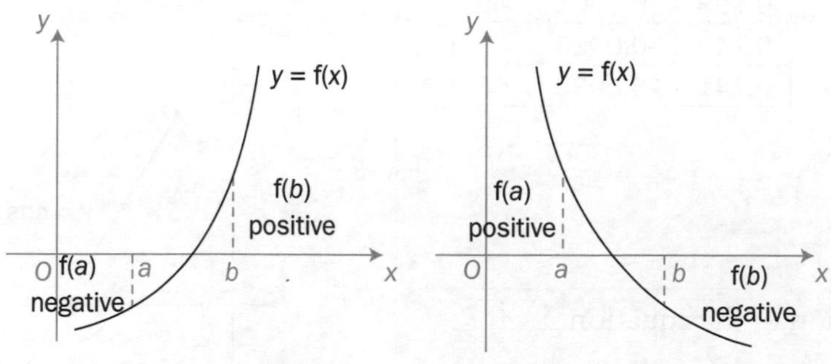

If $f(a)$ and $f(b)$ have different signs for distinct values a and b, then there must be at least one root of $f(x) = 0$ in the interval from a to b.

C3

Be aware that f(x) might cross the x-axis more than once in the interval $[a, b]$.

$[a, b]$ means the interval $a \leqslant x \leqslant b$

Also be aware of the following three possibilities:

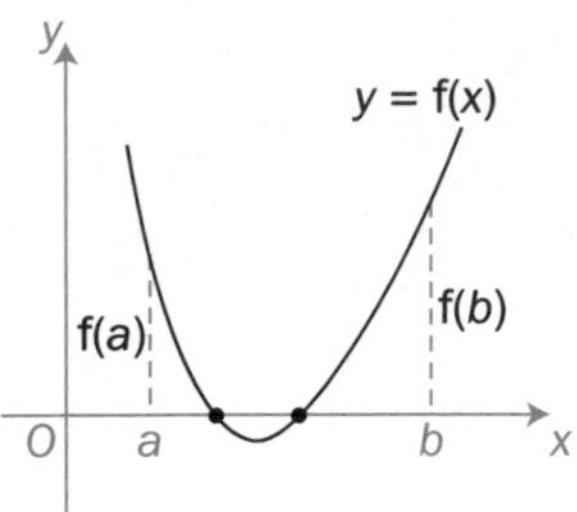

f(a) and f(b) are **both** positive, but there are **two** roots between $x = a$ and $x = b$

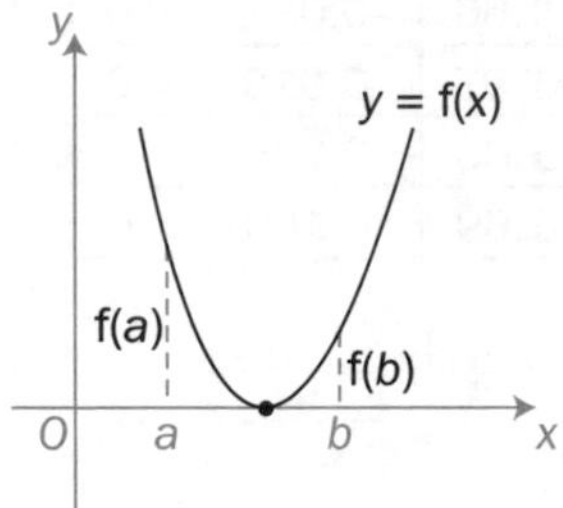

f(a) and f(b) are **both** positive but there is a **single** root when the curve touches the x-axis.

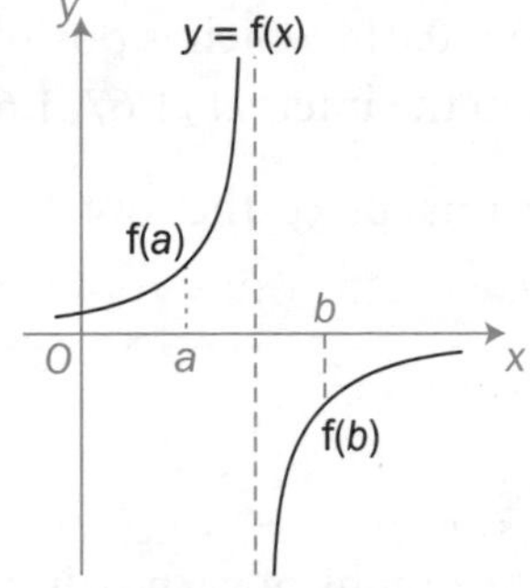

f(a) and f(b) have different signs. However there is **no** root between $x = a$ and $x = b$ because f(x) has a discontinuity.

Sketching a graph will help you to visualise what is happening.

When you have found an interval $[a, b]$ which contains a root, you can then reduce the width of the interval to find the solution to the required accuracy.

C3

EXAMPLE 4

Find all the solutions (to 2 d.p.) of the equation $e^x + x - 7 = 0$

The graph of $y = e^x + x - 7$ cannot be sketched quickly.

Rearrange the equation $e^x + x - 7 = 0$ as $e^x = 7 - x$ and sketch the graphs of $y = e^x$ and $y = 7 - x$:

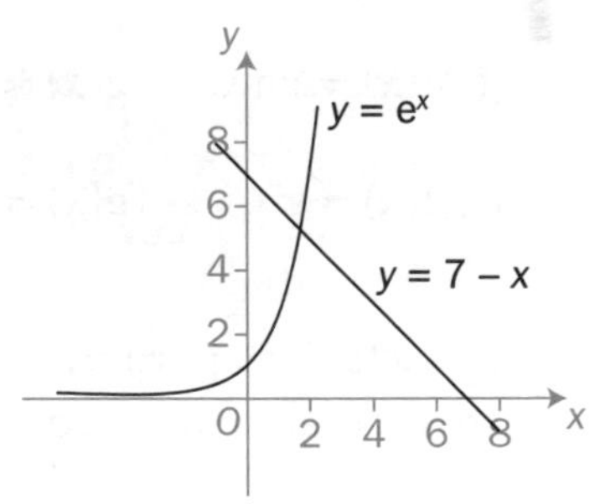

The graphs intersect only once so there is only one root.

The sketch gives its approximate value as $x = 1.5$

Consider $f(x) = e^x + x - 7$

Take values for x between 1.4 and 1.8 with steps of 0.1:

$f(1.6) < 0$ and $f(1.7) > 0$

x	$e^x + x - 7$	
1.4	-1.545	< 0
1.5	-1.018	< 0
1.6	-0.447	< 0
1.7	0.174	> 0
1.8	0.850	> 0

The root is in the interval $[1.6, 1.7]$
The root is nearer to 1.7 because 0.174 is nearer to 0 than is –0.447.

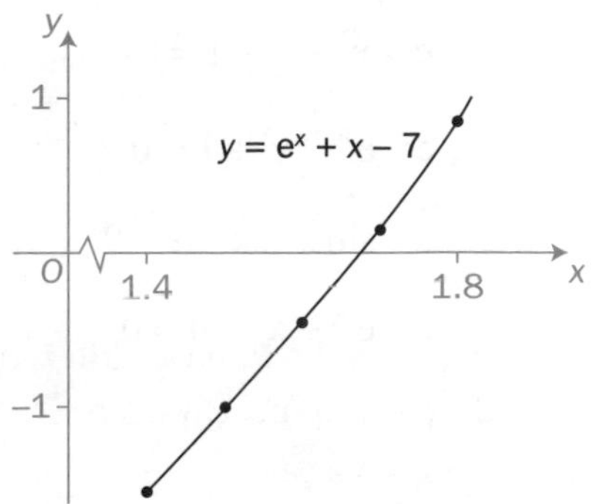

Example 4 is continued on the next page.

EXAMPLE 4 (CONT.)

Explore within this interval in steps of $x = 0.01$:

$f(1.67) < 0$ and $f(1.68) > 0$

The root is in the interval [1.67, 1.68]

As $f(1.675) = 0.014$ which is positive, the root is in the interval [1.67, 1.675]

So, to 2 decimal place, the root is $x = 1.67$

x	$e^x + x - 7$	
1.65	-0.143	< 0
1.66	-0.081	< 0
1.67	-0.018	< 0
1.68	0.046	> 0
1.69	0.109	> 0

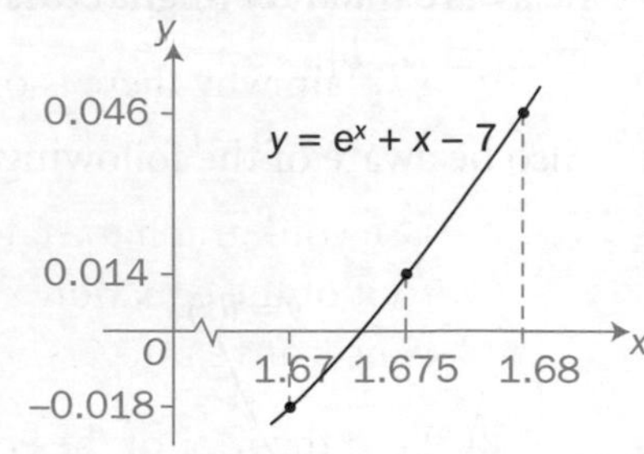

Exercise 5.1

1 Sketch the graphs of each pair of functions $f(x)$ and $g(x)$. State how many roots there are to the equation $f(x) - g(x) = 0$

a $f(x) = x^2$ $\quad g(x) = x + 2$

b $f(x) = x^3$ $\quad g(x) = x + 2$

c $f(x) = \frac{1}{x}$ $\quad g(x) = x + 2$

d $f(x) = \frac{1}{x}$ $\quad g(x) = x^3$

e $f(x) = x^2 - 4$ $\quad g(x) = \frac{1}{x}$

f $f(x) = \sin x$ $\quad g(x) = x + 2$

g $f(x) = \sin x$ $\quad g(x) = \frac{1}{2}x$

h $f(x) = e^x$ $\quad g(x) = 4 - x^2$

In parts **f** and **g**, x is in radians.

2 By sketching appropriate graphs, find how many roots there are for each of these equations.

a $x^3 + x - 5 = 0$

b $x^3 - \frac{1}{x} + 1 = 0$

c $e^x - x^2 + 1 = 0$

d $\sin x + x - 1 = 0$

e $e^{-x} + x^2 - 4 = 0$

f $x^2 + 1 - \sin x = 0$

C3

3 a Sketch the graphs of $y = x^3$ and $y = 9 - x$ on the same axes.

b Explain why there is only one root of the equation $x^3 + x - 9 = 0$

c Use your graph to state the interval $[a, b]$ within which the root of the equation $x^3 + x - 9 = 0$ lies, where a and b are consecutive integers.

d Find the root of the equation $x^3 + x - 9 = 0$ correct to 1 decimal place.

4 a Sketch the graphs of $y = x^2$ and $y = e^{-x}$ on the same axes.

b Show that the only root of the equation $x^2 - e^{-x} = 0$ lies in the interval $[0.70, 0.71]$.

c Find the root correct to 3 decimal places.

5 a Sketch the graphs $y = \frac{1}{x}$ and $y = 5 - x^2$ on the same axes.

b How many roots has the equation $\frac{1}{x} + x^2 - 5 = 0$?

c Show that one root lies in the interval $[2, 3]$ and find its value correct to 2 decimal places.

6 Without drawing any graph, show that each of these equations has a root within the given interval.

a $x^3 - 5x^2 + 6x - 1 = 0$ $[0, 1]$

b $x^3 - 2x^2 + x - 3 = 0$ $[2, 3]$

c $x^3 - \frac{1}{x} - 6 = 0$ $[1, 2]$

d $e^x - x - 8 = 0$ $[2, 3]$

e $\ln x - x^2 + 5 = 0$ $[2.4, 2.5]$

f $\sin x + x - 5 = 0$ $[5.6, 5.7]$

7 Show that the equation $x \sin x + 2 = 0$ has a root α such that $3 < \alpha < 4$. Find the value of α correct to 1 decimal place.

8 The equation $e^{-x} - x + 1 = 0$ has a root in the interval $[a, a + 1]$ where a is a positive integer.
Find the value of a and the value of the root correct to 2 decimal places.

9 **a** Copy and complete this table for the function $f(x) = x^3 - 4x^2 - x + 5$

x	-2	-1	0	1	2	3	4	5
f(x)								

b Sketch the curve $y = x^3 - 4x^2 - x + 5$ for $-2 \leqslant x \leqslant 5$

c Write down the three intervals $[a, b]$ where a, b are consecutive integers within which roots of the equation $x^3 - 4x^2 - x + 5 = 0$ lie.

d Find the largest root of $x^3 - 4x^2 - x + 5 = 0$ correct to 2 decimal places.

10 **a** Copy and complete this table for the function $f(x) = \frac{1}{2}x^3 - 2x^2 - x + 1$

x	-2	-1	0	1	2	3	4	5
f(x)								

b Write down the three intervals $[a, a+1]$, where a is integer, within which roots of the equation $\frac{1}{2}x^3 - 2x^2 - x + 1 = 0$ lie.

c Find the smallest root of $x^3 - 4x^2 - x + 5 = 0$ correct to 2 decimal places.

11 **a** Find the interval $[a, a+1]$, where a is a positive integer, such that the only root of the equation $e^x + \frac{1}{2}x - 10 = 0$ lies within it.

b Find the root correct to 2 decimal places.

12 **a** Show that the function $f(\theta) = \sqrt{\sin\theta} - \theta + 2$ for $0 \leqslant \theta \leqslant \pi$ has a solution in the interval $\left[\frac{3}{4}\pi, \frac{7}{8}\pi\right]$.

b Find the solution correct to 2 decimal places.

C3

13 a By sketching the graphs of $y = 2\sqrt{x}$ and $y = x^3$ on the same axes, find how many solutions there are to the equation $2\sqrt{x} - x^3 = 0$

b Find the non-zero solution of the equation correct to 2 decimal places.

14 a Sketch the graphs of $y = \ln x$ and $y = 12 - x^2$ on the same axes. Explain why the equation $\ln x + x^2 - 12 = 0$ has only one root.

b Find the root of the equation correct to 2 decimal places.

15 a Choose two functions $f(x)$ and $g(x)$ and sketch the graphs of $y = f(x)$ and $y = g(x)$ to find the number of roots of the equation $\cos x - x + 6 = 0$

b Show that a root α exists such that $6 \leqslant \alpha \leqslant 7$

c Find α correct to 3 decimal places.

16 Solve the equation $3^x - x^3 = 0$
giving solutions correct to 3 decimal places where necessary.

17 Find all the roots of the equation $x\cos x + x = 0$
Sketch the graph of $y = x\cos x + x$

C3

INVESTIGATION

18

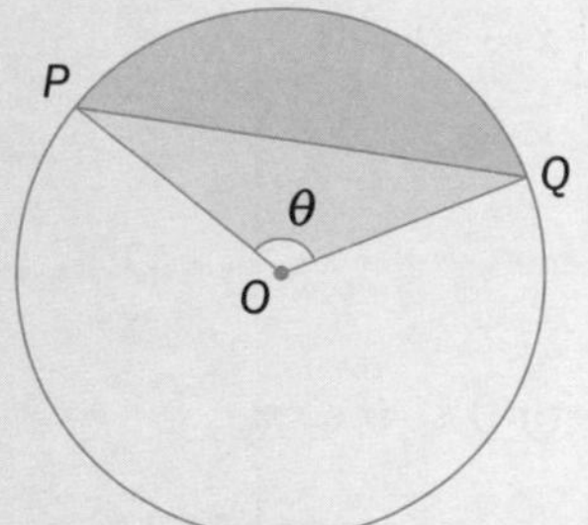

In this diagram, the area of triangle OPQ is half the area of the sector OPQ.
Show that the angle, θ, in radians, must be a solution of the equation

$$2\sin\theta - \theta = 0$$

Solve this equation to find θ.

5.2 Iterative methods

The root of the equation $x - f(x) = 0$ is at the point of intersection of the graphs of $y = x$ and $y = f(x)$ where $x = f(x)$

You can show the steps of the iterative process of finding a root of the equation $x - f(x) = 0$

An iterative method is a repetitive process which uses a succession of approximations. Each approximation builds on the preceding approximation until the required degree of accuracy is achieved.

Numerical process	Graphical process
Choose a value x_0 close to the root α	Locate the value x_0 on the x-axis
Calculate $f(x_0)$	Rise vertically at x_0 to meet the curve at a height of $f(x_0)$
Let $f(x_0) = x_1$	Go horizontally to the line $y = x$ where $f(x_0) = x_1$
Let x_1 be the next approximation to the root	Locate x_1 on the x-axis
Perform the next iteration and calculate x_2	Go vertically to the curve and then go horizontally to the line to find x_2
Perform more iterations.	Go vertically to the curve and go horizontally to the line for each iteration.

EXAMPLE 1

Find the root of the equation $\sqrt[3]{x+1} - x = 0$ using an iterative method, starting with $x_0 = 1$ as the first approximation.

Rearrange the equation:

$$x = \sqrt[3]{x+1}$$

Define the iterative formula:

$$x_{n+1} = \sqrt[3]{x_n + 1} \text{ with } x_0 = 1$$

Carry out the iterations:

$$x_1 = \sqrt[3]{x_0 + 1} = \sqrt[3]{1+1} = \sqrt[3]{2} = 1.259921\ldots$$

$$x_2 = \sqrt[3]{x_1 + 1} = \sqrt[3]{1.259921\ldots + 1} = 1.312293\ldots$$

$$x_3 = \sqrt[3]{x_2 + 1} = \sqrt[3]{1.312293\ldots + 1} = 1.322353\ldots$$

$$x_4 = \sqrt[3]{x_3 + 1} = \sqrt[3]{1.322353\ldots + 1} = 1.324268\ldots \text{ and so on.}$$

The iterations seem to be converging.

After four iterations, the root is correct to 2 decimal places and has a value of 1.32

More iterations will improve the accuracy.

You can do these calculations efficiently on a spreadsheet or using the 'Ans' key on a scientific or graphical calculator.

After 8 iterations, the root is correct to 5 decimal places.

n	x	$\sqrt[3]{x+1}$
0	1	1.25992105
1	1.25992105	1.31229384
2	1.31229384	1.32235382
3	1.32235382	1.32426874
4	1.32426874	1.32463263
5	1.32463263	1.32470175
6	1.32470175	1.32471488
7	1.32471488	1.32471737
8	1.32471737	1.32471785

Whether a sequence converges or diverges depends on the gradient of $f(x)$ at the point of intersection where $x = f(x)$

There are four possibilities:

At the point of intersection, $x = \alpha$ is the root of $x = f(x)$ in each case.

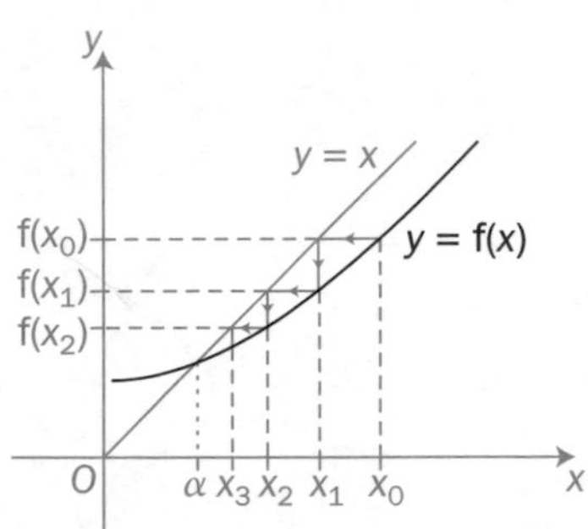

$y = f(x)$ is a rising curve and $0 < f'(\alpha) < 1$
The iterations converge.

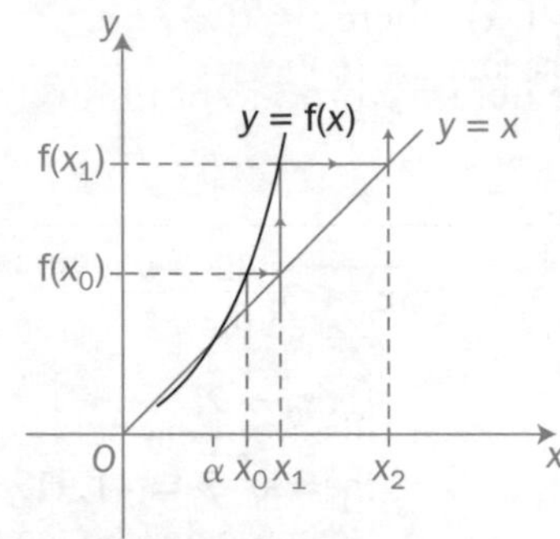

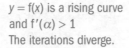

$y = f(x)$ is a rising curve and $f'(\alpha) > 1$
The iterations diverge.

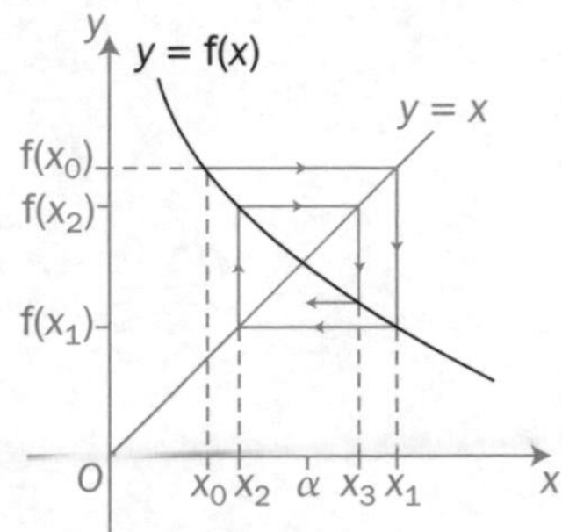

$y = f(x)$ is a falling curve and $-1 < f'(\alpha) < 0$
The iterations converge.

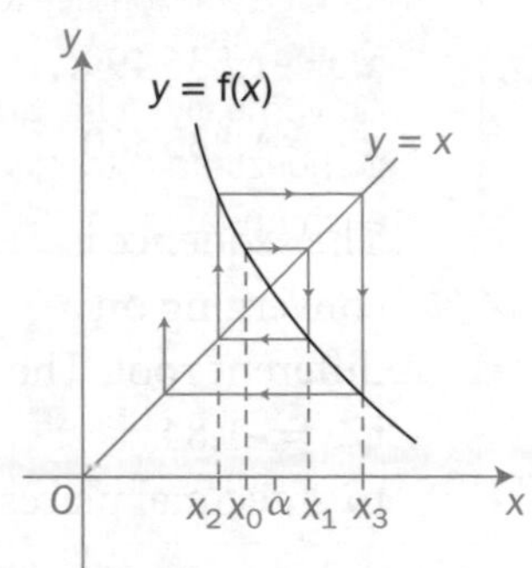

$y = f(x)$ is a falling curve and $f'(\alpha) < -1$
The iterations diverge.

Overall, the iterative process converges provided that $-1 < f'(\alpha) < 1$ where α is the root.

This condition can also be written as $|f'(\alpha)| < 1$

Try to choose a starting value x_0 which is close to the root. Start by looking for an interval which contains the root and then choosing x_0 from within that interval.

A different iterative formula with the same starting value can lead to a different root.
The same iterative formula with a different starting value can lead to a different root or even a diverging sequence.

C3

EXAMPLE 2

Explore the roots of the equation $x^3 - 5x - 3 = 0$ using iterative methods.

First method

Rearrange $x^3 - 5x - 3 = 0$ into the form $x = f(x)$:

$$x^3 = 5x + 3$$
$$x = \sqrt[3]{5x + 3}$$

Define the iterative formula: $x_{n+1} = \sqrt[3]{5x_n + 3}$

Let $x_0 = 2$:

$x_1 = \sqrt[3]{13} = 2.351\,334\ldots$

$x_2 = 2.452\,803\ldots$

$x_3 = 2.480\,597\ldots$

$x_4 = 2.488\,102\ldots$

$x_5 = 2.490\,121\ldots$

$x_6 = 2.490\,664\ldots$

The sequence is converging. The root is $x = 2.49$ to 2 decimal places.

Let $x_0 = -2$:

$x_1 = \sqrt[3]{-7} = -1.912\,931\ldots$

$x_2 = -1.872\,423\ldots$

$x_3 = -1.852\,964\ldots$

$x_4 = -1.843\,470\ldots$

$x_5 = -1.838\,803\ldots$

$x_6 = -1.836\,499\ldots$

$x_7 = -1.835\,360\ldots$

The sequence is converging on a different root. The root is $x = -1.84$ to 2 decimal places.

Second method

Rearrange $x^3 - 5x - 3 = 0$ into the form $x = f(x)$:

$5x = x^3 - 3$ giving $\quad x = \dfrac{x^3 - 3}{5}$

Define the iterative formula: $x_{n+1} = \dfrac{x_n^3 - 3}{5}$

Let $x_0 = 2$:

$x_1 = \dfrac{2^3 - 3}{5} = 1$

$x_2 = \dfrac{1^3 - 3}{5} = -0.4$

$x_3 = -0.6128$

$x_4 = -0.64\,602\ldots$

$x_5 = -0.65\,392\ldots$

$x_6 = -0.65\,593\ldots$

$x_7 = -0.65\,644\ldots$

This series is converging on a third root which is $x = -0.66$ to 2 decimal places.

Let $x_0 = 3$:

$x_1 = \dfrac{3^3 - 3}{5} = 4.8$

$x_2 = \dfrac{4.8^3 - 3}{5} = 21.51\ldots$

$x_3 = 1992.18\ldots$

This sequence is diverging very fast. No root is found.

The equation $x^3 - 5x - 3 = 0$ is a cubic. This example shows that it has three roots $x = 2.49$, -1.84 and -0.66 (to 2 d.p.).

Exercise 5.2

1 a Show, without drawing a graph, that the equation
$x^2 - 5x + 2 = 0$
has a root in the interval $[4, 5]$.

b Show that $x^2 - 5x + 2 = 0$ can be rewritten as $x = 5 - \frac{2}{x}$

c Using the iterative formula $x_{n+1} = 5 - \frac{2}{x_n}$ and $x_0 = 4$
find a root of the equation $x^2 - 5x + 2 = 0$
correct to 2 decimal places.

2 a Show that the equation $x^3 - 4x - 5 = 0$
has a root in the interval $[2, 3]$.

b Show that the equation $x^3 - 4x - 5 = 0$
can be rearranged as $x = \sqrt[3]{4x + 5}$

c Use the iterative formula $x_{n+1} = \sqrt[3]{4x_n + 5}$ with $x_0 = 2$
to find a root of the equation $x^3 - 4x - 5 = 0$
correct to 2 decimal places.

d Show that the equation $x^3 - 4x - 5 = 0$
can also produce the iterative formula $x_{n+1} = \frac{x_n^3 - 5}{4}$.
Take $x_0 = 2$ and find whether this iterative formula
creates a converging or diverging sequence of
approximations to the root.

3 a Show that the equation $x^3 - x^2 - 1 = 0$
has a root in the interval $[1, 2]$.

b Show that the equation $x^3 - x^2 - 1 = 0$
can produce the iterative formula $x_{n+1} = \sqrt[3]{x_n^2 + 1}$
Take $x_0 = 1$ and find this root correct to 3 decimal places.

4 Show that the iterative formula $x_{n+1} = \frac{1}{2}e^{-x_n}$
can be derived from the equation $e^{-x} - 2x = 0$
Taking $x_0 = 1$ find a root of this equation to 2 decimal places.

5 To find the value of $\sqrt[4]{50}$, show that the equation $x^4 = 50$ can be
rearranged to give the iterative formula $x_{n+1} = \sqrt[3]{\frac{50}{x_n}}$
Use this formula with $x_0 = 3$ to calculate $\sqrt[4]{50}$
correct to 4 significant figures.

C3

6 a Find the values of λ and μ such that the iterative formula

$$x_{n+1} = \sqrt{\lambda x_n + \frac{\mu}{x_n}}$$

can be used to solve the equation $x^3 - 5x^2 - 7 = 0$

b Show that a root of the equation lies between $x = 5$ and $x = 6$

c Take $x_0 = 5$ and find the root correct to 3 significant figures.

7 a Evaluate f(1.1) and f(1.4) for the function $f(\theta) = 6\theta - 5\sin\theta - 3$ where θ is in radians.

Explain why there is a root of the equation $6\theta - 5\sin\theta - 3 = 0$ in the range $1.1 < \theta < 1.4$

b Find the values a, b and c such that the iterative formula

$$\theta_{n+1} = \frac{a\sin\theta_n}{b} + \frac{1}{c}$$

can be used to solve the equation $6\theta - 5\sin\theta - 3 = 0$

c Taking $\theta_0 = 1.1$ find a root of equation $6\theta - 5\sin\theta - 3 = 0$ correct to 3 decimal places.

C3

8 The iterative formula $x_{n+1} = \sqrt{6 - \frac{3}{x_n}}$ is used to find the root of the equation $f(x) = 0$

Find the function $f(x)$ and the root of the equation correct to 3 decimal places, given that $x_0 = 2$

9 Find the equation $f(x) = 0$ which each of these iterative formulae can be used to solve.

a $x_{n+1} = \frac{5x_n - 7}{x_n^2}$ **b** $x_{n+1} = \sqrt[3]{5x_n + 1}$ **c** $x_{n+1} = \sqrt{8 - \frac{2}{x}}$

d $x_{n+1} = \sqrt{\frac{50}{x_n^2} - \frac{x_n^2}{2}}$ **e** $x_{n+1} = \frac{4x_n^3 - 1}{3x_n^2 + 2}$ **f** $x_{n+1} = \frac{6 - 2e^{-x_n}}{x_n}$

10 a Find the constants a and b which would allow these two iterative formulae to be used to solve the equation $x^4 = 20$

$$x_{n+1} = \sqrt{\frac{a}{x_n^2} + \frac{x_n^2}{2}} \quad \text{and} \quad x_{n+1} = \sqrt[3]{\frac{b}{x_n}}$$

b Taking $x_0 = 2$ find which one of these two iterative formulae converges faster to the root of the equation.

c Find the value of $\sqrt[4]{20}$ correct to 5 decimal places.

11 **a** Show that the equation $2x - e^x + 3 = 0$ has a root between -1 and -2.

b Show that the iterative formula $x_{n+1} = \frac{e^{x_n} + x_n}{3} - 1$ can be used to solve this equation.

c Take $x_0 = -2$ and find the root of $2x - e^x + 3 = 0$ correct to 4 decimal places.

d Hence find a root of the equation $2\tan\theta - e^{\tan\theta} + 3 = 0$ in degrees correct to 1 decimal place.

12 **a** Show that the equation $x^3 - 4x - 3 = 0$ has a root in the interval $[2,3]$.

b Show that $x_{n+1} = \frac{3(x_n^3 + x_n + 1)}{4x_n^2 - 1}$ is an iterative formula for the equation $x^3 - 4x - 3 = 0$

c Take $x_0 = 3$ and find a root of the equation correct to 3 decimal places.

d Hence, find a root of the equation $8^y - 2^{y+2} = 3$ correct to1 decimal place.

INVESTIGATION

13 Create several of your own iterative formulae to solve the cubic equation

$$ax^3 + bx^2 + cx + d = 0$$

for your choice of *a*, *b*, *c* and *d*.

Use a computer spreadsheet to calculate the iterations for different starting values x_0.
See if your iterations converge or diverge.

Which one of your formulae converges most quickly to a root?

Review 5

1 Sketch the graphs of each pair of functions f(x) and g(x). State how many roots there are to the equation $f(x) - g(x) = 0$

a $f(x) = x^2 \quad g(x) = x + 2$

b $f(x) = e^x \quad g(x) = 4 - x$

2 By sketching appropriate graphs, find how many roots there are for each of these equations.

a $x^3 + 2 - \frac{1}{x} = 0$

b $\sin x - x^3 = 1$

In part **b**, x is in radians.

c $\ln x - x^2 + 2 = 0$

d $|x + 2| - x^2 = 1$

3 a Sketch the graphs of $y = 3 - x^2$ and $y = e^{-x}$ on the same axes.

b Explain why there are only two roots of the equation $x^2 + e^{-x} = 3$
Show that one root lies in the interval [1.6, 1.7] and find this root correct to 2 decimal places.

c Find the other root correct to 2 decimal places.

4 Without drawing any graph, show that each of these equations has a root within the given interval. In each case, find the root correct to 2 decimal places.

a $x^3 - \frac{1}{x^2} - 2 = 0 \quad [1.3, 1.4]$

b $\ln x - e^{-x} = 0 \quad [1.0, 1.5]$

c $\cos x + 2x - 4 = 0 \quad [2.3, 2.4]$

d $e^x + x^2 - 8 = 0 \quad [-3, -2]$

5 a Show that the equation $x^3 - 3x - 4 = 0$ has a root in the interval [2, 3].

b Show that the equation $x^3 - 3x - 4 = 0$ can be rearranged as $x = \sqrt[3]{3x + 4}$

c Use the iterative formula $x_{n+1} = \sqrt[3]{3x_n + 4}$ with $x_0 = 2$ to find the root of the equation $x^3 - 3x - 4 = 0$ correct to 2 decimal places.

d Show that the equation $x^3 - 3x - 4 = 0$ can also produce the iterative formula $x_{n+1} = \frac{x_n^3 - 4}{3}$. Take $x_0 = 2$ and find whether this iterative formula creates a converging or diverging sequence of approximations to the root.

6 $f(x) = x^3 - 2 - \frac{1}{x}, \quad x \neq 0$

a Show that the equation $f(x) = 0$ has a root between 1 and 2.

An approximation for this root is found using the iteration formula

$$x_{n+1} = \left(2 + \frac{1}{x_n}\right)^{\frac{1}{3}} \quad \text{with } x_0 = 1.5$$

b By calculating the values of x_1, x_2, x_3 and x_4, find an approximation to this root, giving your answer to 3 decimal places.

c By considering the change of sign of $f(x)$ in a suitable interval, verify that your answer to part **b** is correct to 3 decimal places. [(c) Edexcel Limited 2004]

7 a Sketch, on the same set of axes, the graphs of $y = 2 - e^{-x}$ and $y = \sqrt{x}$

[It is not necessary to find the coordinates of any points of intersection with the axes.]

Given that $f(x) = e^{-x} + \sqrt{x} - 2, \quad x \geqslant 0$

b explain how your graphs show that the equation $f(x) = 0$ has only one solution

c show that the solution of $f(x) = 0$ lies between $x = 3$ and $x = 4$

The iterative formula $x_{n+1} = (2 - e^{-x_n})^2$ is used to solve the equation $f(x) = 0$

d Taking $x_0 = 4$, write down the values of x_1, x_2, x_3 and x_4, and hence find an approximation to the solution of $f(x) = 0$, giving your answer to 3 decimal places. [(c) Edexcel Limited 2003]

8 $f(x) = x^3 + x^2 - 4x - 1$

The equation $f(x) = 0$ has only one positive root, α.

a Show that $f(x) = 0$ can be rearranged as $x = \sqrt{\left(\frac{4x+1}{x+1}\right)}, \quad x \neq -1$

The iterative formula $x_{n+1} = \sqrt{\left(\frac{4x_n + 1}{x_n + 1}\right)}$ is used to find an approximation to α.

b Taking $x_1 = 1$, find, to 2 decimal places, the values of x_2, x_3 and x_4.

c By choosing values of x in a suitable interval, prove that $\alpha = 1.70$, correct to 2 decimal places.

d Write down a value of x_1 for which the iteration formula $x_{n+1} = \sqrt{\left(\frac{4x_n + 1}{x_n + 1}\right)}$ does not produce a valid value for x_2.

Justify your answer. [(c) Edexcel Limited 2004]

C3

5 Exit

Summary

Refer to

- **Graphical methods**
  - To solve an equation of the type $f(x) = g(x)$, draw the graphs of $y = f(x)$ and $y = g(x)$ and find the x-values of any points of intersection.
  - To solve an equation of the type $f(x) = 0$, draw the graph of $y = f(x)$ and find the x-values of the points at which the graph intersects (or touches) the x-axis.
  - If, between $x = a$ and $x = b$, the graph of $y = f(x)$ is continuous and $f(x)$ changes its sign, then there is at least one root of the equation $f(x) = 0$ between $x = a$ and $x = b$. 5.1
- **Iterative methods**

  To solve an equation of the form $g(x) = 0$
  - rearrange $g(x)$ into the form $x = f(x)$
  - choose a value x_0 which is close to the root
  - use the iterative formula $x_{n+1} = f(x_n)$ to generate the sequence $x_0, x_1, x_2, x_3, \ldots$
  - decide if this sequence of x-values is converging or diverging
  - if it is converging, continue until you have the root to the required accuracy. 5.2

C3

Links

In cases where a real-life problem cannot be solved analytically, a numerical method is applied to find an approximate soluton.

Applied problems can arise from diverse areas such as engineering, economics and biological sciences.

Solutions to such problems often require scientific computation involving advanced iterative methods.

Revision 2

1 Express as a single fraction in its simplest form.

$$\frac{3x^2 - x}{(3x-1)(x+2)} - \frac{12}{x^2 - 2x - 8}$$

2 a Simplify $\frac{3x^2 - x - 14}{x^2 - 4}$

b Hence, express $\frac{3x^2 - x - 14}{x^2 - 4} + \frac{2}{x(x-2)}$ as simply as possible.

3 a Express as a fraction in its simplest form

$$\frac{2}{x-3} + \frac{1}{x^2 - 8x + 15}$$

b Hence, solve the equation $\frac{2}{x-3} + \frac{1}{x^2 - 8x + 15} = 1$

4 Given that $\frac{2x^4 - 3x^2 + x + 1}{(x^2 - 1)} \equiv (ax^2 + bx + c) + \frac{dx + e}{(x^2 - 1)}$,

find the values of the constants a, b, c, d and e. [(c) Edexcel Limited 2008]

5 a Express $2 - \frac{1}{x-4}$ as a single fraction.

b The function f is defined by $\mathrm{f}(x) = 2 - \frac{1}{x-4}$, $x \in \mathbb{R}$, $x \neq 4$

Find an expression for the inverse function $\mathrm{f}^{-1}(x)$.

c Write down the domain of f^{-1}.

6 The functions f and g are defined by

$\mathrm{f}: x \rightarrow 2x + \ln 2$, $x \in \mathbb{R}$

$\mathrm{g}: x \rightarrow e^{2x}$, $x \in \mathbb{R}$

a Prove that the composite function gf is $\mathrm{gf}: x \rightarrow 4e^{4x}$, $x \in \mathbb{R}$

b Sketch the curve with equation $y = \mathrm{gf}(x)$, and show the coordinates of the point where the curve cuts the y-axis.

c Write down the range of gf.

d Find the value of x for which $\frac{\mathrm{d}}{\mathrm{d}x}[\mathrm{gf}(x)] = 3$,

giving your answer to 3 significant figures. [(c) Edexcel Limited 2006]

7 The function f is defined by

$f: x \to |2x - a|, \quad x \in \mathbb{R}, \quad$ where a is a positive constant.

a Sketch the graph of $y = f(x)$, showing the coordinates of the points where the graph cuts the axes.

b On a separate diagram, sketch the graph of $y = f(2x)$, showing the coordinates of the points where the graph cuts the axes.

c Given that a solution of the equation $f(x) = \frac{1}{2}x$ is $x = 4$, find the two possible values of a. [(c) Edexcel Limited 2002]

8 The functions f and g are defined by

$f: x \to \ln(2x - 1), \quad x \in \mathbb{R}, \quad x > \frac{1}{2} \qquad g: x \to \frac{2}{x - 3}, \quad x \in \mathbb{R}, \quad x \neq 3$

a Find the exact value of $fg(4)$.

b Find the inverse function $f^{-1}(x)$, stating its domain.

c Sketch the graph of $y = |g(x)|$. Indicate clearly the equation of the vertical asymptote and the coordinates of the point at which the graph crosses the y-axis.

d Find the exact values of x for which $\left|\frac{2}{x - 3}\right| = 3$ [(c) Edexcel Limited 2008]

9 The functions f and g are defined by

$f: x \to |x - a| + a, \quad x \in \mathbb{R}, \qquad g: x \to 4x + a, \quad x \in \mathbb{R}$

where a is a positive constant.

a On the same diagram, sketch the graphs of f and g, showing clearly the coordinates of any points at which your graphs meet the axes.

b Use algebra to find, in terms of a, the coordinates of the point at which the graphs of f and g intersect.

c Find an expression for $fg(x)$.

d Solve, for x in terms of a, the equation $fg(x) = 3a$ [(c) Edexcel Limited 2003]

10 Prove that $\frac{1 - \tan^2 x}{1 + \tan^2 x} \equiv \cos 2x$. Hence, prove that $\tan^2 \frac{\pi}{12} = 7 - 4\sqrt{3}$

11 a By writing $\sin 3\theta$ as $\sin(2\theta + \theta)$, show that $\sin 3\theta = 3\sin\theta - 4\sin^3\theta$

b Given that $\sin\theta = \frac{\sqrt{3}}{4}$, find the exact value of $\sin 3\theta$. [(c) Edexcel Limited 2007]

12 a Given that $2\sin(\theta + 30)° = \cos(\theta + 60)°$, find the exact value of $\tan\theta°$.

b i Using the identity $\cos(A + B) \equiv \cos A\cos B - \sin A\sin B$, prove that $\cos 2A \equiv 1 - 2\sin^2 A$

ii Hence solve, for $0 \leqslant x \leqslant 2\pi$, $\cos 2x = \sin x$, giving your answers in terms of π.

iii Show that $\sin 2y \tan y + \cos 2y \equiv 1 \quad$ for $0 \leqslant y < \frac{1}{2}\pi$ [(c) Edexcel Limited 2005]

13 **a** Using $\sin^2\theta + \cos^2\theta \equiv 1$, show that $\operatorname{cosec}^2\theta - \cot^2\theta \equiv 1$

b Hence, or otherwise, prove that $\operatorname{cosec}^4\theta - \cot^4\theta \equiv \operatorname{cosec}^2\theta + \cot^2\theta$

c Solve, for $90° < \theta < 180°$, $\operatorname{cosec}^4\theta - \cot^4\theta = 2 - \cot\theta$ [(c) Edexcel Limited 2006]

14 **a** Show that

i $\dfrac{\cos 2x}{\cos x + \sin x} \equiv \cos x - \sin x, \quad x \neq \left(n - \dfrac{1}{4}\right)\pi, \; n \in \mathbb{Z}$

ii $\frac{1}{2}(\cos 2x - \sin 2x) \equiv \cos^2 x - \cos x \sin x - \frac{1}{2}$

b Hence, or otherwise, show that the equation

$$\cos\theta\left(\frac{\cos 2\theta}{\cos\theta + \sin\theta}\right) = \frac{1}{2}$$

can be written as $\sin 2\theta = \cos 2\theta$

c Solve, for $0 \leqslant \theta \leqslant 2\pi$, $\sin 2\theta = \cos 2\theta$
giving your answers in terms of π. [(c) Edexcel Limited 2006]

15 $f(x) = 5\cos x + 12\sin x$
Given that $f(x) = R\cos(x - \alpha)$, where $R > 0$ and $0 < \alpha < \frac{\pi}{2}$,

a find the value of R and the value of α to 3 decimal places

b hence, solve the equation $5\cos x + 12\sin x = 6$ for $0 \leqslant x \leqslant 2\pi$

c i Write down the maximum value of $5\cos x + 12\sin x$.

ii Find the smallest positive value of x for which this maximum value occurs. [(c) Edexcel Limited 2008]

16 **a** The graph of $y = e^x$ is transformed into the graph of each of the following equations.
Name the transformations involved in each case and give the order in which they occur. Sketch the graph of $y = e^x$ and each of its images.

i $y = 1 + 2e^x$ ii $y = 2 + e^{-x}$ iii $y = 3e^{x-2}$

b Name the transformations (and the order in which they occur) that transform the graph of $y = \ln x$ into the graphs of each of these equations. Sketch the graph of $y = \ln x$ and each of its images.

i $y = 3 - \ln x$ ii $y = 1 + 2\ln x$ iii $y = \frac{1}{2}\ln(x + 2)$

17 **a** A population P of individuals increases over a time t (days) from an initial value of P_0, according to the relation $P = P_0 e^{\frac{t}{10}}$
How many days have to elapse for the population to double in size?

b A number of cells are being infected with a virus. The number N of uninfected cells reduces with time t (hours) as given by

$N = 80 + 25e^{-\frac{t}{2}}$

i How many uninfected cells were there initially?

ii What is the limiting value of N as time increases?

C3

18 A savings account earns interest on the money invested in it at a constant rate of 5% each year. An initial investment of £1 thus has a value of £y after t years where $y = 1.05^t$

a Sketch the graph of $y = 1.05^t$ for $t \geqslant 0$

b Find the total value of an initial investment of £500 after it has been in the account for 6 years. Give your answer to the nearest £.

c How many years does it take for any investment to double in value?

d What must be the interest rate (to 1 decimal place) if an investment is to double in value after 10 years?

19 A particular species of orchid is being studied. The population p at time t years after the study started is assumed to be

$$p = \frac{2800ae^{0.2t}}{1 + ae^{0.2t}} \quad \text{where } a \text{ is a constant.}$$

Given that there were 300 orchids when the study started,

a show that $a = 0.12$

b use the equation with $a = 0.12$ to predict the number of years before the population of orchids reaches 1850.

c Show that $p = \dfrac{336}{0.12 + e^{-0.2t}}$

d Hence show that the population cannot exceed 2800. [(c) Edexcel Limited 2005]

20 A heated metal ball is dropped into a liquid. As the ball cools, its temperature, T°C, t minutes after it enters the liquid, is given by

$$T = 400\,e^{-0.05t} + 25, \quad t \geqslant 0$$

a Find the temperature of the ball as it enters the liquid.

b Find the value of t for which $T = 300$, giving your answer to 3 significant figures.

c Find the rate at which the temperature of the ball is decreasing at the instant when $t = 50$. Give your answer in °C per minute to 3 significant figures.

d From the equation for temperature T in terms of t, given above, explain why the temperature of the ball can never fall to 20 °C. [(c) Edexcel Limited 2006]

21 Solve the equations

a $6 + 3e^{-x} = 8$ **b** $\ln(x + 3)^2 = 4$ **c** $e^{2x} = 3e^x - 2$

C3

22 Differentiate with respect to x.

a $e^x \sin x$ b $x^3 \ln x$ c $e^{-x}(x^2 - 3)$ d $\frac{e^x}{\sin x}$

e $\frac{\ln x}{x^2 - 1}$ f $\frac{x^3 - 1}{x^3 + 1}$ g $\ln(\tan x)$ h e^{x^3}

i $\sqrt[3]{x^2 - 2x + 5}$ j $\frac{3}{\sqrt{x^3 - 1}}$ k $\frac{e^x}{e^x - 1}$ l $\cos^2 x \sin x$

23 a Differentiate with respect to x.

i $x^2 e^{3x+2}$ ii $\frac{\cos(2x^3)}{3x}$

b Given that $x = 4\sin(2y + 6)$, find $\frac{dy}{dx}$ in terms of x. [(c) Edexcel Limited 2006]

24 a Differentiate with respect to x

i $e^{3x}(\sin x + 2\cos x)$ ii $x^3 \ln(5x + 2)$

Given that $y = \frac{3x^2 + 6x - 7}{(x+1)^2}$, $x \neq -1$,

b show that $\frac{dy}{dx} = \frac{20}{(x+1)^3}$

c Hence find $\frac{d^2y}{dx^2}$ and the real values of x for which $\frac{d^2y}{dx^2} = -\frac{15}{4}$ [(c) Edexcel Limited 2008]

25 a The curve C has equation $y = \frac{x}{9 + x^2}$

Use calculus to find the coordinates of the turning points of C.

b Given that $y = (1 + e^{2x})^{\frac{3}{2}}$, find the value of $\frac{dy}{dx}$ at $x = \frac{1}{2}\ln 3$ [(c) Edexcel Limited 2007]

26 Find the stationary points on these curves and determine their nature.

a $y = \frac{e^x}{x}$ b $y = xe^x$

c $y = \frac{x}{x^2 + 1}$ d $y = e^{-2x}\sin x$ where $-\pi \leqslant x \leqslant \pi$

27 A curve C has equation $y = x^2 e^x$

a Find $\frac{dy}{dx}$, using the product rule for differentiation.

b Hence, find the coordinates of the turning points of C.

c Find $\frac{d^2y}{dx^2}$

d Determine the nature of each turning point of the curve C. [(c) Edexcel Limited 2008]

C3

28 a Find $\frac{dy}{dx}$ for the curve C where $y = \frac{x}{x^2 - 1}$ and explain why the graph of C is a falling curve for all values of x.

b Find the equation of the tangent to C at the point where $x = 2$. Also find the equation of the normal to C at the point where $x = -2$.

c Find, to 2 decimal places, the coordinates of the point where this tangent and normal intersect.

29 A curve C has equation $y = e^{2x} \tan x$, $\quad x \neq (2n+1)\frac{\pi}{2}$

a Show that the turning points on C occur where $\tan x = -1$

b Find the equation of the tangent to C at the point where $x = 0$ [(c) Edexcel Limited 2008]

30 a Given that $y = \log_a x$, $x > 0$, where a is a positive constant,

i express x in terms of a and y

ii deduce that $\ln x = y \ln a$

b Show that $\frac{dy}{dx} = \frac{1}{x \ln a}$

The curve C has equation $y = \log_{10} x$, $\quad x > 0$
The point A on C has x-coordinate 10.
Using the result in part **b**,

c find an equation for the tangent to C at A.

d The tangent to C at A crosses the x-axis at the point B. Find the exact x-coordinate of B. [(c) Edexcel Limited 2004]

31 a Sketch the graphs of $y = x^3$ and $y = 6 - x^2$ on the same axes.

b Explain why there is only one root of the equation $x^3 + x^2 - 6 = 0$

c Use your graph to state the interval $[a, b]$ within which the root of the equation $x^3 + x^2 - 6 = 0$ lines, where a and b are consecutive integers.

d Find the root of the equation $x^3 + x^2 - 6 = 0$ correct to 1 decimal place.

32 a Find the interval $[a, a + 1]$, where a is a positive integer, such that the only root of the equation $e^x - x^2 - 2 = 0$ lies within this interval.

b Find the root correct to 2 decimal places.

33 a Show that the equation $\frac{1}{10}x^4 - x - 3 = 0$ has a root in the interval $[2,3]$.

b Show that the equation $\frac{1}{10}x^4 - x - 3 = 0$ can produce the iterative formula $x_{n+1} = \sqrt[4]{10x_n + 30}$

c Take $x_0 = 1$ and find the root correct to 3 decimal places.

34 Find the equation in the form $f(x) = 0$ which each of these iterative formulae can be used to solve.

a $x_{n+1} = \dfrac{3 - x_n}{x_n^2}$ **b** $x_{n+1} = \sqrt[3]{5(x_n - 2)}$ **c** $x_{n+1} = \sqrt{3 + \dfrac{2}{x_n}}$

35 a Show that the iterative formula $x_{n+1} = \sqrt{\dfrac{\alpha}{x_n} - \dfrac{x_n}{\beta}}$ can be used to solve the equation $2x^3 + x^2 - 24 = 0$

Find the values of α and β in this case.

b Show that a root of the equation lies between $x = 2$ and $x = 3$

c Take $x_0 = 2$ and find the root correct to 3 significant figures.

36 $f(x) = x^3 - 2 - \frac{1}{x}, \quad x \neq 0$

a Show that the equation $f(x) = 0$ has a root between 1 and 2.

b An approximation for this root is found using the iteration formula

$$x_{n+1} = \left(2 + \frac{1}{x_n}\right)^{\frac{1}{3}} \quad \text{with } x_0 = 1.5$$

By calculating the values of x_1, x_2, x_3 and x_4, find an approximation to this root, giving your answer to 3 decimal places.

c By considering the change of sign of $f(x)$ in a suitable interval, verify that your answer to part **b** is correct to 3 decimal places. [(c) Edexcel Limited 2004]

C3

37 This diagram shows part of the curve C with equation $y = f(x)$, where $f(x) = 0.5e^x - x^2$
The curve C cuts the y-axis at A and there is a minimum at the point B.

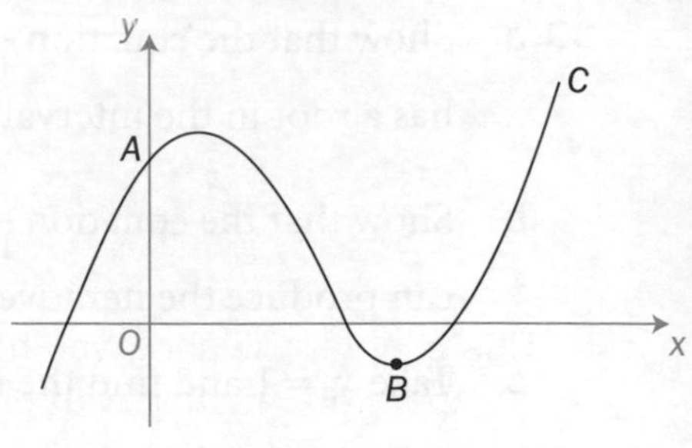

a Find an equation of the tangent to C at A.

The x-coordinate of B is approximately 2.15. A more exact estimate is to be made of this coordinate using iterations $x_{n+1} = \ln g(x_n)$

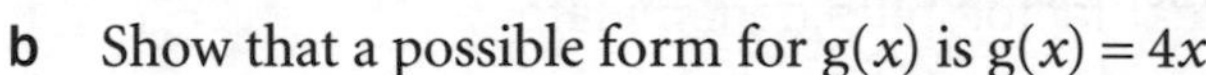

b Show that a possible form for $g(x)$ is $g(x) = 4x$

c Using $x_{n+1} = \ln 4x_n$, with $x_0 = 2.15$, calculate x_1, x_2 and x_3. Give the value of x_3 to 4 decimal places.

[(c) Edexcel Limited 2002]

38 $f(x) = \frac{1}{2x} - 1 + \ln\frac{x}{2}, \quad x > 0$

This diagram shows part of the curve with equation $y = f(x)$. The curve crosses the x-axis at the points A and B, and has a minimum at the point C.

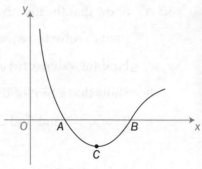

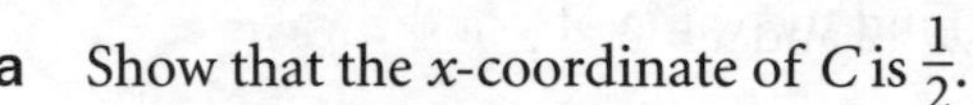

a Show that the x-coordinate of C is $\frac{1}{2}$.

b Find the y-coordinate of C in the form $k\ln 2$, where k is a constant.

c Verify that the x-coordinate of B lies between 4.905 and 4.915

d Show that the equation $\frac{1}{2x} - 1 + \ln\frac{x}{2} = 0$ can be rearranged into the form $x = 2e^{\left(1-\frac{1}{2x_n}\right)}$

The x-coordinate of B is to be found using the iterative formula

$$x_{n+1} = 2e^{\left(1-\frac{1}{2x_n}\right)} \quad \text{with } x_0 = 5$$

e Calculate, to 4 decimal places, the values of x_1, x_2 and x_3.

[(c) Edexcel Limited 2005]

39 $f(x) = 3e^x - \frac{1}{2}\ln x - 2, \quad x > 0$

a Differentiate to find $f'(x)$.

b The curve with equation $y = f(x)$ has a turning point at P. The x-coordinate of P is α. Show that $\alpha = \frac{1}{6}e^{-\alpha}$

The iterative formula $x_{n+1} = \frac{1}{6}e^{-x_n}$, $x_0 = 1$, is used to find an approximate value for α.

c Calculate the values of x_1, x_2, x_3 and x_4, giving your answers to 4 decimal places.

d By considering the change of sign of $f'(x)$ in a suitable interval, prove that $\alpha = 0.1443$ correct to 4 decimal places.

[(c) Edexcel Limited 2005]

C3

6 Partial fractions

This chapter will show you how to
- separate a fraction with different linear factors in its denominator into partial fractions
- separate a fraction with a repeated linear factor in its denominator into partial fractions
- separate an improper fraction into partial fractions
- use the methods of equating coefficients and substitution, including the cover-up rule.

Before you start

You should know how to:

1 Substitute into formulae.

e.g. Find $z = \frac{3x-1}{(x-1)^2}$ when $x = \frac{3}{4}$

$$z = \frac{\frac{9}{4} - 1}{\left(\frac{3}{4} - 1\right)^2} = \frac{5}{4} \times \frac{16}{1} = 20$$

2 Factorise expressions.

e.g. $6x^2 + 11x - 10$

$= (2x + 5)(3x - 2)$

3 Create and use identities.

e.g. Find A and B if $3x + 8 \equiv A(x + 2) + Bx$

Equate constants: $8 = 2A$ so $A = 4$

Equate coefficients of x:

$3 = A + B$ so $B = -1$

4 Add and subtract algebraic fractions.

e.g. $\frac{2}{x+3} + \frac{4}{(x+1)^2} = \frac{2(x+1)^2 + 4(x+3)}{(x+3)(x+1)^2}$

$= \frac{2x^2 + 8x + 14}{(x+3)(x+1)^2}$

5 Divide algebraic expressions.

e.g. $(x^2 + 4x + 7) \div (x + 3)$

$= x + 1 + \frac{4}{x+3}$

Check in:

1 Find the value of y and z when $x = \frac{1}{4}$

if **a** $y = \frac{2x^2 + 1}{x^2(1 - x)}$

b $z = \frac{(2 - x)^2}{x^2(1 + 3x)}$

2 Factorise

a $4x^3 - 9x$

b $x^4 - 1$

3 Find A and B if

a $x(5x + 3) + 6x \equiv x(Ax + B)$

b $(Ax + 3)(2x + B) \equiv 8x^2 + 10x + 3$

4 Work out

a $\frac{3}{x+1} + \frac{4}{(x+1)^2}$

b $\frac{1}{x} + \frac{2}{x+1} + \frac{3}{x-1}$

5 Divide

a $(x^3 - 3x^2 - x + 3)$ by $(x + 1)$

b $(x^2 + 6x - 1)$ by $(x - 2)$

6.1 Separating fractions

You can express a proper fraction of the type

$$\frac{f(x)}{(ax+b)(cx+d)(ex+f)}$$

as **partial fractions** of the type $\frac{A}{ax+b}+\frac{B}{cx+d}+\frac{C}{ex+f}$

where A, B and C are constants.

The initial fraction must be a proper fraction. That is, the numerator must be of a lower degree than the denominator.

You can use the **method of equating coefficients** to separate a fraction into partial fractions.

EXAMPLE 1

Express $\frac{5x-1}{(x-2)(x+1)}$ in partial fractions.

Form an identity (which is true for all values of x):

Let $$\frac{5x-1}{(x-2)(x+1)} \equiv \frac{A}{x-2}+\frac{B}{x+1}$$

$$\equiv \frac{A(x+1)+B(x-2)}{(x-2)(x+1)}$$

$$\equiv \frac{(A+B)x+A-2B}{(x-2)(x+1)}$$

See C3 for revision of adding fractions.

Equate the numerators: $5x-1 \equiv (A+B)x+A-2B$

The two sides of this identity must be identical.

Equate the coefficients of x: $5 = A+B$

Equate the constants: $-1 = A-2B$

You now have two simultaneous equations in A and B.

Subtract these two equations to eliminate A:

$$5-(-1) = B-(-2B)$$
$$6 = 3B$$
$$B = 2$$

Substitute into $5 = A + B$: $A = 3$

So, in partial fractions, $\frac{5x-1}{(x-2)(x+1)} \equiv \frac{3}{x-2}+\frac{2}{x+1}$

Check your answer by adding $\frac{3}{x-2}$ and $\frac{2}{x-1}$

C4

You can use the **method of substitution** to separate a fraction into partial fractions.

EXAMPLE 2

Express $\frac{5x-1}{(x-2)(x+1)}$ in partial fractions.

This is the same as Example 1, using a different method.

Let $\frac{5x-1}{(x-2)(x+1)} \equiv \frac{A}{x-2} + \frac{B}{x+1} \equiv \frac{A(x+1)+B(x-2)}{(x-2)(x+1)}$

Equate the numerators: $5x-1 \equiv A(x+1)+B(x-2)$

In particular, let $x = -1$ to eliminate A:

$-5-1 = 0 + B(-1-2)$ so $B = 2$

An identity is true for *all* values of x.

Now let $x = 2$ to eliminate B:

$10 - 1 = 3A$ so $A = 3$

So, $\frac{5x-1}{(x-2)(x+1)} \equiv \frac{3}{x-2} + \frac{2}{x+1}$

Check your answer by adding $\frac{3}{x-2}$ and $\frac{2}{x-1}$

You can use either of these methods or a mixture of the two to keep your working to a minimum.

C4

EXAMPLE 3

Express as partial fractions $\frac{x^2-11x-6}{(x+2)(x-2)(2x-1)}$

Let $\frac{x^2-11x-6}{(x+2)(x-2)(2x-1)} \equiv \frac{A}{x+2} + \frac{B}{x-2} + \frac{C}{2x-1}$

$$\equiv \frac{A(x-2)(2x-1)+B(x+2)(2x-1)+C(x+2)(x-2)}{(x+2)(x-2)(2x-1)}$$

A, B and C are numerical constants. The Core 4 specification does *not* extend to algebraic numerators.

Consider the numerators.

Let $x = 2$ to eliminate A and C: $4-22-6 = 0 + B \times 4 \times 3 + 0$

so $-24 = 12B$ and $B = -2$

$x^2 - 11x - 6 = 4 - 22 - 6$

Let $x = -2$ to eliminate B and C: $4+22-6 = A \times (-4) \times (-5) + 0 + 0$

so $20 = 20A$ and $A = 1$

Equate coefficients of x^2: $1 = 2A + 2B + C$

Substitute the values of A and B: $C = 3$

To find the coefficients, expand the brackets:
$A(x-2)(2x-1) = A(2x^2-5x+2)$
$B(x+2)(2x-1) = B(2x^2-2x-2)$
$C(x+2)(x-2) = C(x^2-4)$

So, $\frac{x^2-11x-6}{(x+2)(x-2)(2x-1)} \equiv \frac{1}{x+2} - \frac{2}{x-2} + \frac{3}{2x-1}$

You can use the cover-up rule when a fraction has only linear factors. It is a shortened form of the method of substitution.

EXAMPLE 4

Use the cover-up rule to express $\dfrac{3x^2+16x-10}{(x-1)(x+2)(2x-1)}$ in partial fractions.

Let $\dfrac{3x^2+16x-10}{(x-1)(x+2)(2x-1)} \equiv \dfrac{A}{x-1}+\dfrac{B}{x+2}+\dfrac{C}{2x-1}$

To find A, cover up $(x-1)$ and substitute $x=1$ in the rest of the fraction:

$$A=\frac{3+16-10}{(1+2)\times(2-1)}=\frac{9}{3\times 1}=3$$

To find B, cover up $(x+2)$ and substitute $x=-2$ in the rest of the fraction:

$$B=\frac{3\times 4+16\times(-2)-10}{(-2-1)\times(-4-1)}=\frac{-30}{(-3)\times(-5)}=-2$$

To find C, cover up $(2x-1)$ and substitute $x=\frac{1}{2}$ in the rest of the fraction:

$$C=\frac{3\left(\frac{1}{4}\right)+\frac{16}{2}-10}{\left(\frac{1}{2}-1\right)\times\left(\frac{1}{2}+2\right)}=\frac{-1\frac{1}{4}}{\left(-\frac{1}{2}\right)\times\left(\frac{5}{2}\right)}=\frac{5}{4}\times\frac{4}{5}=1$$

Hence $\dfrac{3x^2+16x-10}{(x-1)(x+2)(2x-1)} \equiv \dfrac{3}{x-1}-\dfrac{2}{x+2}+\dfrac{1}{2x-1}$

$$\frac{3x^2+16x-10}{\quad)(x+2)(2x-1)}$$

$$\frac{3x^2+16x-10}{(x-1)(\quad)(2x-1)}$$

$$\frac{3x^2+16x-10}{(x-1)(x+2)(\quad)}$$

EXAMPLE 5

Express as partial fractions $\dfrac{x+1}{8x^2-2x-3}$

Factorise the denominator: $8x^2-2x-3 \equiv (2x+1)(4x-3)$

Let $\dfrac{x+1}{(2x+1)(4x-3)} \equiv \dfrac{A}{2x+1}+\dfrac{B}{4x-3}$

To find A, let $x=-\frac{1}{2}$: $A=\dfrac{-\frac{1}{2}+1}{-2-3}=-\dfrac{1}{10}$

Cover up $(2x+1)$ and substitute $x=-\frac{1}{2}$

To find B, let $x=\frac{3}{4}$: $B=\dfrac{\frac{3}{4}+1}{\frac{3}{2}+1}=\dfrac{7}{10}$

Cover up $(4x-3)$ and substitute $x=\frac{3}{4}$

So $\dfrac{x+1}{8x^2-2x-3} \equiv -\dfrac{1}{10(2x+1)}+\dfrac{7}{10(4x-3)}$

Exercise 6.1

1 Use the method of equating coefficients to express these in partial fractions.

a $\dfrac{4x+5}{(x+2)(x+1)}$ **b** $\dfrac{x+25}{(x-3)(x+4)}$ **c** $\dfrac{2}{(x-3)(x+5)}$

d $\dfrac{4x-3}{x(x-1)}$ **e** $\dfrac{7x+2}{(x+5)(2x-1)}$ **f** $\dfrac{2x-5}{x^2-6x+8}$

2 Use the method of substitution to express these in partial fractions.

a $\dfrac{4x-7}{(x-3)(x-2)}$ **b** $\dfrac{5x+11}{(x+1)(x+4)}$ **c** $\dfrac{x+4}{x(x+1)}$

d $\dfrac{x-11}{x^2-7x+6}$ **e** $\dfrac{3x}{(x-1)(x-2)(x-3)}$ **f** $\dfrac{x^2-x+1}{(x^2-1)(2-x)}$

3 Use the cover-up rule to express these in partial fractions.

a $\dfrac{9}{(x-2)(x+1)}$ **b** $\dfrac{3x+7}{(x-1)(x+4)}$ **c** $\dfrac{1}{(x-5)(x-3)}$

d $\dfrac{x-3}{x(x-2)}$ **e** $\dfrac{4-x}{x(x+1)(x+2)}$ **f** $\dfrac{1}{4x^3-x}$

4 Express these in partial fractions.

a $\dfrac{2x}{x^2-6x+8}$ **b** $\dfrac{3}{(x-3)(x^2+x-2)}$ **c** $\dfrac{6x}{(2x-1)(3x-2)}$

d $\dfrac{3-x}{1-x^2}$ **e** $\dfrac{2x+3}{4x^3-x}$ **f** $\dfrac{2x+5}{9-4x^2}$

g $\dfrac{x^2+4}{2x-x^2-x^3}$ **h** $\dfrac{2x}{(1+2x)(4-x^2)}$ **i** $\dfrac{6}{x^4-5x^2+4}$

INVESTIGATION

5 **a** Examples 1 and 2 showed that $\dfrac{5x-1}{(x-2)(x+1)} \equiv \dfrac{3}{x-2} + \dfrac{2}{x+1}$

Use a graphical calculator or a graphical computer package to draw the three graphs of

$y = \dfrac{5x-1}{(x-2)(x+1)}$, $y = \dfrac{3}{x-2}$ and $y = \dfrac{2}{x+1}$

See how the graphs of the two partial fractions add together to give the graph of the original fraction. Pay particular attention to the asymptotes.

b Repeat this graphical investigation using a fraction which has a denominator with three factors,

such as $\dfrac{3x}{(x-1)(x-2)(x-3)}$

6.2 More partial fractions

A proper fraction of the type $\frac{f(x)}{(ax+b)^2}$, where f(x) is a polynomial in x, will produce two partial fractions of the type

$$\frac{A}{ax+b}+\frac{B}{(ax+b)^2}$$

where A and B are constants.

$(ax + b)^2$ is a **repeated** linear factor.

A repeated factor $(ax + b)^3$ will produce partial fractions

$$\frac{A}{ax+b}+\frac{B}{(ax+b)^2}+\frac{C}{(ax+b)^3}$$

EXAMPLE 1

Express as partial fractions $\frac{4-7x}{(x+3)(x-2)^2}$

Let $\frac{4-7x}{(x+3)(x-2)^2} \equiv \frac{A}{x+3}+\frac{B}{x-2}+\frac{C}{(x-2)^2}$

$$\equiv \frac{A(x-2)^2+B(x+3)(x-2)+C(x+3)}{(x+3)(x-2)^2}$$

Equate the numerators: $4-7x \equiv A(x-2)^2+B(x+3)(x-2)+C(x+3)$

Let $x = 2$: $4-14 = 0+0+C\times(2+3)$

$C = -2$

Let $x = -3$: $4+21 = A\times(-5)^2+0+0$

$A = 1$

No other choice of x-value reduces a bracket to zero.

There are now two ways forward:

either

Use $A = 1$, $C = -2$ and let $x = 1$ in the numerators:

$4-7 = 1\times(-1)^2+B\times 4\times(-1)+(-2)\times 4$

$-3 = 1-4B-8$

$B = -1$

You could choose any value of x. $x = 1$ and $x = 0$ are generally simple to use.

or

Equate the coefficients of x^2 in the numerators:

$0 = A + B$

$B = -A$

$B = -1$

Equating coefficients is usually the most efficient method at this point.

So $\frac{4-7x}{(x+3)(x-2)^2} \equiv \frac{1}{x+3}-\frac{1}{x-2}-\frac{2}{(x-2)^2}$

EXAMPLE 2

Express as partial fractions $\frac{2x^2 - 2x + 14}{(x+4)(x-2)^2}$

Let $\frac{2x^2 - 2x + 14}{(x+4)(x-2)^2} \equiv \frac{A}{x+4} + \frac{B}{x-2} + \frac{C}{(x-2)^2}$ (1)

Use the cover-up rule.

To find A, cover up $(x + 4)$ and let $x = -4$:

So $A = \frac{2x^2 - 2x + 14}{(x-2)^2} = \frac{2\times(-4)^2 - 2\times(-4) + 14}{(-4-2)^2} = \frac{54}{36} = \frac{3}{2}$

To find C, cover up $(x - 2)^2$ and let $x = 2$:

So $C = \frac{2x^2 - 2x + 14}{(x+4)} = \frac{2\times 2^2 - 2\times 2 + 14}{2+4} = \frac{18}{6} = 3$

To find B

either

$$\frac{2x^2 - 2x + 14}{(x+4)(x-2)^2} \equiv \frac{A(x-2)^2 + B(x+4)(x-2) + C(x+4)}{(x+4)(x-2)^2}$$

Equate coefficients of x^2:

$$2 = A + B$$

so $B = 2 - A = 2 - \frac{3}{2} = \frac{1}{2}$

or

Let $x = 0$ and substitute in (1):

$$\frac{14}{4\times 4} = \frac{3}{2\times 4} + \frac{B}{-2} + \frac{3}{4}$$

Any x-value would do, but $x = 0$ is the easiest.

Multiply by 8:

$$7 = 3 - 4B + 6$$

$$B = \frac{1}{2}$$

So $\frac{2x^2 - 2x + 14}{(x+4)(x-2)^2} \equiv \frac{3}{2(x+4)} + \frac{1}{2(x-2)} + \frac{3}{(x-2)^2}$

A fraction with a numerator of degree higher than (or equal to) the degree of the denominator is an improper fraction.

$\frac{x^3+1}{x^2-1}$ is an example of an improper fraction.

Before you can separate an improper fraction into partial fractions, you must first change it to a mixed fraction, consisting of a quotient and a remainder, by

- *either* algebraic long division
- *or* rearranging the numerator and finding the quotient and remainder by inspection.

A numerical equivalent is changing $\frac{9}{4}$ to $2\frac{1}{4}$.

C4

EXAMPLE 3

Express $\frac{2x^2+7}{x^2-x-6}$ as partial fractions.

Numerator and denominator are both of degree 2, so it is an improper fraction. You must change it to a mixed fraction.

either

Use long division:

$$x^2-x-6\overline{)\,2x^2 \qquad +7}$$ with quotient 2:

$$\begin{array}{r} 2x^2-2x-12 \\ \hline 2x+19 \end{array}$$

So $\frac{2x^2+7}{x^2-x-6} \equiv 2 + \frac{2x+19}{x^2-x-6}$

or

Rearrange the numerator:

$$\frac{2x^2+7}{x^2-x-6} \equiv \frac{2(x^2-x-6)+2x+19}{x^2-x-6}$$

The numerator is now 2 × denominator + compensating terms

$$\equiv 2 + \frac{2x+19}{x^2-x-6}$$

The quotient is 2 and the remainder is $\frac{2x+19}{x^2-x-6}$

Let $\frac{2x+19}{x^2-x-6} \equiv \frac{2x+19}{(x+2)(x-3)}$

Factorise the denominator.

$$\equiv \frac{A}{x+2} + \frac{B}{x-3}$$

Use the cover-up rule.

Cover up $(x+2)$ and let $x=-2$: $A = \frac{-4+19}{-2-3} = \frac{15}{-5} = -3$

Cover up $(x-3)$ and let $x=3$: $B = \frac{6+19}{3+2} = \frac{25}{5} = 5$

So $\frac{2x^2+7}{x^2-x-6} \equiv 2 - \frac{3}{x+2} + \frac{5}{x-3}$

EXAMPLE 4

Express as partial fractions $\frac{x^3-2x^2-1}{x^2-3x+2}$

This is an improper fraction.

Use long division:

$$\begin{array}{r} x+1 \\ x^2-3x+2\overline{)\,x^3-2x^2 \qquad -1} \\ \underline{x^3-3x^2+2x} \\ x^2-2x-1 \\ \underline{x^2-3x+2} \\ x-3 \end{array}$$

So $\frac{x^3-2x^2-1}{x^2-3x+2} \equiv x+1+\frac{x-3}{x^2-3x+2}$

Let $\frac{x-3}{x^2-3x+2} \equiv \frac{x-3}{(x-1)(x-2)} \equiv \frac{A}{x-1} + \frac{B}{x-2}$

Use the cover-up rule.

Cover up $(x-1)$ and let $x=1$: $A = \frac{1-3}{1-2} = 2$

Cover up $(x-2)$ and let $x=2$: $B = \frac{2-3}{2-1} = -1$

Hence $\frac{x^3-2x^2-1}{x^2-3x+2} \equiv x+1+\frac{2}{x-1} - \frac{1}{x-2}$

Exercise 6.2

1 These fractions have repeated factors in the denominators.
Express the fractions as partial fractions.

a $\frac{x+1}{(x-1)(x-2)^2}$ b $\frac{16}{(x+1)(x-3)^2}$ c $\frac{x^2-7}{(2x-1)(x+1)^2}$

d $\frac{x}{(x-4)^2}$ e $\frac{x-10}{x^2(x-2)}$ f $\frac{x^2-1}{x^3-2x^2}$

g $\frac{1}{x(3x-1)^2}$ h $\frac{x-4}{x(x+2)^3}$

2 Express these improper fractions as partial fractions.

a $\frac{x}{x+2}$ b $\frac{x^2}{x^2-1}$ c $\frac{x^2-2}{(x-1)(x+3)}$

d $\frac{x^2+1}{x(x+1)}$ e $\frac{x^2}{x-1}$ f $\frac{x^3}{x^2-1}$

g $\frac{x^3-3x^2+5}{x^2+x-2}$ h $\frac{9+x^2}{9-x^2}$ i $\frac{x^3}{4x^2-1}$

3 Express $\frac{x^3+2}{x(x-2)^2}$ in partial fractions of the form $A+\frac{B}{x}+\frac{C}{x-2}+\frac{D}{(x-2)^2}$

C4

4 Express in partial fractions.

a $\frac{4}{(2x-3)(x+1)}$ b $\frac{6}{(x-2)(x+1)^2}$ c $\frac{2x^2+1}{(x+1)(2x-1)}$

d $\frac{2x^2}{x^2-1}$ e $\frac{x^2+5}{x^2+2x-3}$ f $\frac{2x+1}{(2x-1)(3x+1)}$

g $\frac{x^3+2}{x(x+1)}$ h $\frac{x^3}{x^2-x-2}$ i $\frac{x^3+1}{x^2(x-1)}$

j $\frac{3}{x^2(x+1)^2}$ k $\frac{x^3+1}{x(x+1)^2}$ l $\frac{(x^2+1)^2}{x^2(x^2-1)}$

INVESTIGATION

5 Explore the graph of $y=f(x)$ if $f(x)$ is
- a proper fraction with repeated linear factors in the denominator
- an improper fraction.

Consider some of the fractions in this exercise, draw their graphs, and take particular notice of any asymptotes, both vertical and horizontal.

Review 6

1 Use the method of equating coefficients to express these in partial fractions.

a $\frac{x+3}{(x-2)(x-1)}$ b $\frac{8-x}{x(x+4)}$ c $\frac{6}{x^2-9}$

2 Use the method of substitution to express these in partial fractions.

a $\frac{x+1}{(x-1)(x+3)}$ b $\frac{x^2-x+5}{x(x-1)(x-5)}$ c $\frac{x^2-2x-2}{(x^2-1)(x+2)}$

3 Use the cover-up rule to express these in partial fractions

a $\frac{8}{(x-1)(x+3)}$ b $\frac{x^2+1}{(x-2)(x+1)}$ c $\frac{x^2-x-4}{x(x+1)(x-2)}$

The fraction in part **b** is improper.

4 Express these in partial fractions.

a $\frac{4x}{(2x+1)(2x-1)}$ b $\frac{2}{(3x-2)(3x-1)}$

c $\frac{x+1}{(x-2)(x-1)^2}$ d $\frac{x+6}{x^2(x-3)}$

e $\frac{9}{x(2x-3)^2}$ f $\frac{x^2+4x-1}{(x^2-1)(x-1)}$

5 Show that $\frac{x^2+4}{x^2-4}$ can be expressed as $A+\frac{B}{x-2}+\frac{C}{x+2}$
Find the values of A, B and C.

6 Express these in partial fractions.

a $\frac{3x^2-3}{(x-1)(x+2)}$ b $\frac{2x^2-3x-24}{x^2-x-6}$

c $\frac{x^3-2}{x^2(x+1)}$ d $\frac{x^3+2x^2-4}{x^2-4}$

7 Show that $\frac{1}{(x+1)(x-k)}$ can be expressed as partial fractions in the form
$\frac{1}{a}\left(\frac{1}{x-k}-\frac{1}{x+1}\right)$. Find a in terms of k.

8 a Show that, if $f(x)=x^3-2x^2-x+2$, then $f(2)=0$
Hence, factorise $f(x)$ completely.

b Express $\frac{1}{f(x)}$ in partial fractions.

c The line $x=\alpha$ is a vertical asymptote to the curve $y=\frac{1}{f(x)}$
State all possible values of α.

9 a Express $f(x) = \dfrac{3x+2}{(x+4)(x-1)}$ in partial fractions.

b Find $f'(x)$ and deduce that the graph of $y = f(x)$ has a negative gradient at all points on the curve.

10 The function f is given by

$$f(x) = \frac{3(x+1)}{(x+2)(x-1)}, \quad x \in \mathbb{R}, x \neq -2, x \neq 1$$

a Express $f(x)$ in partial fractions.

b Hence, or otherwise, prove that $f'(x) < 0$ for all values of x in the domain. [(c) Edexcel Limited 2003]

11 a Write $\dfrac{1-3x}{(x-2)(x+3)}$ in partial fractions.

b Find the gradient at the point where $x = 1$ on the graph of

$$y = \frac{1-3x}{(x-2)(x+3)}$$

c Explain why the graph is a curve which is always rising.

12 a Express $\dfrac{1}{r(r+1)}$ in partial fractions.

b Deduce that $\dfrac{1}{1\times 2} + \dfrac{1}{2\times 3} + \dfrac{1}{3\times 4} + \dots + \dfrac{1}{n(n+1)} = \dfrac{n}{n+1}$

c Find the value of $\displaystyle\sum_{r=1}^{r=\infty} \frac{1}{r(r+1)}$

13 a Express $\dfrac{1}{(r+1)(r+2)}$ in partial fractions.

b Hence show that $\displaystyle\sum_{r=0}^{r=n} \frac{1}{(r+1)(r+2)} = 1 - \frac{1}{n+2}$

c Show that, as $n \to \infty$, $\displaystyle\sum_{r=0}^{r=n} \frac{1}{(r+1)(r+2)}$ converges.

State the sum to infinity.

C4

6 Exit

Summary

Refer to

- For a proper fraction of the type $\frac{f(x)}{(ax+b)(cx+d)(ex+f)}$ where the factors of the denominator are all different and $f(x)$ is a polynomial in x, the **partial fractions** are of the type $\frac{A}{ax+b} + \frac{B}{cx+d} + \frac{C}{ex+f}$ where A, B and C are constants. 6.1
- For a proper fraction which has a repeated linear factor $(ax+b)^2$ in its denominator, there will be two partial fractions of the type $\frac{A}{ax+b} + \frac{B}{(ax+b)^2}$ where A and B are constants. 6.2
- You can find the constants $A, B, C, \ldots$ in the partial fractions by using
  - the method of equating coefficients
  - the method of substitution

  The cover-up rule is a shorter version of the method of substitution. 6.2
- You must change an **improper fraction** to a mixed fraction, consisting of a quotient and a remainder, before you can create partial fractions by
  - *either* algebraic long division
  - *or* rearranging the numerator and finding the quotient and remainder by inspection. 6.2

Links

Partial fractions allow you to express complicated fractions as the sum of simpler fractions. These simplified expressions can be applied to find antiderivatives as well as inverses of transforms, such as the Laplace transform.

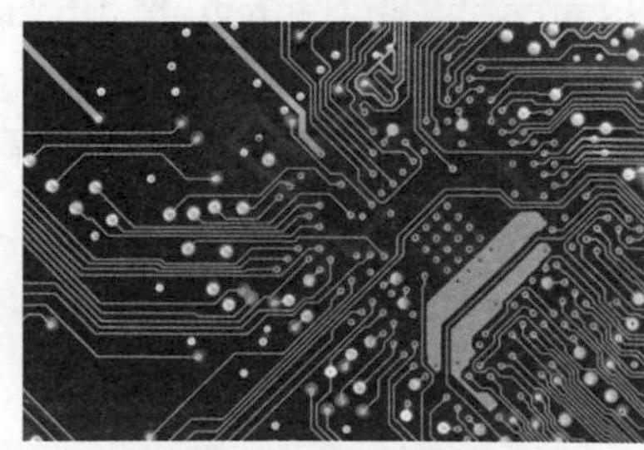

Engineers use Laplace Transforms to simplify problems by converting relationships that are dependent on time t to a set of equations expressed in terms of the Laplace operator s. They can then use Inverse Laplace Transforms to return to the time domain.

Laplace Transforms are particularly useful in analyzing electronic circuits.

7 Parametric equations

This chapter will show you how to

- sketch curves using their parametric equations
- convert parametric equations to Cartesian equations
- find points of intersection of curves and lines using parametric equations
- differentiate parametric equations to find equations of tangents and stationary values
- integrate parametric equations to find areas under curves.

Before you start

You should know how to:

1 Substitute into formulae.

e.g. If $a = 2x + 3$ and $b = 1 - 4x$, find y when $y = a^2 - b$

Substitute for a and b:

$$y = (2x + 3)^2 - (1 - 4x)$$
$$= 4x^2 + 12x + 9 - 1 + 4x$$
$$= 4x^2 + 16x + 8$$

2 Solve simultaneous equations.

e.g. Solve $y = x + 1$ and $y + 5 = x^2$

Eliminate y: $(x + 1) + 5 = x^2$

$$x^2 - x - 6 = 0$$
$$(x + 2)(x - 3) = 0$$

So, $x = -2$ or 3 and $y = -1$ or 4

The solutions are $(-2, -1)$ and $(3, 4)$.

3 Differentiate and integrate functions.

e.g. Find $\frac{dy}{dx}$ and $\int y\,dx$ if $y = \left(1 + \frac{1}{x^2}\right)^2$

Expand the brackets:

$$y = 1 + \frac{2}{x^2} + \frac{1}{x^4} = 1 + 2x^{-2} + x^{-4}$$

Hence $\frac{dy}{dx} = 2(-2)x^{-3} + (-4)x^{-5}$

$$= -\frac{4}{x^3} - \frac{4}{x^5}$$

and $\int y\,dx = x + 2\frac{x^{-1}}{-1} + \frac{x^{-3}}{-3} + c$

$$= x - \frac{2}{x} - \frac{1}{3x^3} + c$$

Check in:

1 If $m = 2(x + 1)$ and $n = \frac{3x - 2}{4}$ find y when

a $y = \frac{1}{4}m + 2n$

b $y = 2m^2 + 16n^2$

2 Solve these simultaneous equations.

a $2x + 3y = 1,\ y + 3x = 5$

b $x^2 + y^2 = 3,\ x + 2y = 1$

3 Find $\frac{dy}{dx}$ and $\int y\,dx$ when

a $y = x^2 + 1 + \frac{1}{x^2}$

b $y = x(1 + x)\left(1 + \frac{1}{x}\right)$

c $y = \frac{x^2 + 1}{x^2}$

C4

7.1 Parametric equations and curve sketching

A **Cartesian equation** has the form $y = f(x)$ — e.g. $y = x^2 + 2x + 1$

Some relationships between x and y involve a third variable. This third variable is called a **parameter**.

The equations $x = f(t)$, $y = g(t)$ are called **parametric equations**. — t is the parameter.

You can sketch a curve described by parametric equations by finding points on the graph for a range of values of t. Each point on the graph has a value of t associated with it.

EXAMPLE 1

Sketch the graph of the curve with the parametric equations $x = 2t - 1$, $y = t^2$ for $-4 \leqslant t \leqslant 4$

Construct a table of values:

t	-4	-3	-2	-1	0	1	2	3	4
x	-9	-7	-5	-3	-1	1	3	5	7
y	16	9	4	1	0	1	4	9	16

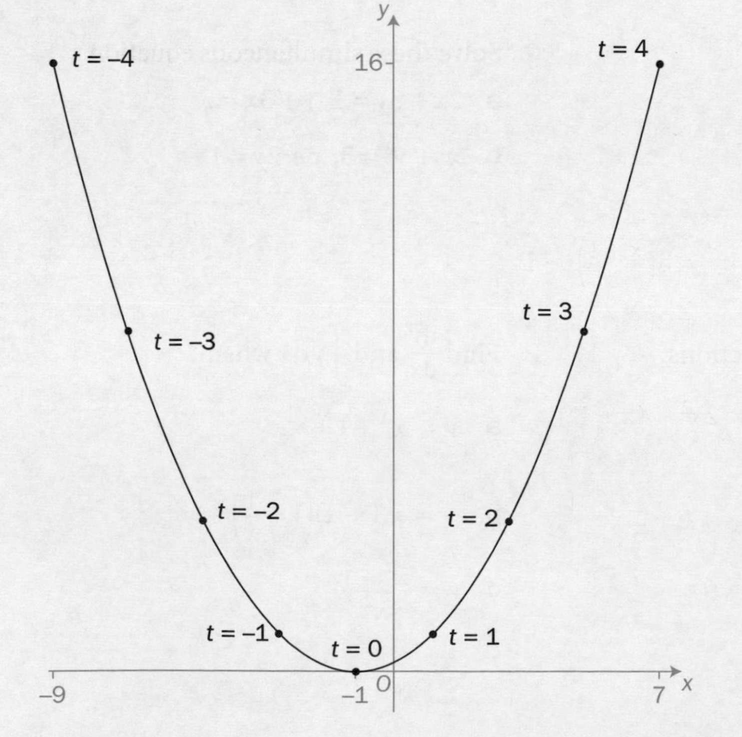

It is useful to label each point with its value of t. You can then see how the curve takes shape as t varies.

The curve is a parabola.

EXAMPLE 2

Sketch the curve given parametrically by the equations
$x = \sin\theta,\ y = \sin 2\theta \quad \text{for } 0 \leqslant \theta \leqslant 2\pi$

θ is the parameter.

Construct a table of values for $0 \leqslant \theta \leqslant 2\pi$:

θ is in radians.

θ	0	$\frac{\pi}{4}$	$\frac{\pi}{2}$	$\frac{3\pi}{4}$	π	$\frac{5\pi}{4}$	$\frac{3\pi}{2}$	$\frac{7\pi}{4}$	2π
x	0	0.707	1	0.707	0	-0.707	-1	-0.707	0
y	0	1	0	-1	0	1	0	-1	0

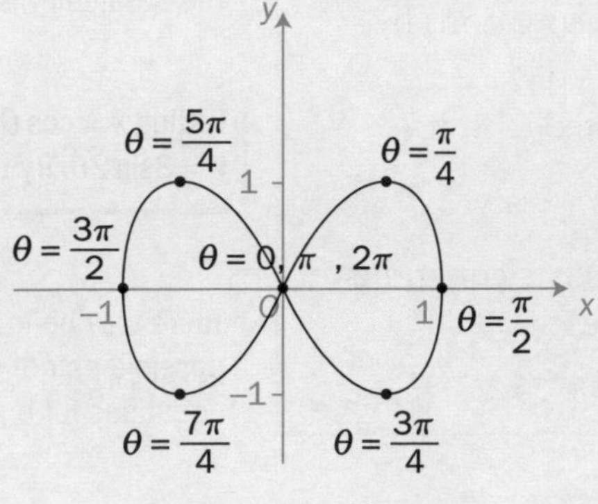

As θ increases from 0, the curve traces the two loops of a figure-of-eight.

When a curve is expressed using parametric equations, you can find the Cartesian equation by eliminating the parameter t (or θ).

EXAMPLE 3

Find the Cartesian equation of the curves which have these parametric equations.

a $x = 2t - 1,\ y = 8t^2 + 3$ **b** $x = 2\sin\theta + 3,\ y = 2\cos\theta - 5$

a Substitute t from $x = 2t - 1$ into the equation for y:

From $x = 2t - 1$, $t = \frac{x+1}{2}$, so $y = 8\left(\frac{x+1}{2}\right)^2 - 1$

$$= 2(x+1)^2 - 1$$
$$= 2x^2 + 4x + 1$$

The Cartesian equation is $y = 2x^2 + 4x + 1$

$2x^2 + 4x + 1$ is a quadratic expression, which indicates that the curve is a parabola.

b Find $\sin\theta$ and $\cos\theta$ in terms of x and y:

From $x = 2\sin\theta + 3$, $\sin\theta = \frac{x-3}{2}$

From $y = 2\cos\theta - 5$, $\cos\theta = \frac{y+5}{2}$

Substitute into $\sin^2\theta + \cos^2\theta = 1$:

$$\left(\frac{x-3}{2}\right)^2 + \left(\frac{y+5}{2}\right)^2 = 1$$
$$(x-3)^2 + (y+5)^2 = 4$$

The Cartesian equation is $(x-3)^2 + (y+5)^2 = 4$

This equation represents a circle, centre (3,–5) and radius $\sqrt{4} = 2$. See C2 for revision.

C4

You can also find parametric equations of a curve represented by a Cartesian equation.

EXAMPLE 4

Find parametric equations for the curve with the Cartesian equation $y = 6x\sqrt{1 - x^2}$

You need to find a parameter which will simplify $\sqrt{1 - x^2}$

Recall that $1 - \sin^2\theta = \cos^2\theta$

Let $x = \sin\theta$:

So $y = 6\sin\theta\sqrt{1 - \sin^2\theta} = 6\sin\theta\cos\theta$
$= 3(2\sin\theta\cos\theta)$
$= 3\sin 2\theta$

Hence, parametric equations for the curve are
$x = \sin\theta,\ y = 3\sin 2\theta$

The parameter is θ.

Letting $x = \cos\theta$ will also give $y = 3\sin 2\theta$. Try this yourself.

There may be more than one possible pair of parametric equations for a given curve.

Exercise 7.1

1 A curve has the parametric equations $x = 3t,\ y = t^2 - 3$
Copy and complete this table.
Hence sketch the graph of the curve for $-3 \leqslant t \leqslant 3$

t	-3	-2	-1	0	1	2	3
x							
y							

2 The parametric equations of a curve are $x = 3t^2,\ y = t^3$
Copy and complete this table.
Hence sketch the graph of the curve for $-2 \leqslant t \leqslant 2$

t	-2	-1	0	1	2
x					
y					

3 Construct your own tables of values to sketch the graphs of the curves with these parametric equations for the range of values given in each case.

a $x = t^2 - 4,\quad y = \frac{1}{2}t^3$ for $-3 \leqslant t \leqslant 3$

b $x = t^3 - 2t + 4,\quad y = t - 1$ for $-2 \leqslant t \leqslant 2$

c $x = t^2,\quad y = \frac{1}{t}$ for $-3 \leqslant t \leqslant 3$

d $x = 4\sin\theta,\quad y = 4\cos\theta$ for $0 \leqslant \theta \leqslant 2\pi$

e $x = 5\cos\theta,\quad y = 3\sin\theta$ for $0 \leqslant \theta \leqslant 2\pi$

f $x = \sec\theta,\quad y = \tan\theta$ for $0 \leqslant \theta \leqslant 2\pi$

C4

4 Find the Cartesian equation for each of the curves given parametrically by these equations.

a $x = t + 4 \quad y = 1 - 2t$

b $x = \frac{3}{t} \quad y = 4t$

c $x = t + 1 \quad y = t^2 - 2$

d $x = t^2 \quad y = t^3$

e $x = t^2 - 1 \quad y = t^3 + 2$

f $x = t^2 \quad y = \frac{2}{t}$

g $x = \frac{1-t}{t} \quad y = \frac{1+t}{t}$

h $x = 3\cos\theta \quad y = 4\sin\theta$

i $x = \sin\theta \quad y = \cos 2\theta$

j $x = 3\cos\theta \quad y = 5\cos 2\theta$

k $x = 2\sec\theta \quad y = 3\tan\theta$

l $x = \frac{1+t}{1-t} \quad y = \frac{2+t}{1-t}$

5 Point P lies on the curve $x = 2t - 4, y = t + 1$
If the y-coordinate of P is 6, find its x-coordinate.

6 Point Q lies on the curve $x = \frac{2+t}{2-t}, y = \frac{3-2t}{2-t}$
If the x-coordinate of Q is 4, find its y-coordinate.

7 The point $(4, k)$ lies on the curve $x = t^2 - 5, y = t - 1$
Find the possible values of k.

8 The variable point $P(at, t^2 - 1)$ meets the line $y = 8$ at the point $(6, 8)$.
Find the possible values of a and the Cartesian equation of the curve along which P moves.

9 Find the coordinates of the points where these curves meet the x-axis.

a $x = t^2 + 1 \quad y = t - 3$

b $x = 1 + t^3 \quad y = 2 - t$

c $x = 5t + 3 \quad y = t^2 - 4$

d $x = 3\cos\theta \quad y = \sin\theta$

10 Find the coordinates of the points where these curves meet the y-axis.

a $x = t - 5 \quad y = t^2 - 2$

b $x = t^2 - 3t + 2 \quad y = t + 4$

c $x = t^3 - t \quad y = t^2$

d $x = \tan\theta \quad y = \sec\theta$

11 The curve $x = at^2 - 3, y = a(t - 2)$ contains the point $(17, 0)$.
Find the value of a.

12 The point $(20, 40)$ lies on the curve $x = at^2, y = 4at$
Find the value of a.

C4

13 The curve $x = 2a\sin\theta$, $y = 1 + a\cos 2\theta$ intersects the y-axis at the point $(0, 4)$.
Find the value of a.

14 Points A and B lie on the curve $x = t^2 - 3$, $y = 2t + 3$ where $t = 2$ and $t = 3$ respectively.
Find

a the distance between A and B

b the gradient of the chord AB

c the equation of the chord AB.

15 Show that these two pairs of parametric equations represent the same straight line. Find the Cartesian equation of the line.

a $x = 1 - t \qquad y = 3 - 2t$

b $x = \dfrac{1}{t+1} \qquad y = \dfrac{t+3}{t+1}$

16 Find parametric equations of the curve with the Cartesian equation

$$y = x\sqrt{4 - x^2}$$

if θ is the parameter such that $x = 2\cos\theta$

17 Use the identity $1 + \tan^2\theta = \sec^2\theta$ to find parametric equations for the curve with the Cartesian equation $y = \dfrac{x}{\sqrt{1 + x^2}}$

18 The Cartesian equation of a curve is $y = \dfrac{3\sqrt{1 - x^2}}{x}$
Find parametric equations for this curve if

a $x = \sin\theta$ **b** $x = \dfrac{1}{t}$

19 The equation of a circle is $x^2 + y^2 - 6x - 4y + 12 = 0$ Refer to Example 3.

a The equation is written in the form $(x - \alpha)^2 + (y - \beta)^2 = 1$
Find the values of α and β.

b Hence, find parametric equations for the circle in terms of the parameter θ.

20 A hyperbola has the equation $9x^2 - 4y^2 - 18x + 16y - 43 = 0$

a The equation is written in the form $\dfrac{(x - a)^2}{b^2} - \dfrac{(y - c)^2}{d^2} = 1$

Find the values of a, b, c and d.

b Hence, find parametric equations for the hyperbola in terms of θ.

21 A curve has parametric equations
$x = t - 2\sin t, \quad y = 1 - 2\cos t, 0 \leqslant t \leqslant 2\pi$

a Find the values of t, in terms of π, at the two points where the curve crosses the x-axis.

b The curve crosses the y-axis at two points where $t = \alpha$ and $t = \beta$
Show that one of these points has $\alpha = 0$. Find, by trial-and-improvement, the value of β to 1 decimal place. Find the coordinates of these two points on the y-axis.

22 By substituting $y = tx$, find parametric equations for the curves with these Cartesian equations.

a $y^3 = x^2$

b $y = x^2 - 2x$

c $x^3 - y^3 = x^2$

d $x - y = xy$

23 A curve has the Cartesian equation $x^3 + y^3 = 3xy$

a Show, by substituting $y = tx$, that the curve can be represented by the parametric equations

$$x = \frac{3t}{1+t^3}, \; y = \frac{3t^2}{1+t^3}$$

b **i** Find the points where $t = 0$ and $t = \infty$
ii Investigate the curve when t is close to -1.

c Hence, sketch the curve and find the equation of any asymptote.

INVESTIGATION

24 **a** Use a computer's graphical package to check your answers to questions **1**, **2** and **3**.
You can also check answers to other questions by drawing appropriate graphs.

b Investigate how changing the values of constants A, B, m and n in these parametric equations alters the graphs of the curves.

$$x = A\sin m\theta, \quad y = B\cos n\theta \quad \text{for } 0 \leqslant \theta \leqslant 2\pi$$

Your computer software may need to have θ in degrees, that is, $0° \leqslant \theta \leqslant 360°$

C4

7.2 Points of intersection

You can use simultaneous equations to find the points of intersection when a curve is expressed in parametric equations.

EXAMPLE 1

Find the points of intersection of the curve with parametric equations $x = 2t - 1, y = t^2$ and the straight line $y = 3x - 2$

Solve the equations $\left.\begin{array}{l} x = 2t - 1, y = t^2 \\ y = 3x - 2 \end{array}\right\}$ simultaneously.

There are three unknowns, x, y and t.

Substitute $x = 2t - 1$, $y = t^2$ into $y = 3x - 2$ to eliminate x and y:

$$t^2 = 3(2t - 1) - 2$$

$$t^2 - 6t + 5 = 0$$

$$(t - 1)(t - 5) = 0 \quad \text{so} \quad t = 1 \text{ or } 5$$

When $t = 1$, $x = 2(1) - 1 = 1$ and $y = 1^2 = 1$

When $t = 5$, $x = 2(5) - 1 = 9$ and $y = 5^2 = 25$

The points of intersection are $(1,1)$ and $(9,25)$.

EXAMPLE 2

Find the points A, B and C where the curve given parametrically by $x = t^2 - 4, y = t - 1$ intersects the two coordinate axes. Hence, find the area of triangle ABC.

When $y = 0$, $t - 1 = 0$ giving $t = 1$

When $t = 1$, $x = 1^2 - 4 = -3$,

giving the point of intersection $A(-3,0)$.

The curve meets the x-axis when $y = 0$

When $x = 0$, $t^2 - 4 = 0$ giving $t = \pm 2$

When $t = 2$, $y = 2 - 1 = 1$

and, when $t = -2$, $y = -2 - 1 = -3$,

giving the points of intersection $B(0,1)$ and $C(0,-3)$.

The area of triangle ABC is

$\frac{1}{2}BC \times OA = \frac{1}{2} \times 4 \times 3 = 6$ square units

The curve meets the y-axis when $x = 0$

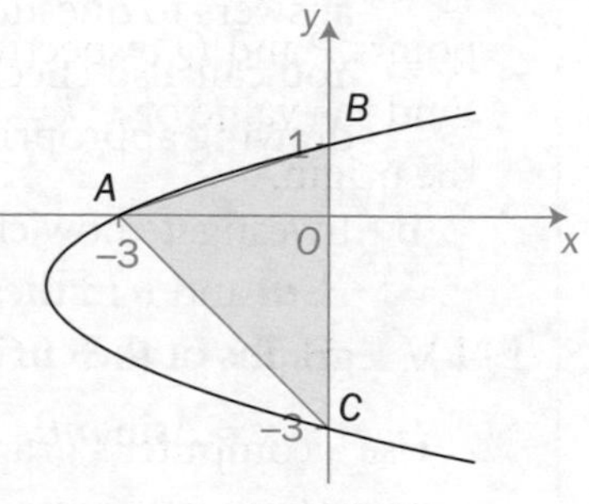

Exercise 7.2

1 Find the points of intersection of the parabola $x = t^2$ $y = 2t$ and the straight lines

a $x + y = 3$

b $4x + 2y = 15$

2 Find the points of intersection of each curve and the given line.

a $x = t^2 - 1, y = 2t + 1 \quad y = x + 2$

b $x = t^3, y = t^2 + 2t \quad y = 2x + 1$

c $x = t^2 - 1, y = t^2 + t + 1 \quad 2y - x - 3 = 0$

3 Find the points of intersection of the curve with parametric equations $x = 2t^2, y = 3t$ and the circle $x^2 + y^2 - 6x - 1 = 0$

4 Find the points of intersection of the parabola $x + y^2 = 9$ and the curve $x = (t - 3)^2, y = 2t$

5 Find the points where these curves cross the coordinate axes.

a $x = t - 1 \quad y = t - 4$

b $x = t - 2 \quad y = t^2 - 9$

c $x = t^2 + 1 \quad y = t - 3$

d $x = t^3 - 1 \quad y = t^2 - 4$

e $x = 1 - \frac{1}{t} \quad y = \frac{t + 1}{t^2 - 4}$

f $x = \pi - 2t \quad y = 1 - \sin t$

6 The variable point $P(t^2, 2t)$ moves along a locus.
Find the points where the locus crosses the straight line $y = 2x - 4$

7 The point $P(t^2, 4t)$ moves as t varies. Q is the midpoint of OP where O is the origin. Write down the coordinates of Q. Find the Cartesian equation of the locus of Q.

8 The point $P(2t^2, 6t)$ lies on a curve. The foot of the perpendicular from P to the x-axis is Q. The midpoint of PQ is M. Find

a the coordinates of Q and M in terms of t

b the Cartesian equation of the locus of M as P moves.

9 Find the points of intersection of the curve
$x = 1 - 5t, y = t^3 + t^2$ and the line $x + y + 2 = 0$

10 The curve $x = t + 1, y = t^2 - k$ intersects the x- and y-axes at points P and Q respectively.
Find the value of $k\,(k \neq 1)$ such that $OP = 2OQ$ where O is the origin.

INVESTIGATION

11 Use a computer's graphical software to draw graphs using their parametric equations.
Check your answers to the problems in this exercise where you have found points of intersection.

C4

7.3 Differentiation

You can differentiate parametric equations to obtain $\frac{dy}{dx}$.

If $x = f(t)$ and $y = g(t)$
the chain rule gives $\frac{dy}{dt} = \frac{dy}{dx} \times \frac{dx}{dt}$ or $\frac{dy}{dx} = \frac{\frac{dy}{dt}}{\frac{dx}{dt}}$

Once you know $\frac{dy}{dx}$, you can use it to find equations of tangents and normals to a curve and to find stationary values.

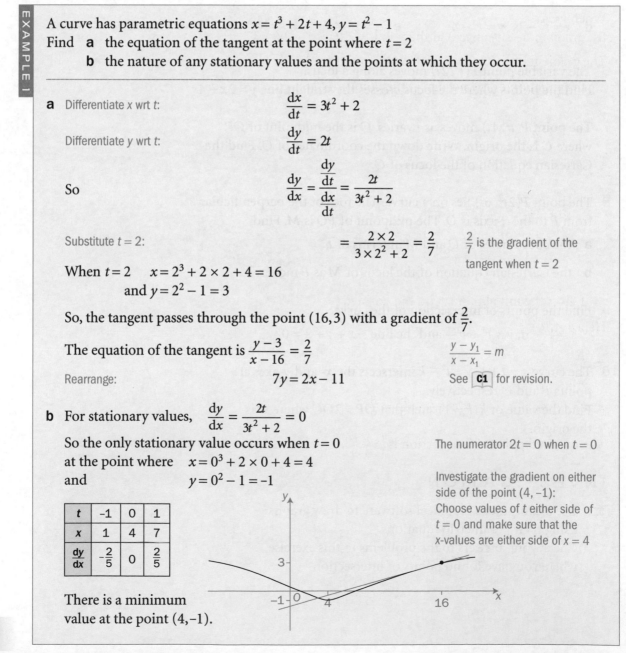

EXAMPLE 1

A curve has parametric equations $x = t^3 + 2t + 4$, $y = t^2 - 1$
Find **a** the equation of the tangent at the point where $t = 2$
b the nature of any stationary values and the points at which they occur.

a Differentiate x wrt t: $\frac{dx}{dt} = 3t^2 + 2$

Differentiate y wrt t: $\frac{dy}{dt} = 2t$

So $\frac{dy}{dx} = \frac{\frac{dy}{dt}}{\frac{dx}{dt}} = \frac{2t}{3t^2 + 2}$

Substitute $t = 2$: $= \frac{2 \times 2}{3 \times 2^2 + 2} = \frac{2}{7}$ — $\frac{2}{7}$ is the gradient of the tangent when $t = 2$

When $t = 2$ $\quad x = 2^3 + 2 \times 2 + 4 = 16$
and $y = 2^2 - 1 = 3$

So, the tangent passes through the point (16, 3) with a gradient of $\frac{2}{7}$.

The equation of the tangent is $\frac{y - 3}{x - 16} = \frac{2}{7}$ — $\frac{y - y_1}{x - x_1} = m$

Rearrange: $7y = 2x - 11$ — See C1 for revision.

b For stationary values, $\frac{dy}{dx} = \frac{2t}{3t^2 + 2} = 0$

So the only stationary value occurs when $t = 0$ — The numerator $2t = 0$ when $t = 0$
at the point where $x = 0^3 + 2 \times 0 + 4 = 4$
and $y = 0^2 - 1 = -1$

Investigate the gradient on either side of the point (4, −1): Choose values of t either side of $t = 0$ and make sure that the x-values are either side of $x = 4$

t	-1	0	1
x	1	4	7
$\frac{dy}{dx}$	$-\frac{2}{5}$	0	$\frac{2}{5}$

There is a minimum value at the point (4, −1).

EXAMPLE 2

A curve is defined parametrically by $x = 3 - t$, $y = 4 - t - t^2$
A normal is drawn to the curve at the point A where $t = 1$
Find another point B at which this normal intersects the curve.

Let $t = 1$: $x = 3 - 1 = 2$ and $y = 4 - 1 - 1^2 = 2$
So, the normal is drawn at the point A (2,2).

Find $\frac{dy}{dx}$ and substitute $t = 1$:

$$\frac{dy}{dx} = \frac{\frac{dy}{dt}}{\frac{dx}{dt}} = \frac{-1 - 2t}{-1}$$

$$= 1 + 2t$$

When $t = 1$,
the gradient of the tangent at the point $A(2,2)$ is $1 + 2(1) = 3$
The gradient of the normal at the same point is $-\frac{1}{3}$.
So, the equation of the normal at the point $A(2,2)$
is
$$\frac{y-2}{x-2} = -\frac{1}{3}$$
$$3y + x = 8$$

If the gradients of tangent and normal are m and m', then $m' = -\frac{1}{m}$

To find the intersections of the normal and curve, substitute the parametric equations into $3y + x = 8$:

$$3(4 - t - t^2) + (3 - t) = 8$$
$$3t^2 + 4t - 7 = 0$$
$$(t - 1)(3t + 7) = 0$$
$$t = 1 \text{ or } -\frac{7}{3}$$

You know that the curve and normal intersect when $t = 1$

$t = 1$ gives the initial point A on the normal,
Hence at B, $t = -\frac{7}{3}$

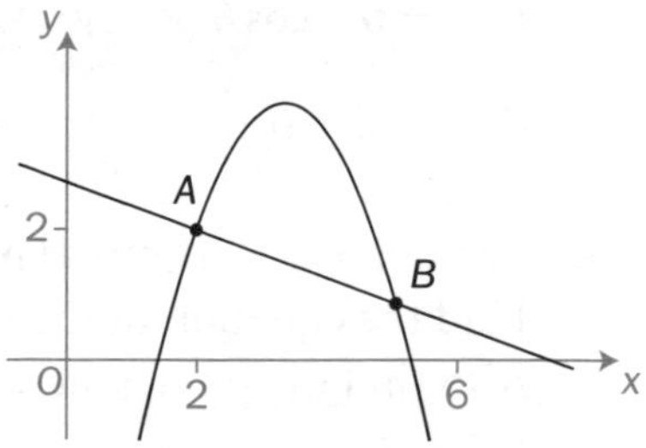

Substitute $t = -\frac{7}{3}$ into the parametric equations:

$$x = 3 + \frac{7}{3} = 5\frac{1}{3} \qquad y = 4 + \frac{7}{3} - \frac{49}{9} = \frac{8}{9}$$

The required point of intersection is $\left(5\frac{1}{3}, \frac{8}{9}\right)$.

C4

Exercise 7.3

1 Find the gradient of each curve with these parametric equations at the point with the given value of t (or θ).

a $x = t + 1$ $\quad y = 3t^2$ $\quad$ when $t = 2$

b $x = t^2 - 2$ $\quad y = t^3 + 1$ $\quad$ when $t = 4$

c $x = t^3 - t^2$ $\quad y = (t + 3)^2$ $\quad$ when $t = 1$

d $x = 3\sin\theta$ $\quad y = 5\cos\theta$ $\quad$ when $\theta = \frac{\pi}{4}$

e $x = 1 + t^2$ $\quad y = 1 + \frac{1}{t}$ $\quad$ when $t = -2$

f $x = \sin 2\theta$ $\quad y = \theta\cos\theta$ $\quad$ when $\theta = 0$

2 Find the equation of the tangent and the normal to the curves with these equations at the point where t (or θ) has the given value.

a $x = 2t^2$ $\quad y = 4t$ $\quad$ when $t = \frac{1}{2}$

b $x = t^2 + 1$ $\quad y = t^3 - 1$ $\quad$ when $t = 1$

c $x = 2\cos\theta$ $\quad y = \cos 2\theta$ $\quad$ when $\theta = \frac{\pi}{4}$

d $x = \frac{2t}{1 + t^2}$ $\quad y = \frac{1 - t^2}{1 + t^2}$ $\quad$ when $t = 2$

3 Find the stationary points on these curves.

a $x = t$ $\quad y = t^3 - t$

b $x = t^2$ $\quad y = t + \frac{1}{t}$

c $x = \theta - \cos\theta$ $\quad y = \sin\theta$ $\quad$ for $0 < \theta < 2\pi$

d $x = 3\sin\theta + 2$ $\quad y = 3\cos\theta + 5$ $\quad$ for $0 \leqslant \theta \leqslant 2\pi$

4 The curve $x = 2t^2, y = 4t$ has a normal at point $P(8, 8)$.
Find the equation of the normal.
Also find the point where the normal meets the curve a second time.

5 Find the equation of the normal to the curve $x = 6t$, $y = \frac{6}{t}$ at the point where $t = 2$
Also find the point where the normal intersects the curve again.

C4

6 The tangent at point $P(1, 1)$ to the curve $x = \frac{1}{t}, y = t^2$ intersects the curve at point Q.

Find the equation of the tangent at P and the coordinates of Q.

7 The point P lies on the curve $x = 5\cos\theta, y = 4\sin\theta$
A and B are the points $(-3, 0)$ and $(3, 0)$ respectively.

a Find the distances AP and BP in terms of θ.

b Show that the sum of the distances AP and BP is constant for all points P.

8 The line from the variable point $P\left(t, \frac{1}{t}\right)$ to the origin O intersects the line $x = 1$ at the point Q.

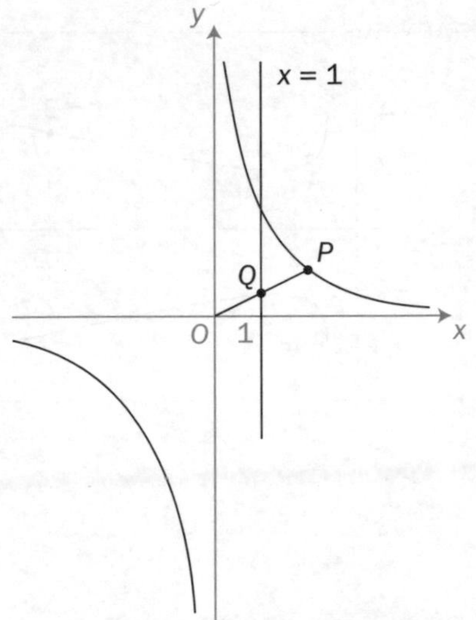

Find

a the gradient and the equation of the line OP

b the coordinates of Q in terms of t

c the Cartesian equation of the locus of the midpoint of PQ.

INVESTIGATION

9 Investigate how to draw tangents and normals to curves using a computer's graphical package. Hence, check some of your answers to the problems in this exercise.

C4

Integration

You can modify the expression $\int_a^b f(x)\,dx$ for the area under a curve using the chain rule.

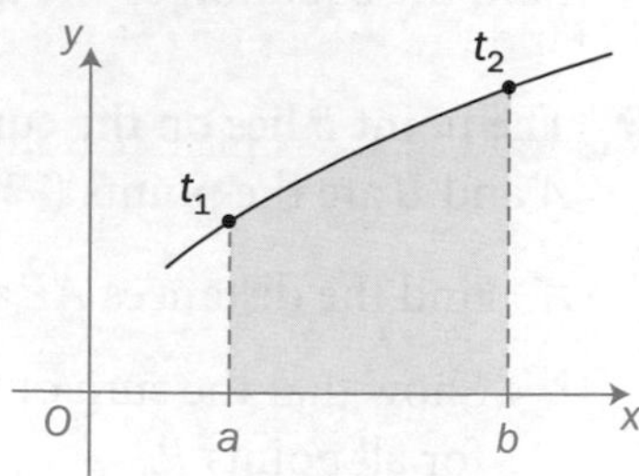

For parametric equations $x = f(t)$, $y = g(t)$, the area under the curve between the points where $t = t_1$ and $t = t_2$ is given by

$$\int_{t_1}^{t_2} y\frac{dx}{dt}\,dt$$

The **limits of this integral** are $t = t_1$ and $t = t_2$, as the independent variable is now t and not x.

EXAMPLE 1

This sketch shows the curve with parametric equation

$x = t^2 + 1,\ y = t + \frac{1}{t}$

Find the shaded area between the curve and the x-axis from $x = 2$ to $x = 10$

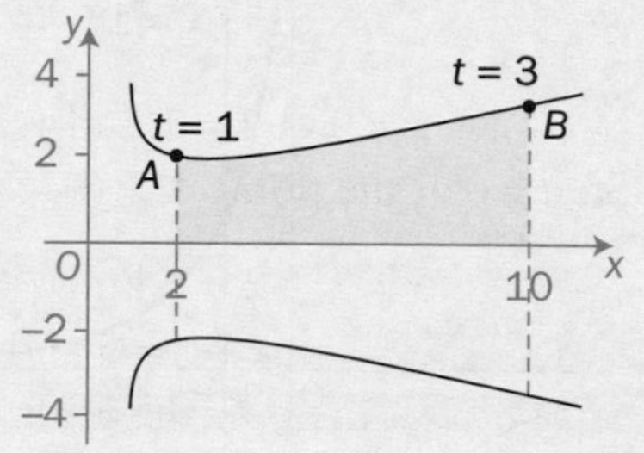

Let x = 2: $t^2 = 1, t = \pm1$ so $y = 2$ or -2.
So, point A is $(2, 2)$.

Let x = 10: $t^2 = 9, t = \pm3$ so $y = 3\frac{1}{3}$ or $-3\frac{1}{3}$.

So, point B is $\left(10, 3\frac{1}{3}\right)$.

You want the area under the curve from $A(2, 2)$ where $t = 1$ to $B\left(10, 3\frac{1}{3}\right)$ where $t = 3$

You are calculating the area above the x-axis only.

Integrate:

$$\text{Required area} = \int_2^{10} y\,dx = \int_1^3 y\frac{dx}{dt}\,dt \text{ where } y = t + \frac{1}{t} \text{ and } \frac{dx}{dt} = 2t$$

$$= \int_1^3 \left(t + \frac{1}{t}\right)(2t)\,dt$$

$$= 2\int_1^3 (t^2 + 1)\,dt$$

$$= 2\left[\frac{t^3}{3} + t\right]_1^3$$

$$= 2\left(9 + 3 - \frac{1}{3} - 1\right) = 21\frac{1}{3} \text{ square units}$$

Notice the change in the limits as the independent variable changes from x to t.

EXAMPLE 2

This diagram shows the curve with parametric equations $x = t^2 + 1, y = t^3 - 4t$

Find the values of t at the points $A(1,0)$ and $B(5,0)$.

Find the area of the region enclosed by the loop of the curve.

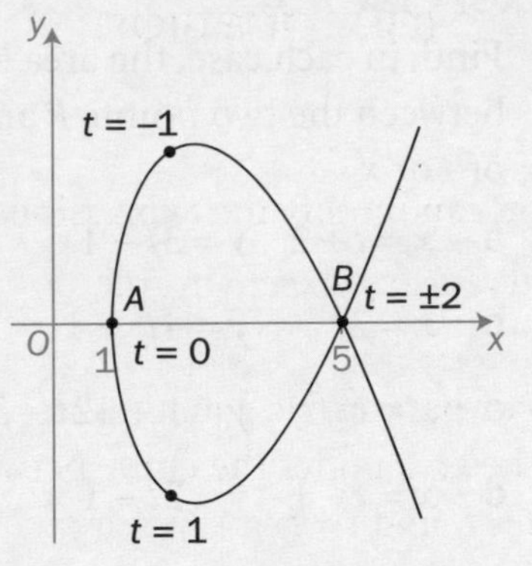

Find t at A and B:

At points A and B, $y = 0$

$$t^3 - 4t = 0$$
$$t(t-2)(t+2) = 0$$
$$t = 0 \text{ or } \pm 2$$

Let $t = 0$: $x = 0^2 + 1 = 1$ and $y = 0 - 0 = 0$ — Find x and y when $t = 0$

So, $t = 0$ at the point $A(1,0)$.

Let $t = 2$: $x = 2^2 + 1 = 5$ and $y = 2^3 - 4 \times 2 = 0$

Let $t = -2$: $x = (-2)^2 + 1 = 5$ and $y = (-2)^3 - 4 \times (-2) = 0$

Find x and y when $t = +2$ and $t = -2$.

So, $t = \pm 2$ at the point $B(5,0)$.

You want the area under the curve from $A(1,0)$ where $t = 0$ to $B(5,0)$ where $t = \pm 2$

When $t = 1$, $x = 2$, $y = 1 - 4 = -3$, giving the point $(2,-3)$ below the x-axis. So, integrating from $t = 0$ to $t = 2$ will give the area **below** the x-axis.

Similarly, $t = -1$ gives the point $(2,3)$ and integrating from $t = 0$ to $t = -2$ gives the area **above** the x-axis.

area of loop = 2 × area enclosed by the curve above the x-axis

The curve is symmetrical about the x-axis.

$$= 2 \times \int_1^5 y\,dx = 2\int_0^{-2} y\frac{dx}{dt}dt \text{ where } y = t^3 - 4t \text{ and } \frac{dx}{dt} = 2t$$

$$= 2\int_0^{-2} (t^3 - 4t)(2t)dt$$

$$= 4\int_0^{-2} (t^4 - 4t^2)dt$$

$$= 4\left[\frac{t^5}{5} - \frac{4t^3}{3}\right]_0^{-2}$$

$$= 4\left(-\frac{32}{5} + \frac{32}{3} - 0\right)$$

$$= 17\frac{1}{15} \text{ square units}$$

You could find the answer directly by integrating the whole way from $t = -2$ to $t = 2$ using

$$\int_2^{-2} 2t^4 - 8t^2\,dt$$

Try this yourself.

C4

Exercise 7.4

1 Find, in each case, the area between the x-axis and the curve between the two points P and Q defined by the given values of t or x.

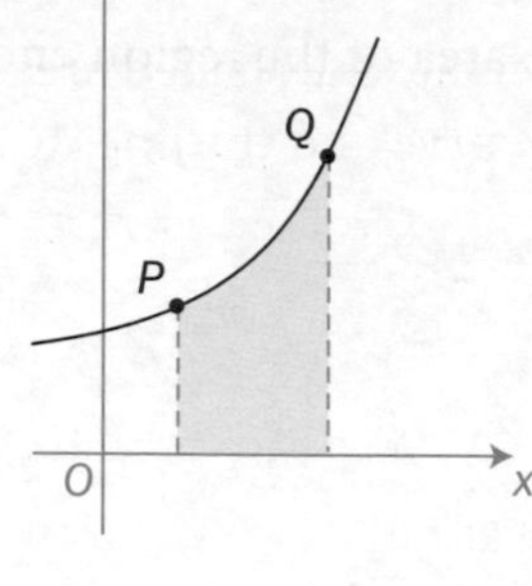

a $x = t + 2 \quad y = 3t - 1$ for $t = 1$ to $t = 4$

b $x = 2t \quad y = 4t^2 + 1$ for $t = 0$ to $t = 2$

c $x = t^2 \quad y = 1 + 12t - 3t^2$ for $t = 0$ to $t = 4$

d $x = t - 1 \quad y = 2t - 1$ for $x = 1$ to $x = 5$

e $x = 2t \quad y = t^2 + 2$ for $x = 2$ to $x = 4$

f $x = 4t \quad y = \frac{2}{t}$ for $x = 2$ to $x = 4$

2 A and B are the points on the curve $x = t^3, y = t^2$ where $t = 0$ and $t = 2$
Find the shaded area on the diagram.
Also find the area of the region labelled R.

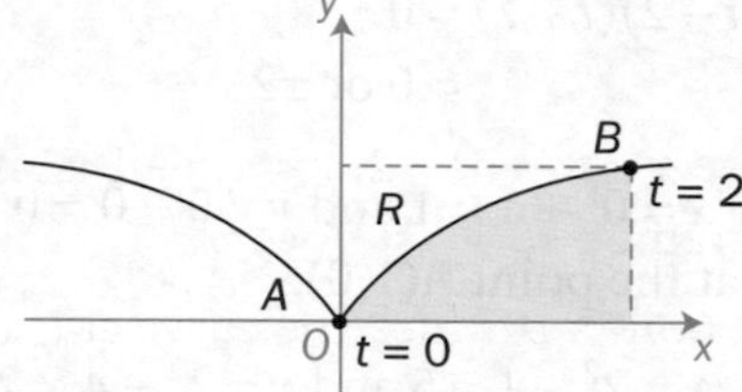

3 **a** The curve $x = t^2 - 1, y = 2t(2 - t)$ cuts the coordinate axes at the points A, B, C and D. Find the positions of these points and their associated t-values.

b Calculate the area of the region R in the first quadrant enclosed by the curve and the two coordinate axes.

c Calculate the total shaded area.

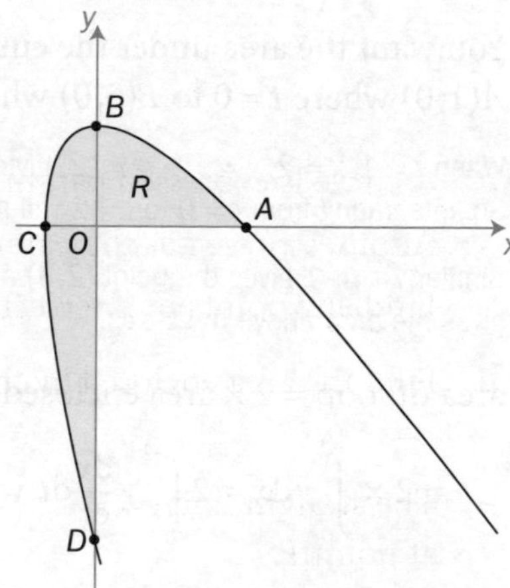

4 This diagram shows the curve $x = t^2, y = t(4 - t^2)$
Find the values of t at the two points where the curve cuts the x-axis.
Hence, find the area enclosed by the loop.

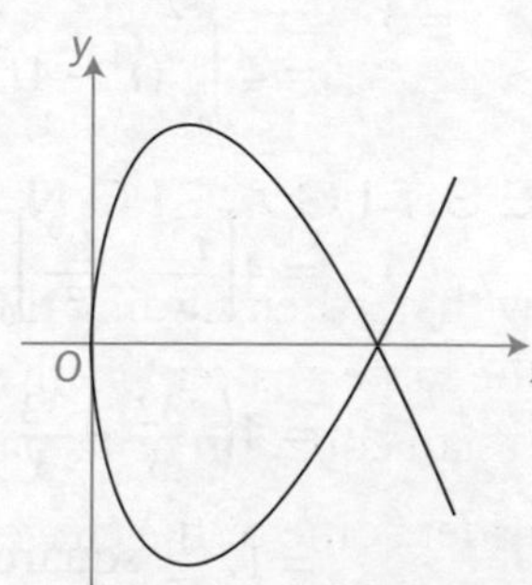

5 **a** This diagram shows the curve $x = t^2 + \frac{1}{t^2}$, $y = 2t$

Find the coordinates of the points A, B and C on the curve where $t = \frac{1}{2}$, 1 and 2 respectively.

b Calculate the shaded area of the diagram bounded by the curve and the line AC.

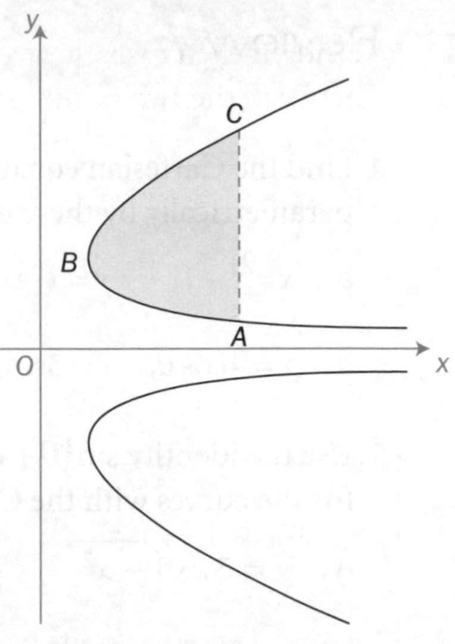

6 The curve $x = t^2 - 3t + 2$, $y = 4t^2 + 1$ is shown on this diagram.

Find the points at which the curve cuts the y-axis.

Also find the area bounded by the curve and the line $y = 5$

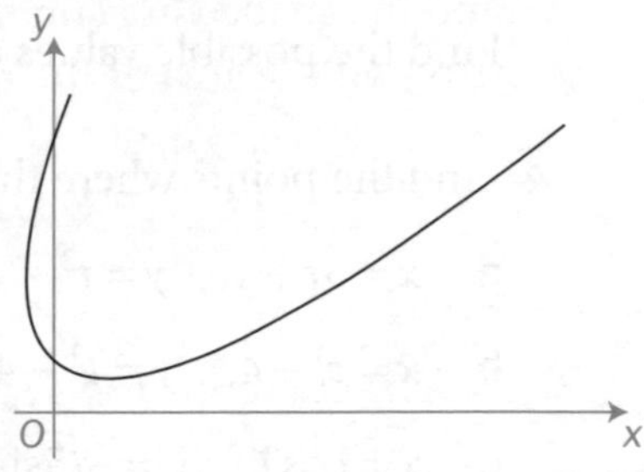

7 **a** A curve is expressed parametrically by $x = t^2 + 1$, $y = 2t$

Another curve has the parametric equations $x = 2s$, $y = \frac{2}{s}$

Find any points of intersection.

b Find the area enclosed by the two curves and the line $x = 5$

8 **a** The straight line $y = c - 2x$ touches the curve $x = t$, $y = \frac{1}{t}$ at point P.

Find the possible values of c and the coordinates of P.

b Find the area enclosed by the curve $x = t$, $y = \frac{1}{t}$ and the line $y = 3 - 2x$

INVESTIGATION

9 Show that, when finding the area under a curve, you get the same answer whether you use the curve's Cartesian equation or its parametric equations.

Consider some of the curves in questions **1** and **2** of this exercise.

C4

Review 7

1 Find the Cartesian equation for each of the curves given parametrically by these equations.

a $x = t - 1,\quad y = t^2 + 1$ b $x = \frac{t-1}{t},\quad y = \frac{t+1}{t}$

c $x = 4\cos\theta,\quad y = 3\sin\theta$ d $x = 2\cos\theta,\quad y = \cos 2\theta$

2 Use the identity $\sin^2\theta + \cos^2\theta \equiv 1$ to find parametric equations for the curves with the Cartesian equation

a $y = 3x\sqrt{1 - x^2}$ b $y = \frac{5x}{\sqrt{4 - x^2}}$

3 The point $(5, a)$ lies on the curve $x = t^2 + 1,\ y = \frac{1}{3}(t - 1)$
Find the possible values of a.

4 Find the points where the curve given by these parametric equations

a $x = 3t + 1,\quad y = t^2 - 1$ intersects the straight line $2x + y = 6$

b $x = t^3 - 4,\quad y = t^2 - 4$ intersects the x-axis

c $x = \cos t,\quad y = \sqrt{5}\sin t$ intersects the circle $x^2 + y^2 = 2$

5 Find the equation of the tangent and the normal to the curves with these parametric equations at the point where t has the given value.

a $x = 2t - 1,\quad y = t^3 + 1$ when $t = 1$ b $x = t^3 - 1,\quad y = t^2 + t + 1$ when $t = 2$

c $x = 1 + t,\quad y = \frac{1}{1-t}$ when $t = 2$ d $x = 2\sin t,\quad y = \sin 2t$ when $t = \frac{\pi}{6}$

6 Find, in each case, the area between the x-axis and the curve between the two points defined by the given values of t or x.

a $x = t + 2$, $y = 3t - 1$ from $t = 1$ to $t = 4$ b $x = 2t + 1$, $y = \sqrt{t}$ from $x = 3$ to $x = 9$

c $x = \ln t$, $y = t\sin t$ from $t = 2$ to $t = 3$ d $x = e^{-t}$, $y = e^{2t} + 1$ from $x = 1$ to $x = 2$

7 The curve C is defined parametrically, for $0 \leqslant \theta \leqslant \pi$, by the equations
$x = 3\cos\theta,\quad y = 3\cos 2\theta + 6$

a Find $\frac{dy}{dx}$ in terms of θ.
Explain why the gradient at any point on the curve C is never greater than 4.

b Find the Cartesian equation of C and sketch the graph of C.

C4

8 A curve has parametric equations $x = t^2$, $y = 2t$
R is the point on the curve where $t = r$

- **a** Show that the normal to the curve at point R has a gradient of $-r$.
- **b** If S is the point $(s^2, 2s)$, find the gradient of the chord RS in terms of r and s.
- **c** If the chord RS is normal to the curve at R, show that $r^2 + rs + 2 = 0$
- **d** At what point does the normal at the point (9, 6) meet the curve again?

9 The trajectory of a cricket ball is given parametrically by the equations

$$x = 10t, \quad y = 2 + 10t - 5t^2$$

where x and y are the horizontal and vertical distances travelled (in metres) after a time of t seconds from the ball being struck.

- **a** Find the Cartesian equation of the trajectory.
- **b** Find the time taken before the ball hits the ground.
- **c** What is the horizontal distance travelled by the ball before it hits the ground?

10 This diagram shows a sketch of part of the curve C with parametric equations

$$x = 1 + \frac{3}{t}, \quad y = t^2 \sin t, \quad \frac{\pi}{2} \leqslant t \leqslant \pi$$

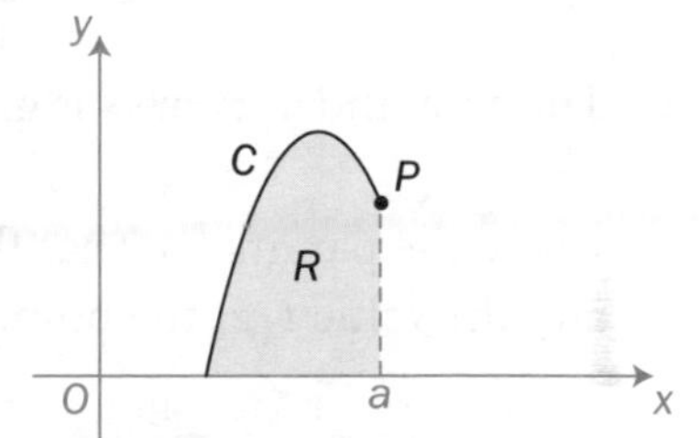

- **a** The point $P\left(a, \frac{\pi^2}{4}\right)$ lies on C. Find the value of a.
- **b** Region R is enclosed by C, the x-axis and the line $x = a$ as shown in the diagram.
  Show that the area of region R is given by $3\int_{\frac{\pi}{2}}^{\pi} \sin t \, dt$

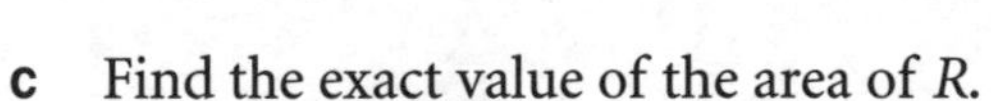

- **c** Find the exact value of the area of R.

11 A curve has parametric equations

$$x = 2\cot\theta, \quad y = 2\sin^2\theta, \quad 0 < \theta \leqslant \frac{\pi}{2}$$

- **a** Find an expression for $\frac{dy}{dx}$ in terms of the parameter θ.
- **b** Find an equation of the tangent to the curve at the point where $\theta = \frac{\pi}{4}$
- **c** Find a Cartesian equation of the curve in the form $y = f(x)$
  State the domain on which the curve is defined. [(c) Edexcel Limited 2005]

7

Exit

Summary

Refer to

- You can sketch the graph of a curve given by the parametric equations $x = f(t)$, $y = g(t)$ by using a table of values showing the values of x and y as t varies. 7.1
- To convert the parametric equations $x = f(t)$, $y = g(t)$ of a curve into a Cartesian equation, eliminate the parameter t from the two equations. 7.1
- You can find the points of intersection of two curves by solving the equations of the curves simultaneously.
For example, you can substitute the parametric equations of one curve into the Cartesian equation of the other curve. 7.2
- To differentiate $x = f(t)$, $y = g(t)$ to find $\frac{dy}{dx}$,
the chain rule gives: $\frac{dy}{dx} = \frac{\frac{dy}{dt}}{\frac{dx}{dt}} = \frac{g'(t)}{f'(t)}$ 7.3
- The area under a curve is given by $\int_{x=a}^{x=b} y\,dx = \int_{t=t_1}^{t=t_2} y\frac{dx}{dt}\,dt$
where the parameter t has the value t_1 at the point where $x = a$ and the value t_2 at the point where $x = b$ 7.4

Links

The path of any projectile is the result of two independent motions, horizontal and vertical, which can be expressed in terms of a parameter time t.

You can model the path of a projectile at any time t by the parametric equations

$x(t) = (v_0 \cos\theta)t$, $y(t) = (v_0 \sin\theta)t - \frac{1}{2}gt^2$

where θ is the angle at which the projectile is launched at time $t = 0$, v_0 is the initial velocity of the projectile, and g is the acceleration due to gravity.

This model can be used to analyse the motion of a specific projectile, which is useful in areas such as sports science to study, for example, the flight of a golf ball.

8 The binomial series

This chapter will show you how to

- find the binomial expansion of $(a + b)^n$ and $(1 + x)^n$ when n is a positive integer
- find the binomial expansion of $(1 + x)^n$ when n is a fraction or a negative number and write the condition for which the expansion is valid
- use partial fractions to express certain kinds of algebraic fraction as a binomial series
- find numerical and algebraic approximations using binomial expansions.

Before you start

You should know how to:

1 Use the laws of indices.

e.g. $\dfrac{1}{\sqrt[3]{(1+x)^2}} = \dfrac{1}{(1+x)^{\frac{2}{3}}} = (1+x)^{-\frac{2}{3}}$

2 Manipulate surds.

e.g. $\sqrt{0.99} = \sqrt{\dfrac{9\times 11}{100}} = \dfrac{3}{10}\sqrt{11}$

3 Use Pascal's triangle and calculate $\dbinom{n}{r}$.

e.g. $\dbinom{10}{3} = {}^{10}C_3 = \dfrac{10\times 9\times 8}{1\times 2\times 3} = 120$

4 Find terms when multiplying brackets.

e.g. Find the term in x^2 in the expansion of

$\left(1 + 2x + 3x^2 + \ldots\right)\left(1 - \frac{1}{2}x - \frac{1}{3}x^2 - \ldots\right)$

The term is $(1)\left(-\frac{1}{3}x^2\right) + (2x)\left(-\frac{1}{2}x\right) + (3x^2)(1)$

$= \left(-\frac{1}{3} - 1 + 3\right)x^2 = \frac{5}{3}x^2$

5 Express a fraction in partial fractions.

e.g. Let $\dfrac{7}{(1-x)(5+2x)} \equiv \dfrac{A}{1-x} + \dfrac{B}{5+2x}$

The cover-up rule uses $x = 1$ to give $A = 1$

and uses $x = -2\frac{1}{2}$ to give $B = 2$

So $\dfrac{7}{(1-x)(5+2x)} \equiv \dfrac{1}{1-x} + \dfrac{2}{5+2x}$

Check in:

1 Write in the form $(1 + x)^n$

a $\sqrt{(1+x)^5}$ **b** $\sqrt{\dfrac{1}{(1+x)^3}}$

2 Write in terms of the root of an integer

a $\sqrt{1.21}$ **b** $\sqrt{4\times 3 + \frac{4}{9}}$ **c** $\dfrac{1}{\sqrt{1.2}}$

3 a Use Pascal's triangle to expand $(2 + x)^3$

b Find the value of ${}^8C_3$ and expand $(2 + x)^8$

4 In the expansion of

$\left(1 - 3x - 5x^2 + \ldots\right)\left(1 + \frac{1}{2}x + \frac{1}{4}x^2 - \ldots\right)$

find

a the term in x

b the term in x^2.

5 Express in partial fractions

a $\dfrac{3}{(1-2x)(1+x)}$

b $\dfrac{3}{(1-2x)(1+x)^2}$

8.1 The binomial series

For n a positive integer, the **binomial series** is

$$(a+b)^n = \binom{n}{0}a^n + \binom{n}{1}a^{n-1}b + \binom{n}{2}a^{n-2}b^2 + \cdots + \binom{n}{r}a^{n-r}b^r + \cdots \binom{n}{n}b^n$$

This is a finite series with $n + 1$ terms.

You can find the coefficients $\binom{n}{0}, \binom{n}{1}, \ldots$

either
from the nth row of
Pascal's triangle when n is small

```
            1               Row
         1     1          ← 1st
      1     2     1       ← 2nd
   1     3     3     1    ← 3rd
1     4     6     4     1 ← 4th
```

or
by using

$$\binom{n}{r} = {}^nC_r = \frac{n!}{(n-r)!r!}$$

$n!$ means factorial n $= n \times (n-1) \times (n-2) \ldots \times 1$

$$= \frac{n(n-1)(n-2)\ldots(n-r+1)}{1 \times 2 \times 3 \ldots \times r}$$

There are r terms in both the numerator and denominator.

The special case when $a = 1$ and $b = x$ gives the binomial series for $(1 + x)^n$

$$(1+x)^n = 1 + nx + \frac{n(n-1)}{1 \times 2}x^2 + \frac{n(n-1)(n-2)}{1 \times 2 \times 3}x^3 + \cdots + \frac{n(n-1)(n-2)\ldots(n-r+1)}{1 \times 2 \times 3 \ldots \times r}x^r + \cdots + x^n$$

This is valid for all x when n is a positive integer.

EXAMPLE 1

Find the first four terms in the expansion of $(1 + 2x)^{20}$

$$(1+2x)^{20} = 1 + 20(2x) + \frac{20 \times 19}{1 \times 2}(2x)^2 + \frac{20 \times 19 \times 18}{1 \times 2 \times 3}(2x)^3 + \ldots$$

$$= 1 + 40x + 760x^2 + 9120x^3 + \ldots$$

which are the first four terms of the series.

n is large, so use $\binom{n}{r}$ to find the coefficients. Using Pascal's triangle is not an appropriate method in this case.

The binomial expansion of $(1 + x)^n$ is also valid for all negative or fractional values of n provided that $-1 < x < 1$.

You can write $-1 < x < 1$ as $|x| < 1$. This restriction on the value of x ensures that the series converges.

When n is negative or fractional

- the binomial series obtained is an infinite series – it does not terminate after $n+1$ terms
- the series can be written as a series of ascending powers of x
- Pascal's triangle and ${}^nC_r$ have no meaning when n is negative or fractional and cannot be used to find coefficients
- the condition $-1 < x < 1$ which restricts the range of values of x must always be stated
- to expand $(a+x)^n$ as a series when n is **not** a positive integer, you must first rearrange $(a+x)^n$ as $a^n\left(1+\frac{x}{a}\right)^n$

EXAMPLE 2

Find the first four terms in the expansion of $\frac{1}{\sqrt{1-x}}$

State the range of values of x for which the expansion is valid.

As $\frac{1}{\sqrt{1-x}} = (1-x)^{-\frac{1}{2}}$, let $n = -\frac{1}{2}$ and replace x by $-x$ in the expansion of $(1+x)^n$:

$$\frac{1}{\sqrt{1-x}} = (1-x)^{-\frac{1}{2}} = 1 + \left(-\frac{1}{2}\right)(-x) + \frac{\left(-\frac{1}{2}\right)\left(-\frac{3}{2}\right)}{1\times 2}(-x)^2 + \frac{\left(-\frac{1}{2}\right)\left(-\frac{3}{2}\right)\left(-\frac{5}{2}\right)}{1\times 2\times 3}(-x)^3 + \dots$$

$$= 1 + \frac{1}{2}x + \frac{3}{8}x^2 + \frac{5}{16}x^3 + \dots$$

Write out the expansion in full with brackets to avoid errors with the fractions and negative signs.

The expansion is valid for $|-x| < 1$ which gives $|x| < 1$

EXAMPLE 3

Find the first four terms in the expansion of $(8+3x)^{\frac{2}{3}}$

Give the range of values of x for which the expansion is valid.

Rearrange so that the first term inside the bracket is 1.

Then let $n = \frac{2}{3}$ and replace x by $\frac{3}{8}x$ in the expansion of $(1+x)^n$:

$$(8+3x)^{\frac{2}{3}} = 8^{\frac{2}{3}}\left(1+\frac{3x}{8}\right)^{\frac{2}{3}}$$

$$= 4\left[1 + \left(\frac{2}{3}\right)\left(\frac{3x}{8}\right) + \frac{\frac{2}{3}\left(\frac{2}{3}-1\right)}{1\times 2}\left(\frac{3x}{8}\right)^2 + \frac{\frac{2}{3}\left(\frac{2}{3}-1\right)\left(\frac{2}{3}-2\right)}{1\times 2\times 3}\left(\frac{3x}{8}\right)^3 + \dots\right]$$

$8^{\frac{2}{3}} = 4$

$$= 4\left[1 + \frac{x}{4} - \frac{x^2}{64} + \frac{x^3}{384} - \dots\right] = 4 + x - \frac{x^2}{16} + \frac{x^3}{96} - \dots$$

The expansion is valid for $\left|\frac{3}{8}x\right| < 1$

which gives $|x| < \frac{8}{3}$ or $-2\frac{2}{3} < x < 2\frac{2}{3}$

As a check, substitute a small value of x (such as $x = 0.01$) into the series and compare with the value of $8.03^{\frac{2}{3}}$ from your calculator.

EXAMPLE 4

Find the coefficient of x^2 in the expansion of $\dfrac{(1+x)^4}{\sqrt[3]{1-3x}}$

For what values of x is the expansion valid?

$$\frac{(1+x)^4}{\sqrt[3]{1-3x}} = (1+x)^4(1-3x)^{-\frac{1}{3}}$$

$$= (1+4x+6x^2+\dots)\left(1+\left(-\frac{1}{3}\right)(-3x)+\frac{\left(-\frac{1}{3}\right)\left(-\frac{1}{3}-1\right)}{1\times 2}(-3x)^2+\dots\right)$$

$$= (1+4x+6x^2+\dots)(1+x+2x^2+\dots)$$

The term in x^2 is $(1)(2x^2)+(4x)(x)+(6x^2)(1)=12x^2$

The coefficient of the x^2 term is 12.

The expansion of $(1+x)^4$ is valid for all values of x because the index 4 is a positive integer.

The expansion of $(1-3x)^{-\frac{1}{3}}$ is valid for $|-3x|<1$ or $-\frac{1}{3}<x<\frac{1}{3}$

The whole expansion is thus valid for $-\frac{1}{3}<x<\frac{1}{3}$

You can find the coefficients of $(1+x)^4$ from Pascal's triangle.

C4

Exercise 8.1

1 Expand as a series of ascending powers of x up to and including x^3. State the range of values for which each series is valid.

a $\dfrac{1}{1+2x}$ **b** $\sqrt{1+x}$ **c** $\sqrt{1+2x}$ **d** $\dfrac{1}{(1-x)^3}$

e $\sqrt[3]{1+3x}$ **f** $\dfrac{1}{\sqrt{1-x}}$ **g** $\dfrac{1}{(1-3x)^2}$ **h** $\sqrt[4]{1+2x}$

i $\dfrac{1}{\sqrt{1-2x}}$ **j** $\dfrac{1}{\left(1-\frac{1}{2}x\right)^2}$ **k** $\sqrt{(1+2x)^3}$ **l** $\left(1-\frac{1}{2}x\right)^{\frac{2}{3}}$

2 **a** Expand **i** $\dfrac{1}{1+x}$ **ii** $\dfrac{1}{1-x}$
as a series of ascending powers of x as far as x^4

b Identify each series as a geometric progression and so find the sum of each series using the formula for a geometric progression.

3 Find the first three terms in the binomial expansion of each expression. Give the values of x for which each expansion is valid.

a $\sqrt{4+x}$ **b** $\dfrac{1}{2+x}$ **c** $\dfrac{1}{(2-3x)^2}$ **d** $\dfrac{1}{\sqrt{9-x}}$

4 Expand each expression as far as the term in x^2.
Find the values of x for which each expansion is valid.

a $(1+x)\sqrt{1-x}$

b $\dfrac{1-2x}{\sqrt{1+x}}$

c $\dfrac{2+x}{\left(1-\frac{1}{2}x\right)^2}$

d $\dfrac{x+3}{x-1}$

e $\sqrt{\dfrac{1+x}{1-x}}$

f $(3-x)(1+2x)^{\frac{2}{3}}$

g $(2+x)^2\sqrt{1-2x}$

5 Find the coefficient of x^2 in the expansion of $\sqrt{1+x+x^2}$

6 Find the first four terms in the binomial expansion of $\sqrt{1+\dfrac{2}{x}}$ in descending powers of x.
For what values of x is the expansion valid?

7 The coefficient of x^2 in the expansion of $\sqrt{1+ax}$ is -2.
Find two possible values of a and the first four terms of each possible expansion.

8 The first three terms in the expansion of $(1+\alpha x)^n$ are 1, $+2x$, and $-2x^2$.
Find α, n and the coefficient of x^3 in the expansion.

9 The second and fourth terms in the expansion of $(1+kx)^n$ are x and $\frac{5}{3}x^3$.
Given that $k, n > 0$, find k and n. Also find the third term of the expansion.
For what values of x is the expansion valid?

INVESTIGATION

10 Express $f(x) = \sqrt{1+\dfrac{1}{x}}$ as a series of descending powers of x.

Show that $f(x) = \dfrac{1}{\sqrt{x}}\sqrt{1+x}$ and so express $f(x)$ as a series of ascending powers of x. Use computer software to draw the graphs of $f(x)$ and these two series.
Notice the importance on the graphs of the values of x for which the two expansions are valid.

C4

8.2 Using partial fractions

You can express some fractions as a series of ascending powers of x by firstly expressing them as partial fractions.

EXAMPLE 1

Find the first four terms of the expansion of $\frac{1+5x}{(1-x)(2+x)}$ as a series of ascending powers of x.

Find the range of values of x for which the expansion is valid.

Let $$\frac{1+5x}{(1-x)(2+x)} \equiv \frac{A}{1-x} + \frac{B}{2+x}$$

$$\equiv \frac{A(2+x)+B(1-x)}{(1-x)(2+x)}$$

Equate the numerators:

$$1+5x \equiv A(2+x) + B(1-x)$$

When $x = -2$, $1 - 10 = 0 + 3B$

$B = -3$

When $x = 1$, $1 + 5 = 3A + 0$

$A = 2$

The cover-up rule gives the same results with less written working.

So $$\frac{1+5x}{(1-x)(2+x)} = \frac{2}{1-x} - \frac{3}{2+x} = \frac{2}{1-x} - \frac{3}{2\left(1+\frac{1}{2}x\right)}$$

$$= 2(1-x)^{-1} - \frac{3}{2}\left(1+\frac{1}{2}x\right)^{-1}$$

$$= 2\left[1 + (-1)(-x) + \frac{(-1)(-2)}{1\times 2}(-x)^2 + \frac{(-1)(-2)(-3)}{1\times 2\times 3}(-x)^3 + \dots\right]$$

$$-\frac{3}{2}\left[1 + (-1)\left(\frac{1}{2}x\right) + \frac{(-1)(-2)}{1\times 2}\left(\frac{1}{2}x\right)^2 + \frac{(-1)(-2)(-3)}{1\times 2\times 3}\left(\frac{1}{2}x\right)^3 + \dots\right]$$

$$= 2(1 + x + x^2 + x^3 + \dots) - \frac{3}{2}\left(1 - \frac{1}{2}x + \frac{1}{4}x^2 - \frac{1}{8}x^3 + \dots\right)$$

$$= \frac{1}{2} + \frac{11}{4}x + \frac{13}{8}x^2 + \frac{35}{16}x^3 + \dots$$

The expansion of $(1-x)^{-1}$ is valid for $|x| < 1$; that is $-1 < x < 1$

The expansion of $\left(1+\frac{1}{2}x\right)^{-1}$ is valid for $\left|\frac{1}{2}x\right| < 1$; i.e. $-2 < x < 2$

$-1 < x < 1$ is a stricter condition than $-2 < x < 2$.

Both these conditions must apply, so the whole expansion is valid for $-1 < x < 1$.

Exercise 8.2

1 Write each expression in partial fractions and so expand each expression as a series of ascending powers of x as far as x^3.
Find the range of values of x for which each series is valid.

a $\dfrac{3}{(1-2x)(1-x)}$ b $\dfrac{4}{(1+3x)(1-x)}$ c $\dfrac{5}{\left(1-\frac{1}{2}x\right)(1+2x)}$

d $\dfrac{3}{(2-x)(1+x)}$ e $\dfrac{6}{(3+x)(2+x)}$ f $\dfrac{2}{(3-2x)(1-2x)}$

g $\dfrac{4}{(1+x)(1-x)^2}$

2 a Express $\dfrac{2x+3}{(x-4)^2}$ in the form $\dfrac{A}{x-4}+\dfrac{B}{(x-4)^2}$, where A and B are constants.

b Hence, or otherwise, express $\dfrac{2x+3}{(x-4)^2}$ as a binomial series up to and including the term in x^4. Give the range of values of x for which the series is valid.

c Compare the value of $\dfrac{2x+3}{(x-4)^2}$ with the value of the series up to and including the term in x^4 when

i $x=1$ ii $x=1.5$ iii $x=2$

What do you notice?

3 $f(x)=\dfrac{1+14x}{(1-x)(1+2x)}$, $|x|<\dfrac{1}{2}$

a Express $f(x)$ in partial fractions.

b Hence find the exact value of $\displaystyle\int_{\frac{1}{6}}^{\frac{1}{3}} f(x)\,dx$, giving your answer in the form $\ln p$, where p is rational.

c Use the binomial theorem to expand $f(x)$ in ascending powers of x, up to and including the term in x^3, simplifying each term. [(c) Edexcel Limited 2003]

INVESTIGATION

4 Use a computer's graphical software to draw the graph of $y=f(x)$ where $f(x)$ is one of the fractions in this exercise.
Also draw the graph of the equivalent binomial expansion.
Explore the graphical significance of the range of valid values of x. Repeat for other functions $f(x)$ selected from this exercise, especially for Question 2.

C4

8.3 Approximations

The terms of the binomial expansion of $(1 + x)^n$ have ascending powers of x.
If the value of x is small, such as 0.01, then successive terms in the expansion have smaller and smaller values.

EXAMPLE 1

Find the value of 0.999^8 correct to 6 significant figures by letting $x = 0.001$ in the expansion of $(1 - x)^8$.

$$(1-x)^8 = 1 + 8(-x) + \frac{8\times7}{1\times2}(-x)^2 + \frac{8\times7\times6}{1\times2\times3}(-x)^3 + \cdots + x^8$$
$$= 1 - 8x + 28x^2 - 56x^3 + \cdots + x^8$$

Let $x = 0.001$
so $(1 - x)^8 = 0.999^8$
Then $0.999^8 = 1 - 0.008 + 0.000\,028 - 0.000\,000\,056 + \ldots$
$\approx 0.992\,027\,944$
$= 0.992\,028$ correct to 6 significant figures.

You can check this value on a calculator.
By including more terms of the series, the value could be calculated beyond the accuracy of a calculator.

Keep your working to at least 7 significant figures if you require accuracy to 6 s.f.

EXAMPLE 2

Find the value of $\sqrt{2}$ correct to 5 decimal places by letting $x = -0.02$ in the expansion of $\sqrt{1 + x}$.

$$\sqrt{1+x} = (1+x)^{\frac{1}{2}} = 1 + \left(\frac{1}{2}\right)x + \frac{\left(\frac{1}{2}\right)\left(-\frac{1}{2}\right)}{1\times2}x^2 + \frac{\left(\frac{1}{2}\right)\left(-\frac{1}{2}\right)\left(-\frac{3}{2}\right)}{1\times2\times3}x^3 + \ldots$$
$$= 1 + \frac{x}{2} - \frac{x^2}{8} + \frac{x^3}{16} - \ldots$$

This expansion is valid for $|x| < 1$ and is thus valid for $x = -0.02$

Let $x = -0.02$, so $\sqrt{0.98} = 1 - 0.01 - 0.000\,05 - 0.000\,000\,5$
$= 0.989\,949\,5$ to 7 decimal places

Now $\sqrt{0.98} = \sqrt{\frac{49\times2}{100}} = \frac{7}{10}\times\sqrt{2}$

So $\sqrt{2} = \frac{10}{7} \times 0.989\,949\,5 = 1.41421$
correct to five decimal places.

You can compare the graph of $y = f(x)$ with the graph of its binomial expansion.

In example 2, you found that $\sqrt{1+x} = 1 + \frac{x}{2} - \frac{x^2}{8} + \frac{x^3}{16} - \ldots$ for $|x| < 1$

The graph of $y = \sqrt{1+x}$ and the graph of this infinite series are identical for $|x| < 1$.

However, you can also compare the graph of $y = \sqrt{1+x}$ with the graphs of just parts of the series; for instance

$y = 1 + \frac{x}{2}$ $y = 1 + \frac{x}{2} - \frac{x^2}{8}$ and $y = 1 + \frac{x}{2} - \frac{x^2}{8} + \frac{x^3}{16}$

These are approximations to $\sqrt{1+x}$ and are valid when x is small and high powers of x are negligible in size. That is, they are valid when x is close to $x = 0$ and the graph is in the neighbourhood of the point (0, 1).

The approximation $y = 1 + \frac{x}{2}$ gives a straight line graph. It is a linear approximation.

The line $y = 1 + \frac{1}{2}x$ is, in fact, the tangent to the curve $y = \sqrt{1+x}$ at the point (0, 1).

The approximation $y = 1 + \frac{x}{2} - \frac{x^2}{8}$ is a quadratic approximation.

The approximation $y = 1 + \frac{x}{2} - \frac{x^2}{8} + \frac{x^3}{16}$ is a cubic approximation.

This diagram shows that the more terms there are in the approximation, the closer the graph of the approximation is to the graph of $y = \sqrt{1+x}$

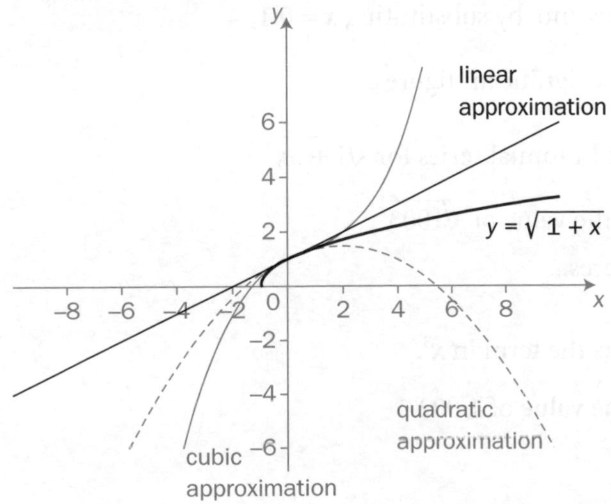

C4

EXAMPLE 3

Find linear and quadratic approximations to $\dfrac{1}{(1+x)^2}$

$$\frac{1}{(1+x)^2} = (1+x)^{-2} = 1 + (-2)x + \frac{(-2)\times(-3)}{1\times 2}x^2 - \ldots$$

$$= 1 - 2x + 3x^2 - \ldots \text{ for } |x| < 1$$

The linear approximation is $\dfrac{1}{(1+x)^2} \approx 1 - 2x$

The quadratic approximation is $\dfrac{1}{(1+x)^2} \approx 1 - 2x + 3x^2$

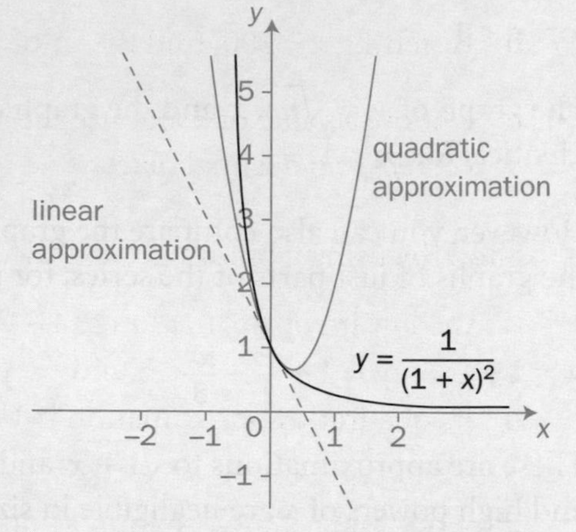

Exercise 8.3

You can check many of the answers in this exercise on a calculator. However, you should not use a calculator to work out the answers.

1 Substitute $x = 0.03$ in the expansion of $(1 - x)^{10}$ to find the value of 0.97^{10} correct to 5 decimal places.

2 Expand $(3 + x)^6$ and show, by substituting $x = 0.02$, that $3.02^6 = 758.650$ correct to 3 decimal places.

3 By expanding $\sqrt{1+x}$ as a binomial series, find

   a the value of $\sqrt{5}$ correct to 5 decimal places by letting $x = \frac{1}{4}$

   b the value of $\sqrt{3}$ correct to 5 decimal places by letting $x = -\frac{1}{4}$

4 Expand $\dfrac{1}{\sqrt{1-x}}$ as a binomial series and, by substituting $x = 0.1$, find the value of $\sqrt{10}$ correct to six significant figures.

5 a Find the first four terms of the binomial series for $\sqrt[3]{1+3x}$

   b By substituting $x = \dfrac{1}{1000}$ find the value of $\sqrt[3]{1003}$ correct to eight significant figures.

6 Expand $\dfrac{1}{(1+2x)^3}$ as a series as far as the term in x^3.
Substitute $x = 0.001$ and so find the value of 1.002^{-3} correct to 8 decimal places.

C4

7 a Find the first four terms in the binomial expansion of $\sqrt{1-\frac{1}{x}}$ in descending powers of x.
For what range of x-values is the expansion valid?

b By letting $x = 100$, find the value of $\sqrt{99}$ correct to 6 decimal places.

c Choose a value of x and use the same series to find $\sqrt{101}$ correct to 6 decimal places.

8 Show, by using a binomial expansion, that

a the linear approximation to $\frac{1}{(1+x)^3}$ is $1-3x$

b its quadratic approximation is $1-3x+6x^2$

9 Find the cubic approximation of the function $\sqrt{\left(1+\frac{x}{2}\right)^3}$

10 Find

a a quadratic approximation to the function $\frac{1+2x}{\sqrt{1-2x}}$

b a cubic approximation to the function $\frac{5}{(1-2x)(2+x)}$

11 A quadratic approximation to $\frac{1}{(A+Bx)^3}$ is $\frac{1}{8}-\frac{3}{4}x+Cx^2$
Find the values of A, B and C.

INVESTIGATION

12 Use a computer's graphical package to explore the graph of $y=\frac{1}{(1+x)^3}$ and the linear, quadratic and cubic approximations found from its binomial expansion.

Explore other functions and their approximations from questions in this exercise.

See question 8 in this exercise.

C4

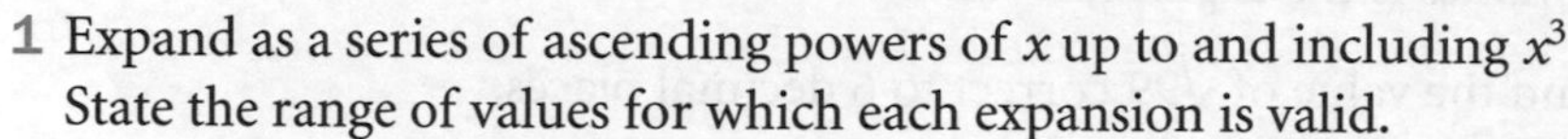

Review 8

1 Expand as a series of ascending powers of x up to and including x^3
State the range of values for which each expansion is valid.

a $\frac{1}{1-3x}$ **b** $\sqrt{1+2x}$

c $\left(1-\frac{1}{2}x\right)^{-\frac{2}{3}}$ **d** $\sqrt{9+2x}$

e $\frac{1}{2-x}$ **f** $\frac{1}{\sqrt{4-x}}$

g $(1-x)\sqrt{1+x}$ **h** $\frac{x+3}{x-1}$

2 Find the constants A, B and C where $\frac{\sqrt{1-x}}{2-x} = A + Bx + Cx^2 + \ldots$
Find the range of values of x for which the expansion is valid.

3 Find the coefficient of x^2 in the expansion of $\sqrt{1-x-x^2}$

4 Find the first four terms in the binomial expansion of $\sqrt{1-\frac{2}{x}}$ in descending powers of x.
For what values of x is the expansion valid?

5 When $(1+ax)^n$ is expanded as a series in ascending powers of x, the coefficients of x and x^2 are –6 and 27 respectively.

a Find the value of a and the value of n.

b Find the coefficient of x^3.

c State the set of values of x for which the expansion is valid. [(c) Edexcel Limited 2004]

6 Write each expression in partial fractions and so expand it as a series of ascending powers of x as far as x^3.
Find the range of values of x for which each expansion is valid.

a $\frac{3}{(1-x)(1+2x)}$

b $\frac{3x+1}{(1-x)(1+x)^2}$

C4

7 Given that $\frac{3+5x}{(1+3x)(1-x)} \equiv \frac{A}{1+3x} + \frac{B}{1-x}$,

a find the values of the constants A and B.

b Hence, or otherwise, find the series expansion in ascending powers of x, up to and including the term in x^2, of $\frac{3+5x}{(1+3x)(1-x)}$

c State, with a reason, whether your series expansion in part **b** is valid for $x = \frac{1}{2}$

[(c) Edexcel Limited 2004]

8 Expand $(2+x)^5$ and show, by substituting $x = 0.03$, that $2.03^5 = 34.473$ correct to 3 decimal places.

9 By expanding $\sqrt{1+x}$ as a binomial series, find the value of $\sqrt{10}$ correct to 5 decimal places by letting $x = \frac{1}{9}$

10 Show, by using a binomal expansion, that

a the linear approximation to $\frac{1}{(1+x)^3}$ is $1 - 3x$

b the quadratic approximation to $\frac{1}{(1+x)^3}$ is $1 - 3x + 6x^2$

11 The binomial expansion of $(1+12x)^{\frac{3}{4}}$ in ascending powers of x up to and including the term in x^3 is $1 + 9x + px^2 + qx^3$, $|12x| < 1$

a Find the value of p and the value of q.

b Use this expansion with your values of p and q together with an appropriate value of x to obtain an estimate of $(1.6)^{\frac{3}{4}}$.

c Obtain $(1.6)^{\frac{3}{4}}$ from your calculator and hence make a comment on the accuracy of the estimate you obtained in part **b**.

[(c) Edexcel Limited 2003]

12 $f(x) = \frac{3x^2+16}{(1-3x)(2+x)^2} = \frac{A}{(1-3x)} + \frac{B}{(2+x)} + \frac{C}{(2+x)^2}$, $|x| < \frac{1}{3}$

a Find the values of A and C and show that $B = 0$

b Hence, or otherwise, find the series expansion of $f(x)$, in ascending powers of x, up to and including the term in x^3. Simplify each term.

[(c) Edexcel Limited 2006]

C4

8 Exit

Summary

Refer to

- When n is a positive integer, the binomial expansion of $(a+b)^n$ is a finite series, valid for all values of x, where

$$(a+b)^n = \binom{n}{0}a^n + \binom{n}{1}a^{n-1}b + \binom{n}{2}a^{n-2}b^2 + \dots + \binom{n}{r}a^{n-r}b^r + \dots \binom{n}{n}b^n$$ 8.1

- You can find the coefficients $\binom{n}{r}$ either from Pascal's triangle or by using $\frac{n!}{(n-r)!r!}$ 8.1
- When n is negative or fractional, the binomial expansion of $(1+x)^n$ is an infinite series, valid only for $|x| < 1$, where

$$(1+x)^n = 1 + nx + \frac{n(n-1)}{1\times 2}x^2 + \frac{n(n-1)(n-2)}{1\times 2\times 3}x^3 + \dots + \frac{n(n-1)(n-2)\dots(n-r+1)}{1\times 2\times 3\dots\times r}x^r + \dots$$ 8.1

- When n is negative or fractional, you must rearrange $(a+x)^n$ as $a^n\left(1+\frac{x}{a}\right)^n$ to obtain its binomial expansion. 8.1
- You can rewrite some algebraic fractions using partial fractions before expressing them as binomial series. 8.2
- You can use binomial expansions to find numerical and algebraic approximations when x is small and terms containing high powers of x are negligible in size and can be ignored. 8.3

Links

The binomial series is useful for approximations. There are many applications of this in physics.

For example, in special relativity, which studies space and time, the parameter γ is defined as

$$\gamma = \left(1 - \frac{v^2}{c^2}\right)^{-\frac{1}{2}}$$

where v is the velocity of a particle and c is the speed of light. This is of the form $(1+x)^n$ and, with v very much smaller than c, can be approximated by the first few terms of its series expansion

$$\gamma = 1 + \frac{1}{2}\left(\frac{v}{c}\right)^2 + \frac{3}{8}\left(\frac{v}{c}\right)^4$$

Using this approximation can make calculations easier and allow related equations to be defined.

9 Differentiation

This chapter will show you how to

- differentiate implicit functions and parametric functions
- use implicit and parametric functions in problems of coordinate geometry
- apply exponential functions to problems involving growth and decay
- find rates of change and explore how different rates of change relate to each other in practical situations.

Before you start

You should know how to:

1 Find the equation of a tangent and a normal to a curve.

e.g. Find the equation of the normal to the curve $y = x + \frac{1}{x}$ at $\left(2, 2\frac{1}{2}\right)$.

$\frac{dy}{dx} = 1 - \frac{1}{x^2}$

At the point $\left(2, 2\frac{1}{2}\right)$, $\frac{dy}{dx} = 1 - \frac{1}{4} = \frac{3}{4}$

So, gradient of normal is $-\frac{4}{3}$

The equation of the normal is

$\frac{y - 2\frac{1}{2}}{x - 2} = -\frac{4}{3}$

which gives $6y + 8x = 31$

Check in:

1 a Find the equation of the tangent to the curve $y = 2x^2 - \frac{1}{x^2}$ at the point where $x = 1$.

b Find the equation of the normal to the curve $y = \sin x - \cos x$ at the point where $x = \frac{\pi}{4}$.

You should know how to:

2 Use the chain rule and product rule.

e.g. Differentiate $y = \sin^2 x + \sin x \cos x$

$\frac{dy}{dx} = (2\sin x \cos x) + (\sin x \times -\sin x + \cos x \cos x)$

$= 2\sin x \cos x + \cos^2 x - \sin^2 x$

$= \sin 2x + \cos 2x$

Check in:

2 Differentiate each expression with respect to x.

a $y = \tan^3 x$

b $y = \sqrt{x^3 + 1}$

c $y = e^x \sin x$

d $y = x^3 \ln x$

You should know how to:

3 Manipulate logarithmic expressions.

e.g. Solve the equation $3^x = 2^{x+4}$

Take natural logarithms:

$x\ln 3 = (x + 4)\ln 2$

$x(\ln 3 - \ln 2) = 4\ln 2$

$x = \frac{4\ln 2}{\ln 3 - \ln 2} = \frac{\ln 16}{\ln 1.5}$

Check in:

3 Solve these equations, giving answers to 2 s.f.

a $\ln 3^x = x\ln 2 + 1$

b $e^{2x-3} = 20$

c $e^{1-0.02x} = 1.5$

9.1 Differentiating implicit functions

Equations such as $y^2 + xy + y = 8$, which are not easily rearranged into the form $y = f(x)$, are known as **implicit functions.**

You usually need to use the chain rule or the product rule (or both) to differentiate an implicit function.

See Chapter 4 for revision.

The chain rule gives $\frac{d(y^n)}{dx} = \frac{d(y^n)}{dy} \times \frac{dy}{dx} = ny^{n-1}\frac{dy}{dx}$

and the product rule gives $\frac{d(xy)}{dx} = x\frac{dy}{dx} + y\frac{dx}{dx} = x\frac{dy}{dx} + y$

EXAMPLE 1

Find $\frac{dy}{dx}$ when **a** $2x^3 + 4y^3 = 3$ **b** $x^2 + 3xy + y^2 = 6$ **c** $x\ln y = y^2 + 1$ **d** $x^3y^2 = \sin(x - y)$

a Differentiate wrt x: $6x^2 + 4 \times 3y^2 \times \frac{dy}{dx} = 0$

$12y^2\frac{dy}{dx} = -6x^2$ and $\frac{dy}{dx} = -\frac{x^2}{2y^2}$

Use the chain rule for $4y^3$ to give $\frac{d(4y^3)}{dy} \times \frac{dy}{dx} = 4 \times 3y^2 \times \frac{dy}{dx}$

b Differentiate wrt x: $2x + 3\left(x\frac{dy}{dx} + y \times 1\right) + 2y\frac{dy}{dx} = 0$

$$\frac{dy}{dx}(3x + 2y) = -(2x + 3y)$$

$$\frac{dy}{dx} = -\frac{2x + 3y}{3x + 2y}$$

Use the product rule for $3xy$ and the chain rule for y^2.

c Differentiate wrt x: $x \times \left(\frac{1}{y} \times \frac{dy}{dx}\right) + \ln y \times 1 = 2y\frac{dy}{dx} + 0$

$$\ln y = 2y\frac{dy}{dx} - \frac{x}{y}\frac{dy}{dx}$$

$$\frac{y\ln y}{2y^2 - x} = \frac{dy}{dx}$$

Use the product rule to differentiate $x\ln y$ wrt x. Within the product rule, use the chain rule to differentiate $\ln y$ wrt x.

d Differentiate with respect to x:

$$x^3 \times \left(2y\frac{dy}{dx}\right) + y^2 \times 3x^2 = \cos(x - y) \times \left(1 - \frac{dy}{dx}\right)$$

$$\frac{dy}{dx}(\cos(x - y) + 2x^3y) = \cos(x - y) - 3x^2y^2$$

$$\frac{dy}{dx} = \frac{\cos(x - y) - 3x^2y^2}{\cos(x - y) + 2x^3y}$$

Use the chain rule to differentiate $\sin(x - y)$ wrt x:

$\frac{d(\sin(x - y))}{dx}$

$= \cos(x - y) \times \frac{d(x - y)}{dx}$

$= \cos(x - y) \times \left(1 - \frac{dy}{dx}\right)$

C4

Implicit functions and coordinate geometry

You can use the approaches from the Core 1 and Core 2 units with curves expressed by implicit functions.

Refer to C1 and C2 for revision.

EXAMPLE 2

Find the equation of the tangent to the curve $x^3 - y^3e^x + 8 = 0$ at the point (0,2). Find the equation of the normal at this point.

Differentiate with respect to x:

$$3x^2 - \left(y^3 \times e^x + e^x \times 3y^2\frac{dy}{dx}\right) + 0 = 0$$

$$\text{so} \quad \frac{dy}{dx} = \frac{3x^2 - y^3e^x}{3y^2e^x}$$

At the point (0,2), $\dfrac{dy}{dx} = \dfrac{0 - 8\times 1}{3\times 4\times 1} = -\dfrac{2}{3}$

Find the equation of the tangent at (0,2):

$\dfrac{y-2}{x-0} = -\dfrac{2}{3}$ so the equation of the tangent is $3y + 2x = 6$

The gradient of the normal at $(0, 2) = +\dfrac{3}{2}$

Find the equation of the normal at (0,2):

$\dfrac{y-2}{x-0} = \dfrac{3}{2}$ so the equation of the normal is $2y = 3x + 4$

The gradient of a tangent is given by the value of $\frac{dy}{dx}$ at the point of contact with the curve.

$e^0 = 1$

Use $\dfrac{y - y_1}{x - x_1} = m$

The gradient of the normal, $m' = -\dfrac{1}{m}$

EXAMPLE 3

Find the stationary points on the curve $x^2 + y^2 = 12x$

Differentiate with respect to x:

$$2x + 2y\frac{dy}{dx} = 12$$

$$\frac{dy}{dx} = \frac{12 - 2x}{2y} = \frac{6 - x}{y}$$

$$\frac{dy}{dx} = 0 \text{ when } x = 6$$

When $x = 6$, $y^2 = 12 \times 6 - 6^2 = 36$ and $y = \pm 6$

So, there are stationary points at (6,6) and (6,−6).

The gradient is zero at stationary points.

You can determine if a stationary point is a maximum or minimum by

either investigating the change in sign of $\dfrac{dy}{dx}$ near to the stationary point

or finding whether $\dfrac{d^2y}{dx^2}$ is negative or positive at the stationary point.

C4

Exercise 9.1

1 Find $\frac{dy}{dx}$ when

a $x^2 - 3y^2 = 5$
b $y^3 - x^3 = 2$
c $x^2 + y^2 - 3x + 4 = 0$
d $y^3 + 3y^2 = x$
e $xy = 2$
f $x + xy + y = 1$
g $x^2 + xy + y^2 = 1$
h $x^2y^2 + x^2 + y^2 = 2$
i $3x^2 + 2y^2 = x^2y^2$
j $x^3 + 3x^2y + 3xy^2 + y^3 = 1$
k $\frac{2}{x} - \frac{2}{y} = 1$
l $\frac{3}{x^2} + \frac{4}{y^2} = 1$
m $\cos x \sin y = 1$
n $x + y = \tan y$

2 Find $\frac{dy}{dx}$ when

a $y^2 = \ln x$
b $y \ln x = \ln y$
c $\sqrt{x} + \sqrt{y} = 1$
d $\sin(x + y) = \sin x$
e $\cos 3x \sin 2y = 1$
f $e^{x+y} = x$
g $xe^y + ye^x = 1$
h $x = e^x \ln y$

3 Find the gradient of the tangent to each of these curves at the given point.

a $x^2 - y^2 = 7$ $(4,3)$
b $x^2y = 12$ $(2,3)$
c $x^2 + 3xy + y^2 = 1$ $(3,-1)$
d $\cos x = \sin y + 1$ $(0,0)$
e $\cos(x - y) = x - \frac{\pi}{2}$ $\left(\frac{\pi}{2}, 0\right)$
f $xy + e^x y = 1$ $(0,1)$
g $y^2 = 12x$ $(3,6)$
h $\frac{x^2}{9} + \frac{y^2}{16} = 1$ $(3,0)$
i $x^3 + xy^2 + y^3 = 11$ $(2,1)$
j $xe^y = 2$ $(2,0)$

4 Find the equations of the tangent and the normal to these curves at the given point.

a $x(y - 3) = y^2$ $(-4,2)$
b $2\sin x \cos y = 1$ $\left(\frac{\pi}{4}, \frac{\pi}{4}\right)$
c $\ln(xy) = 2y - 1$ $\left(2, \frac{1}{2}\right)$
d $e^{2x} + e^{2y} = x + y + 2$ $(0,0)$

5 Find the equation of the normal to the curve $y^2 = \frac{8}{1 + x^2}$ at the point $(1,2)$.

6 The tangent to the curve $y^2 = x^3$ at the point $(4,8)$ intersects the x-axis at the point P.
The normal to the curve at the same point intersects the x-axis at point Q.
Find the length PQ.

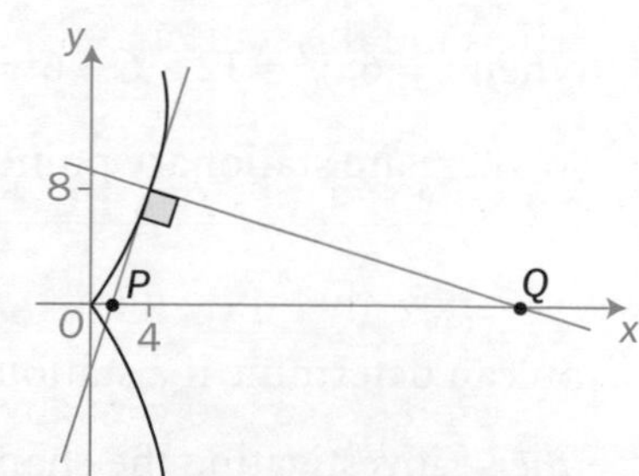

7 a Find the equation of the normal to the curve $y^2 + 3xy - 2x^2 + 1 = 0$ at the point $A(2,1)$.

b The normal intersects the y-axis at the point B.
If O is the origin, find the area of triangle OAB.

C4

8 Tangents to the curve $y^2 = 4x$ at the points (1,2) and (4, 4) meet at the point P. Find the coordinates of P.

9 The gradients of two tangents to the curve $y^2 = 16x$ are 2 and $\frac{1}{3}$. Find the point at which the two tangents intersect.

10 The curve $y(x+y)^2 + 15 = 3x^3$ has a tangent at the point (2, 1). Find the point where the tangent intersects the x-axis. Also find the angle that the tangent makes with the x-axis.

11 Find the two points of intersection of the curves $y^2 = x$ and $x^2 = 8y$
Find the gradients of the curves at both points of intersection.
Hence, find the angles at which the curves intersect.

12 Find the equations of the tangents to the curve $y^2 = 3x + \frac{12}{x^2}$ which are parallel to the x-axis.

13 Find the x-coordinates of the turning points on the curve
$x^3 - y^3 - 2x^2 + 3y + x - 4 = 0$

14 Find the points on each of these curves where the gradient of the curve is zero.

a $x^2 + 3y^2 - 8x - 4y + 17 = 0$ **b** $2x^2 + 2y^2 - 4x + 5y + 4 = 0$

15 Find the turning points on each of these curves and determine whether they are maximum or minimum points.

a $x^2 - y^2 + 10x - 5y + 19 = 0$ **b** $x^2 - 2y^2 + 6x - 3y + 18 = 0$

16 Find the turning points on the curve $x^3 - 3xy^2 - y^3 + 3 = 0$

17 Find the maximum and minimum values of y when

a $3(x-2)^2 + 4(y-1)^2 = 16$ **b** $3(y-1)^2 - 2(x+1)^2 = 12$

18 **a** Find the values of x which give stationary values of y on the curve $x^3 + y^3 = 3xy$

b Find the stationary values and determine whether they are maxima or minima.

INVESTIGATION

19 Show that the equation $xy - x^2 + 4y + x = 0$ can also be written as $y = \frac{x(x-1)}{x+4}$

Find $\frac{dy}{dx}$ in two ways: by using implicit differentiation of the first equation and by using the quotient rule in the second equation.
Show that your two answers are equivalent.

C4

9.2 Differentiating parametric functions

You can find $\frac{dy}{dx}$ from the **parametric equations** $x = f(t)$, $y = g(t)$ by using the chain rule to give

The chain rule gives $\frac{dy}{dt} = \frac{dy}{dx} \times \frac{dx}{dt}$

$$\frac{dy}{dx} = \frac{dy}{dt} \div \frac{dx}{dt} = \frac{g'(t)}{f'(t)}$$

EXAMPLE 1

The curve shown in the diagram has parametric equations $x = 3\cos\theta + \cos 3\theta$ and $y = 3\sin\theta + \sin 3\theta$
Find the gradient of the tangent to the curve at the point P where $\theta = \frac{\pi}{6}$

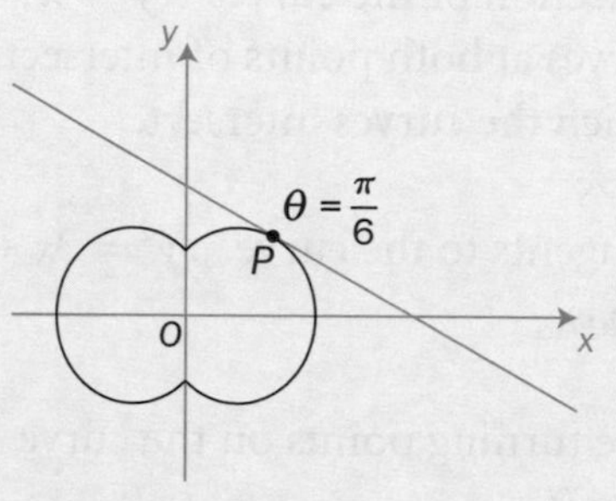

$$\frac{dx}{d\theta} = -3\sin\theta - 3\sin 3\theta \qquad \frac{dy}{d\theta} = 3\cos\theta + 3\cos 3\theta$$

$$\frac{dy}{dx} = \frac{dy}{d\theta} \div \frac{dx}{d\theta} = \frac{3\cos\theta + 3\cos 3\theta}{-(3\sin\theta + 3\sin 3\theta)} = -\frac{\cos\theta + \cos 3\theta}{\sin\theta + \sin 3\theta}$$

When $\theta = \frac{\pi}{6}$, $\frac{dy}{dx} = -\frac{\cos\frac{\pi}{6} + \cos\frac{\pi}{2}}{\sin\frac{\pi}{6} + \sin\frac{\pi}{2}} = -\frac{\frac{\sqrt{3}}{2} + 0}{\frac{1}{2} + 1} = -\frac{\sqrt{3}}{3} = -\frac{1}{\sqrt{3}}$

The gradient of the tangent at P is $-\frac{1}{\sqrt{3}}$

Exercise 9.2

1 A hyperbola has parametric equations $x = \sec\theta$, $y = \tan\theta$
Prove that $\frac{dy}{dx} = \operatorname{cosec}\theta$
Find the equations of the two tangents to the curve which are parallel to the y-axis.

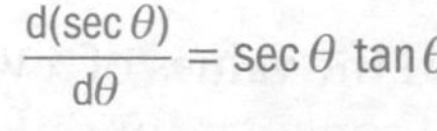

2 The astroid in this diagram has parametric equations $x = 4\cos^3\theta$, $y = 4\sin^3\theta$

Prove that $\frac{dy}{dx} = -\tan\theta$ and find the equations of the four tangents to the astroid which are equally inclined at 45° to both axes. Also find the area of the square that they enclose.

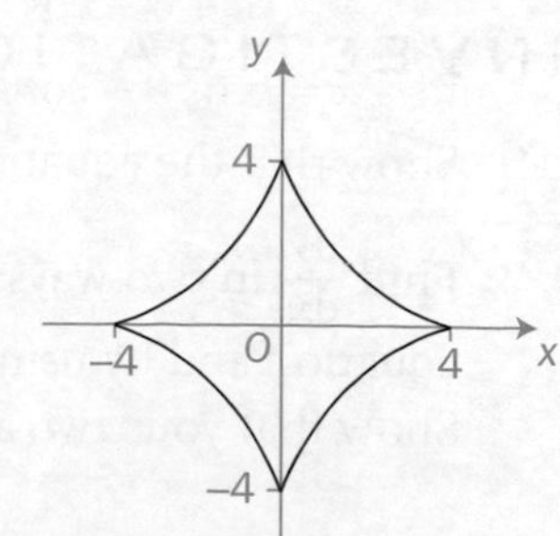

C4

3 The strophoid in this diagram has parametric equations

$$x = \frac{t^2 - 1}{t^2 + 1},\ y = t\left(\frac{t^2 - 1}{t^2 + 1}\right)$$

Find the positions of its stationary points.

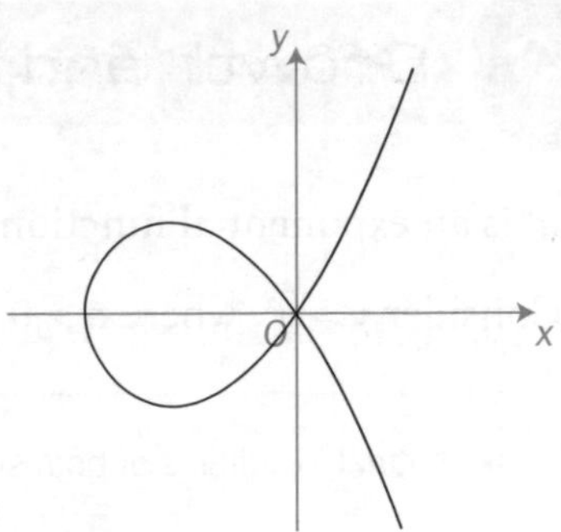

4 The point $P\left(4, 2\frac{2}{5}\right)$ lies on the ellipse with parametric equations
$x = 5\cos\theta$, $y = 4\sin\theta$ and S is the point $(3, 0)$.
The normal to the ellipse at P meets the x-axis at Q.
Prove that the length $QS = \frac{3}{5}PS$

5 A curve has parametric equations

$$x = \frac{t^2}{1+t^2},\quad y = \frac{t^3}{1+t^2}$$

The parameter $t = 1$ at the point P.
Show that the tangent to the curve at P has the equation $y = 2x - \frac{1}{2}$
Find the point Q at which the tangent meets the curve again.

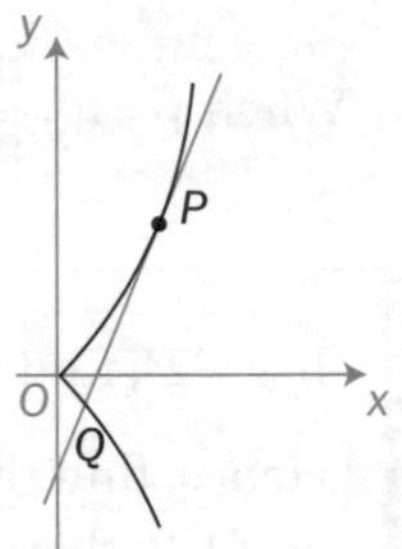

INVESTIGATION

6 When a curve (such as a circle) rolls along a fixed curve, a point on the rolling curve traces out a locus called a *roulette.*

If a moveable circle rolls on the *outside* of a fixed circle, a point on its circumference traces out a special roulette called an *epicycloid.*
If the moveable circle rolls on the *inside* of a fixed circle, the point traces out a *hypocycloid.*

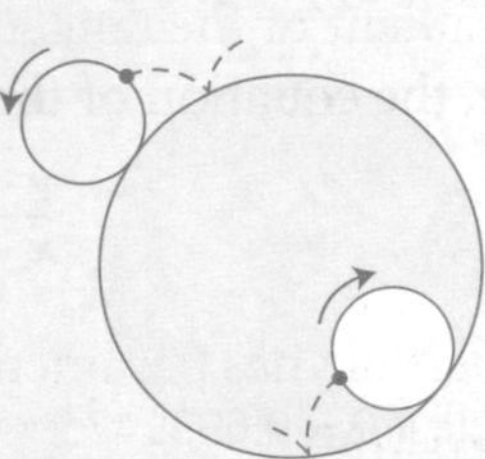

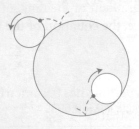

Use a computer's graphical software to investigate these special roulettes for different values of k.

Their parametric equations are:

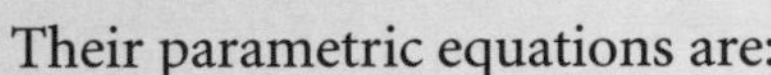

$x = k\cos\theta + \cos k\theta$, $\quad y = k\sin\theta - \sin k\theta$ for hypocycloids

and $x = k\cos\theta - \cos k\theta$, $\quad y = k\sin\theta - \sin k\theta$ for epicycloids.

9.3 Growth and decay

a^x is an exponential function for all values of a.

When $a = e$, the function e^x is called **the** exponential function.

Consider $y = a^x$ where $a > 0$

Take natural logarithms of both sides:

$$\ln y = \ln a^x = x \ln a$$

Differentiate with respect to x:

$$\frac{1}{y}\frac{dy}{dx} = 1 \times \ln a$$

$$\frac{dy}{dx} = \ln a \times y$$

Use the chain rule to differentiate $\ln y$.
$\ln a$ is a constant.

When $y = a^x$, $\frac{dy}{dx} = \ln a \times a^x$

When $a = e$, $\frac{dy}{dx} = \ln e \times e^x = e^x$

C4

EXAMPLE 1

If $y = 2^x$, find the value of $\frac{dy}{dx}$ when $x = 3$

Hence, find the equation of the tangent to the curve at the point P where $x = 3$.

For $y = 2^x$, $\frac{dy}{dx} = \ln 2 \times 2^x$

$= \ln 2 \times 2^3 = 8\ln 2 \quad$ when $x = 3$

When $x = 3$, $y = 2^3 = 8$

Hence, the equation of the tangent at the point $(3, 8)$ is

$$\frac{y - 8}{x - 3} = 8\ln 2$$

$$y = (8\ln 2)x - 24\ln 2 + 8$$

Substitute $\ln 2 = 0.693147...$: $\quad y = 5.55x - 8.64 \quad$ to 2 d.p.

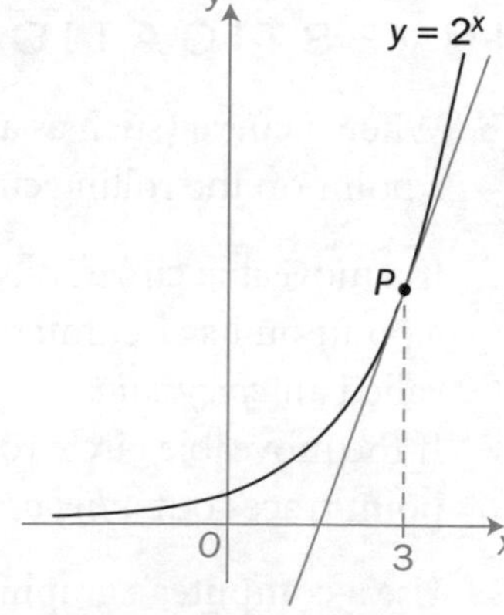

EXAMPLE 2

Find the value of $f'(1)$ for the function $f(x) = 3^{2x+1} + 4 \times 3^x + 1$

Differentiate $f(x) = 3^{2x+1} + 4 \times 3^x + 1$ with respect to x:

$$f'(x) = (\ln 3 \times 3^{2x+1}) \times 2 + 4 \times (\ln 3 \times 3^x) + 0$$
$$= (2\ln 3)3^{2x+1} + (4\ln 3)3^x$$

Use the chain rule to differentiate 3^{2x+1}

So $\quad f'(1) = (2\ln 3) \times 3^3 + (4\ln 3) \times 3$
$$= 54\ln 3 + 12\ln 3$$
$$= 66\ln 3$$

EXAMPLE 3

Cells in a dish grow so that, at a time t (hours), the number of cells, n, is given by $n = 10 \times 8^{\frac{t}{2}}$

Find

a the number of cells when $t = 1$ and $t = 5$

b the average rate of increase over the period $t = 1$ to $t = 5$

c the instantaneous rate of increase when $t = 5$.

---

a When $t = 1$, $n = 10 \times 8^{0.5} = 28$
When $t = 5$, $n = 10 \times 8^{2.5} = 1810$

n is given to the nearest whole number.

b The average rate of increase $= \dfrac{1810 - 28}{5 - 1}$
$= 445.5$ cells per hour

c The instantaneous rate of increase is given by the gradient of the curve at the instant when $t = 5$.

For $n = 10 \times 8^{\frac{t}{2}}$, $\dfrac{dn}{dt} = 10 \times \left(\ln 8 \times 8^{\frac{t}{2}}\right) \times \dfrac{1}{2}$
$= (5 \ln 8) \times 8^{2.5}$ when $t = 5$
$= 1882$

The instantaneous rate of increase is 1882 cells per hour.

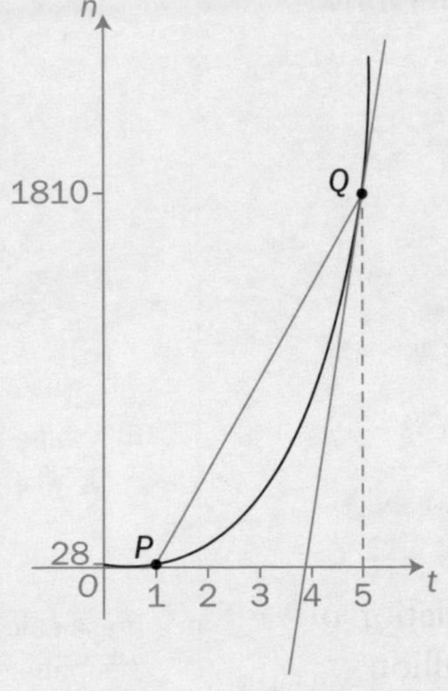

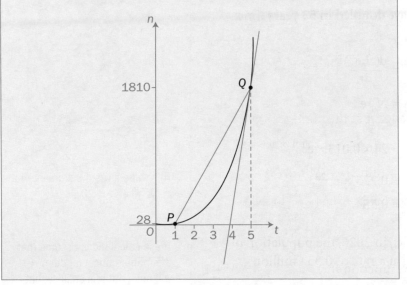

The gradient of the chord PQ gives the average rate of increase.

The gradient of the tangent at the point Q gives the rate of increase at the instant when $t = 5$.

Exponential functions of time of the type a^{kt} and a^{-kt} are used in real life to develop models of **exponential growth and decay.**

When $a = e$, the models use functions involving e^{kt} for exponential growth and e^{-kt} for exponential decay, where $k > 0$

C4

EXAMPLE 4

Sri Lanka had a population N_0 in 2006 of 21 million people.
Its population N in another t years is predicted to be
$N = N_0 e^{rt}$ where $r = 0.013$

Find

a the predicted population of Sri Lanka in 2046

b how long it will take to double the 2006 population

c the rate of growth of the population in 2026.

---

a In 2046, $t = 40$

and the predicted population $N = 21 \times e^{0.013\times40}$

$$= 21 \times e^{0.52}$$

$$= 35.3 \text{ million}$$

b When $N = 2N_0$

$$e^{rt} = 2$$

$$rt = \ln 2$$

$$t = \frac{\ln 2}{r}$$

$$= \frac{0.6931..}{0.013}$$

$$= 53.3$$

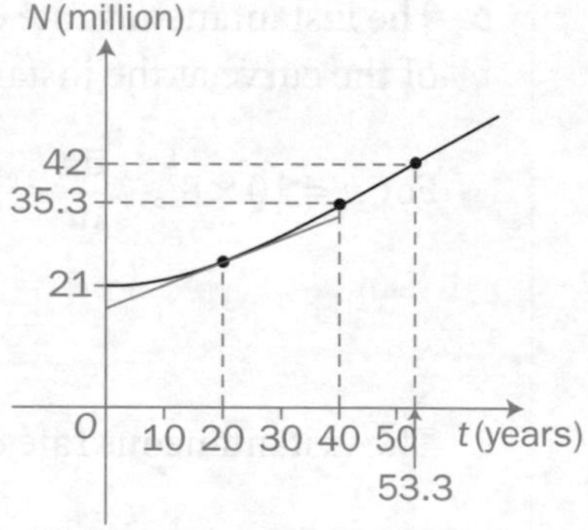

So, the population will have doubled in 53 years time; that is, by 2059.

c The rate of growth $\frac{dN}{dt} = \frac{d(N_0 e^{rt})}{dt}$

$$= N_0 r e^{rt}$$

In 2026, $t = 20$ and $\frac{dN}{dt} = 21 \times 0.013 \times e^{0.013\times20}$

$$= 0.273 \times e^{0.26}$$

$$= 0.354$$

This value gives the gradient of the curve when $t = 20$.

So, according to this model, in 2026, the population of Sri Lanka will be growing at a rate of 0.354 million = 354 000 people per year.

These calculations assume that the mathematical model $N = N_0 e^{rt}$ will hold, but it is unlikely that r will stay constant over many years.

C4

EXAMPLE 5

Radioactive fallout in the atmosphere contains the isotope strontium-90.
The mass m grams after a time t years is given by $m = m_0 e^{-kt}$ where $k = 0.024$ and m_0 is the mass at time $t = 0$.
Find

a the mass after 10 years if $m_0 = 5$

b the initial rate of decay and the rate of decay after 10 years if $m_0 = 5$

c the time taken for m to reduce to a value of $\frac{1}{2}m_0$

---

a When $t = 10$ and $m_0 = 5$, $m = 5 \times e^{-0.024 \times 10}$

$$= 5 \times 0.786\ldots$$

$$= 3.93 \text{ grams (to 3 s.f.)}$$

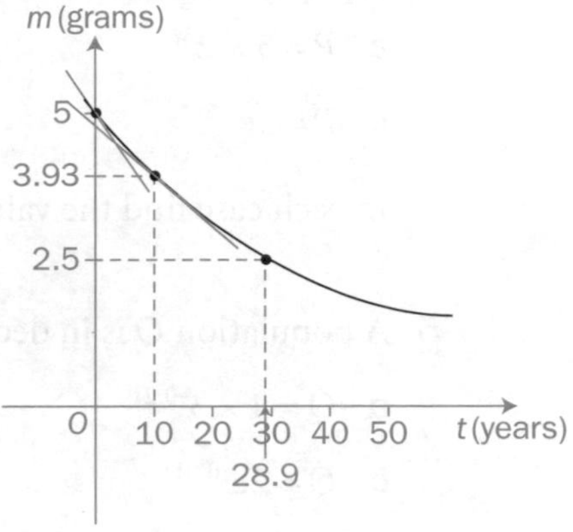

b The rate of decay $= \frac{dm}{dt} = \frac{d(m_0 e^{-kt})}{dt}$

$$= m_0 \times (-k) \times e^{-kt}$$

$$= -km_0 e^{-kt}$$

When $t = 0$, $\frac{dm}{dt} = -0.024 \times 5 \times e^0$

$$= -0.12$$

The initial rate of decay is 0.12 grams per year.

This value of –0.12 is the gradient of the curve when $t = 0$.
In the final answer, the negative sign is implied by the word 'decay'.

When $t = 10$, $\frac{dm}{dt} = -0.024 \times 5 \times e^{-0.024 \times 10}$

$$= -0.094$$

The rate of decay after 10 years is 0.094 grams per year.

c When $m = \frac{1}{2}m_0$ $\quad \frac{1}{2} = e^{-0.024 \times t}$

Take the reciprocal of both sides.

$$2 = e^{0.024 \times t}$$

$$\ln 2 = 0.024 \times t$$

$$t = \frac{0.6931\ldots}{0.024} = 28.9$$

It takes almost 29 years for half the initial mass to decay.

This time of 29 years is called the *half-life* of strontium-90.

Exercise 9.3

1 Find $\frac{dy}{dx}$ when

a $y = 3^x$ **b** $y = 3^{2x-1}$ **c** $y = 4 \times 3^{5x+2}$

d $y = \frac{1}{3^{2x}}$ **e** $y = 10^{2x+5}$ **f** $y = 3 \times 5^{1-2x}$

C4

2 Find $f'(x)$ and $f'(2)$ when $f(x) = 2^{\frac{1}{2}x+2} + 3(2^x) - 4$

3 Find the equation of the tangent and the equation of the normal to the curve

a $y = 2 \times 3^x$ when $x = 1$

b $y = 3 \times 2^{x+1}$ when $x = 0$.

4 A population P grows over a time t according to

a $P = 5 \times 2^{3t}$

b $P = 2^{0.4t+2}$

c $P = 3e^{1.5t}$

d $P = e^{0.4t-1}$

In each case find the value of P and $\frac{dP}{dt}$ when $t = 1$.

5 A population Q is in decline over time according to

a $Q = 4 \times 3^{-0.2t}$

b $Q = 3^{2-0.4t}$

c $Q = 2e^{-0.01t}$

d $Q = e^{6-2t}$

In each case, find the value of Q and $\frac{dQ}{dt}$ when $t = 1$.

C4

6 A number of cells, n, grows over a time t such that $n = n_0e^{0.2t}$ where $n_0 = 5$.
Find

a the number of cells when $t = 0$ and when $t = 10$

b the average rate of growth over the period $t = 0$ to $t = 10$

c the instantaneous rate of growth when $t = 5$.

7 The population, P, of Manchester was 126 000 in 1821 and 236 000 in 1841.

a If t is the number of years after 1821, model the population as $P = P_0e^{kt}$ and find the constants P_0 and k.

b Find $\frac{dP}{dt}$ and evaluate the rate of growth in 1831.

c Estimate the population in 1851. Find the percentage error in this estimate compared to the actual population in 1851 of 303 000.

8 The number, N, of cells infected with a virus changes over t hours according to $N = 200 - 50e^{-2t}$

a Find the value of N and $\frac{dN}{dt}$ when $t = 0$.

b Find how many infected cells there are after 4 hours and the rate of change in the number of infected cells at this time.

9 A mass m grams of a substance decays exponentially over a time t hours where $m = m_0e^{-kt}$

a If $m = 20$ when $t = 0$, find m_0.

If $m = 15$ when $t = 5$, find k.

b Find the time taken for mass of the substance to decay to

i half its original mass ii 10% of its original mass.

c Find the rate of decay when $t = 0$ and when $t = 5$.

10 A hot liquid cools such that the difference, θ, between its temperature and that of its surroundings at a time t minutes is given by $\theta = ke^{-\alpha t}$

a If the liquid's temperature is 70 °C after 1 minute, find the values of k and α. Room temperature is constant at 10 °C and the liquid's initial temperature is 80 °C.

b Calculate the initial rate of cooling and the rate of cooling after 5 minutes.

c Write an expression for the temperature T of the liquid at time t.

d Calculate the time taken for the temperature of the liquid to drop to 40 °C.

C4

INVESTIGATION

11 The population of London was 1 950 000 in 1841 and 2 800 000 in 1861.
Model the population in two ways as:

- $P = P_0a^t$
- $P = P_0e^{kt}$

where P_0, a and k are constants.

The actual population in 1871 was 3 250 000.
Do the two models give the same estimates?
Explain your answer.

9.4 Rates of change

The rate at which water is flowing out of a tap affects the rate at which the depth of water in the bath increases.

This is an example of a **rate of change** in which a change in one variable over a given time produces a change in the other variable over that time.

If V is the volume of water running into the bath, then the rate of change of V is $\frac{dV}{dt}$.

Similarly, if h is the depth of water in the bath, then the rate of change of h is $\frac{dh}{dt}$.

These two rates of change are linked by the chain rule

$$\frac{dV}{dt} = \frac{dV}{dh} \times \frac{dh}{dt}$$

The differential $\frac{dV}{dh}$ is independent of time t and can be derived from some physical or geometrical relationship between V and h.

C4

EXAMPLE 1

A cylindrical water tank of radius 2 m holds water which is being pumped in from the top of the tank at a rate of 3 m³ per hour.

Find the rate at which the depth h of water is increasing.

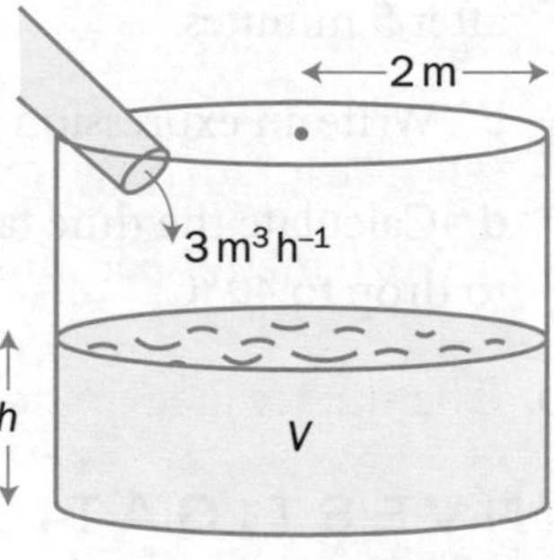

You have to find $\frac{dh}{dt}$

If the volume of water in the tank is V, then $\frac{dV}{dt} = 3\,\text{m}^3\,\text{h}^{-1}$

The chain rule gives $\frac{dV}{dt} = \frac{dV}{dh} \times \frac{dh}{dt}$

From the geometry of the cylinder, $V = \pi r^2 h$

$= 4\pi h \quad \text{for } r = 2$

$$\frac{dV}{dh} = 4\pi$$

Substitute into the chain rule:

$$3 = 4\pi \times \frac{dh}{dt}$$

The rate of change of the depth of water, $\frac{dh}{dt} = \frac{3}{4\pi}$

$= 0.24\,\text{m}\,\text{h}^{-1}$

Make sure you use the correct units in your answer.

EXAMPLE 2

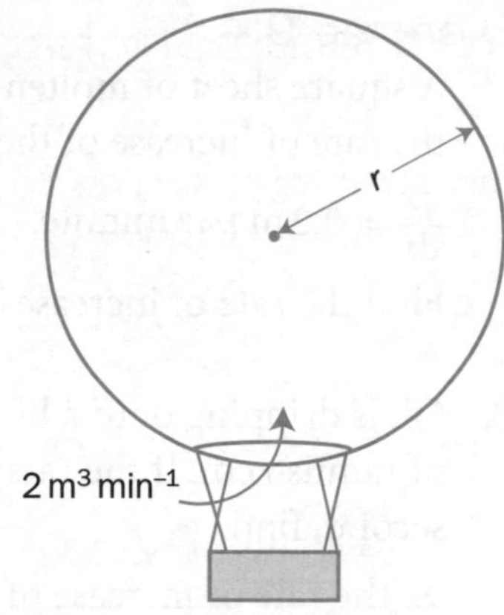

A hot-air balloon is being blown up at a rate of $2\,\text{m}^3$ per minute.
Assuming the balloon is spherical, find

a the rate of increase in its radius r when $r = 2.5$ metres

b the rate of increase in its surface area A when $r = 2.5$ metres.

a If the volume of the balloon is $V\,\text{m}^3$, then $\frac{dV}{dt} = 2\,\text{m}^3\,\text{min}^{-1}$

To find $\frac{dr}{dt}$, use the chain rule $\frac{dV}{dt} = \frac{dV}{dr} \times \frac{dr}{dt}$

To find $\frac{dV}{dr}$, use the geometry of the sphere.

$$V = \frac{4}{3}\pi r^3$$

$$\frac{dV}{dr} = \frac{4}{3}\pi \times 3r^2 = 4\pi r^2$$

Substitute into the chain rule with $r = 2.5$:

$$\frac{dV}{dt} = \frac{dV}{dr} \times \frac{dr}{dt}$$

$$2 = 4\pi \times 2.5^2 \times \frac{dr}{dt}$$

The rate of increase in the radius, $\frac{dr}{dt} = \frac{2}{25\pi}$
$= 0.025$ metres per minute to 2 s.f.

b The surface area of the balloon $A = 4\pi r^2$

$$\frac{dA}{dr} = 8\pi r$$
$$= 8\pi \times 2.5 = 20\pi$$
when $r = 2.5$

The chain rule gives $\frac{dA}{dt} = \frac{dA}{dr} \times \frac{dr}{dt}$ where, from part **a**,
$\frac{dr}{dt} = \frac{2}{25\pi}$

Substitution gives $\frac{dA}{dt} = 20\pi \times \frac{2}{25\pi} = 1.6$

When $r = 2.5\,\text{m}$, the rate of increase in the surface area is $1.6\,\text{m}^2$ per minute.

C4

Exercise 9.4

1 A square sheet of molten glass is increasing in size such that the rate of increase of the length, x, of its sides is given by
$\frac{dx}{dt} = 0.2$ m per minute.
Find the rate of increase of its area, A, when $x = 0.5$ metres.

2 Oil is dripping onto a horizontal floor forming a circular pool of radius r cm. If the rate of increase of the radius is 0.5 cm per second, find

a the rate of increase of the area of the pool when $r = 15$ cm

b the rate of increase of the circumference of the pool.

3 A crystal forms in the shape of cube of edge length x mm. If the rate of increase of its edges is 0.3 mm per minute, find

a the rate of increase of its volume when $x = 6$ mm

b the rate of increase of its surface area when $x = 6$ mm.

4 An ice cube melts uniformly on all its faces. When its edges are 2 cm long, the rate of decrease of its surface area is 4 cm^2 per hour. Find the rate of decrease in its volume at this instant.

5 A spherical bubble under water is rising to the surface and its radius r is increasing at a rate of 0.2 mm per second.

a When its radius is 4 mm, find the rate of increase in its volume.

b Find the rate of increase in its surface area when its volume is 36π mm^3.

6 Molten plastic is extruded from a nozzle at a speed of 20 cm s^{-1} and forms a cylindrical shape. The nozzle has a circular cross-section of area 0.75 cm^2.
Find the rate of change in the volume of the extruded plastic.

7 Air is leaking from a spherical balloon at a rate of 2 cm^3 per second. When its radius is 12 cm, find the rate of decrease of

a its radius

b its surface area.

8 A hollow cone, with a semi-vertical angle of 60°, is held vertex downwards with its axis vertical. Water drips into the cone at a constant rate of 4 cm^3 per minute. Find the rate at which the depth of water is increasing when the water is 4 cm deep.

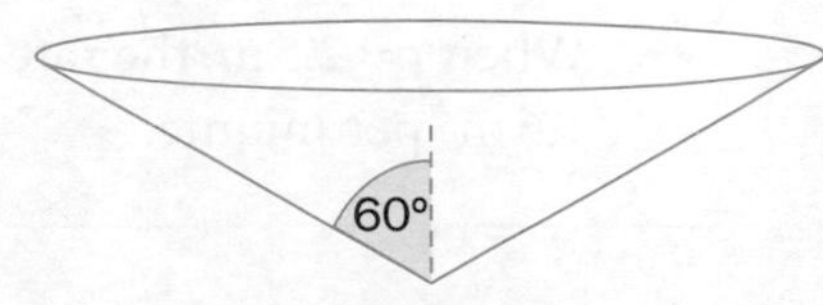

C4

9 A spillage of coffee on to a horizontal table forms a circular stain with a radius increasing at a rate of 2 mm s^{-1}. Find the rate at which the area of the stain is increasing after 5 seconds.

10 Sand falls onto horizontal ground at a constant rate of 20 cm^3 s^{-1} to form a pile in the shape of a circular-based cone with a semi-vertical angle of 45°. Calculate the rate at which the vertical height of the conical pile is increasing after 5 seconds.

You will need to use the formulae for volume and height of a cone.

11 A boy 1.5 m tall runs directly away from a light which is fixed 2 m above a horizontal road. If he runs at a speed of 3 m s^{-1}, find the rate at which his shadow is lengthening.

12 Robert Boyle (1627–1691) discovered that, for a fixed amount of gas at a constant temperature, its pressure varies inversely as its volume varies. A quantity of gas has an initial volume of 0.25 m^3 and an initial pressure of 2 N m^{-2}. Its volume is allowed to increase at a constant rate of 0.05 m^3 s^{-1}. Find the rate at which its pressure is changing at the instant when its volume is double its initial value.

C4

INVESTIGATION

13 As a piston moves into the cylinder of an engine, the length L of the cylindrical space changes. This change in length causes a change in the volume V of the gas in the cylinder, which causes the pressure P of the gas to change.

Find an expression for the rate of change of pressure in terms of radius r, length L and time t.

Have you made any assumptions?

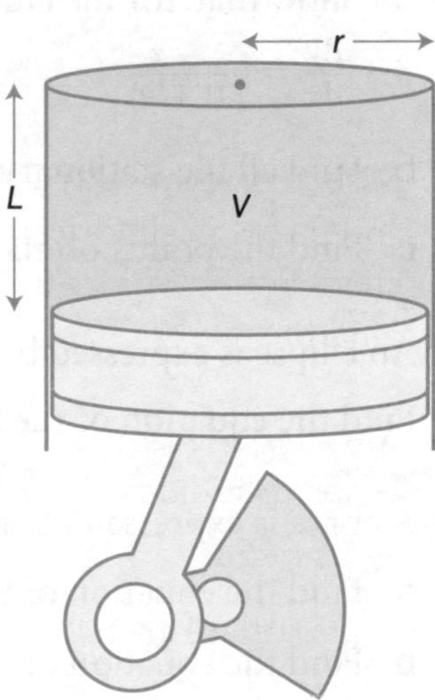

1 Find $\frac{dy}{dx}$ when

a $3x^2 + 4y^2 = 9$ b $x^2 + 3xy + y^2 = 2$ c $\frac{2}{x^2} - \frac{3}{y^2} = 4$

d $\cos 2x \sin 3y = 1$ e $y^3 = \ln x$ f $x^2e^y - y^2e^x = 2$

2 Find $\frac{dy}{dx}$ for the curve $x^3 - 2x^2y - 3xy^2 + y^3 = 9$

Find the gradient of the tangent to this curve at the point (2, –1).

3 Find the equations to the tangent to these curves at the given points.

a $y^3 = 2x^2$ (2,2) b $x^3 - 2x^2y - y^3 = 8$ (3,1) c $ye^{2x} = 3$ (0,3)

4 The normal to the curve $x^2 + 3xy + y^2 = 11$ at the point (2,1) meets the axes at the points P and Q. Given that O is the origin, show that the area of triangle OPQ is $\frac{81}{112}$ square units.

5 A curve has equation $x^3 - 2xy - 4x + y^3 - 51 = 0$
Find the equation of the normal to the curve at the point (4,3).
Give your answer in the form $ax + by + c = 0$, where a, b and c are integers.

6 a Show that, for the curve $x^2 + 2y^2 - 4x + 4y = 26$

$$\frac{dy}{dx} = \frac{2 - x}{2(1 + y)}$$

b Find all the stationary points on the curve.

c Find the points on the curve where the tangents are parallel to the y-axis.

7 An ellipse is expressed by the parametric equations $x = 4\sin\theta$, $y = 3\cos\theta$
Find the equation of the tangent to the ellipse at the point where $\theta = \frac{\pi}{6}$

8 A curve is expressed parametrically by the equations $x = t^2 + 1$, $y = t^3$

a Find the equation of the tangent at the point where $t = 2$

b Find the equation of the normal at the point (2,–1).

9 The curve C has parametric equation $x = a\sec t$, $y = b\tan t$, $0 < t < \frac{\pi}{2}$, where a and b are positive constants.

Prove that $\frac{dy}{dx} = \frac{b}{a}\operatorname{cosec} t$

Find the equation in the form $y = px + q$ of the tangent to C at the point where $t = \frac{\pi}{4}$

10 The parametric equations of a curve are

$x = 2\cos\theta + \cos 2\theta$ and $y = 2\sin\theta + \sin 2\theta$

Show that stationary values occur on this curve when $\cos\theta = \frac{1}{2}$
Find two stationary points for $0 \leqslant \theta \leqslant 2\pi$.

11 A curve is given by the parametric equations $x = t^2$, $y = \frac{1}{t}$

a Find the equation of the tangent at the point $A\left(9, \frac{1}{3}\right)$.

b The tangent at A intersects the curve at point B.
Find the value of the parameter t at B.

12 The normal to the curve with parametric equations $x = \frac{1}{t-1}$, $y = t^2$ at the point $P(1, 4)$ meets the curve again at points Q and R.

a Show that the gradient of the normal is $\frac{1}{4}$ and find its equation.

b Prove that the values of t at points Q and R are given by $t = -\frac{1}{2} \pm \sqrt{2}$

13 The curve C is given parametrically by $x = t - \frac{1}{t}$, $y = t + \frac{1}{t}$, $t \neq 0$

a Find the coordinates of the points on C at which the gradient is zero.

b Find the equation of the normal to C at the point $(0, 2)$.

14 a Find $\frac{dy}{dx}$ when **i** $y = 4^x + 4$ **ii** $y = 4^{2x} + 2 \times 4^x + 1$

b Find the equation of the tangent to the graphs of both these equations at the points where $x = 0$

15 Bacteria grow so that, at a time t, the number n of bacteria is given by $n = 5 \times 4^{\frac{t}{3}}$

Find

a the number of bacteria when $t = 2$ and $t = 5$

b the average rate of increase in the number of bacteria over the period from $t = 2$ to $t = 5$

c the instantaneous increase in the number of bacteria when $t = 2$

d the value of t at which the value of n is double its initial value.

16 The value £V of a car t years after the 1st January 2001 is given by the formula $V = 10\,000 \times (1.5)^{-t}$

a Find the value of the car on 1st January 2005

b Find the value of $\frac{dV}{dt}$ when $t = 4$

c Explain what the answer to part **b** represents. [(c) Edexcel Limited 2005]

9 Exit

Summary

Refer to

- You use implicit differentiation when the relation between x and y cannot be expressed explicitly by $y = f(x)$

  The chain rule gives $\dfrac{d(y^n)}{dx} = ny^{n-1}\dfrac{dy}{dx}$

  The product rule gives $\dfrac{d(xy)}{dx} = x\dfrac{dy}{dx} + y\dfrac{dx}{dx} = x\dfrac{dy}{dx} + y$ 9.1

- For parametric equations $x = f(t), y = g(t)$,

  the chain rule gives $\dfrac{dy}{dt} = \dfrac{dy}{dx} \times \dfrac{dx}{dt}$ 9.2

- If $y = a^x$, then $\dfrac{dy}{dx} = \ln a \times a^x$ 9.3

  When $a = e$, $y = e^x$ gives $\dfrac{dy}{dx} = \ln e \times e^x = e^x$

- You can model
  - exponential growth by $y = Ae^{kt}$
  - exponential decay by $y = Ae^{-kt}$

  where $k > 0$ 9.3

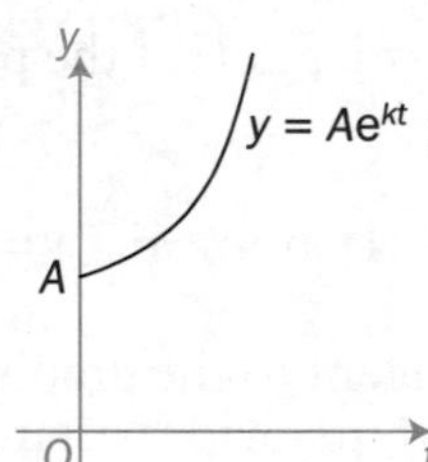

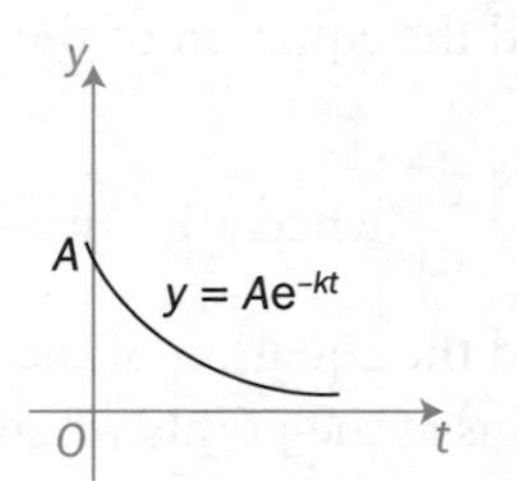

- When changes in x effect changes in y over a period of time, their rates of change are connected by the chain rule where

  $\dfrac{dy}{dt} = \dfrac{dy}{dx} \times \dfrac{dx}{dt}$ 9.4

Links

Differentiation is a versatile tool that can be used in many fields, particularly within industry.

Derivatives can be used to express the rate of decay of a radioactive substance in a chemical power plant.

Engineers can use differentiation to calculate rates of change of variables when designing systems to ensure efficiency.

Managers can solve maximum and minimum problems to determine how to maximize profit or minimize waste.

C4

Revision 3

1 Express these as partial fractions.

a $\dfrac{x+4}{(x-2)(x+1)}$ b $\dfrac{x+3}{x(x^2-1)}$ c $\dfrac{2x}{x^2-5x+6}$

d $\dfrac{x^2+2}{x(x+1)^2}$ e $\dfrac{2}{x^2(2x-1)}$ f $\dfrac{x+2}{(1-2x)(x-3)^2}$

2 Show that $\dfrac{x^3+3}{x(x^2-1)}$ can be expressed as $A+\dfrac{B}{x}+\dfrac{C}{x-1}+\dfrac{D}{x+1}$
Find the values of A, B, C and D.

3 Express these as partial fractions.

a $\dfrac{x^2+2}{x^2-1}$ b $\dfrac{x^2-2}{(x-1)(x+2)}$ c $\dfrac{x^3-2x^2+2}{x(x-1)^2}$

d $\dfrac{x^3+1}{x(x-2)}$ e $\dfrac{4x^3+1}{x^2(2x-1)}$ f $\dfrac{2x^3-1}{(x+1)^2(2x-1)}$

4 If $\dfrac{x^4+2x^2+3x+1}{x^2-2}\equiv px^2+qx+r+\dfrac{sx+t}{x^2-2}$,
find the values of the constants p, q, r, s and t.

5 a Point P lies on the curve with parametric equations $x=t^2-4$, $y=t+1$
If the y-coordinate of P is 6, find its x-coordinate.

b The point $(4, k)$ lies on the curve with parametric equations $x=\dfrac{1}{t^2}-5$, $y=t-1$
Find the possible values of k.

6 The variable point $P(\lambda\sqrt{t}, 3t-1)$ meets the line $y=11$ at the point $(6, 11)$.

a Find the value of λ.

b Find the Cartesian equation of the curve along which P moves in the form $y=\mathrm{f}(x)$

7 Find the points where the curve expressed parametrically
by the equations $x=\dfrac{1+t}{1-t}$, $y=2t+1$ intersects

a the x-axis b the y-axis c the line $y=x-1$

8 A curve is defined parametrically by $x=2t+1$, $y=t^2-4t+1$
A normal is drawn to the curve at the point where $t=4$.

Find

a the equation of the normal

b the point at which the normal intersects the curve a second time.

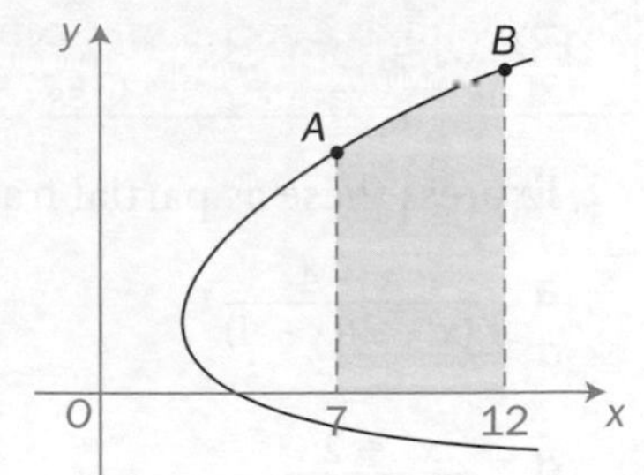

9 a A curve has parametric equations $x = t^2 + 3$, $y = 1 + t$
Find the two points A and B on the curve at which $x = 7$ and $x = 12$.

b Find the area between the curve, the x-axis and the ordinates $x = 7$ and $x = 12$.

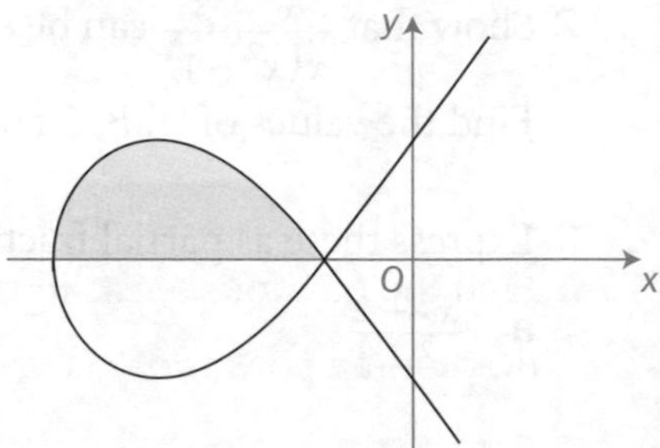

10 The diagram shows the graph of the curve with parametric equations $x = t^2 - 12$, $y = t^3 - 9t$

a Find the values of t at the points where the curve intersects the x-axis.

b Find the shaded area on the diagram.
Hence, find the total area of the loop.

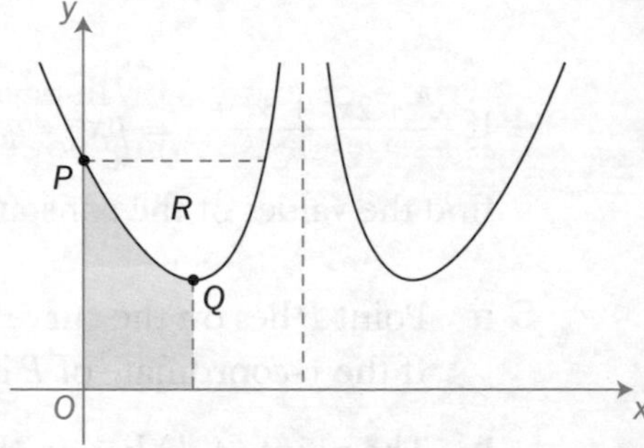

11 The curve expressed parametrically by

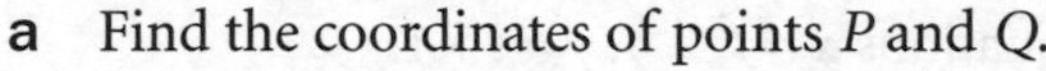

$$x = 4 - 2t,\ y = t^2 + \frac{1}{t^2}$$

is shown on this diagram. The curve cuts the y-axis at the point P and has a minimum value at the point Q.

a Find the coordinates of points P and Q.

b Find the shaded area on the diagram.

c Find the area of the region labelled R.

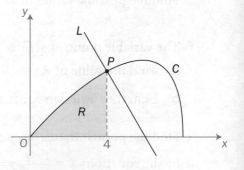

12 This diagram shows the curve C with parametric equations

$$x = 8\cos t,\quad y = 4\sin 2t,\quad 0 \leqslant t \leqslant \frac{\pi}{2}$$

The point P lies on C and has coordinates $(4, 2\sqrt{3})$.

a Find the value of t at the point P.

The line L is a normal to C at P.

b Show that an equation for L is $y = -x\sqrt{3} + 6\sqrt{3}$

The finite region R is enclosed by the curve C, the x-axis and the line $x = 4$, as shown shaded in the diagram.

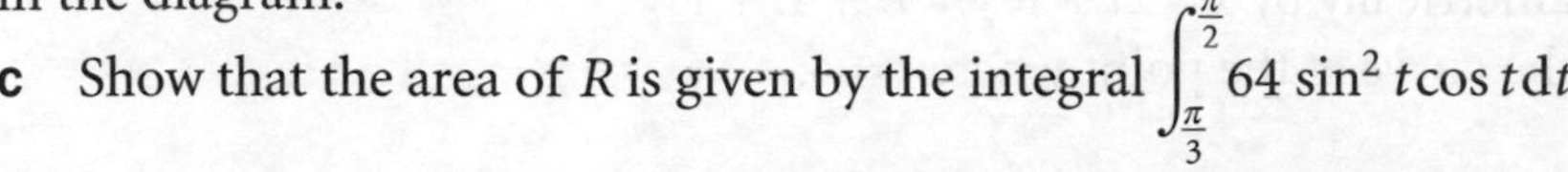

c Show that the area of R is given by the integral $\int_{\frac{\pi}{3}}^{\frac{\pi}{2}} 64\sin^2 t\cos t\,\mathrm{d}t$

d Use this integral to find the area of R, giving your answer in the form $a + b\sqrt{3}$, where a and b are constants to be determined.

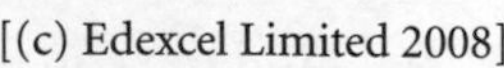

[(c) Edexcel Limited 2008]

13 Expand as a series of ascending powers of x up to and including x^3
State the range of values of x for which each expansion is valid.

a $\dfrac{1}{1+3x}$ **b** $\sqrt{1+5x}$ **c** $\sqrt{9+2x}$

d $\dfrac{x}{\sqrt{1-\frac{1}{4}x}}$ **e** $\dfrac{1}{4+3x}$ **f** $\dfrac{1}{\sqrt{4-x}}$

14 Find the first three terms in the binominal expansion of each expression.
Give the values of x for which each expansion is valid.

a $(1-2x)\sqrt{1+x}$ **b** $\dfrac{1+x}{\sqrt{1+3x}}$

15 Find the first four terms in the binomial expansion of $\sqrt{1-\dfrac{3}{x}}$ in descending powers of x. For what values of x is the expansion valid?

16 The coefficient of x^2 in the expansion of $\sqrt{4+ax}$ is -1. Find two possible values of a and the first three terms of each possible expansion.

17 Write each expression in partial fractions and so expand it as a series of ascending powers of x as far as x^3.
Find the range of values of x for which each expansion is valid.

a $\dfrac{2-x}{(1-2x)(1+x)}$ **b** $\dfrac{3-2x}{(2+x)(1-3x)}$

18 $f(x) = \dfrac{3x-1}{(1-2x)^2}$, $|x| < \dfrac{1}{2}$

Given that, for $+x \neq \dfrac{1}{2}$, $\dfrac{3x-1}{(1-2x)^2} \equiv \dfrac{A}{(1-2x)} + \dfrac{B}{(1-2x)^2}$
where A and B are constants,

a find the values of A and B.

b Hence, or otherwise, find the series expansion of $f(x)$, in ascending powers of x, up to and including the term in x^3, simplifying each term.

19 Find $\dfrac{dy}{dx}$ when

a $x^2 - 5xy + y^2 = 1$ **b** $xe^y = y^3 - 1$ **c** $xy = 1 + x\sin y$

d $\sin x \cos y = 1$ **e** $\sqrt{y} - \sqrt{x} = 2$ **f** $x^2y = \tan(x+y)$

20 **a** Find the equation of the tangent to the curve $e^{xy} + 1 = x$ at $(2,0)$.

b Find the equation of the normal to the curve $\ln(xy) = y^2 - 1$ at $(1,1)$.

C4

21 A curve is expressed parametrically by $x = \frac{t^2+1}{t}$, $y = t^2 - 1$ $(t \neq 0)$
Show that there are tangents to the curve which are parallel to the *y*-axis and that they meet the curve at the points (2, 0) and (–2, 0).

22 **a** Find $\frac{dy}{dx}$ for an ellipse with parametric equations $x = 3\sin\theta$, $y = 2\cos\theta$, $0 \leqslant \theta \leqslant \pi$

b Prove that the equation of the tangent to the ellipse at the point *P* where $\theta = \alpha$ is given by $3y\cos\alpha + 2x\sin\alpha = 6$

c The tangent at *P* intersects the coordinate axes at the points *A* and *B*. Find, in terms of α, the area of triangle *OAB*, where *O* is the origin.

d Find the value of α which gives the smallest possible value of the area of triangle *OAB*. State this area.

23 The tangent and normal to the curve $y = 5 \times 2^x$, at the point where $x = 1$, intersect the *y*-axis at the points *P* and *Q* respectively. Find the distance *PQ* as a decimal, correct to 1 decimal place.

24 The population N_0 of a town in 2008 is 56 000. A model of its growth predicts that its population *N* in *t* years after 2008 will be $N = N_0e^{0.008t}$

Find

a its population in 2018

b the rate of growth of its population in 2018

c how long it takes for its population to be double the 2008 figure.

25 A spherical balloon is being inflated at a constant rate of 0.2 m^3 per minute.

Find

a the rate of increase in its radius *r* when $r = 0.5$ metres

b the rate of increase in its surface area when $r = 0.5$ metres.

26 This diagram shows a right circular cylindrical metal rod which is expanding as it is heated. After *t* seconds, the radius of the rod is *x* cm and the length of the rod is 5*x* cm.

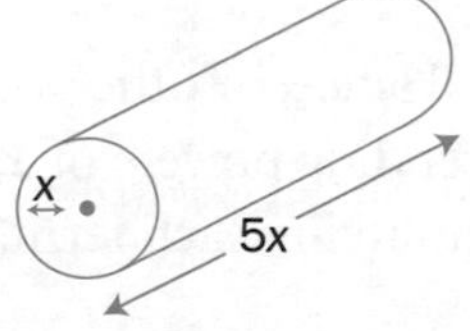

The cross-sectional area of the rod is increasing at the constant rate of 0.032 cm^2 s^{-1}

a Find $\frac{dx}{dt}$ when the radius of the rod is 2 cm, giving your answer to 3 significant figures.

b Find the rate of increase of the volume of the rod when $x = 2$.

C4

10 Integration

This chapter will show you how to

- find the area under a curve to a specified accuracy using a numerical method
- find the exact area under a curve by integration
- integrate a variety of functions by using standard integral forms, substitution, trigonometric identities and partial fractions, and integration by parts
- be systematic in your approach to integration
- use integration to find volumes of revolution.

Before you start

You should know how to:

1 Differentiate various functions using the product, quotient and chain rules.

e.g. Differentiate $y = x^2 \sin 3x$

$$\frac{dy}{dx} = 2x\sin 3x + (3\cos 3x)x^2$$
$$= 2x\sin 3x + 3x^2\cos 3x$$

2 Manipulate trigonometric identities.

e.g. Prove that $\frac{\sin 2A}{1-\cos 2A} \equiv \cot A$

Use the double-angle formulae:

$$\frac{\sin 2A}{1-\cos 2A} \equiv \frac{2\sin A\cos A}{1-(1-2\sin^2 A)}$$
$$\equiv \frac{2\sin A\cos A}{2\sin^2 A} \equiv \frac{\cos A}{\sin A} \equiv \cot A$$

3 Find partial fractions.

e.g. Express $\frac{2}{x(x-2)}$ in partial fractions.

Let $\frac{2}{x(x-2)} \equiv \frac{A}{x} + \frac{B}{x-2}$

$$\equiv \frac{A(x-2)+Bx}{x(x-2)}$$

Hence $2 \equiv A(x-2) + Bx$

Letting $x = 0$ gives $A = -1$

Equate coefficients of x: $A + B = 0$ so $B = 1$

So $\frac{2}{x(x-2)} \equiv \frac{1}{x-2} - \frac{1}{x}$

Check in:

1 Differentiate

a $x^2 \ln x$

b $x^3 e^x$

c $e^x \tan x$

d $x\sqrt{x^2+1}$

e $\frac{\ln x}{x}$

f $\ln(\sin x)$

2 Prove these identities.

a $\tan A + \cot A \equiv 2\operatorname{cosec} 2A$

b $\frac{2\tan A}{1+\tan^2 A} \equiv \sin 2A$

c $\cos 3A \equiv 4\cos^3 A - 3\cos A$

3 Express these in partial fractions.

a $\frac{x}{x^2-25}$

b $\frac{6}{x(x-1)(x+2)}$

c $\frac{1}{x(x-1)^2}$

d $\frac{x^2+3}{x(x-2)}$

10.1 The trapezium rule

When the area enclosed by a graph, the x-axis and two **ordinates** $x = a$ and $x = b$ is split into n equal intervals of width h, then the **trapezium rule** gives

For n equal intervals, use $(n + 1)$ x-values. x-values are sometimes called 'ordinates'.

$$\int_a^b f(x)\,dx \approx \frac{1}{2}h[y_0 + y_n + 2(y_1 + y_2 + \cdots + y_{n-1})] \quad \text{where } h = \frac{b-a}{n}$$

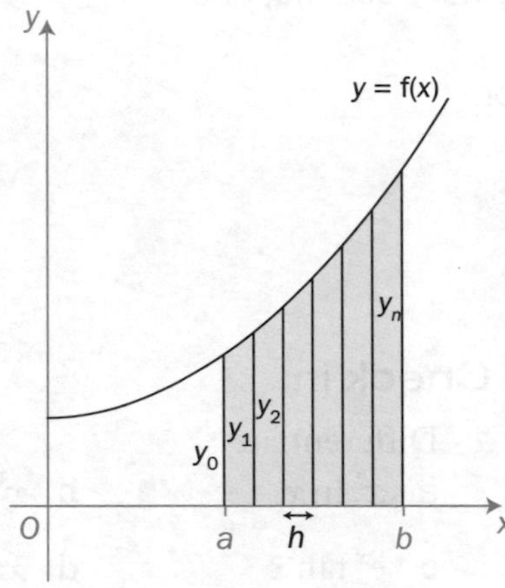

Using more strips involves more calculation but gives a more accurate approximation.

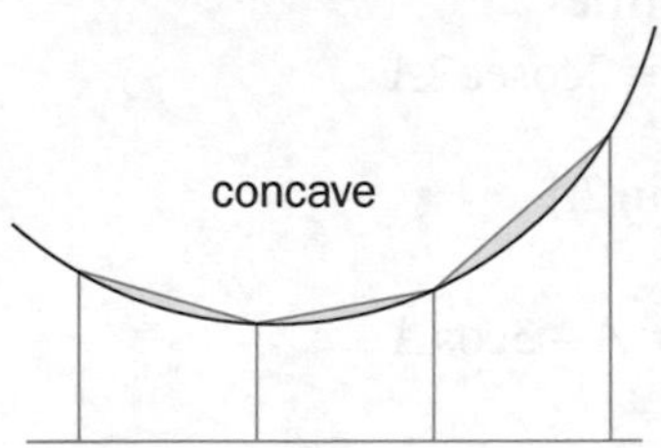

If the graph is a concave curve, the trapezium rule overestimates the actual area.

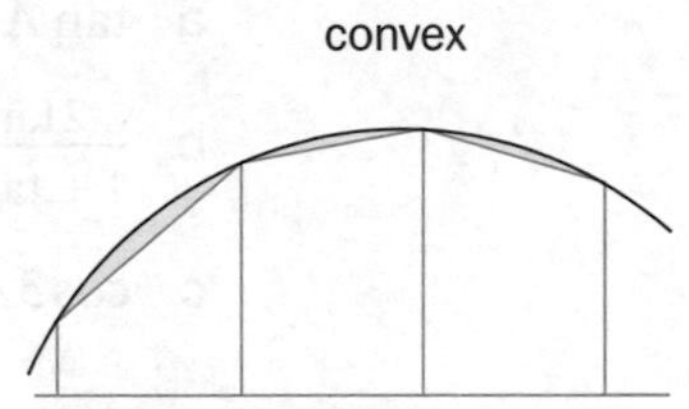

If the graph is convex, the trapezium rule underestimates the actual area.

See C1 for revision.

The shaded areas are the errors in each approximation.

C4

EXAMPLE 1

a Estimate (to 3 significant figures) the area under the curve $y = \sec^2 x$ from $x = 0$ to $x = \frac{\pi}{3}$ by using the trapezium rule with

i 4 equal strips ii 8 equal strips.

b Calculate the exact value of $\int_0^{\frac{\pi}{3}} \sec^2 x\,dx$ as a surd.

Find the percentage error in each of the two estimated values of area found using the trapezium rule.

---

a i With 4 strips, the width of each strip, $h = \dfrac{\frac{\pi}{3} - 0}{4} = \dfrac{\pi}{12}$

Record the values in a table:

x	0	$\frac{\pi}{12}$	$\frac{2\pi}{12}$	$\frac{3\pi}{12}$	$\frac{4\pi}{12}$
$y = \sec^2 x$	1	1.0718	1.3333	2	4

Use radians on your calculator.

Estimate of the area required $= \frac{1}{2} \times \frac{\pi}{12} \times [1 + 4 + 2(1.0718 + 1.3333 + 2)]$

$= \frac{1}{2} \times \frac{\pi}{12} \times [1 + 4 + 8.8102]$

$= 1.8078 = 1.81$ units^2 (to 3 s.f.)

ii With 8 strips, the width of each strip, $h = \dfrac{\frac{\pi}{3} - 0}{8} = \dfrac{\pi}{24}$

x	0	$\frac{\pi}{24}$	$\frac{2\pi}{24}$	$\frac{3\pi}{24}$	$\frac{4\pi}{24}$	$\frac{5\pi}{24}$	$\frac{6\pi}{24}$	$\frac{7\pi}{24}$	$\frac{8\pi}{24} = \frac{\pi}{3}$
y	1	1.0173	1.0718	1.1716	1.3333	1.5888	2	2.6984	4

Estimate of the area required $= \frac{1}{2} \times \frac{\pi}{24} \times [1 + 4 + 2 \times 10.8812]$

All the 'middle values' add together to give 10.8812

$= \frac{1}{2} \times \frac{\pi}{24} \times 26.7624$

$= 1.7516 = 1.75$ units^2 (to 3 s.f.)

b $\int_0^{\frac{\pi}{3}} \sec^2 x\,dx = [\tan x]_0^{\frac{\pi}{3}}$

$= \sqrt{3} - 0 = \sqrt{3}$

$\frac{d(\tan x)}{dx} = \sec^2 x$

i The percentage error with 4 strips $= \dfrac{1.8078 - \sqrt{3}}{\sqrt{3}} \times 100$

$= 4.4\%$ (to 2 s.f.)

ii The percentage error with 8 strips $= \dfrac{1.7516 - \sqrt{3}}{\sqrt{3}} \times 100$

$= 1.1\%$ (to 2 s.f.)

The percentage error is smaller when more strips are used (but more calculation is involved).

C4

Exercise 10.1

1 Use the trapezium rule to estimate the value of these integrals correct to 3 significant figures using the number of strips given.

a $\int_0^1 2^x \, dx$ 5 strips

b $\int_1^5 \ln(1 + x^2) \, dx$ 4 strips

c $\int_0^{\frac{\pi}{2}} (\sqrt{\sin\theta}) \, d\theta$ 6 strips

d $\int_1^2 e^{-x^2} \, dx$ 5 strips

e $\int_{-\frac{\pi}{2}}^{\frac{\pi}{2}} \cos^2 \theta \, d\theta$ 6 strips

f $\int_2^4 1 + e^{-x} \, dx$ 6 strips

C4

2 a Estimate the value of the integral $I = \int_0^{\frac{\pi}{2}} \cos x \, dx$ using the trapezium rule by dividing the interval from 0 to $\frac{\pi}{2}$ into six strips.

b Find the exact value of I by integration.

c Calculate the percentage error in the estimated value of I.

d Explain, using a suitable diagram, why the answer to part **a** is an underestimate of the exact value of I.

3 Find an approximate value for $I = \int_0^1 e^x \sin x \, dx$ using the trapezium rule with

a six ordinates

b eleven ordinates.

Give reasons why one of these values is more accurate than the other.

4 The semicircle $y = +\sqrt{36 - x^2}$ is split into twelve vertical strips. Find an estimate of the area of the semicircle (to 3 s.f.) using the trapezium rule. Hence, obtain an approximate value of π.

5 Use the trapezium rule with seven ordinates to estimate the value of $I = \int_0^3 \sqrt{1 - \frac{x^2}{9}} \, dx$ to 2 decimal places.

Sketch the graph of the ellipse $\frac{x^2}{9} + \frac{y^2}{4} = 1$ and use your value of I to estimate its area.

6 a Estimate the value of the integral $I = \int_0^8 (\sqrt[3]{x} + 1)^2 \, dx$ using the trapezium rule with 8 strips.

b Calculate the exact value of I.

c Find the percentage error in the estimated value to 1 decimal place.

INVESTIGATION

7 Do some research to find a formula for the area of an ellipse.

How does this formula also give you the area of a circle?

Use your answer to question **5** to find another estimate for the value of π.

10.2 Integration as summation

You can find the exact area under a curve by summing an infinite number of infinitely thin rectangles.

See C2 for revision.

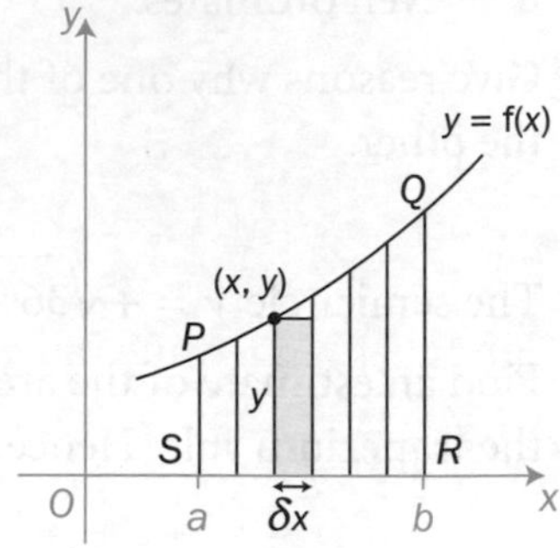

$$\text{Area } PQRS = \sum_a^b \delta A \approx \sum_a^b y\delta x$$

Each strip of area δA is approximated by a rectangle of height y and width δx, so its area $\delta A \approx y\delta x$

The Greek letter sigma, Σ, indicates the sum of many of these rectangles.

In the limit, as $\delta x \to 0$,

$$\text{area } PQRS = \lim_{\delta x \to 0} \sum_a^b y\delta x$$

$$= \int_a^b y\,\mathrm{d}x$$

As $\delta x \to 0$, the number of rectangles $\to \infty$.
In the limit, the Greek letters δx and Σ become the English letters dx and $\int$ respectively.

$\int$ is an elongated letter S (for Sum) invented by Leibnitz in about 1680.

Integration is also the reverse of differentiation, so you should already recognise these basic results:

$\frac{\mathrm{d}(\sin x)}{\mathrm{d}x} = \cos x$	$\int \cos x\,\mathrm{d}x = \sin x + c$	$\frac{\mathrm{d}(\mathrm{e}^x)}{\mathrm{d}x} = \mathrm{e}^x$	$\int \mathrm{e}^x\,\mathrm{d}x = \mathrm{e}^x + c$
$\frac{\mathrm{d}(\cos x)}{\mathrm{d}x} = -\sin x$	$\int \sin x\,\mathrm{d}x = -\cos x + c$	$\frac{\mathrm{d}(\ln x)}{\mathrm{d}x} = \frac{1}{x}$	$\int \frac{1}{x}\,\mathrm{d}x = \ln x + c$
$\frac{\mathrm{d}(\tan x)}{\mathrm{d}x} = \sec^2 x$	$\int \sec^2 x\,\mathrm{d}x = \tan x + c$	c is the constant of integration.	

EXAMPLE 1

Find the area enclosed by the graphs of $y = \cos x$ and $y = \mathrm{e}^x$ and the line $x = 1$

From the sketch, the area required is the area under $y = \cos x$ subtracted from the area under $y = \mathrm{e}^x$

$$\text{Area} = \int_0^1 \mathrm{e}^x\,\mathrm{d}x - \int_0^1 \cos x\,\mathrm{d}x$$

$$= \left[\mathrm{e}^x\right]_0^1 - \left[\sin x\right]_0^1$$

$$= (\mathrm{e} - 1) - (\sin 1 - 0) = 2.718 - 1 - 0.841 = 0.88$$

The area required is 0.88 units^2 to 2 d.p.

A sketch graph will help you to visualise the problem.

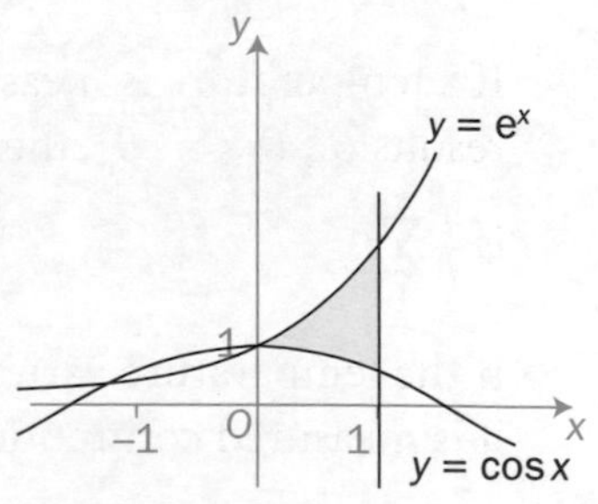

You could write this as one integral $\int_0^1 (\mathrm{e}^x - \cos x)\,\mathrm{d}x$

sin 1 means the sine of 1 radian (not 1°).

Exercise 10.2

1 Find the values of these definite integrals.

a $\int_0^{\frac{\pi}{2}} 2\cos x\,dx$ **b** $\int_{\frac{\pi}{4}}^{\frac{\pi}{2}} 5\sin x\,dx$ **c** $\int_{-1}^{1} e^x\,dx$ **d** $\int_1^2 \frac{1}{x}\,dx$

e $\int_{-\frac{\pi}{4}}^{\frac{\pi}{4}} \sin x - \cos x\,dx$ **f** $\int_1^4 \frac{x+1}{x}\,dx$ **g** $\int_1^4 \frac{(x+1)^2}{x}\,dx$ **h** $\int_0^{\frac{\pi}{3}} \frac{\cos x + \sec x}{\cos x}\,dx$

2 Find these indefinite integrals.

a $\int \frac{1}{2}\sec^2\theta\,d\theta$ **b** $\int x - \frac{2}{x}\,dx$ **c** $\int 5\cos x - 3\sin x\,dx$

3 **a** Find the area bounded by the graph of $y = e^x$, the two coordinate axes and the line $x = 2$

b The region bounded by the graph of $y = e^x$, the two coordinate axes and the line $x = k$ has an area of 2 units^2. Find the value of k.

4 **a** Find the point of intersection of the graphs $y = \cos x$ and $y = \sin x$ for $0 \leqslant x \leqslant \frac{\pi}{2}$

b Find the area, A, bounded by the graphs of $y = \cos x$, $y = \sin x$ and the x-axis as shown in this diagram.

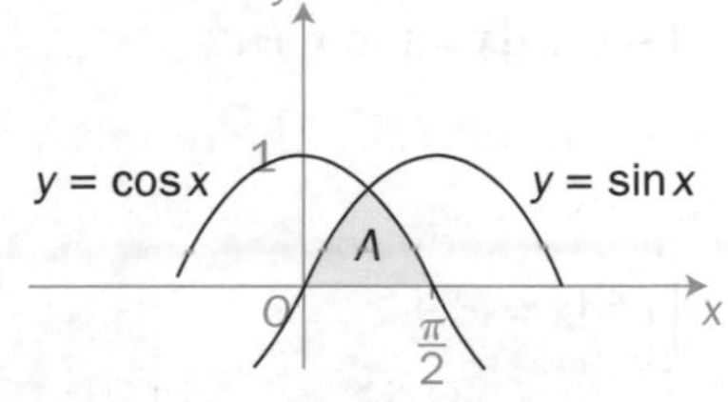

c Find the area bounded by the two graphs and the y-axis.

5 Find the area between the graphs of $y = e^x$ and $y = \frac{1}{x}$ from $x = 1$ to $x = 2$.

6 If the region under the graph of $y = \frac{1}{x}$ from $x = 1$ to $x = \alpha$ has an area of 4 units^2, find the value of α.

INVESTIGATION

7 If a temperature is measured discretely n times with the results $\theta_1, \theta_2, \ldots, \theta_n$, then the average (mean) temperature is $\frac{1}{n}\sum_1^n \theta_i$

If the temperature varies with time t such that $\theta = f(t)$ and it is measured continuously over a period T, then the mean temperature is $\frac{1}{T}\int_0^T f(\theta)\,dt$

Find the mean temperature if $\theta = 5\sin t$ from $t = 0$ to $t = \pi$

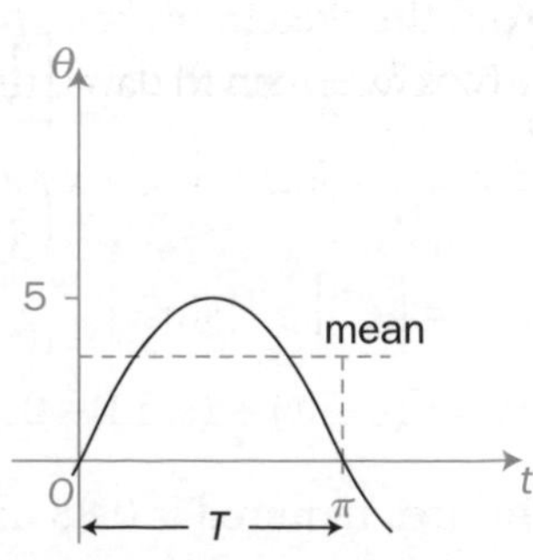

10.3 Integration using standard forms

The Core 3 and 4 specifications list the derivatives and integrals that you are expected to remember.
The formulae booklet lists derivatives and integrals that are provided for you in your examinations.

Here is a list of standard integral forms where a, b, c and n are constants:

Try to derive these results yourself.

$$\int x^n \,dx = \frac{1}{n+1}x^{n+1} + c \ (n \neq -1)$$

$$\int (ax+b)^n \,dx = \frac{1}{a} \times \frac{1}{n+1}(ax+b)^{n+1} + c$$

$$\int \cos x \,dx = \sin x + c$$

$$\int \sin x \,dx = -\cos x + c$$

$$\int \sec^2 x \,dx = \tan x + c$$

$$\int \cos(ax+b) \,dx = \frac{1}{a}\sin(ax+b) + c$$

$$\int \sin(ax+b) \,dx = -\frac{1}{a}\cos(ax+b) + c$$

$$\int \sec^2(ax+b) \,dx = \frac{1}{a}\tan(ax+b) + c$$

$$\int e^x \,dx = e^x + c$$

$$\int e^{ax+b} \,dx = \frac{1}{a}e^{ax+b} + c$$

$$\int \frac{1}{x} \,dx = \ln x + c$$

$$\int \frac{1}{ax+b} \,dx = \frac{1}{a}\ln|ax+b| + c$$

You can always check your integration by differentiating your answer.

See the formulae booklet for other standard integrals.

EXAMPLE 1

Evaluate $\int_0^{\frac{\pi}{3}} (\cos 3x + 4\sin x) \,dx$

$$\int_0^{\frac{\pi}{3}} (\cos 3x + 4\sin x) \,dx = \left[\frac{1}{3}\sin 3x - 4\cos x\right]_0^{\frac{\pi}{3}}$$

$$= \left(\frac{1}{3}\sin \pi - 4\cos\frac{\pi}{3}\right) - \left(\frac{1}{3}\sin 0 - 4\cos 0\right)$$

$$= 0 - 4 \times \frac{1}{2} - 0 + 4 = 2$$

Exercise 10.3

1 Integrate with respect to x.

a $\cos 5x$ b $\sin 4x$ c $\sec^2 3x$

d $\cos \frac{1}{2}x$ e $\operatorname{cosec}^2 4x$ f e^{4x-3}

g $(3x+2)^4$ h $\tan 3x$ i $\frac{1}{3x-1}$

j $\frac{1}{(3x-1)^2}$ k $\cot \frac{2x}{5}$ l $\cos(2x+3)$

m $\sec^2(4x+1)$ n $\sec 4x \tan 4x$ o $\sec 4x$

p $\operatorname{cosec} 4x$ q e^{-2x} r $\cos 3x + \sin \frac{1}{3}x$

s $\operatorname{cosec} 2x \cot 2x$ t $(e^x - e^{-x})^2$

2 Evaluate these integrals.

a $\int_0^{\frac{\pi}{3}} \cos 3x \, dx$ b $\int_0^{\frac{\pi}{8}} \sec^2 2x \, dx$ c $\int_{\frac{\pi}{2}}^{\pi} \sin \frac{x}{2} \, dx$

d $\int_0^{\frac{\pi}{3}} \sin \frac{3x}{4} \, dx$ e $\int_{-1}^{1} e^{2x+1} \, dx$ f $\int_0^2 \frac{1}{3x+1} \, dx$

3 Integrate with respect to x.

a $\frac{1}{x^2} + \frac{1}{\cos^2 x}$ b $\frac{1}{x} + \frac{1}{\sin^2 x}$

c $e^{2x-1} + e^{1-2x}$ d $\frac{\sin 3x}{\cos^2 3x}$

e $\cos 3x \operatorname{cosec}^2 3x$

INVESTIGATION

4 An AC current i varies with time t such that
$i = 5\sin \omega t$ where $\omega = 3$
The current is rectified so that $i = |5\sin \omega t|$
Draw the graph of the rectified current for $0 \leqslant t \leqslant 2\pi$

Use the ideas of the investigation in Section 10.2 to find the mean value of the rectified current for $t = 0$ to $t = \frac{\pi}{3}$

AC means 'alternating current'. 'Rectified' means that the current always flows in the same direction; that is, $i > 0$ at all times.

10.4 Further use of standard forms

There are two special cases which use standard forms and the chain rule in reverse.

Consider $y = (3x^2 + 2x + 1)^5$

Apply the chain rule: $\frac{dy}{dx} = 5(3x^2 + 2x + 1)^4 \times (6x + 2)$

$= 5(6x + 2)(3x^2 + 2x + 1)^4$

In reverse, $\int (6x + 2)(3x^2 + 2x + 1)^4 \, dx = \frac{1}{5}(3x^2 + 2x + 1)^5 + c$

> In general, you can perform an integration of the form
>
> $\int f'(x) \times g[f(x)] \, dx$ by sight.

Check your answers mentally by differentiating them using the chain rule.

EXAMPLE 1

Integrate $\int x \cos(x^2 + 1) \, dx$

$$\int x \cos(x^2 + 1) \, dx = \frac{1}{2}\int 2x \cos(x^2 + 1) \, dx = \frac{1}{2}\sin(x^2 + 1) + c$$

The 'inside' function $f(x)$ is $x^2 + 1$ and its derivative is $2x$. Introduce $\frac{1}{2} \times 2$ to make a $2x$ on the 'outside'.

EXAMPLE 2

Evaluate $\int_0^1 \frac{x^2}{(x^3 - 2)^2} \, dx$

$$\int_0^1 \frac{x^2}{(x^3 - 2)^2} \, dx = \frac{1}{3}\int_0^1 \frac{3x^2}{(x^3 - 2)^2} \, dx = \frac{1}{3}\int_0^1 3x^2(x^3 - 2)^{-2} \, dx$$

$$= \frac{1}{3}\left[\frac{(x^3 - 2)^{-1}}{-1}\right]_0^1 = -\frac{1}{3}\left[\frac{1}{x^3 - 2}\right]_0^1$$

$$= -\frac{1}{3}\left(\frac{1}{-1} - \frac{1}{-2}\right) = -\frac{1}{3} \times -\frac{1}{2} = \frac{1}{6}$$

The 'inside' function $f(x)$ is $x^3 - 2$ and its derivative is $3x^2$. Introduce $\frac{1}{3} \times 3$ to make a $3x^2$ on the 'outside'.

Now consider $y = \ln|3x^2 + 2x + 1|$

Apply the chain rule: $\frac{dy}{dx} = \frac{1}{3x^2 + 2x + 1} \times (6x + 2) = \frac{6x + 2}{3x^2 + 2x + 1}$

In reverse, $\int \frac{6x + 2}{3x^2 + 2x + 1} \, dx = \ln|3x^2 + 2x + 1| + c$

In general, $\int \frac{f'(x)}{f(x)}\,dx = \ln|f(x)| + c$

EXAMPLE 3

Integrate **a** $\int \frac{x}{x^2+1}\,dx$ **b** $\int \frac{\cos x}{3\sin x - 4}\,dx$

a $\int \frac{x}{x^2+1}\,dx = \frac{1}{2}\int \frac{2x}{x^2+1}\,dx = \frac{1}{2}\ln(x^2+1) + c$

The derivative of $x^2 + 1$ is $2x$. Modify the integral so that you have $2x$ 'on the top'.

b $\int \frac{\cos x}{3\sin x - 4}\,dx = \frac{1}{3}\int \frac{3\cos x}{3\sin x - 4}\,dx = \frac{1}{3}\ln|3\sin x - 4| + c$

The derivative of $3\sin x - 4$ is $3\cos x$. You need to have $3\cos x$ 'on the top'.

Exercise 10.4

1 Integrate each expression with respect to x.

a $\frac{3x^2}{x^3-1}$ **b** $3x^2(x^3-1)^5$ **c** $\frac{2x+3}{x^2+3x-1}$

d $(2x+3)(x^2+3x-1)^4$ **e** $\frac{x-2}{x^2-4x+1}$ **f** $(x-2)(x^2-4x+1)^3$

g $\frac{\cos x}{\sin x + 1}$ **h** $x\cos(x^2+1)$ **i** $x\sqrt{x^2-1}$

j $\frac{x}{\sqrt{x^2-1}}$ **k** $\frac{x}{x^2-1}$ **l** xe^{x^2}

m xe^{-x^2} **n** $(x+1)\sqrt{x^2+2x+3}$ **o** $\frac{x+1}{x^2+2x+3}$

p $\cos x e^{\sin x}$ **q** $x^2e^{x^3}$ **r** $\frac{1}{x\ln x}$

2 By writing $\cot\theta \equiv \frac{\cos\theta}{\sin\theta}$, find $\int \cot\theta\,d\theta$ and show that $\int_{\frac{\pi}{4}}^{\frac{\pi}{2}} \cot\theta\,d\theta = \frac{1}{2}\ln 2$

3 By writing $\tan\theta \equiv \frac{\sin\theta}{\cos\theta}$, prove that $\int \tan\theta\,d\theta = \ln|\sec\theta| + c$

Find the value of $\int_0^{\frac{\pi}{4}} \tan\theta\,d\theta$

INVESTIGATION

4 By writing $x = x - 1 + 1$, find $\int \frac{x}{x-1}\,dx$

Use long division to rewrite $\frac{x^2}{x-1}$, $\frac{x^3}{x-1}$,..., $\frac{x^n}{x-1}$ and so integrate each of them.

C4

10.5 Integration by substitution

You can simplify an **indefinite integral** using a substitution.

EXAMPLE 1

Consider the indefinite integral $\int (3x+2)^5\,dx$

Change the variable x by substituting $u = 3x + 2$:

You now have $\int u^5\,dx$

You cannot integrate a function of u with respect to x.

Find a substitution for the operator 'dx' in terms of u.

Differentiate $u = 3x + 2$ wrt x: $\frac{du}{dx} = 3$

Separate the operators du and dx: $\frac{du}{3} = dx$

You now have

$$\int (3x+2)^5\,dx = \int u^5 \times \frac{du}{3} = \frac{1}{3}\int u^5\,du$$

$$= \frac{1}{3} \times \frac{1}{6}u^6 + c$$

$$= \frac{1}{18}u^6 + c$$

Substitute $u = 3x + 2$: $\int (3x+2)^5\,dx = \frac{1}{18}(3x+2)^6 + c$

You must give the final answer in terms of x.

EXAMPLE 2

Perform this integration $\int \frac{x}{\sqrt{x+1}}\,dx$ using the substitution $u = \sqrt{x+1}$

This substitution is chosen to simplify the denominator in the integral.

Rearrange $u = \sqrt{x+1}$: $u^2 = x + 1$

Differentiate wrt x using the chain rule: $2u\frac{du}{dx} = 1$

Separate the operators: $2u\,du = dx$

Rewrite the integral in terms of u:

$$\int \frac{x}{\sqrt{x+1}}\,dx = \int \frac{u^2-1}{u} \times 2u\,du = 2\int (u^2-1)\,du$$

$$= 2\left(\frac{1}{3}u^3 - u\right) + c = \frac{2}{3}u^3 - 2u + c$$

Substitute for u and return to x:

$$\int \frac{x}{\sqrt{x+1}}\,dx = \frac{2}{3}\sqrt{(x+1)^3} - 2\sqrt{x+1} + c$$

You could also use the substitution $u = x + 1$. Show that it gives the same answer.

To evaluate a definite integral, you can

either change back to the variable x and use the original limits of x
or stay with the new variable u, provided you change the limits on the integral to the corresponding values of u.

EXAMPLE 3

Evaluate $\displaystyle\int_0^{\frac{\pi}{2}} \frac{\cos x}{4+\sin x}\,dx$

Let $u = 4 + \sin x$, so $\dfrac{du}{dx} = \cos x$ and $du = \cos x\,dx$

Either

$$\int_0^{\frac{\pi}{2}} \frac{\cos x}{4+\sin x}\,dx = \int_{u_1}^{u_2} \frac{1}{u}\,du$$

The actual values u_1 and u_2 are not needed.

$$= [\ln u]_{u_1}^{u_2}$$

Change back to x and use the limits of x:

$$= \left[\ln(4+\sin x)\right]_0^{\frac{\pi}{2}}$$
$$= \ln(4+1) - \ln(4+0)$$
$$= \ln(1.25)$$

or

When $x = 0$, $\quad u = 4 + \sin 0 = 4$

When $x = \frac{\pi}{2}$, $\quad u = 4 + \sin\frac{\pi}{2} = 5$

$$\int_0^{\frac{\pi}{2}} \frac{\cos x}{4+\sin x}\,dx = \int_4^5 \frac{1}{u}\,du$$
$$= [\ln u]_4^5$$

Stay with u and use the limits of u.

$$= \ln 5 - \ln 4$$
$$= \ln(1.25)$$

C4

Exercise 10.5

1 Use the given substitutions to find these integrals.

a $\displaystyle\int x^2(1+x^3)^4\,dx \quad u = 1 + x^3$

b $\displaystyle\int x^2\sqrt{1+x^3}\,dx \quad u = 1 + x^3$

c $\displaystyle\int \cos x \sin^4 x\,dx \quad u = \sin x$

d $\displaystyle\int \sec^2 x(1+\tan x)\,dx \quad u = \tan x$

e $\displaystyle\int \frac{1}{(3x-1)^4}\,dx \quad u = 3x - 1$

f $\displaystyle\int \frac{x}{(x+5)^2}\,dx \quad u = x + 5$

g $\displaystyle\int \left(\frac{x}{x-1}\right)^2 dx \quad u = x - 1$

h $\displaystyle\int \frac{e^x}{(e^x+2)^3}\,dx \quad u = e^x + 2$

2 Evaluate each integral using the given substitution.

a $\int e^x\sqrt{1+e^x}\,dx$ $\quad u = 1 + e^x$

b $\int \sin x\sqrt{1-\cos x}\,dx$ $\quad u = 1 - \cos x$

c $\int \frac{x}{\sqrt{x-1}}\,dx$ $\quad u^2 = x - 1$

d $\int x^3\sqrt{1+x^4}\,dx$ $\quad u^2 = 1 + x^4$

e $\int \frac{(\ln x)^2}{x}\,dx$ $\quad u = \ln x$

f $\int \frac{1}{2+\sqrt{x}}\,dx$ $\quad u^2 = x$

g $\int \frac{1}{1+\sqrt{x+1}}\,dx$ $\quad u^2 = x + 1$

h $\int \frac{x}{\sqrt{1-x^2}}\,dx$ $\quad x = \sin\theta$

i $\int \frac{1}{\sqrt{1-x^2}}\,dx$ $\quad x = \sin\theta$

j $\int \sec^3 x\tan x\,dx$ $\quad u = \sec x$

k $\int \frac{dx}{e^x - e^{-x}}$ $\quad u = e^x$

l $\int \frac{dx}{x^2\sqrt{4-x^2}}$ $\quad x = 2\sin\theta$

m $\int \frac{x+4}{x\sqrt{x-2}}\,dx$ $\quad u^2 = x + 1$

3 Calculate the values of these definite integrals using the given substitutions.

a $\int_0^{\frac{\pi}{2}} 2\cos x\, e^{\sin x}\,dx$ $\quad u = \sin x$

b $\int_2^3 (x-1)(x-2)^3\,dx$ $\quad u = x - 2$

c $\int_0^1 \frac{x}{\sqrt{1+x^2}}\,dx$ $\quad u^2 = 1 + x^2$

d $\int_{\frac{1}{2}}^3 2x\sqrt{2x+3}\,dx$ $\quad u^2 = 2x + 3$

e $\int_0^{\frac{\pi^2}{4}} \frac{1}{\sqrt{x}}\cos\sqrt{x}\,dx$ $\quad u^2 = x$

f $\int_1^2 \frac{x\,dx}{(2x-1)^4}$ $\quad u = 2x - 1$

g $\int_0^1 xe^{x^2-1}\,dx$ $\quad u = x^2 - 1$

h $\int_1^5 \frac{x^2-1}{\sqrt{x-1}}\,dx$ $\quad u^2 = x - 1$

4 The x-axis is a tangent to the curve $y=(x-1)(x\ 2)^4$ at the point P.

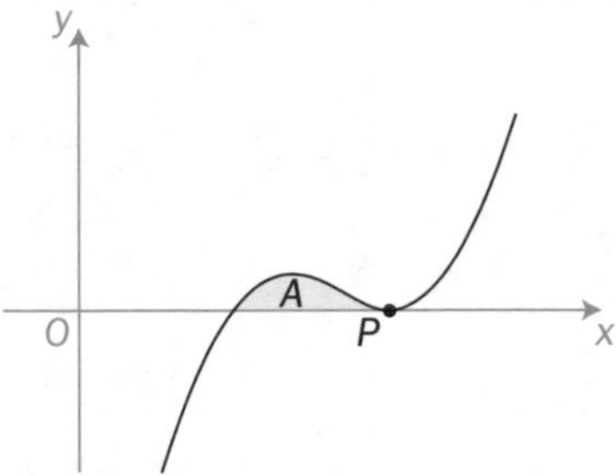

a Write down the coordinates of P.

b Let $u=x-2$ and find the area A enclosed between the curve and the x-axis.

5 Find the points where each of these curves meets the x-axis. For each curve find the area enclosed between the curve and the x-axis.

a $y=x\sqrt{4-x}$

b $y=x(x-2)^2$

c $y=5\cos x\sin^3 x$ for $0\leqslant x\leqslant\pi$

INVESTIGATION

6 All these integrations can be performed by choosing appropriate substitutions.
Some of them, however, can be done immediately on sight using $\int f'(x)\times g[f(x)]\,dx$

Decide which integrals can be written down on sight.
Find all the integrals.

a $\int 3x^2(x^3+1)^4\,dx$ **b** $\int 2x^3(x^2+1)^2\,dx$

c $\int x\sqrt{x+1}\,dx$ **d** $\int \sec^2 x\sqrt{1+\tan x}\,dx$

e $\int \frac{x}{\sqrt{x^2+1}}\,dx$ **f** $\int \frac{x^2}{\sqrt{x+1}}\,dx$

g $\int xe^{x^2+1}\,dx$ **h** $\int e^x\sqrt{e^x-1}\,dx$

10.6 Integration using trigonometric identities

You can integrate some trigonometric expressions after rearranging them into one of the standard integral forms.

EXAMPLE 1

Integrate **a** $\displaystyle\int \frac{\operatorname{cosec} x}{\sec x}\,dx$ **b** $\displaystyle\int \sec^2 x \sin x\,dx$

a $\displaystyle\int \frac{\operatorname{cosec} x}{\sec x}\,dx = \int \frac{1}{\sin x} \times \frac{\cos x}{1}\,dx = \int \frac{\cos x}{\sin x}\,dx = \int \cot x\,dx$

$= \ln|\sin x| + c$

The formula book gives $\int \cot x\,dx = \ln|\sin x|$ and $\frac{d(\sec x)}{dx} = \sec x \tan x$

b $\displaystyle\int \sec^2 x \sin x\,dx = \int \frac{\sec x \sin x}{\cos x}\,dx = \int \sec x \tan x\,dx$

$= \sec x + c$

You can integrate powers of sine and cosine by first changing the powers to multiples.

EXAMPLE 2

Integrate **a** $\displaystyle\int \sin^2 x\,dx$ **b** $\displaystyle\int \cos^2 3x\,dx$ **c** $\displaystyle\int (1 + \tan x)^2\,dx$

a $\displaystyle\int \sin^2 x\,dx = \frac{1}{2}\int (1 - \cos 2x)\,dx = \frac{1}{2}\left(x - \frac{1}{2}\sin 2x\right) + c = \frac{1}{2}x - \frac{1}{4}\sin 2x + c$

$\sin^2 x = \frac{1}{2}(1 - \cos 2x)$

b $\displaystyle\int \cos^2 3x\,dx = \frac{1}{2}\int (1 + \cos 6x)\,dx = \frac{1}{2}\left(x + \frac{1}{6}\sin 6x\right) + c = \frac{1}{2}x + \frac{1}{12}\sin 6x + c$

$\cos^2 3x = \frac{1}{2}(1 + \cos 6x)$

c $\displaystyle\int (1 + \tan x)^2\,dx = \int (1 + 2\tan x + \tan^2 x)\,dx = \int (2\tan x + \sec^2 x)\,dx$

$= 2\ln|\sec x| + \tan x + c$

$1 + \tan^2 x = \sec^2 x$

$\int \tan x\,dx$ and $\int \sec^2 x\,dx$ are both standard forms.

For higher powers of sine and cosine, the method you use depends on whether the power is even or odd.

For **even** powers, use the double-angle formulae as many times as is needed.
For **odd** powers, use $\sin^2 A + \cos^2 A = 1$ as shown in Example 3 part **b**.

EXAMPLE 3

Integrate **a** $\int \cos^4 x\,dx$ **b** $\int \cos^5 x\,dx$

a $$\int \cos^4 x\,dx = \int \left(\frac{1}{2}(1+\cos 2x)\right)^2 dx$$
$$= \frac{1}{4}\int (1 + 2\cos 2x + \cos^2 2x)\,dx$$
$$= \frac{1}{4}\int \left(1 + 2\cos 2x + \frac{1}{2}(1+\cos 4x)\right) dx$$
$$= \frac{1}{4}\int \left(\frac{3}{2} + 2\cos 2x + \frac{1}{2}\cos 4x\right) dx$$
$$= \frac{1}{4}\left(\frac{3}{2}x + \sin 2x + \frac{1}{2}\times\frac{1}{4}\sin 4x\right) + c$$
$$= \frac{3}{8}x + \frac{1}{4}\sin 2x + \frac{1}{32}\sin 4x + c$$

$\cos^2 x = \frac{1}{2}(1 + \cos 2x)$

b Write $\cos^5 x$ as

$\cos x(\cos^4 x) = \cos x(\cos^2 x)^2 = \cos x\,(1 - \sin^2 x)^2$

$$\int \cos^5 x\,dx = \int \cos x(1-\sin^2 x)^2\,dx$$
$$= \int \cos x(1 - 2\sin^2 x + \sin^4 x)\,dx$$
$$= \int (\cos x - 2\cos x\sin^2 x + \cos x\sin^4 x)\,dx$$
$$= \sin x - \frac{2}{3}\sin^3 x + \frac{1}{5}\sin^5 x + c$$

You could use the substitution method with $u = \sin x$. Try this method to show that it gives the same result.

$\int \cos x\sin^2 x\,dx = \frac{1}{3}\sin^3 x$ is of the form $\int f'(x) \times g[f(x)]\,dx$, as is $\int \cos x \sin^4 x\,dx$

The simplest examples of a *product* of sine and cosine involve the *same* multiple of the angle.

EXAMPLE 4

Find $\int \sin 3x\cos 3x\,dx$

Use the double-angle formula $\sin 2A = 2\sin A\cos A$:

$$\sin 3x\cos 3x = \frac{1}{2}\times 2\sin 3x\cos 3x = \frac{1}{2}\sin 6x$$

Hence, $$\int \sin 3x\cos 3x\,dx = \frac{1}{2}\int \sin 6x\,dx$$
$$= \frac{1}{2}\times\left(-\frac{1}{6}\cos 6x\right) + c = -\frac{1}{12}\cos 6x + c$$

C4

If the product involves *different* multiples of the angle, you can write the product as the sum or difference of sines and cosines and then integrate.

EXAMPLE 5

Find $\int \sin 5x \cos 3x \, dx$

Consider the expansions of $\sin(A \pm B)$ where $A = 5x$ and $B = 3x$:

$$\sin(5x + 3x) = \sin 5x \cos 3x + \cos 5x \sin 3x$$

and $$\sin(5x - 3x) = \sin 5x \cos 3x - \cos 5x \sin 3x$$

Add the expressions together:

$$\sin 8x + \sin 2x = 2\sin 5x \cos 3x$$

Hence, $$\int \sin 5x \cos 3x \, dx = \frac{1}{2}\int (\sin 8x + \sin 2x)\, dx$$

$$= \frac{1}{2}\left(-\frac{1}{8}\cos 8x - \frac{1}{2}\cos 2x\right) + c$$

$$= -\frac{1}{16}\cos 8x - \frac{1}{4}\cos 2x + c$$

Refer to Section 2.7 on the CD-ROM.

EXAMPLE 6

Find $\int_0^{\frac{\pi}{8}} \sin 6x \sin 2x \, dx$

Consider the expansions of $\cos(A \pm B)$ where $A = 6x$ and $B = 2x$:

$$\cos(6x - 2x) = \cos 6x \cos 2x + \sin 6x \sin 2x$$

and $$\cos(6x + 2x) = \cos 6x \cos 2x - \sin 6x \sin 2x$$

Subtract:

$$\cos 4x - \cos 8x = 2\sin 6x \sin 2x$$

Hence, $$\int_0^{\frac{\pi}{8}} \sin 6x \sin 2x \, dx = \frac{1}{2}\int_0^{\frac{\pi}{8}} (\cos 4x - \cos 8x)\, dx$$

$$= \frac{1}{2}\left[\frac{1}{4}\sin 4x - \frac{1}{8}\sin 8x\right]_0^{\frac{\pi}{8}}$$

$$= \frac{1}{2}\left(\frac{1}{4} \times 1 - 0 - 0 + 0\right) = \frac{1}{8}$$

Exercise 10.6

1 Integrate with respect to x.

a $\cos 3x$ b $\sin 4x$

c $\cos\left(\frac{1}{2}x\right)$ d $\sin\left(\frac{3x}{2}\right)$

e $\cos(2x+1)$ f $\sin(3x-2)$

g $\sec^2 4x$ h $\sec^2(2x-3)$

2 Find these integrals by rearranging into standard forms.

a $\int \frac{1}{\cos^2 3x}\,dx$ b $\int \tan 3x \cos 3x\,dx$

c $\int \frac{1}{\sin^2 4x}\,dx$ d $\int \tan^2 x \operatorname{cosec}^2 x\,dx$

e $\int \cot 2x \sec 2x\,dx$ f $\int \frac{\sec^2 x}{\operatorname{cosec} x}\,dx$

g $\int \tan x \operatorname{cosec} x\,dx$ h $\int (1+\sec x)^2\,dx$

i $\int \sin 3x(1+\cot 3x)\,dx$ j $\int (\operatorname{cosec} x+2)^2\,dx$

k $\int \frac{(\cos x+1)^2}{\sin^2 x}\,dx$ l $\int \frac{\cot 2x}{\sin 4x}\,dx$

3 Find these integrals.

a $\int \cos^2 x\,dx$ b $\int \sin^2 3x\,dx$

c $\int \cos^2\left(\frac{x}{2}\right)dx$ d $\int \sin^2(3x+1)\,dx$

e $\int (1-\tan x)^2\,dx$ f $\int (1+\sin x)^2\,dx$

g $\int \sec^2 x \tan^4 x\,dx$ h $\int \cos^3 x\,dx$

i $\int \sin^3 x\,dx$ j $\int \sin^4 x\,dx$

k $\int \cos^7 x\,dx$ l $\int \tan^3 x\,dx$

4 Integrate with respect to x.

a $\sin x \cos x$

b $\sin 2x \cos 2x$

c $\tan\frac{x}{2}\cos^2\frac{x}{2}$

d $\sin^2 3x \cot 3x$

e $\tan^2 3x$

f $\cot^2 3x$

g $\dfrac{1-\sin^2 2x}{\sin^2 2x}$

h $\dfrac{1}{1-\sin^2\left(\frac{1}{2}x\right)}$

5 By expanding $\sin(A+B)$ and $\sin(A-B)$, show that

$2\sin A\cos B = \sin(A+B) + \sin(A-B)$

Hence find $\int \sin 6x \cos 2x\,dx$

6 Find

a $\int \sin 4x \cos x\,dx$

b $\int \cos 5x \cos 4x\,dx$

c $\int \sin 3x \sin 2x\,dx$

7 Evaluate these definite integrals.

a $\int_0^{\frac{\pi}{2}} \sin 3x \cos 2x\,dx$

b $\int_0^{\frac{\pi}{4}} \sin 4x \sin 6x\,dx$

c $\int_0^{\frac{\pi}{4}} \cos 2x \cos 3x\,dx$

8 Evaluate

a $\int_{-\frac{\pi}{2}}^{\frac{\pi}{2}} 1+\sin^2 x\,dx$

b $\int_{-\frac{\pi}{4}}^{\frac{\pi}{4}} \operatorname{cosec} x \tan x\,dx$

c $\int_{-\pi}^{\pi} \sin^5 x\,dx$

d Explain your answer to part **c** in terms of the graph of $y = \sin^5 x$

9 Find $\int_{\frac{\pi}{4}}^{\frac{\pi}{2}} \operatorname{cosec}^2 x \cot^2 x\,dx$ using the substitution $u = \cot x$

10 Prove that the area between the graph of $y = \tan x$ and the x-axis from $x = 0$ to $x = \frac{\pi}{3}$ is $\ln 2$.

C4

11 Find the shaded areas in these diagrams where

a $y = 2\sin x\cos^3 x$ **b** $y = 2\cos^3 x$

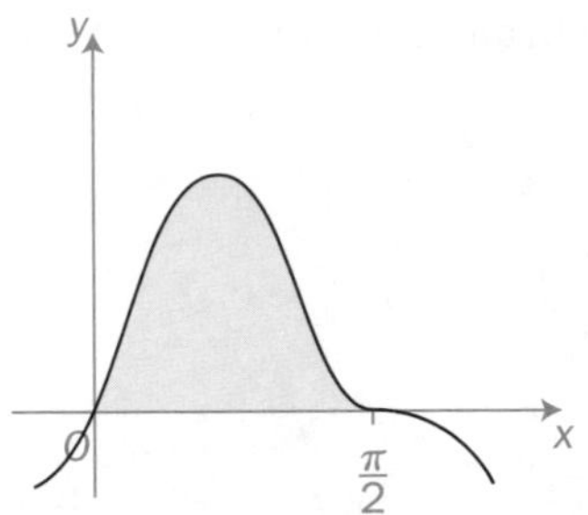

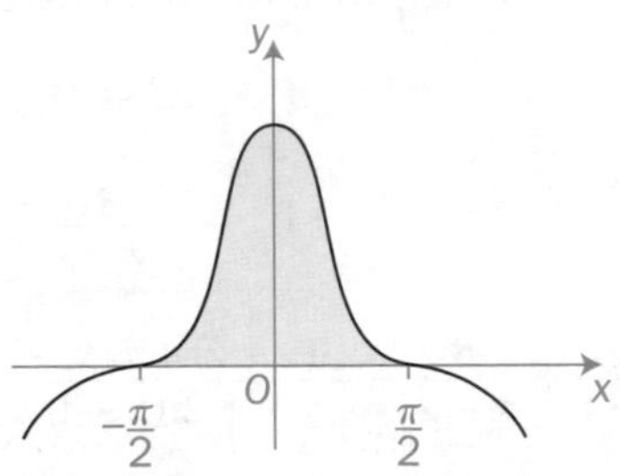

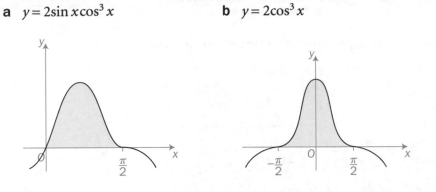

12 Find the area enclosed by the graphs of $y = 2\cos^2 x$ and $y = \frac{1}{2}\cos 3x$ and the y-axis, as shown in this diagram.

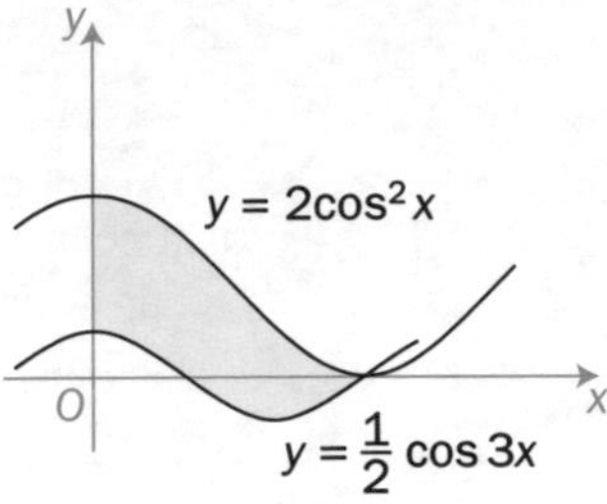

INVESTIGATION

13 Consider the integral $\int_{-\pi}^{\pi} \sin^n x\,dx$ where n is a positive integer.

Use computer software to explore the graph of $y = \sin^n x$ for different values of n.

Give a reason why

a $\int_{-\pi}^{\pi} \sin^n x\,dx = 0$ when n is odd

b $\int_{-\pi}^{\pi} \sin^n x\,dx = 2 \times \int_{0}^{\pi} \sin^n x\,dx$ when n is even.

10.7 Integration using partial fractions

You can use partial fractions to help you to integrate certain algebraic fractions.

EXAMPLE 1

Find $\int \frac{x+5}{(x-1)(x+2)}\,dx$

Let $\frac{x+5}{(x-1)(x+2)} \equiv \frac{A}{x-1} + \frac{B}{x+2} \equiv \frac{A(x+2)+B(x-1)}{(x+2)(x-1)}$

Equate the numerators: $x+5 \equiv A(x+2) + B(x-1)$

Let $x = 1$: $1+5 = A(1+2) + 0$ so $A = 2$

Let $x = -2$: $-2+5 = 0 + B(-2-1)$ so $B = -1$

So $\frac{x+5}{(x-1)(x+2)} \equiv \frac{2}{x-1} - \frac{1}{x+2}$

Hence, $\int \frac{x+5}{(x-1)(x+2}\,dx = \int \frac{2}{x-1} - \frac{1}{x+2}\,dx$

$= 2\ln|x-1| - \ln|x+2| + c$

$= \ln\left|\frac{(x-1)^2}{x+2}\right| + c$

When you work with partial fractions, you often get an answer involving logarithmic functions.

EXAMPLE 2

Evaluate $\int_0^1 \frac{x+1}{(x-2)^2}\,dx$

Let $\frac{x+1}{(x-2)^2} \equiv \frac{A}{x-2} + \frac{B}{(x-2)^2} \equiv \frac{A(x-2)+B}{(x-2)^2}$

Equate numerators: $x+1 \equiv A(x-2) + B$

Let $x = 2$: $2+1 = 0 + B$ so $B = 3$

Equate coefficients of x: $A = 1$

So $\frac{x+1}{(x-2)^2} \equiv \frac{1}{x-2} + \frac{3}{(x-2)^2}$

Hence, $\int_0^1 \frac{x+1}{(x-2)^2}\,dx = \int_0^1 \frac{1}{x-2} + \frac{3}{(x-2)^2}\,dx$

$= \int_0^1 \frac{1}{x-2} + 3(x-2)^{-2}\,dx$

$= \left[\ln|x-2| + \frac{3}{-1}(x-2)^{-1}\right]_0^1 = \left[\ln|x-2| - \frac{3}{x-2}\right]_0^1$

$= \ln 1 - \frac{3}{-1} - \ln 2 + \frac{3}{-2} = \frac{3}{2} - \ln 2$

$\ln 1 = 0$

Exercise 10.7

1 Find these integrals.

a $\int \frac{4}{x(x+2)}\,dx$ b $\int \frac{x+5}{(x+1)(x-3)}\,dx$ c $\int \frac{4x+1}{(2x-1)(x+1)}\,dx$

d $\int \frac{2x}{(x-1)(x+3)}\,dx$ e $\int \frac{2}{4x^2-1}\,dx$ f $\int \frac{8}{x(x^2-1)}\,dx$

g $\int \frac{1}{x^2(x-1)}\,dx$ h $\int \frac{1}{x(x-1)^2}\,dx$ i $\int \frac{9}{(x-2)(x+1)^2}\,dx$

2 Evaluate these integrals.

a $\int_3^4 \frac{3}{(x-2)(x+1)}\,dx$ b $\int_2^3 \frac{x+2}{(x-1)^2}\,dx$ c $\int_3^4 \frac{1+2x}{(3+x)(2-x)}\,dx$

3 Show that the area enclosed by the curve $y = \frac{4}{(x+3)(x-1)}$, the lines $x = -4$ and $x = -5$ and the x-axis is $\ln\frac{5}{3}$ square units.

4 Find the area between the graph of $y = \frac{1}{x^3 - 3x^2 + 2x}$ and the x-axis from the ordinates $x = 3$ to $x = 4$.

5 Either rearrange the numerator or use long division before integrating these expressions using partial fractions.

a $\int \frac{x^2}{x^2-9}\,dx$ b $\int \frac{x^2+1}{x^2-1}\,dx$

c $\int \frac{(x+2)(x-1)}{x(x+1)}\,dx$ d $\int \frac{x^3+x^2+1}{x^2-x-6}\,dx$

INVESTIGATION

6 You can evaluate the integral $\int \frac{x}{x^2-1}\,dx$ by several different methods.

Integrate using

a a logarithmic standard form on sight

b partial fractions

c the substitution $u = x^2 - 1$

d the substitution $x = \sec u$

Show that your four answers are equivalent.

C4

10.8 Integration by parts

You can use integration by parts to integrate the product of two functions.

The method is based on reversing the product rule for differentiation.

If u and v are both functions of x, then the product rule states:

$$\frac{d(uv)}{dx} = u\frac{dv}{dx} + v\frac{du}{dx}$$

Rearrange: $$u\frac{dv}{dx} = \frac{d(uv)}{dx} - v\frac{du}{dx}$$

Integrate with respect to x:

$$\int u\frac{dv}{dx}\,dx = uv - \int v\frac{du}{dx}\,dx$$

The overall aim is to make sure that $v\frac{du}{dx}$ is easier to integrate than the $u\frac{dv}{dx}$ that you started with.

Your first step in choosing which function is u and which is $\frac{dv}{dx}$ is crucial for success.

u must be simple to differentiate. $\frac{dv}{dx}$ must be simple to integrate.

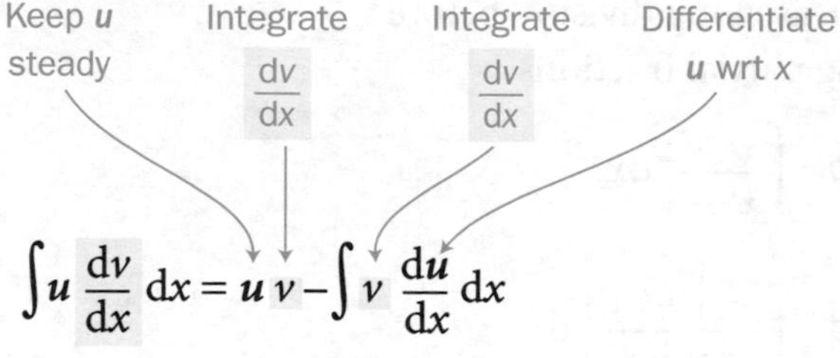

$$\int u\frac{dv}{dx}\,dx = uv - \int v\frac{du}{dx}\,dx$$

EXAMPLE 1

Find $\int x\cos x\,dx$

Let $u = x$ and $\frac{dv}{dx} = \cos x$

Then,

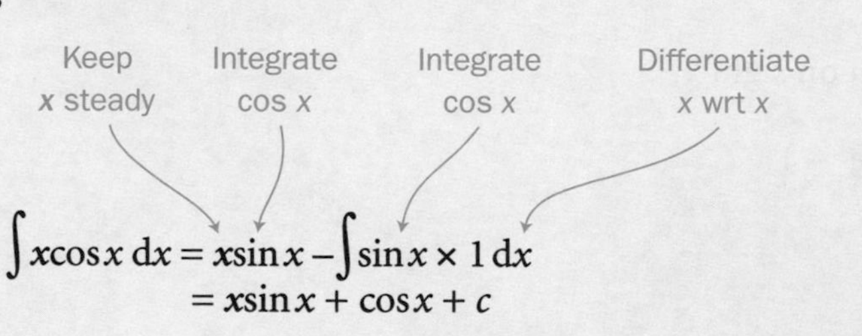

$$\int x\cos x\,dx = x\sin x - \int \sin x \times 1\,dx$$
$$= x\sin x + \cos x + c$$

Choosing $u = x$ gives a simplified final integral because $\frac{du}{dx} = 1$

EXAMPLE 2

Find **a** $\int x^4 \ln x \, dx$ **b** $\int \ln x \, dx$

a Let $u = \ln x$ and $\frac{dv}{dx} = x^4$

$\frac{dv}{dx} = x^4$ so $v = \frac{1}{5}x^5$ and

$u = \ln x$ so $\frac{du}{dx} = \frac{1}{x}$

Then, $\int x^4 \ln x \, dx = \frac{1}{5}x^5 \ln x - \int \frac{1}{5}x^5 \times \frac{1}{x} \, dx$

$$= \frac{1}{5}x^5 \ln x - \frac{1}{5}\int x^4 \, dx$$

$$= \frac{1}{5}x^5 \ln x - \frac{1}{5} \times \frac{1}{5}x^5 + c$$

$$= \frac{1}{25}x^5(5\ln x - 1) + c$$

b Let $u = 1$ and $\frac{dv}{dx} = \ln x$

This example is important.
By thinking of $\ln x$ as the product $1 \times \ln x$, you can use 'integration by parts'.

You can differentiate $\ln x$ by sight but you can not integrate it by sight.

Then, $\int \ln x \, dx = \int 1 \times \ln x \, dx$

$$= x\ln x - \int x \times \frac{1}{x} \, dx$$

$$= x\ln x - \int 1 \, dx$$

$$= x\ln x - x + c$$

$\frac{dv}{dx} = 1$ so $v = x$ and

$u = \ln x$ so $\frac{du}{dx} = \frac{1}{x}$

C4

EXAMPLE 3

Use integration by parts to evaluate the definite integral

$$\int_0^1 xe^x \, dx$$

Let $u = x$ and $\frac{dv}{dx} = e^x$

Then, $\int_0^1 xe^x \, dx = \left[xe^x\right]_0^1 - \int_0^1 e^x \times 1 \, dx$

$$= \left[xe^x\right]_0^1 - \left[e^x\right]_0^1$$

$$= (1 \times e^1 - 0) - (e^1 - 1)$$

$$= 1$$

In some cases, integration by parts gives you an integral which is still not simple enough to integrate.
However, if you integrate the new integral by parts again, you can sometimes solve the problem in two stages.

EXAMPLE 4

Find $\int x^2 e^{3x} dx$

Integrate by parts:

Let $u = x^2$ and $\frac{dv}{dx} = e^{3x}$

$\frac{dv}{dx} = e^{3x}$ so $v = \frac{1}{3}e^{3x}$

and $u = x^2$ so $\frac{du}{dx} = 2x$

$$\int x^2 e^{3x} dx = x^2 \times \frac{1}{3}e^{3x} - \int \frac{1}{3}e^{3x} \times 2x\, dx$$

$$= \frac{1}{3}x^2e^{3x} - \frac{2}{3}\int xe^{3x} dx$$

Now integrate xe^{3x} by parts:

$$= \frac{1}{3}x^2e^{3x} - \frac{2}{3}\left(x \times \frac{1}{3}e^{3x} - \int \frac{1}{3}e^{3x} \times 1\, dx\right)$$

Let $u = x$, so $\frac{du}{dx} = 1$, and integrate e^{3x} using a standard form.

$$= \frac{1}{3}x^2e^{3x} - \frac{2}{3}\left(\frac{1}{3}xe^{3x} - \frac{1}{3} \times \frac{1}{3}e^{3x} + c\right)$$

$$= \frac{1}{3}x^2e^{3x} - \frac{2}{9}xe^{3x} + \frac{2}{27}e^{3x} + c$$

$$= \frac{e^{3x}}{3}\left(x^2 - \frac{2}{3}x + \frac{2}{9}\right) + c$$

EXAMPLE 5

Find $\int e^x \cos x\, dx$

Let $u = e^x$ and $\frac{dv}{dx} = \cos x$

You could also solve this problem using $u = \cos x$ and $\frac{dv}{dx} = e^x$
Try it for yourself.

$$\int e^x \cos x\, dx = e^x \sin x - \int e^x \sin x\, dx$$

Integrate by parts a second time:

$$= e^x \sin x - \left\{e^x(-\cos x) - \int (-\cos x)e^x\, dx\right\}$$

$$= e^x \sin x + e^x \cos x - \int e^x \cos x\, dx$$

So, $2\int e^x \cos x\, dx = e^x \sin x + e^x \cos x$

Rearrange with both integrals on LHS.

$$\int e^x \cos x\, dx = \frac{e^x}{2}(\sin x + \cos x) + c$$

C4

Exercise 10.8

1 Find

a $\int x\sin x\,dx$ b $\int xe^x\,dx$

c $\int x\ln x\,dx$ d $\int xe^{2x}\,dx$

e $\int x\sec^2 x\,dx$ f $\int xe^{-x}\,dx$

g $\int x\sin 2x\,dx$ h $\int \frac{\ln x}{x^3}\,dx$

i $\int xe^{2x+1}\,dx$ j $\int x^2\ln x\,dx$

k $\int \sqrt{x}\ln x\,dx$ l $\int x\cos\left(x-\frac{\pi}{4}\right)dx$

m $\int x\sin\left(x+\frac{\pi}{6}\right)dx$ n $\int x\tan^2 x\,dx$

o $\int (\ln x)^2\,dx$

2 Find

a $\int x\cos nx\,dx$ b $\int xe^{nx}\,dx$

c $\int x^n\ln x\,dx$ d $\int \sin nx\ln(\sec nx)\,dx$

3 Evaluate these definite integrals.

a $\int_0^{\frac{\pi}{2}} x\cos x\,dx$ b $\int_1^2 x^3\ln x\,dx$

c $\int_2^3 \ln x\,dx$ d $\int_1^2 x\log_{10} x\,dx$

e $\int_2^4 \log_{10} x\,dx$ f $\int_{\frac{\pi}{4}}^{\frac{\pi}{2}} x\cot^2 x\,dx$

g $\int_0^1 x^3e^{x^2}\,dx$ h $\int_0^{\frac{\pi}{2}} e^x\sin x\,dx$

C4

4 The graph of $y = x \sin x$ for $0 \leqslant x \leqslant 2\pi$ is shown here.

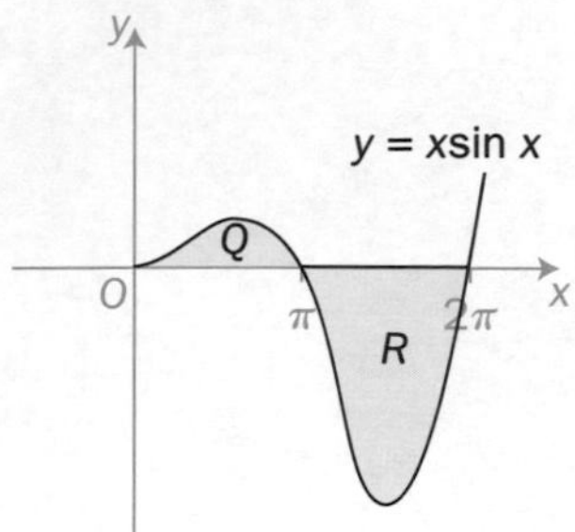

Find area Q and area R.

5 This diagram shows the graph of $y = xe^{-x}$

Find the position of the stationary value.
Find the area between the curve, the x-axis and the ordinates $x = 0$ and $x = 5$.

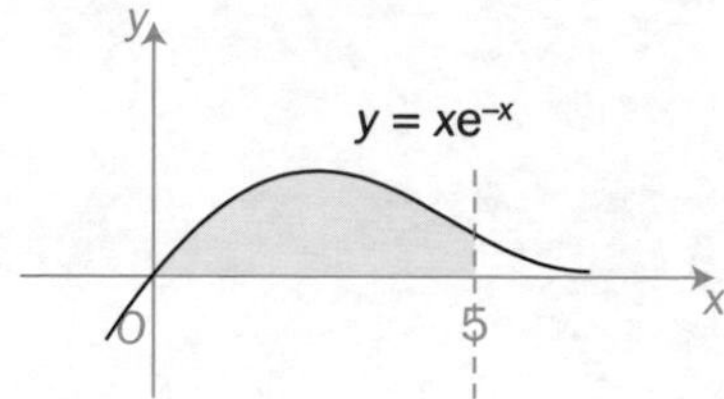

6 **a** Prove that the curve $y = x \ln x$ has a minimum and find this minimum value.

b Find the area enclosed by the curve and the x-axis.

7 Integrate each of these functions either by parts or by using a substitution of your choice.

a $\int x(1 + x)^4 \, dx$ **b** $\int (x + 1)^2 e^x \, dx$ **c** $\int x\sqrt{x - 1} \, dx$

8 Find

a $\int x^2 e^x \, dx$ **b** $\int x^2 \sin x \, dx$ **c** $\int x^2 e^{2x} \, dx$

d $\int x^2 e^{-3x} \, dx$ **e** $\int e^{2x} \cos 2x \, dx$ **f** $\int x^2 \cos 3x \, dx$

g $\int e^{3x} \sin 2x \, dx$

9 Evaluate these definite integrals.

a $\int_0^{\frac{\pi}{2}} e^{2x} \cos x \, dx$

b $\int_0^1 2x^2 e^{-2x} \, dx$

c $\int_1^2 x(\ln x)^2 \, dx$

10 This diagram shows the graph of $y = x^2 e^{2x}$

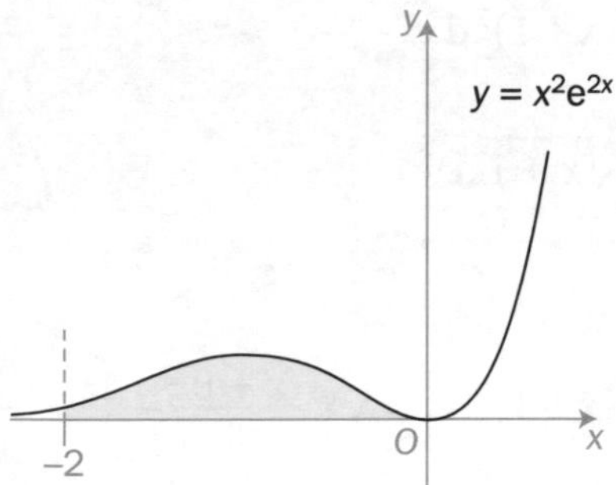

Find

a the y value of the maximum stationary point

b the area enclosed by the curve, the x-axis and the ordinate $x = -2$.

INVESTIGATION

11 Let $I_n = \int \sin^n x \, dx$

By writing $\sin^n x$ as $\sin x \sin^{n-1} x$, use integration by parts to show that $nI_n = -\cos x \sin^{n-1} x + (n-1)I_{n-2}$

Use this formula to find $\int \sin^3 x \, dx$ and $\int \sin^4 x \, dx$

Find the value of $\int_0^{\pi} \sin^6 x \, dx$

C4

10.9 A systematic approach to integration

When integrating a particular function, you should look to use:

- standard integrals to see if the function can be integrated on sight
- the two particular cases $\int \frac{f'(x)}{f(x)}\,dx$ and $\int f'(x)g[f(x)]\,dx$
- the method of substitution
- trigonometric identities
- partial fractions
- the method of integration by parts.

EXAMPLE 1

Find **a** $\int x(x^3+1)^2\,dx$ **b** $\int x(x+1)^7\,dx$

c $\int x^2(x^3+1)^7\,dx$ **d** $\int x\sqrt{x+1}\,dx$

Let $u = \sqrt{x+1}$ in part **d**.

---

a $\int x(x^3+1)^2\,dx = \int x(x^6+2x^3+1)\,dx$

$= \int (x^7+2x^4+x)\,dx$

$= \frac{1}{8}x^8 + \frac{2}{5}x^5 + \frac{1}{2}x^2 + c$

b $\int x(x+1)^7\,dx = x \times \frac{(x+1)^8}{8} - \int \frac{(x+1)^8}{8} \times 1\,dx$

$= \frac{1}{8}x(x+1)^8 - \frac{1}{8} \times \frac{1}{9}(x+1)^9 + c$

$= \frac{1}{72}(x+1)^8(8x-1) + c$

c $\int x^2(x^3+1)^7\,dx = \frac{1}{3}\int 3x^2(x^3+1)^7\,dx$

$= \frac{1}{3} \times \frac{1}{8}(x^3+1)^8 + c$

$= \frac{1}{24}(x^3+1)^8 + c$

d $\int x\sqrt{x+1}\,dx = \int (u^2-1) \times u \times 2u\,du$

$= 2\int (u^4-u^2)\,du$

$= \frac{2}{5}u^5 - \frac{2}{3}u^3 + c$

$= \frac{2}{15}(x+1)^{\frac{3}{2}}(3x-2) + c$

You could also use integration by parts in **d**.
Try it yourself to show that the two methods give the same answer.

Exercise 10.9

Integrate these functions with respect to x.

1 $x^2(x-3)$ **2** $x(x-3)^2$ **3** $(x-6)^6$ **4** $3x^2(x^3-2)^7$

5 $x^2\sqrt{x^3+1}$ **6** $\frac{x^2}{x^3+1}$ **7** $x^2\cos(2x^3)$ **8** $3\cos^2 x$

9 $2\sin^2 3x$ 10 $4\sec^2 x$ 11 $5\cos 4x$ 12 $x\sin 2x$

13 $x^2\ln(2x)$ 14 $\frac{3x}{x^2+7}$ 15 $x(x+1)^4$ 16 $x(x^2+2)^5$

17 xe^{2x^2} 18 xe^{2x} 19 $2\tan^2 2x$ 20 $\frac{\cot 2x}{\operatorname{cosec} 2x}$

21 $\tan\left(\frac{1}{2}x\right)\cot^2\left(\frac{1}{2}x\right)$ 22 $\ln(2x)$ 23 $\ln(x^2)$ 24 $\frac{x+2}{x+1}$

25 $(x+3)(x^2+1)$ 26 $\cos\left(2x+\frac{\pi}{2}\right)$ 27 $\tan(2x-\pi)$ 28 $e^{2x}\sin 2x$

29 $\sin 2x\cos 2x$ 30 $\sin 2x\cos 4x$ 31 $\frac{5}{(x+1)(x-4)}$ 32 $\frac{1}{x^2-2x}$

33 $\frac{x}{\sqrt{x-2}}$ 34 $e^x\sqrt{e^x+1}$ 35 $\frac{e^{2x}}{e^{2x}+1}$ 36 $\frac{\cos x}{\sin^4 x}$

37 $\sin 5x\cos 2x$ 38 $\cos x\sin^7 x$ 39 $(x^2-9)^{-1}$ 40 $\sin x\sqrt{\cos x}$

41 Evaluate

a $\int_0^{\frac{\pi}{4}} \sin^2 2x\,dx$ **b** $\int_1^2 \frac{1}{x(2x-1)}\,dx$ **c** $\int_0^{\frac{\pi}{4}} \frac{\tan x}{\sin x}\,dx$ **d** $\int_0^{\frac{\pi}{12}} 3\sin 3x\cos 3x\,dx$

42 Find

a $\int \frac{\sec^2 x}{\tan^3 x}\,dx$ **b** $\int \tan^4 x\,dx$ **c** $\int x^2 e^{x^3}\,dx$ **d** $\int \cos^5 x\,dx$

e $\int \sin^4 x\,dx$ **f** $\int \frac{1}{x\sqrt{\ln x}}\,dx$ **g** $\int \frac{\sec x\tan x}{3+\sec x}\,dx$ **h** $\int 3^x\,dx$

INVESTIGATION

43 What is a geometrical interpretation of the integral $\int_a^b f(x)\,dx$ for $b > a$?

Without evaluating these integrals, find whether they are positive, negative or zero.

a $\int_0^1 \frac{1}{x-2}\,dx$ **b** $\int_0^1 x^5(1-x^2)\,dx$ **c** $\int_{-\frac{\pi}{4}}^{\frac{\pi}{4}} \sin^3 x\,dx$

C4

10.10 Volumes of revolution

The area under a curve $y = f(x)$ between the ordinates $x = a$ and $x = b$ is given by

$$\text{Area } A = \lim_{\delta x \to 0} \sum_{a}^{b} y\delta x = \int_a^b y\,dx$$

If you rotate a rectangular strip of width δx about the x-axis, the strip generates a thin circular disc.

The volume of this thin disc is $\delta V = \pi y^2 \delta x$.

The disc is a cylinder, radius y and thickness δx.

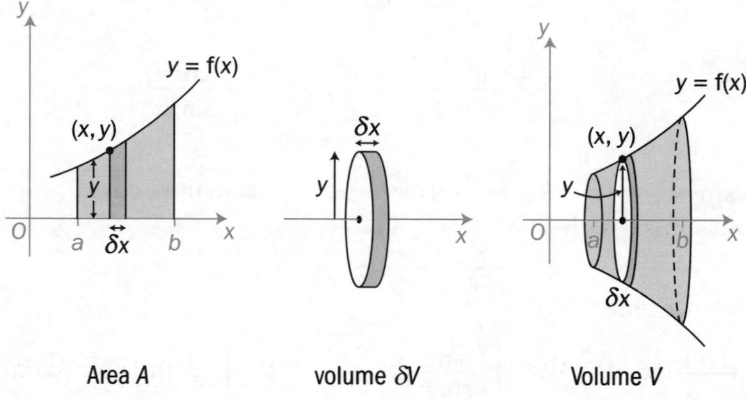

If you rotate the whole of area A about the x-axis, you generate a solid which is formed by summing an infinite number of thin discs.

As the number of discs increases, $\delta x \to 0$ and, in the limit, the summation gives the exact value of the volume, V, of the solid. This is known as a **volume of revolution**.

$$\text{Volume } V = \lim_{\delta x \to 0} \sum_{a}^{b} \pi y^2 \delta x = \int_a^b \pi y^2\,dx$$

EXAMPLE 1

The area enclosed by the curve $y = x^2 + 1$, the x-axis and the ordinates $x = 1$ and $x = 2$ is rotated about the x-axis through 360°.
Find the volume of the solid of revolution.

$$\text{The volume} = \int_1^2 \pi y^2\,dx = \pi\int_1^2 (x^2+1)^2\,dx$$

$$= \pi\int_1^2 (x^4 + 2x^2 + 1)\,dx$$

$$= \pi\left[\frac{1}{5}x^5 + \frac{2}{3}x^3 + x\right]_1^2$$

$$= \pi\left(\frac{32}{5} + \frac{16}{3} + 2 - \frac{1}{5} - \frac{2}{3} - 1\right)$$

$$= \frac{178}{15}\pi \text{ cubic units}$$

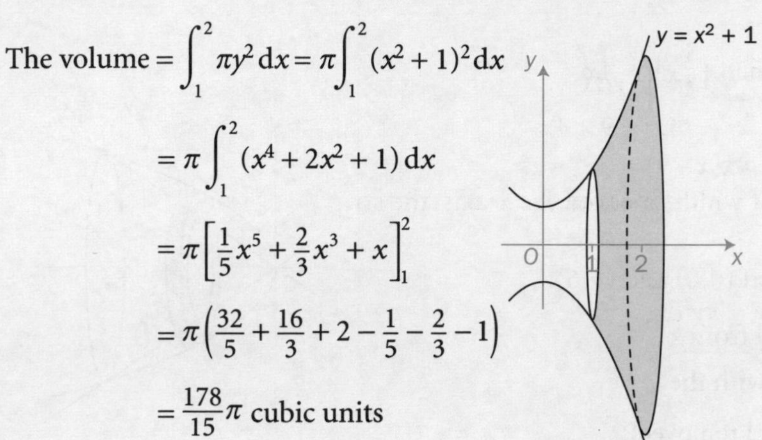

π is a constant. You can take it outside the integral sign.

You can leave your answer as a multiple of π.

EXAMPLE 2

Show that the volume of a sphere of radius r is given by $\frac{4}{3}\pi r^3$.

The circle $x^2 + y^2 = r^2$ has a radius r and a centre (0,0).

The shaded semicircle is rotated about the x-axis through 360° to make a sphere of radius r.

Imagine a thin disc of thickness δx and radius y.

The volume of the disc, $\delta V = \pi y^2 \delta x = \pi(r^2 - x^2)\delta x$

$$\text{So, the volume of the sphere} = \int_{-r}^{r} \pi(r^2 - x^2)\,dx$$

$$= \pi\left[r^2x - \frac{1}{3}x^3\right]_{-r}^{r}$$

$$= \pi\left(r^3 - \frac{1}{3}r^3 - r^2(-r) + \frac{1}{3}(-r)^3\right)$$

$$= \frac{4}{3}\pi r^3$$

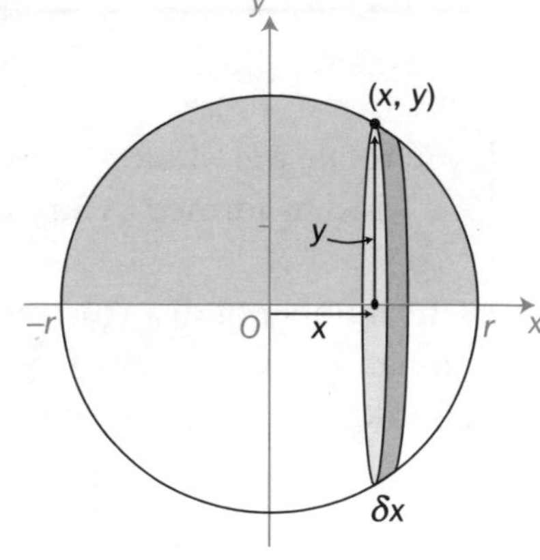

EXAMPLE 3

A hollow bowl is formed by a solid of revolution when the area between the parabola $y^2 = 4x$ and the straight line $y = \frac{2}{3}x$ is rotated 360° about the x-axis.
Find the volume of the bowl.

The curve and line intersect when $\left(\frac{2}{3}x\right)^2 = 4x$

$$4x^2 = 9 \times 4x$$
$$4x(x-9) = 0$$
$$x = 0 \text{ or } 9$$
$$y = 0 \text{ or } 6$$

The points of intersection are at (0,0) and (9,6).

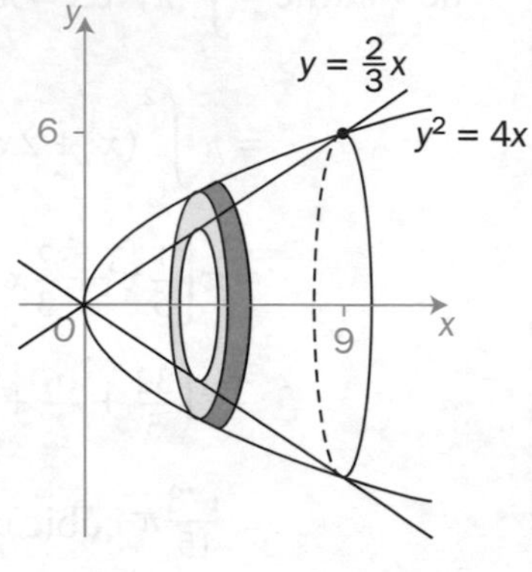

A solid of revolution is formed from thin discs (of radius $y = \sqrt{4x}$) with the central parts $\left(\text{of radius } y = \frac{2}{3}x\right)$ removed.

The volume of the disc, $\delta V = \pi\left(\sqrt{4x}\right)^2 \delta x - \pi\left(\frac{2}{3}x\right)^2 \delta x$

$$= \left(\pi \times 4x - \pi \times \frac{4}{9}x^2\right)\delta x$$

The volume of the solid of revolution

$$= \lim_{\delta x \to 0} \sum_{x=0}^{x=9} \delta V$$

$$= \int_0^9 \left(\pi \times 4x - \pi \times \frac{4}{9}x^2\right) \mathrm{d}x$$

$$= \pi \int_0^9 \left(4x - \frac{4}{9}x^2\right) \mathrm{d}x$$

$$= \pi\left[2x^2 - \frac{4}{27}x^3\right]_0^9$$

$$= \pi\,(162 - 108 - 0)$$

$$= 54 \text{ cubic units}$$

A slightly quicker method is to realise that the 'hollow' in the bowl is a cone of base radius 6 units and height 9 units.
The volume of the cone is thus $\frac{1}{3}\pi \times 36 \times 9 = 108\pi$ and you can subtract it immediately from

$$\int_0^9 \pi \times 4x \,\mathrm{d}x = 162\pi$$

Curves with parametric equations

If a curve is specified by parametric equations $x = \mathrm{f}(t)$, $y = \mathrm{g}(t)$, then the volume V of the solid of revolution is given by

$$V = \int_a^b \pi y^2 \,\mathrm{d}x = \int_{t_1}^{t_2} \pi y^2 \frac{\mathrm{d}x}{\mathrm{d}t} \,\mathrm{d}t$$

The independent variable is t. The limits of this integral are $t = t_1$ and $t = t_2$.

C4

EXAMPLE 4

The curve with parametric equations $x = t^2 + 1$, $y = t + \frac{1}{t}$ is shown in this diagram.

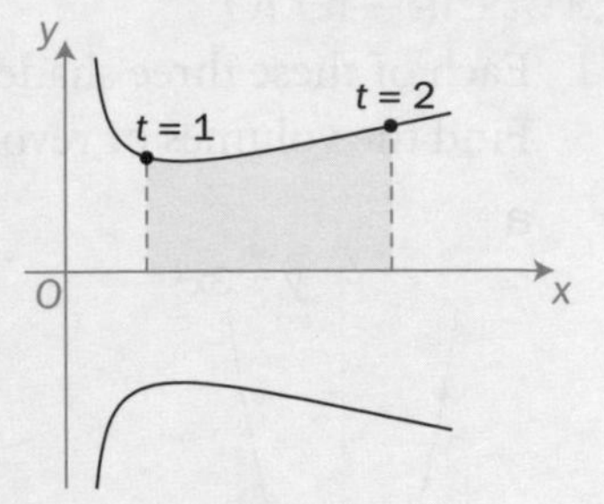

Find the volume of revolution when the shaded area bounded by the curve from $t = 1$ to $t = 2$ is rotated about the x-axis through an angle of 2π.

The volume $V = \int_1^2 \pi y^2 \frac{dx}{dt} dt = \int_1^2 \pi\left(t + \frac{1}{t}\right)^2 \times 2t\, dt$

$\frac{dx}{dt} = 2t$

The limits are values of t (not values of x).

Expand the bracket $\left(t + \frac{1}{t}\right)^2$:

$$= 2\pi\int_1^2 \left(t^3 + 2t + \frac{1}{t}\right) dt$$

$$= 2\pi\left[\frac{1}{4}t^4 + t^2 + \ln t\right]_1^2 = \frac{27}{2}\pi + \pi \ln 4 \text{ cubic units}$$

EXAMPLE 5

A circle of radius r has parametric equations

$x = r\cos\theta$, $y = r\sin\theta$

The shaded semicircle is rotated 360° about the x-axis.

Show that the volume V of the sphere generated is $\frac{4}{3}\pi r^3$

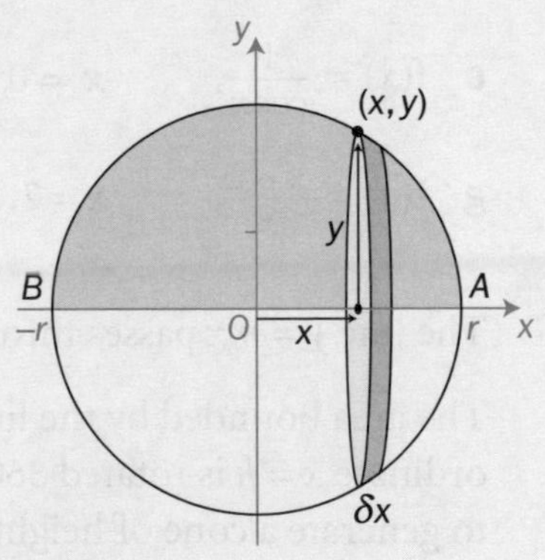

At points A and B, $y = 0$, $\sin\theta = 0$, so $\theta = 0$ and π respectively.

$$V = \int_0^\pi \pi y^2 \frac{dx}{d\theta} d\theta = \int_0^\pi \pi(r\sin\theta)^2 \times (-r\sin\theta)\, d\theta$$

$\frac{dx}{d\theta} = -r\sin\theta$

$$= -\pi r^3 \int_0^\pi (1 - \cos^2\theta)\sin\theta\, d\theta$$

Using $\sin^2\theta + \cos^2\theta = 1$

$$= -\pi r^3 \int_0^\pi (\sin\theta - \sin\theta\cos^2\theta)\, d\theta$$

$$= -\pi r^3\left[-\cos\theta + \frac{1}{3}\cos^3\theta\right]_0^\pi$$

Either recognise the integral of $-\sin\theta\cos^2\theta$ on sight or use the substitution $u = \cos\theta$

$$= -\pi r^3\left(\left(1 - \frac{1}{3}\right) - \left(-1 + \frac{1}{3}\right)\right) = -\frac{4}{3}\pi r^3$$

The negative sign indicates that you have integrated anticlockwise around the circle from A (where $t = 0$) to B (where $t = \pi$).

The volume of the sphere $= \frac{4}{3}\pi r^3$ cubic units

C4

Exercise 10.10

1 Each of these three shaded areas is rotated 360° about the x-axis. Find the volumes of revolution.

a

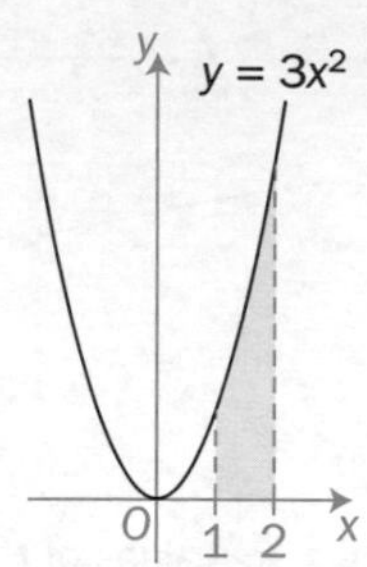

b

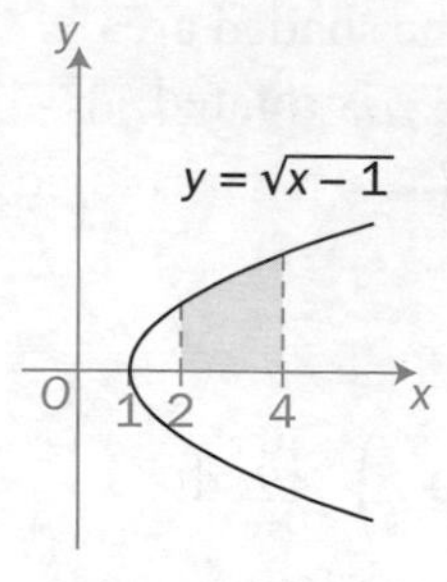

c

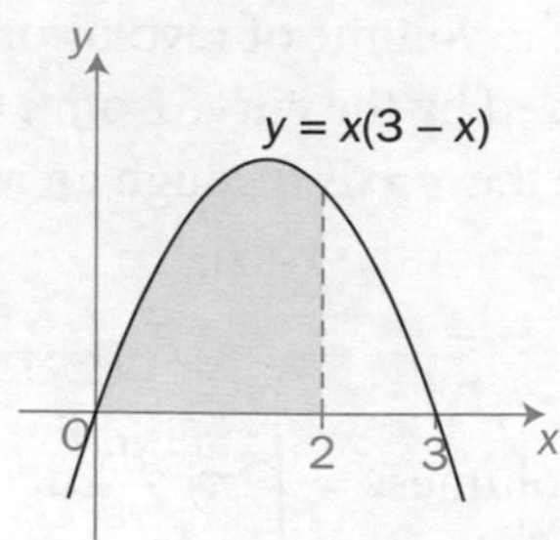

2 In each case, the region R is bounded by the curve $y = \mathrm{f}(x)$, the x-axis and the given ordinates. Find the volume generated when R is rotated through an angle of 2π about the x-axis.

a $\mathrm{f}(x) = x^2 - 1,\quad x = 1, x = 2$ **b** $\mathrm{f}(x) = \dfrac{1}{\sqrt{x}},\quad x = 2, x = 3$

c $\mathrm{f}(x) = x(x - 2),\quad x = 0, x = 1$ **d** $\mathrm{f}(x) = \sin x,\quad x = 0, x = \dfrac{\pi}{2}$

e $\mathrm{f}(x) = \dfrac{1}{\cos x},\quad x = 0, x = \dfrac{\pi}{4}$ **f** $\mathrm{f}(x) = \mathrm{e}^x\sqrt{x},\quad x = 1, x = 2$

g $\mathrm{f}(x) = \dfrac{x - 1}{x},\quad x = 2, x = 3$ **h** $\mathrm{f}(x) = \sqrt{x \ln x},\quad x = 2, x = 3$

3 The line $y = mx$ passes through the point (h, r).

The area bounded by the line, the x-axis and the ordinate $x = h$ is rotated 360° about the x-axis to generate a cone of height h.

Find the value of m in terms of r and h.

Prove that the volume of the cone is $\frac{1}{3}\pi r^2 h$.

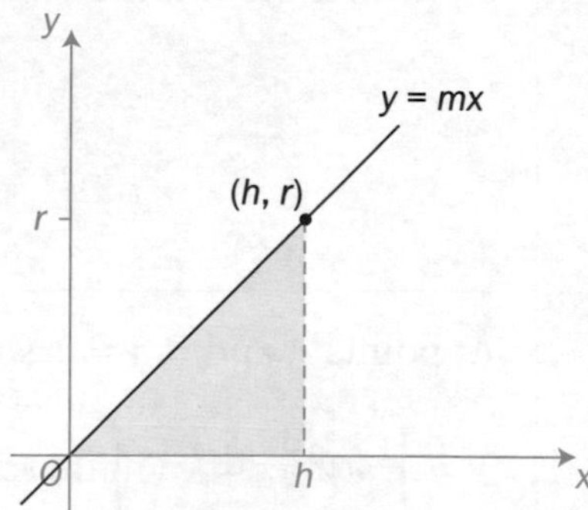

4 A solid of revolution is formed by rotating the curve $y = x\sqrt{4 - x^2}$, where $x > 0$, a full turn about the x-axis. Sketch the graph of the curve and find the volume of the solid.

5 The region R is defined as the area between the curve $y = \tan x$, the x-axis and the line $x = \frac{\pi}{4}$. If R is rotated 180° about the x-axis, find the volume of the solid which is generated.

6 a Find the volume generated when the area enclosed by the curve $y = x^2$, the x-axis and the line $x = 3$ is rotated 360° about the x-axis.

b Find the points of intersection of the curve $y = x^2$ and the line $x + y = 12$. Find the volume generated when the area in the first quadrant enclosed by the curve, the line and the y-axis is rotated 360° about the x-axis.

7 a Find the points of intersection of the line $y = 2x$ and the curve $xy = 8$

b The area in the first quadrant bounded by this line and curve, the x-axis and the ordinate $x = 4$ is rotated through 2π about the x-axis. Find the volume generated.

8 The area between the two curves $y = x^3$ and $y^2 = x$ is rotated through 2π about the x-axis. Find the volume of the solid which is generated.

9 The region R is defined as the area enclosed by the y axis and the two curves $y = \sin x$ and $y = \cos x$. If R is rotated through 2π about the x-axis, find the volume generated.

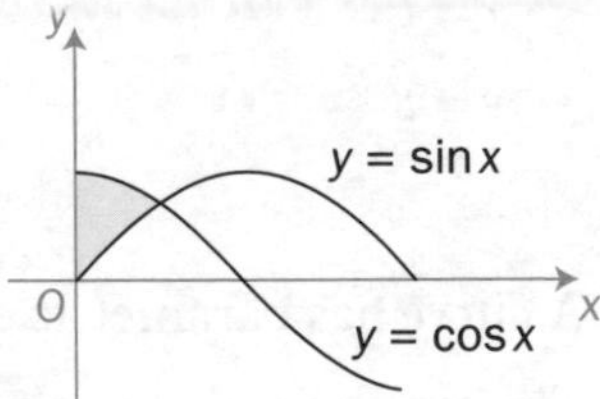

10 A metal washer has the shape of a solid of revolution. The shaded area in this diagram, enclosed between the curve $y = x^2 - x + 4$ and the line $y = 4$, is rotated through an angle of 2π about the x-axis to form the washer.

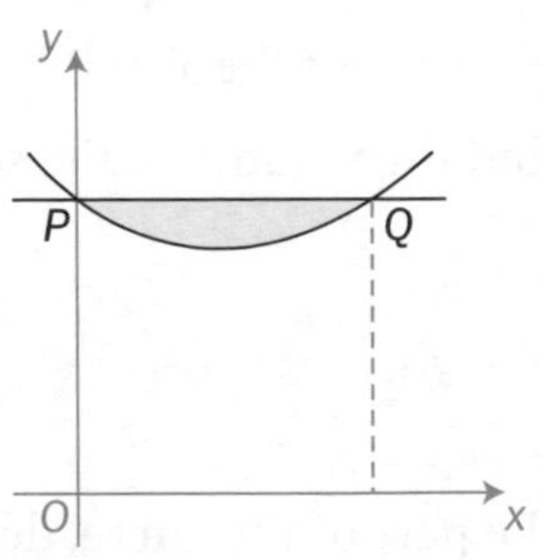

a Write down the coordinates of the points of intersection P and Q.

b Find the volume of revolution.

11 If the area between the curve $y = \ln x$, the x-axis and the ordinate $x = 3$ is rotated through 180° about the x-axis, find the volume generated.

12 A sphere of radius 13 cm is intersected by two parallel planes 7 cm apart.
The larger intersection is 5 cm at its nearest point from the centre of the sphere.
Find that volume of the sphere which lies between the two planes.

13 Find the volumes of revolution when the shaded areas on these diagrams are rotated about the x-axis through 360°.

a

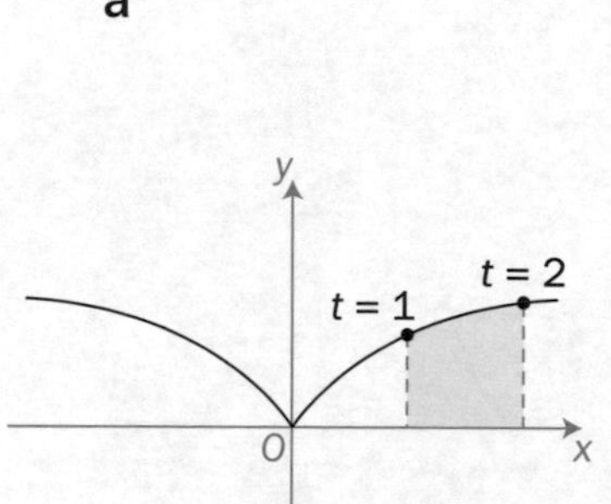

b

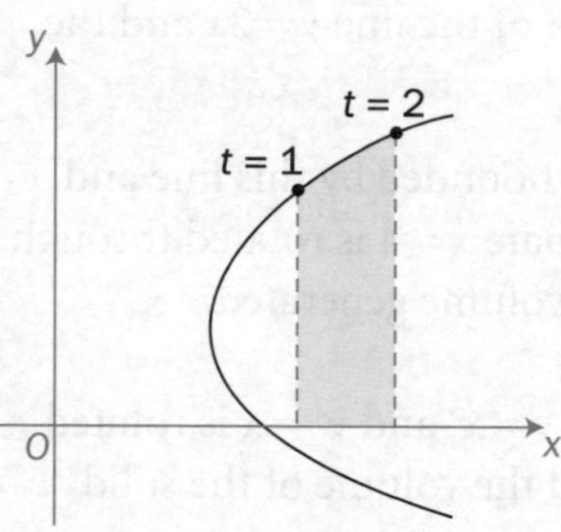

c

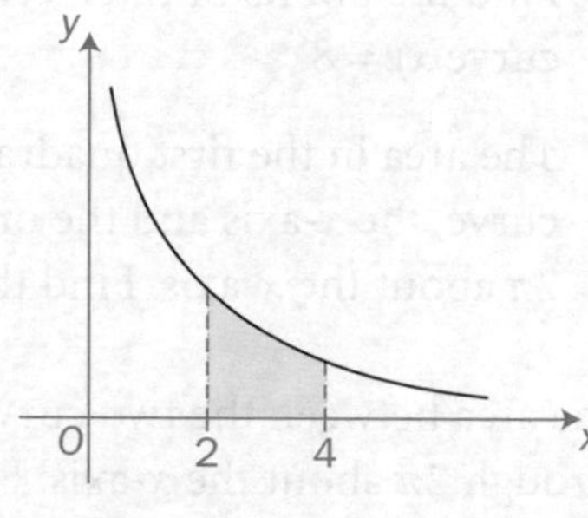

The parametric equations are

a	b	c
$x = t^3$	$x = t^2 + 3$	$x = 4t$
$y = t^2$	$y = 3t + 1$	$y = \frac{2}{t}$

14 A curve has parametric equations $x = \frac{1}{t}$, $y = t^2$
The area bounded by the curve, the x-axis and the ordinates $x = \frac{1}{2}$ and $x = 2$ is rotated through an angle of 2π about the x-axis.
Find the volume of the solid of revolution.

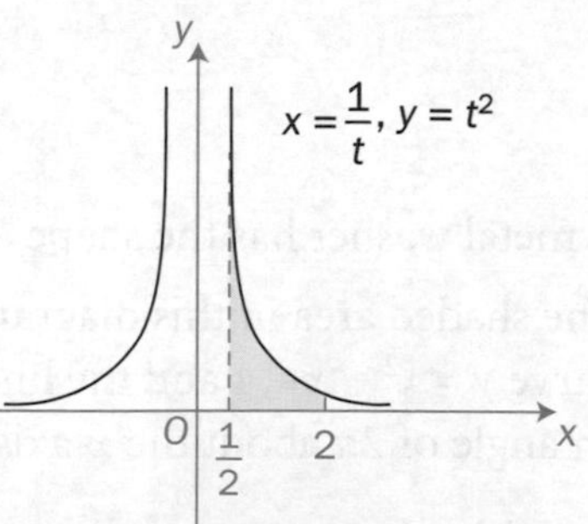

15 The part of the curve, defined parametrically by $x = t^2 - 1$, $y = e^t$, from the ordinate of P where $t = 0$ to the ordinate of Q where $t = -1$ is rotated about the x-axis through 360°.

Find the volume of revolution.

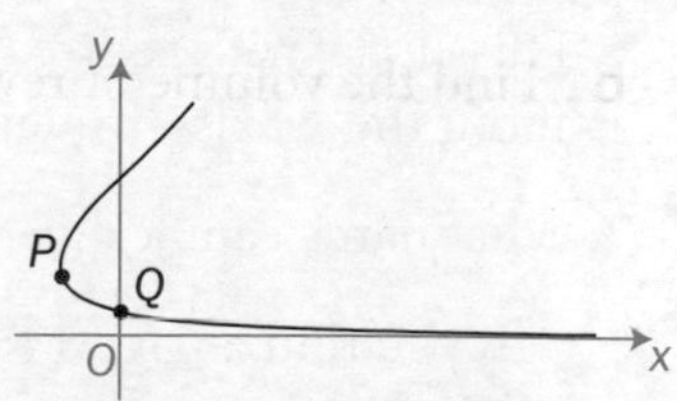

16 That part of the curve, defined parametrically by $x = t^2$, $y = \dfrac{1}{1+t}$, which lies between the ordinates of $R\left(1, \frac{1}{2}\right)$ and $S\left(4, \frac{1}{3}\right)$ is rotated a full turn about the x-axis.

Find the volume of revolution.

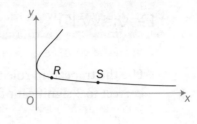

17 The shape of a rugby ball is called a spheroid. It is the solid of revolution when an ellipse is rotated about one of its axes. Find the volume of the spheroid which is generated by rotating the ellipse with parametric equations $x = a\cos\theta$, $y = b\sin\theta$ about the x-axis.

18 This diagram shows the curve with parametric equations $x = \sin\theta$, $y = \sin 2\theta$

Find the volume of revolution when one of the loops is rotated a full turn about the x-axis.

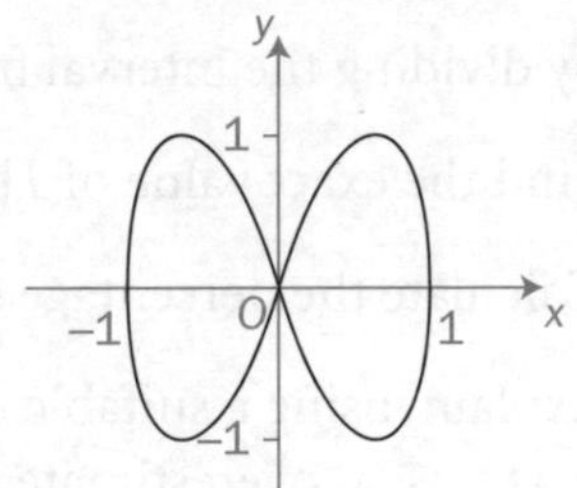

19 Find the volume of the solid of revolution when the area bounded by the curve $x = \tan\theta$, $y = \sin\theta$, the x-axis and the ordinates $x = 1$ and $x = \sqrt{3}$ is rotated through an angle of 2π about the x-axis.

INVESTIGATION

20

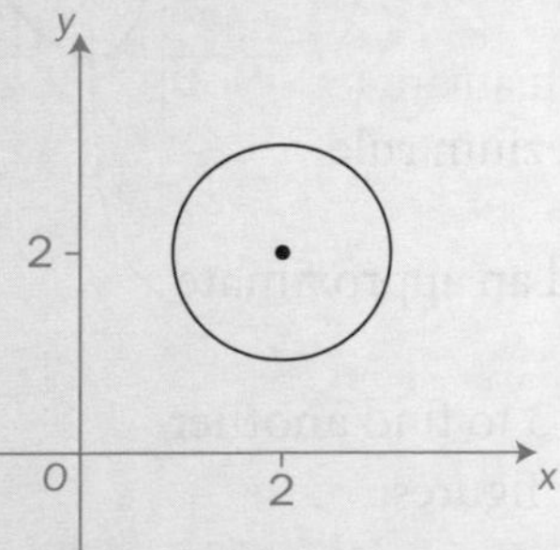

This circle, with centre (2, 2) and radius 1 unit, is rotated about the x-axis through an angle of 2π to generate a solid.

What name can be given to the solid?

Can you find an expression for the volume of the solid?

1 Use the trapezium rule to estimate the value of these integrals correct to 3 significant figures using the number of strips given.

a $\int_0^2 3^x \mathrm{d}x$ 5 strips

b $\int_0^1 e^{x^2} \mathrm{d}x$ 6 strips

c $\int_2^4 \ln(x^2 - 1)\,\mathrm{d}x$ 6 strips

d $\int_0^{\frac{\pi}{4}} \sqrt{\tan\theta}\,\mathrm{d}\theta$ 5 strips

2 a Use the trapezium rule to estimate the value of $I = \int_0^{\frac{\pi}{4}} \sec^2 x\,\mathrm{d}x$ by dividing the interval from 0 to $\frac{\pi}{4}$ into 5 strips.

b Find the exact value of I by integration.

c Calculate the percentage error in the estimated value of I.

d Explain, using a suitable diagram, why the answer to part **a** is an overestimate of the exact value of I.

3 This figure shows a sketch of the curve with equation $y = (x - 1)\ln x$, $x > 0$

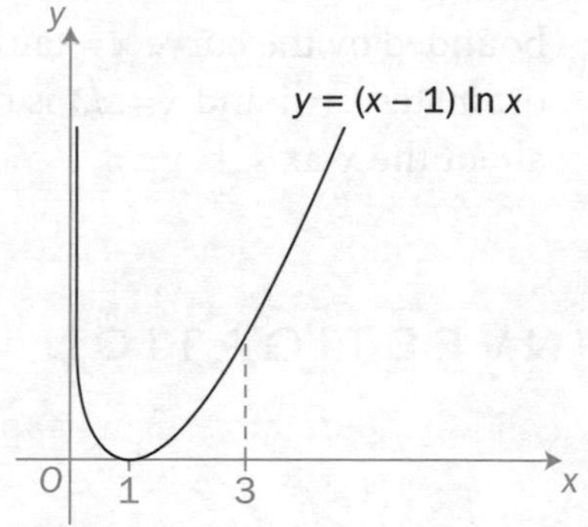

a Copy and complete the table with the values of y corresponding to $x = 1.5$ and $x = 2.5$

x	1	1.5	2	2.5	3
y	0		ln 2		2ln 3

b Given that $I = \int_1^3 (x - 1)\ln x\,\mathrm{d}x$, use the trapezium rule

i with values of y at $x = 1$, 2 and 3 to find an approximate value for I to 4 significant figures

ii with values of y at $x = 1$, 1.5, 2, 2.5 and 3 to find another approximate value for I to 4 significant figures.

c Explain, with reference to the figure shown, why an increase in the number of values improves the accuracy of the approximation.

d Show, by integration, that the exact value of $\int_1^3 (x - 1)\ln x\,\mathrm{d}x$ is $\frac{3}{2}\ln 3$. [(c) Edexcel Limited 2006]

C4

4 Use standard forms to integrate these expressions with respect to x.

a $\cos 3x$ **b** $\sec^2 5x$ **c** $\sin\frac{x}{4}$ **d** $(6x+1)^4$

e $\cot 2x$ **f** $\frac{1}{4x+3}$ **g** $\frac{1}{(4x+3)^2}$ **h** e^{4x}

i $(e^x+1)^2$ **j** $\operatorname{cosec}^2\left(\frac{x}{2}+1\right)$ **k** $\sec 2x$ **l** $\sec 2x\tan 2x$

5 Find

a $\int 2x(x^2+3)^5\,dx$ **b** $\int \frac{x}{x^2+3}\,dx$ **c** $\int x\cos(x^2+1)\,dx$ **d** $\int \cos x\sin^6 x\,dx$

e $\int \frac{x-1}{x^2-2x-1}\,dx$ **f** $\int \frac{\sec^2 x}{1+\tan x}\,dx$ **g** $\int xe^{-x^2}\,dx$ **h** $\int x^2\sqrt{1+x^3}\,dx$

6 Use the given substitutions to find these integrals.

a $\int \frac{1}{(x+3)^2}\,dx \quad u=x+3$ **b** $\int (x-1)(x-4)^3\,dx \quad u=x-4$

c $\int e^x\sqrt{e^x+3}\,dx \quad u=e^x+3$ **d** $\int \frac{x}{\sqrt{x+1}}\,dx \quad u^2=x+1$

7 Calculate the value of $\int_1^2 \left(\frac{x}{x+2}\right)^2 dx$ using the substitution $u=x+2$

8 Use the substitution $x=\sin\theta$ to find the exact value of $\int_0^{\frac{1}{2}} \frac{1}{(1-x^2)^{\frac{3}{2}}}\,dx$ [(c) Edexcel Limited 2005]

9 Find these integrals, using appropriate trigonometric identities where necessary.

a $\int \cos x\operatorname{cosec} x\,dx$ **b** $\int \tan^2 x\operatorname{cosec}^2 x\,dx$ **c** $\int \frac{\tan x}{\sqrt{1-\cos^2 x}}\,dx$

d $\int \cos^2 3x\,dx$ **e** $\int \sin^2 3x\,dx$ **f** $\int (2+\tan x)^2\,dx$

g $\int \cos^4\left(\frac{x}{2}\right)dx$ **h** $\int \cos^3\left(\frac{x}{2}\right)dx$ **i** $\int \sin^5 x\,dx$

j $\int \sin^4 2x\,dx$ **k** $\int (\tan^4 x-\sec^4 x)\,dx$ **l** $\int \left(1-\sin^2\left(\frac{1}{4}x\right)\right)dx$

10 **a** By expanding $\cos(A+B)$ and $\cos(A-B)$, show that
$2\cos A\cos B=\cos(A-B)+\cos(A+B)$
Hence, find $\int \cos 6x\cos 4x\,dx$

b Use a similar method with $\sin(A\pm B)$ to find the value of $\int_0^{\frac{\pi}{4}} \sin 6x\cos 4x\,dx$

C4

11 Integrate using partial fractions.

a $\int \frac{x+3}{(x-1)(x-3)}\,dx$ **b** $\int \frac{5x}{(2x-1)(x+2)}\,dx$ **c** $\int \frac{7}{(x+1)(x-3)^2}\,dx$

12 Find the value of

a $\int_2^3 \frac{1}{(x-1)(2x-1)}\,dx$ **b** $\int_1^2 \frac{9}{x^2(3-x)}\,dx$

13 $g(x) = \frac{5x+8}{(1+4x)(2-x)}$

a Express g(x) in the form $\frac{A}{(1+4x)} + \frac{B}{(2-x)}$, where A and B are constants to be found.

b The finite region R is bounded by the curve with equation $y = g(x)$, the coordinate axes and the line $x = \frac{1}{2}$

Find the area of R, giving your answer in the form $a \ln 2 + b \ln 3$ [(c) Edexcel Limited 2003]

14 Use integration by parts to find

a $\int x\cos x\,dx$ **b** $\int x\cos 3x\,dx$ **c** $\int xe^{3x}\,dx$ **d** $\int x\sin nx\,dx$

e $\int x^2e^{2x}\,dx$ **f** $\int e^{2x}\sin x\,dx$ **g** $\int x^3\ln x\,dx$

15 Evaluate these integrals using integration by parts.

a $\int_0^{\frac{\pi}{4}} x\sin x\,dx$ **b** $\int_0^1 x^2e^x\,dx$ **c** $\int_0^{\frac{\pi}{3}} e^x\cos x\,dx$ **d** $\int_1^2 x^4\ln x\,dx$

16 a Use integration by parts to find $\int x\cos 2x\,dx$

b Hence, or otherwise, find $\int x\cos^2 x\,dx$ [(c) Edexcel Limited 2005]

17 Choose an appropriate method of integration to find each of these integrals.

a $\int x^3(x^2-1)\,dx$ **b** $\int (x-2)(x^2-1)\,dx$ **c** $\int (x-5)^6\,dx$ **d** $\int 2x(x^2-2)^5\,dx$

e $\int x\sqrt{x-2}\,dx$ **f** $\int \frac{x}{x^2-1}\,dx$ **g** $\int \frac{1}{x^2-1}\,dx$ **h** $\int \frac{1}{x^2-x}\,dx$

i $\int xe^{\frac{x^2}{2}}\,dx$ **j** $\int xe^{\frac{x}{2}}\,dx$ **k** $\int x\ln(3x)\,dx$ **l** $\int x^2\ln(3x)\,dx$

18 Integrate

a $\int \sqrt{\frac{1-\sin^2 x}{1-\cos^2 x}}\,dx$ **b** $\int \frac{\sqrt{\sec^2 x - 1}}{\sin x}\,dx$ **c** $\int \cos^3(4x)\,dx$ **d** $\int \sin^4\left(\frac{x}{2}\right)dx$

e $\int (2-\sin^2 x)\,dx$ **f** $\int \sin 7x \sin 3x\,dx$ **g** $\int e^x \sin 3x\,dx$ **h** $\int x^2 \cos 4x\,dx$

19 Evalute

a $\int_1^2 x(x-1)^5\,dx$ **b** $\int_2^3 (x-1)^2 \ln(x-1)\,dx$ **c** $\int_{\frac{\pi}{4}}^{\frac{\pi}{3}} \frac{\sin x}{\sqrt{1+\tan^2 x}}\,dx$

20 The region R is bounded by the curve $y = f(x)$, the x-axis and the given ordinates. In each case, find the volume generated when R is rotated through an angle of 360° about the x-axis.

a $f(x) = 4 - x^2 \quad x = 1, x = 2$ **b** $f(x) = \sin x \quad x = \frac{\pi}{4}, x = \frac{\pi}{2}$

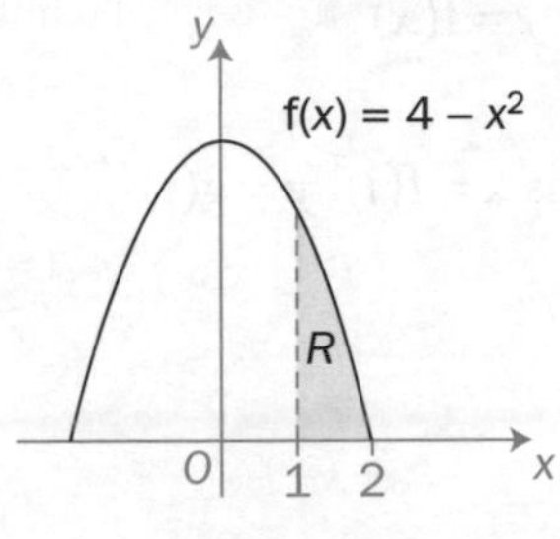

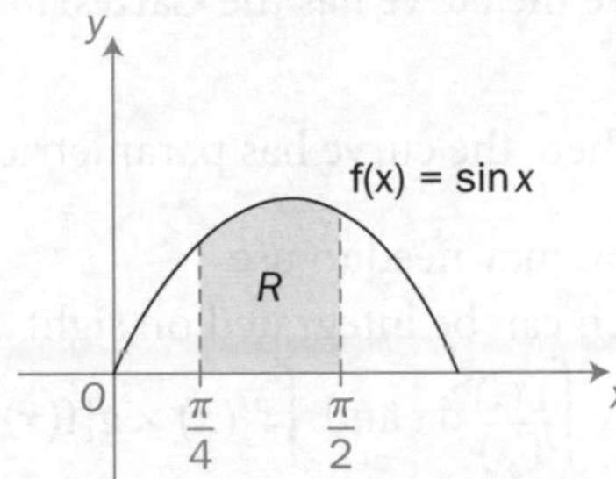

21 The area enclosed by the y-axis, the curve $y = x^2 + 4$ and the straight line $y = 5x$ is rotated about the x-axis through an angle of 2π. Draw a sketch of the area and find the volume of the solid which is generated.

22 Part of a curve is defined by a range of values of the parameter t. In each case, find the volume of revolution when the area between the x-axis and the defined part of the curve is rotated about the x-axis through 360°.

a $x = t^2 + 1, y = t^3 \quad 1 \leqslant t \leqslant 2$ **b** $x = e^t, y = t + 1 \quad 0 \leqslant t \leqslant 1$

23 This diagram shows part of the curve with equation $y = 1 + \frac{1}{2\sqrt{x}}$

The shaded region R, bounded by the curve, the x-axis and the lines $x = 1$ and $x = 4$, is rotated through 360° about the x-axis. Using integration, show that the volume of the solid generated is $\pi\left(5 + \frac{1}{2}\ln 2\right)$

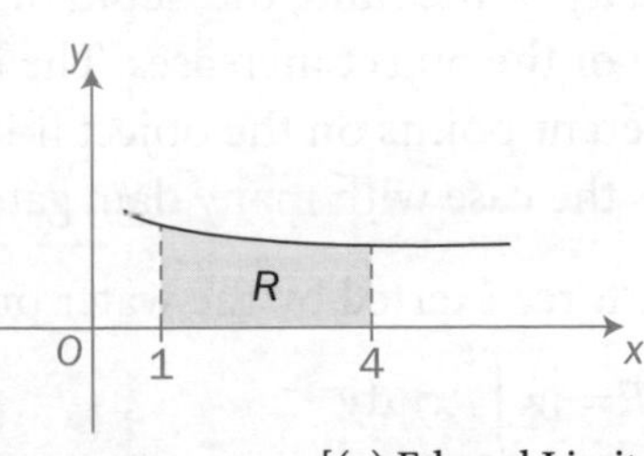

[(c) Edexcel Limited 2003]

10 Exit

Summary

Refer to

- The trapezium rule gives an estimate of the area between a curve and the x-axis. 10.1
- The area A between a curve and the x-axis from $x = a$ to $x = b$ is given by

  either $A = \int_a^b y\,\mathrm{d}x$ where the curve has the Cartesian equation $y = \mathrm{f}(x)$

  or $A = \int_{t_1}^{t_2} y\frac{\mathrm{d}x}{\mathrm{d}t}\,\mathrm{d}t$ where the curve has parametric equations $x = \mathrm{f}(t)$, $y = \mathrm{g}(t)$ 10.2
- The volume of revolution when the area between a curve and the x-axis from $x = a$ to $x = b$ is rotated 360° about the x-axis is given by

  either $V = \int_a^b \pi y^2\,\mathrm{d}x$ where the curve has the Cartesian equation $y = \mathrm{f}(x)$

  or $V = \int_{t_1}^{t_2} \pi y^2\frac{\mathrm{d}x}{\mathrm{d}t}\,\mathrm{d}t$ where the curve has parametric equations $x = \mathrm{f}(t)$, $y = \mathrm{g}(t)$ 10.10
- To integrate a function, you may need to use
  - standard integrals which can be integrated on sight 10.3
  - the two particular cases $\int\frac{\mathrm{f}'(x)}{\mathrm{f}(x)}\,\mathrm{d}x$ and $\int \mathrm{f}'(x) \times \mathrm{g}[\mathrm{f}(x)]\,\mathrm{d}x$ 10.4
  - the method of substitution 10.5
  - trigonometric identities 10.6
  - partial fractions 10.7
  - the method of integration by parts, where $\int u\frac{\mathrm{d}v}{\mathrm{d}x}\,\mathrm{d}x = uv - \int v\frac{\mathrm{d}u}{\mathrm{d}x}\,\mathrm{d}x$ 10.8

Links

Engineers use integration to determine the pressure exerted on the vertical gates of a dam by the water.

Pressure is defined as the force per unit area. In a fluid, the force exerted on a submerged object increases if either the density of the fluid, the depth of the object, or the exposed area of the object increases. The force can also vary at different points on the object if its shape is not uniform, as is the case with many dam gates.

The force exerted by the water on the gate can be modelled by

$$F = w\int_a^b xy\,\mathrm{d}y$$

where w is the density of the water, $(b - a)$ is the vertical length of the gate in the water and x is the horizontal length of the gate at a point at depth y below the surface of the water.

11 Differential equations

This chapter will show you how to

- solve first-order differential equations of the forms
  $\frac{dy}{dx} = f(x)$, $\frac{dy}{dx} = f(y)$ and $\frac{dy}{dx} = f(x)g(y)$
- use the method of 'separating the variables'
- use first-order differential equations to solve problems in practical contexts.

Before you start

You should know how to:

1 Solve simple problems involving gradients of curves.

e.g. A curve passes through the point (1,5) such that $\frac{dy}{dx} = 2x$. Find the equation of the curve.

If $\frac{dy}{dx} = 2x$, then $y = x^2 + c$

To pass through (1,5), $5 = 1^2 + c$

so $c = 4$

The equation of the curve is $y = x^2 + 4$

2 Integrate various functions.

e.g. Find **a** $\int 4e^{-3x}dx$ **b** $\int \frac{1}{x(x+2)}dx$

a $\int 4e^{-3x}dx = 4 \times \frac{e^{-3x}}{-3} + c = -\frac{4}{3}e^{-3x} + c$

b $\int \frac{1}{x(x+2)}dx = \frac{1}{2}\int\left(\frac{1}{x} - \frac{1}{x+2}\right)dx$

$= \frac{1}{2}(\ln x - \ln(x+2)) + c$

$= \frac{1}{2}\ln\left(\frac{x}{x+2}\right) + c$

3 Use exponentials and logarithms.

e.g. If $\ln(x+1) - \ln x = k$, find x in terms of k.

$\ln(x+1) - \ln x = \ln\left(\frac{x+1}{x}\right) = k$

So $\frac{x+1}{x} = e^k$

$1 = x(e^k - 1)$

$x = \frac{1}{e^k - 1}$

Check in:

1 **a** The gradient function of the curve $y = f(x)$ is $f'(x) = x^2 - 5$
If the curve passes through the point (3,−4), find its Cartesian equation.

b The derivative of a curve is $\frac{dy}{dx} = (x+1)^2$
If the curve passes through the point (2,5), find its equation.

2 Integrate

a $(2x+1)^3$

b $\cos 2x + \sec^2 3x$

c $\frac{1}{(x-1)(x+3)}$

d $x\cos x$

e $\frac{x}{x^2+3}$

f $\frac{x}{x-1}$

3 Find x in terms of k when

a $\ln(x-2) = \ln x + \ln 2 + k$

b $2\ln x = \ln(x^2+1) + k$

c $k^2 = 1 + ke^{-x}$

11.1 First-order differential equations

A **differential equation** is a relationship between two (or more) variables and one (or more) of their derivatives.

The order of a differential equation is given by the highest derivative in the equation.

E.g.

$\frac{dy}{dx} = x\sin x$ is a **first-order differential equation** in x and y.

$\frac{d^2y}{dx^2} - 6\frac{dy}{dx} + 9 = 0$ is a **second-order differential equation** in x and y.

This chapter considers only certain types of first-order differential equations.

Consider the differential equation $\frac{dy}{dx} = 2x$

You have $y = \int 2x\,dx$

giving $y = x^2 + c$ where c is an arbitrary constant.

To find the solution of the equation, integrate with respect to x.

This solution is called the **general solution**.

Different values of c give different solutions but the graphs of these different solutions all have the same basic shape. The graphs form a **family of curves**, in which each curve can be formed from any other by a simple translation parallel to the y-axis.

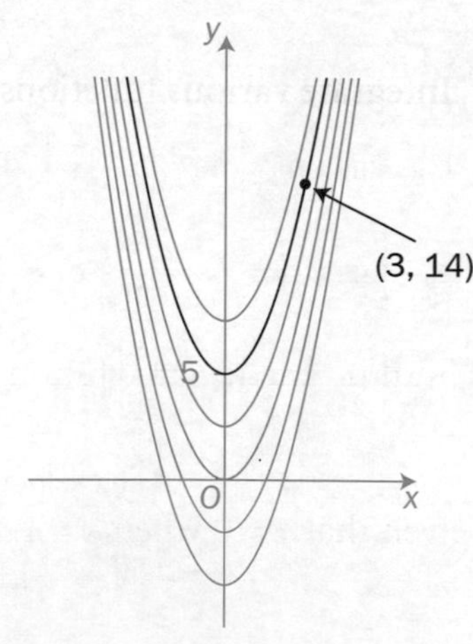

If you choose one member of the family, then that solution is called a **particular solution**.

For example, consider the graph which passes through the point (3, 14).
Substitute $x = 3$ and $y = 14$ into $y = x^2 + c$

$$14 = 3^2 + c \qquad \text{giving } c = 5$$

The particular solution in this case is $y = x^2 + 5$

If $\frac{dy}{dx} = f(x)$, then the general solution is given by

$$y = \int f(x)\,dx = g(x) + c$$

EXAMPLE 1

Find the particular solution of the differential equation $x^2\frac{dy}{dx} - 3x = 1$ if $y = 4$ when $x = 1$.

$$x^2\frac{dy}{dx} - 3x = 1$$

Rearrange:

$$x^2\frac{dy}{dx} = 1 + 3x$$

$$\frac{dy}{dx} = \frac{1+3x}{x^2} = \frac{1}{x^2} + \frac{3}{x}$$

Integrate with respect to x:

$$y = \int\left(\frac{1}{x^2} + \frac{3}{x}\right)dx$$

$$= -\frac{1}{x} + 3\ln x + c$$

This is the general solution.

When $x = 1$, $y = 4$, so $\quad 4 = -1 + 3\ln 1 + c$

$$c = 4 + 1 - 3 \times 0$$
$$= 5$$

The particular solution is $y = -\frac{1}{x} + 3\ln x + 5$

If $\frac{dy}{dx} = f(y)$, then $\frac{dx}{dy} = \frac{1}{f(y)}$
and integrating with respect to y gives the general solution

$$x = \int\frac{1}{f(y)}dy = g(y) + c$$

$$\frac{dx}{dy} = \frac{1}{\frac{dy}{dx}}$$

EXAMPLE 2

Find the particular solution of the differential equation

$$\frac{dy}{dx} = \cos^2 y$$

given that $y = 0$ when $x = 1$.

Rearrange $\frac{dy}{dx} = \cos^2 y$: $\quad \frac{dx}{dy} = \frac{1}{\cos^2 y} = \sec^2 y$

Integrate with respect to y:

The general solution is $\quad x = \int \sec^2 y\, dy = \tan y + c$

Given $y = 0$ when $x = 1$, $\quad 1 = \tan 0 + c$, so $c = 1$

The particular solution is $\quad x = 1 + \tan y$

C4

EXAMPLE 3

Find the particular solution of $(1-y)\frac{dy}{dx} + y^2 = 1$, if $y = 2$ when $x = 0$.

Rearrange: $$\frac{dy}{dx} = \frac{(1+y)(1-y)}{1-y} = 1 + y$$

So $$\frac{dx}{dy} = \frac{1}{1+y}$$

Integrate with respect to y: $$x = \int \frac{1}{1+y}\,dy = \ln(1+y) + c$$

This is the general solution.

When $x = 0$, $y = 2$ so $$0 = \ln(1+2) + c$$ $$c = -\ln 3$$

The particular solution is $$x = \ln(1+y) - \ln 3$$ $$= \ln\left(\frac{1+y}{3}\right)$$ $$e^x = \frac{1+y}{3}$$ $$y = 3e^x - 1$$

If $\frac{dy}{dx} = f(x) \times g(y)$, then the general solution is given by

$$\int \frac{1}{g(y)}\,dy = \int f(x)\,dx$$

You could have used this method on $\frac{dy}{dx} = 1 + y$ in Example 3. Try it yourself to show that it gives the same answer.

This method is called **separating the variables.**

x and y (with the operators dx and dy) are separated onto the two sides of the equation.

EXAMPLE 4

Find the general solution of $xy^2\frac{dy}{dx} = x^2 + 1$

Separate the variables and integrate:

$$\int y^2\,dy = \int \frac{x^2+1}{x}\,dx$$
$$= \int \left(x + \frac{1}{x}\right) dx$$
$$\frac{1}{3}y^3 = \frac{1}{2}x^2 + \ln x + c$$
$$y^3 = \frac{3}{2}x^2 + 3\ln x + c'$$

A new arbitrary constant c' has been introduced where $c' = 3c$

EXAMPLE 5

Find the particular solution of $\frac{dy}{dx} = \frac{y}{x^2 - 1}$ if $y = \sqrt{3}$ when $x = 2$.

Separate the variables and integrate:

$$\int \frac{dy}{y} = \int \frac{dx}{x^2 - 1}$$

$$= \frac{1}{2}\int \left(\frac{1}{x-1} - \frac{1}{x+1}\right) dx$$

Factorise and use partial fractions:

$$\frac{1}{x^2 - 1} = \frac{1}{(x-1)(x+1)}$$

$$\ln y = \frac{1}{2}(\ln(x-1) - \ln(x+1)) + c$$

$$\ln y = \ln\sqrt{\frac{x-1}{x+1}} + \ln A$$

$$= \ln\left(A\sqrt{\frac{x-1}{x+1}}\right)$$

Define a new arbitrary constant A so that $c = \ln A$

You can now incorporate A within the logarithm, using the fact that $\ln p + \ln q = \ln(pq)$

The general solution is $y = A\sqrt{\frac{x-1}{x+1}}$

Substitute for x and y:

$$\sqrt{3} = A\sqrt{\frac{1}{3}}$$

so $A = 3$

The particular solution is $y = 3\sqrt{\frac{x-1}{x+1}}$

Exercise 11.1

1 Find the general solutions of these differential equations.

a $\frac{dy}{dx} = 3x^4$ **b** $\frac{dy}{dx} = \cos 2x$

c $(x^2 + 1)\frac{dy}{dx} = x$ **d** $(x + 1)\frac{dy}{dx} = x$

2 Find the general solutions of these differential equations.

a $\frac{dy}{dx} = 4y^4$ **b** $\frac{dy}{dx} = e^{-2y}$

c $\frac{dy}{dx} - y = 3$ **d** $\frac{dy}{dx} = \cot y$

3 Find the general solutions of these differential equations.

a $\frac{dy}{dx} = \frac{x}{y}$ **b** $\frac{dy}{dx} = \frac{y}{x}$ **c** $\frac{dy}{dx} = xy$

d $x\frac{dy}{dx} - y = 1$ **e** $y\frac{dy}{dt} = e^t$ **f** $t\frac{dy}{dt} = y + ty$

g $t^2\frac{dx}{dt} = x + 1$ **h** $t\frac{dx}{dt} = \cot x$

4 Find the particular solutions of these differential equations.

a $\frac{dy}{dx} = x^2 + x + 1$, given $y = 2$ when $x = 0$

b $\frac{dy}{dx} = \frac{1}{y-1}$, given $y = 2$ when $x = 1$

c $\left(\frac{dy}{dx}\right)^2 = x$, given $y = 4$ when $x = 1$

d $(1+x)^2\frac{dy}{dx} = xy^2$, given $y = 1$ when $x = 0$

e $\frac{dy}{dx} = \frac{\sec y}{\sec x}$, given that $y = 0$ when $x = \frac{\pi}{2}$

f $e^{x+y}\frac{dy}{dx} = 1$, given $y = \ln 2$ when $x = 0$

5 Use a mixture of methods to find the general solutions of these differential equations.

a $(x+1)\frac{dy}{dx} = 1$

b $\frac{dy}{dx} = e^{-3y}$

c $\frac{dy}{dx} = \frac{x+2}{y-2}$

d $\frac{dy}{dx} = \tan y$

e $y\frac{dy}{dx} - x = 1$

f $x\cos y\frac{dy}{dx} = \sin y$

g $3x\frac{dy}{dx} + x = x^2$

h $2y\frac{dy}{dx} + y = 1$

i $\frac{dy}{dx} = \frac{y}{x(x+1)}$

j $\frac{dy}{dx} = \frac{x^2+x}{y^2+y}$

k $\tan x\frac{dy}{dx} = \cot y$

l $(x+1)\frac{dy}{dx} - xy = 0$

m $\cos^2 x\frac{dy}{dx} = \sin^2 x$

n $y\frac{dy}{dx} = \sec y$

o $e^x\frac{dy}{dx} + y^2 = 4$

p $(\cos x - \sin x)\frac{dy}{dx} = 2\sin x$

6 Find the equation of the parabolic curve which passes through the point (–1,2) and for which $(y-1)\frac{dy}{dx} = 4$

7 Show that the curve, which contains the point (3,7) and for which $\frac{dy}{dx} = \frac{1+y}{1+x}$, has the equation $y + 1 = k(x+1)$
Find the value of k.

C4

8 The gradient at the point (x, y) on a curve is given by
$\frac{dy}{dx} = \frac{x}{y\sec x}$
If the curve passes through the point (0,2), find its Cartesian equation.

9 Find the particular solution of the equation $(1 + \cos 2\theta)\frac{dx}{d\theta} = 2$ given that $x = 1$ when $\theta = \frac{\pi}{4}$

10 Find the general solution of the equation $(y^3 + 1)\frac{dy}{dx} - xy = x$

11 The differential equation $xy\frac{dy}{dx} = x^2 + y^2$ can be solved using the substitution $y = xz$

a Find an expression for $\frac{dy}{dx}$ in terms of x and z.

b Eliminate y from $xy\frac{dy}{dx} = x^2 + y^2$ and show that $xz\frac{dz}{dx} = 1$

c Hence, prove that $y^2 = 2x^2\ln(ax)$ where a is a constant.

12 Prove, by using the substitution $y = tz$, that the general solution of the equation $\frac{dy}{dt} = \frac{y(t+y)}{t(y-t)}$ is $Aty = e^{\frac{y}{t}}$, where A is a constant.

INVESTIGATION

13 Oscillations and waves in, for example, simple pendulums and electrical circuits, can be modelled using differential equations. Architects and structural engineers use differential equations when designing buildings and bridges to take account of their natural frequencies when these structures sway in the wind. If the oscillations induced by the wind match the natural oscillations of the structure, then resonance occurs. Use the Internet to investigate

- the early discovery of resonance;
- the conditions under which resonance leads to instability;
- the structural collapse of the Tacoma Narrows Bridge in the USA in 1940.

C4

11.2 Applications of differential equations

You can solve problems involving **rates of change** by forming a differential equation and then using integration.

EXAMPLE 1

The acceleration, $a\,\text{m s}^{-2}$, of a moving particle depends on the time, t seconds, which has elapsed since the start of the motion. When $a = 6\cos 2t$

a find the velocity of the particle, $v\,\text{m s}^{-1}$, in terms of the time t, given that the velocity after $\frac{\pi}{4}$ seconds is $7\,\text{m s}^{-1}$

b find the initial velocity.

---

a Acceleration $\quad a = \dfrac{dv}{dt} = 6\cos 2t$

Acceleration is the rate of change of velocity.

Integrate with respect to t: $\quad v = \int 6\cos 2t\,dt = 3\sin 2t + c$

This is the general solution.

When $t = \frac{\pi}{4}$, $v = 7$, so $\quad 7 = 3\sin\left(2\times\frac{\pi}{4}\right) + c$

$c = 7 - 3 = 4$

The velocity of the particle, $v = 3\sin 2t + 4$

This is the particular solution.

b When $t = 0$, $v = 3\times 0 + 4 = 4$

The initial velocity $= 4\,\text{m s}^{-1}$

EXAMPLE 2

At any given time t, the rate of increase of a population of bacteria is proportional to the size of the population, N. The initial population is 50. If the population has increased to 100 when $t = 1$, find the size of the population when $t = 5$.

The rate of increase of the population is $\dfrac{dN}{dt}$.

So $\dfrac{dN}{dt} \propto N$ giving $\dfrac{dN}{dt} = kN$ where $k > 0$

Separate the variables and integrate:

$$\int \frac{1}{N}\,dN = \int k\,dt$$

$$\ln N = kt + c$$

$$= kt + \ln A$$

Let $c = \ln A$, so that $\ln N - \ln A$ can be combined.

$$\ln\left(\frac{N}{A}\right) = kt$$

$$N = Ae^{kt}$$

You have $N = 50$ at $t = 0$ and also $N = 100$ at $t = 1$

There are two unknown constants, A and k. You need two items of information to find their values.

When $t = 0$, $50 = Ae^0$, so $A = 50$

When $t = 1$, $100 = 50e^k$, so $e^k = 2$

Hence $N = Ae^{kt} = A(e^k)^t = 50\times 2^t$

This is the particular solution.

When $t = 5$, $N = 50\times 2^5 = 1600$

The size of the population when $t = 5$ is 1600.

EXAMPLE 3

The radioactive element strontium-90 has a half-life of 29 years. Find what percentage of the initial amount of radioactive strontium is left after a hundred years.

In radioactive material, atoms disintegrate spontaneously. The rate of disintegration at a given time t is proportional to the amount of radioactive material left in the sample. The amount of radioactive material M is decreasing over time, so the rate of disintegration $\frac{dM}{dt}$ is negative.

You have $\frac{dM}{dt} \propto M$ giving $\frac{dM}{dt} = -kM$ where k is positive.

Separate the variables and integrate:

$$\int \frac{dM}{M} = \int -k\,dt$$

$$\ln M = -kt + c = -kt + \ln A$$

Let $c = \ln A$, so that $\ln M - \ln A$ can be combined.

$$\ln\left(\frac{M}{A}\right) = -kt \text{ giving } M = Ae^{-kt}$$

A half-life of 29 years means that, if the initial amount of strontium is M_0, then $\frac{1}{2}M_0$ remains after 29 years.

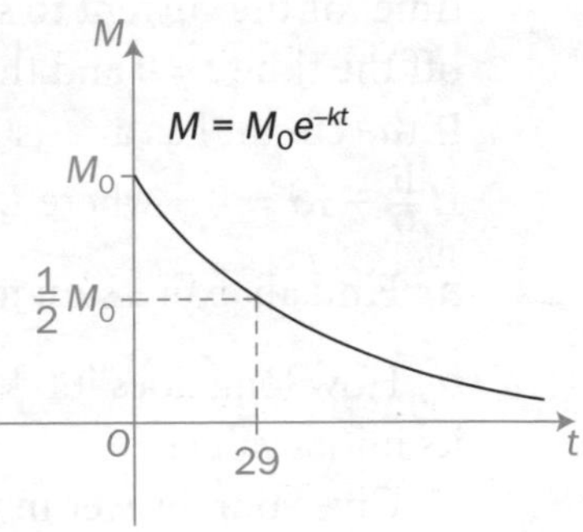

This graph illustrates the half-life of strontium as 29 years.

When $t = 0$, $M_0 = Ae^0$ so $A = M_0$ and $M = M_0e^{kt}$

When $t = 29$, $\frac{1}{2}M_0 = M_0e^{-29k}$ so $e^{-29k} = \frac{1}{2}$

$e^{29k} = 2$ gives $29k = \ln 2$ and $k = \frac{1}{29}\ln 2$

Hence $M = M_0e^{-\left(\frac{\ln 2}{29}\right)t}$

When $t = 100$, $M = M_0e^{-\frac{100\ln 2}{29}} = M_0e^{-2.39}$

$= 0.0916M_0$ (to 3 s.f.)

After 100 years, just over 9% of the strontium is radioactive.

Exercise 11.2

1 As a train enters a tunnel with a velocity of 6 m s^{-1}, its acceleration, $a\text{ m s}^{-2}$, is given by $a = \frac{1}{100}t + \frac{1}{10}$, where t is the time in seconds spent in the tunnel.

a Find an expression for the speed of the train in terms of t.

b If the train leaves the tunnel after 30 seconds, find its speed on exit.

2 The population of a certain kind of insect is growing at a rate which is proportional to the number, N, of insects present at a given time t (in days), where $\frac{dN}{dt} = 0.1N$

The population is monitored over time and $N = 200$ when $t = 0$.

a Find the size of the population one week after monitoring has begun.

b How long will it take for the initial population to increase tenfold?

3 Pine trees in a forest are dying due to a fungal disease. Initially there were 2000 trees, but, with a rate of infection proportional to the number of trees still unaffected by the fungus, the number unaffected after 2 years is 1600.
Find how long it will take before half the trees are infected.

4 A crystal suspended in a chemical solution is increasing in size over time. The rate of increase of volume is inversely proportional to the square of its volume. Initially, the crystal had a volume of 3 cm^3 and, one day later, its volume was 4 cm^3.
How long will it take for its volume to increase to 10 cm^3?

5 When the power is switched off in an electrical circuit, it takes time for the current to stop flowing. At the moment of switching off the time $t = 0$ and the current $i = i_0$.
If the circuit has a resistance R and an inductance L, then
$L\frac{di}{dt} + Ri = 0$ where L and R are constants.

a Find an expression for i as a function of t in terms of L and R.

b How long does it take the current to fall to a value of $\frac{1}{100}$th of its initial value?
Give your answer in terms of R and L.

6 A oil-tank 4 metres tall has a leak. The depth of oil, h metres, in the tank is decreasing over time at a rate, in metres per hour, which is proportional to the square root of the depth.

a Given that the tank is full initially and that the initial rate at which the depth is decreasing is 16 cm per hour, find the depth of oil in the tank after 10 hours.

b How long does it take for the tank to empty?

7 Radium is a radioactive element and it decays at a rate proportional to the mass M of radium which exists in a sample at a time t.

a If the initial mass of radium in a sample is M_0, show that the mass of radium remaining after a time t is given by $M = M_0e^{-kt}$, where k is a constant.

b If the half-life of radium is 1620 years, show that the value of k is approximately 4.28×10^{-4}

c How long will it take for the initial mass of radium to decay by 99%?

8 Newton's law of cooling states that, for an object at a temperature θ°C, the rate of decrease in its temperature is proportional to the difference between its temperature and the ambient temperature. An object is initially at 70 °C in a room at a constant temperature of 10 °C. During the first 10 minutes its temperature falls to 60 °C.

a Prove that $\theta = 10 + 60e^{-kt}$ where $k = \frac{1}{10}\ln\left(\frac{6}{5}\right)$

b Find how much longer elapses before its temperature falls to 50 °C.

c What will its temperature be one hour after it initially started to cool?

9 The growth of a leaf on a plant depends on the water absorbed from the plant and the water lost by evaporation from its surface. If the width of a leaf is w cm, then the rate of increase of the width is equal to $2w - w^2$.

a If $w = 1$ when $t = 0$, show that $w = \frac{2e^{2t}}{1 + e^{2t}}$

b Find the maximum width of the leaf.

10 In a chemical reaction in a solution a new chemical is formed. At a time t seconds, y grams of this chemical are present in the solution and the rate of increase of the chemical is given by $\frac{dy}{dt} = 2(3 - y)(1 - y)$

a Given that $y = 0$ when $t = 0$, find an expression for y in terms of t.

b Find the mass, y grams, which has formed 2 seconds after the start of the reaction.

c Find the limiting value of y as t increases.

INVESTIGATION

11 Shake 100 drawing pins on to a table.

Pins landing point down can be used to simulate radioactive atoms which have decayed in the first second. Remove them.

Repeat over and over again, removing all the pins which land point downwards, until you have only half of the original 100 pins left.

The number of shakes that you have made gives you the half-life in seconds.

Construct a mathematical model and check its validity for predicting the number of shakes required to leave 10 pins.

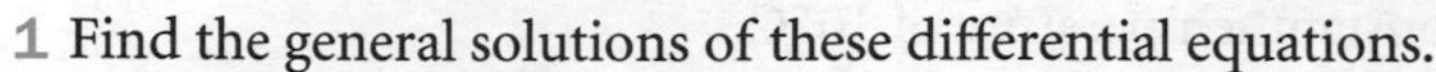

Review 11

1 Find the general solutions of these differential equations.

a $x\frac{dy}{dx} = x^3 + 2$ **b** $(x^3 - 1)\frac{dy}{dx} = 3x^2$

c $\sin x\frac{dy}{dx} = \cos x$ **d** $\frac{dy}{dx} = y^2$

e $\frac{dy}{dx} = e^{-3y}$ **f** $\frac{dy}{dx} + y = 1$

2 Find the particular solutions of these differential equations.

a $\frac{dy}{dx} = 1 - 2x + 2x^2$ given $y = 4$ when $x = 3$

b $y\frac{dy}{dx} = y^2 - 1$ given $y = \sqrt{2}$ when $x = \frac{1}{2}$

3 Separate the variables to find the general solutions of these differential equations.

a $y^2\frac{dy}{dx} = 1 + e^x$ **b** $(x^2 + 1)\frac{dy}{dx} = xy$

c $\sec x\frac{dy}{dx} = \cos^2 y$ **d** $3y\frac{dy}{dx} - y = 1$

e $\frac{dy}{dx} = \frac{y}{x^2 - 3x + 2}$ **f** $y\frac{dy}{dx} = 2x\sec y$

4 Find the particular solutions of these different equations, given that $y = 2$ when $x = 1$.

a $y^3\frac{dy}{dx} = x^2$

b $x\frac{dy}{dx} = y + 1$

c $\operatorname{cosec}(x - 1)\frac{dy}{dx} = \frac{y}{y^2 - 1}$

5 A curve contains the point $(0, \pi)$. Show that, if $\frac{dy}{dx} = x\sec^2 y$, then the curve has the equation $y + \sin y\cos y - x^2 = k$

Find the value of k.

6 Given that $y = \frac{1}{2}$ when $x = \pi$, solve the differential equation

$$\frac{dy}{dx} = y^2x\cos x$$

C4

7 **a** The acceleration, a m s^{-2}, of a moving particle is inversely proportional to the velocity, v m s^{-1}, at which it is travelling after a time of t seconds.

i Find v in terms of t, given that its initial velocity is 10 m s^{-1} and, after 5 seconds, its velocity is 20 m s^{-1}.

ii Find its velocity after a further 5 seconds of motion.

b If the acceleration is directly proportional to the velocity, find its velocity after 10 seconds of motion, given that $v = 10$ when $t = 0$ and $v = 20$ when $t = 5$.

8 Liquid is stored in a cylindrical tank of radius 5 metres. Algae on the surface of the liquid is growing at a rate proportional to the surface area, A, which is covered with algae. The algae was first noticed when it covered 10% of the surface and, after a further 10 days, it covered 20% of the surface.

a Write down a differential equation involving $\frac{dA}{dt}$.

Solve the equation to find the area of the surface covered with algae after another 10 days.

b How many days in total does it take for 75% of the surface to be covered with algae?

9 Newton's law of cooling states that an object at a temperature θ°C cools in such a way that the rate of decrease in its temperature is proportional to the difference between its temperature and room temperature. An object is initially at 70 °C in a room at a constant temperature of 20 °C. During the first 5 minutes, its temperature falls to 60 °C.

a Show that $\theta = 20 + 50e^{-kt}$ and find the value of the constant k.

b How many more minutes elapse before its temperature falls to 50 °C?

10 Chemical A is converted into chemical B during a reaction. At any time during the reaction, the rate at which A is converted into B is proportional to the quantity of A that remains at that time.

a The quantity of A at time t is x and, when $t = 0$, $x = x_0$. Write down a differential equation involving t, x and x_0.

b In a particular experiment, the initial quantity of A is reduced by a half in 5 minutes. Find how many more minutes it takes for A to reduce to only 10% of its initial quantity.

C4

Exit

Summary

Refer to

- $\dfrac{dy}{dx} = f(x)$ has the general solution $y = \int f(x)dx = g(x) + c$ 11.1
- $\dfrac{dy}{dx} = f(y)$ has the general solution $x = \int \dfrac{1}{f(y)}dy = g(y) + c$ 11.1
- $\dfrac{dy}{dx} = f(x) \times g(y)$ has the general solution found from $\int \dfrac{1}{g(y)}dy = \int f(x)dx$ 11.1

  This method is known as **separating the variables.** 11.1
- The **general solution** of a first-order differential equation has just one arbitrary constant. 11.1

  You can represent the general solution graphically by a **family of curves.**
- You can represent the **particular solution** of a first-order differential equation by just one curve selected from the family of curves. 11.1

C4

Links

An example of a differential equation with many applications is the logistic equation

$$\frac{dP}{dt} = P(1 - P)$$

where P is a variable dependent on t.

This differential equation is often used to model population growth, where the rate of reproduction is proportional to both the existing population and the supportability of that population.

In this setting, the equation now takes the form

$$\frac{dP}{dt} = rP\left(1 - \frac{P}{K}\right)$$

where the constant r defines the growth rate of the population, and K is the supportable population within the given environment.

12 Vectors

This chapter will show you how to

- express vectors in different ways and use the components of a vector to calculate its magnitude
- investigate properties of vectors, including how to add and subtract them
- calculate the distance between two points and find their midpoint in 3-dimensional space
- find the scalar product (or dot product) of two vectors
- calculate the angle between two vectors and the intersection of two lines
- find the vector equation of a straight line.

Before you start

You should know how to:

1 Describe a translation using a vector.

e.g. The point (4, 1) maps onto the point (5,3) under a translation given by the vector $\begin{pmatrix}1\\2\end{pmatrix}$.

2 Use Pythagoras' theorem.

e.g. If the two shorter sides of a right-angled triangle are 8 cm and 6 cm, then the longest side is given by $\sqrt{64+36} = 10$ cm

3 Solve simultaneous equations in three unknowns.

e.g. $2x + 3y + z = 8$ (1)

$3x - y + z = 1$ (2)

$x + 4y + z = 9$ (3)

Subtract (2) from (1) to get

$-x + 4y = 7$ (4)

Subtract (3) from (1) to get

$x - y = -1$ (5)

Add (4) and (5): $3y = 6$

$y = 2$

Substitute into (5) and (1):

$x = 1, z = 0$

Check in:

1 **a** Find the vector which maps the point (3,4) onto these points.

i (4,6) **ii** (4,−1) **iii** (2,7)

b Find the image of the square (3,3), (5,5), (3,5), (5,3) under the translation $\begin{pmatrix}6\\-2\end{pmatrix}$.

2 **a** Find the distance from the point (0,0) to the point (4,5).

b A right-angled triangle has two sides of length 5 cm and 7 cm. Calculate two possible values of the length of its third side.

3 Solve these simultaneous equations.

a $3x + 4y - z = 3$

$2x - 2y + z = 7$

$x - 3y + 2z = 8$

b $x + y + 2z = 4$

$2x - 3y + z = 7$

$3x + 2y + 3z = 7$

12.1 Basic definitions and notations

A vector is a quantity which has both a magnitude (size) and a direction. For example, velocity and force are both vectors.

A scalar is a quantity which needs only a magnitude to describe it fully. For example, speed and mass are both scalar quantities.

A displacement vector represents a movement from a point P to a point Q.
The vector is represented by a directed line segment or an arrow.
The length of the arrow represents the magnitude of the vector.

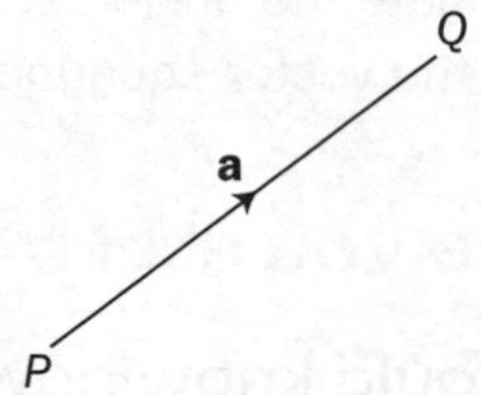

The vector in the diagram is printed as $\overrightarrow{PQ}$, **PQ** or **a**.
When handwritten, this vector can be written as $\overrightarrow{PQ}$, $\underset{\sim}{PQ}$ or $\underset{\sim}{a}$.

The magnitude of the vector $\overrightarrow{PQ}$ (also called its modulus) can

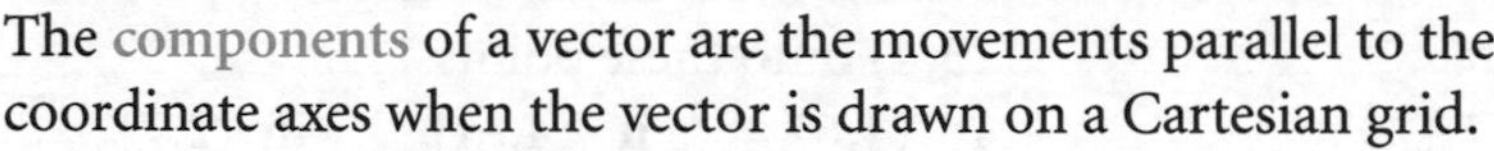

be printed or handwritten as PQ, $|\overrightarrow{PQ}|$, $|\underset{\sim}{a}|$ or just a. It can also be printed as $|\mathbf{a}|$.

The components of a vector are the movements parallel to the coordinate axes when the vector is drawn on a Cartesian grid.

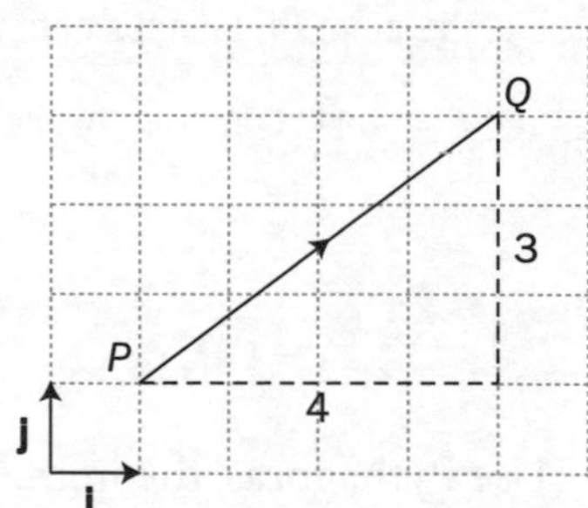

The components of this vector are 4 and 3.

The vector can be written as $\overrightarrow{PQ} = \begin{pmatrix} 4 \\ 3 \end{pmatrix}$

Pythagoras' theorem gives you

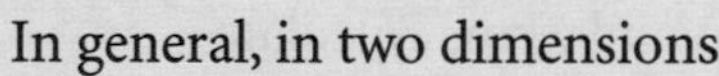

$|\overrightarrow{PQ}|^2 = 4^2 + 3^2 = 25$ and $|\overrightarrow{PQ}| = 5$

In general, in two dimensions

the magnitude of $\overrightarrow{PQ} = \begin{pmatrix} x \\ y \end{pmatrix}$ is $\sqrt{x^2 + y^2}$

and, in three dimensions, the magnitude of

$\overrightarrow{PQ} = \begin{pmatrix} x \\ y \\ z \end{pmatrix}$ is $\sqrt{x^2 + y^2 + z^2}$

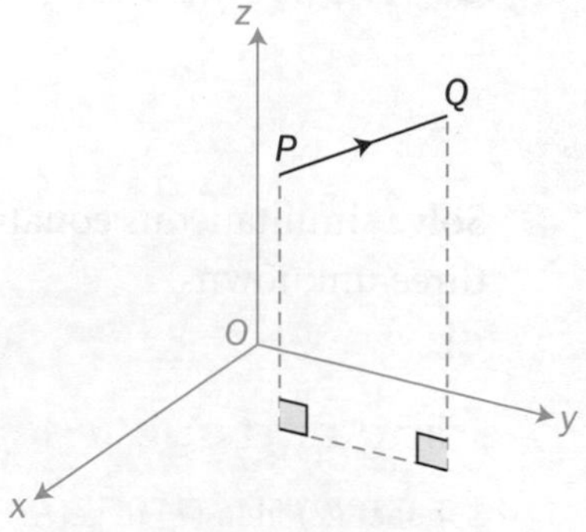

In three dimensions there are three axes for x, y and z.

The zero vector $\underset{\sim}{0}$ has zero magnitude and no direction.

A unit vector is a vector with a magnitude of 1.
The notation for a unit vector uses a 'hat' so that $\hat{\mathbf{a}}$ or $\underset{\sim}{\hat{a}}$ is the unit vector in the direction of vector **a**.

The symbol ~ is known as 'twiddle'.

C4

Unit vectors parallel to the coordinate axes are denoted by

$\mathbf{i} = \begin{pmatrix} 1 \\ 0 \end{pmatrix}$, $\mathbf{j} = \begin{pmatrix} 0 \\ 1 \end{pmatrix}$ in two dimensions

and by $\mathbf{i} = \begin{pmatrix} 1 \\ 0 \\ 0 \end{pmatrix}$, $\mathbf{j} = \begin{pmatrix} 0 \\ 1 \\ 0 \end{pmatrix}$ and $\mathbf{k} = \begin{pmatrix} 0 \\ 0 \\ 1 \end{pmatrix}$ in three dimensions.

The vector $\overrightarrow{PQ} = \begin{pmatrix} 4 \\ 3 \end{pmatrix}$ can be written as $\overrightarrow{PQ} = 4\mathbf{i} + 3\mathbf{j}$

The unit vector parallel to $x\mathbf{i} + y\mathbf{j}$ is $\dfrac{x\mathbf{i} + y\mathbf{j}}{\sqrt{x^2 + y^2}}$

For the vector $\overrightarrow{PQ} = \begin{pmatrix} 4 \\ 3 \end{pmatrix}$ with $|\overrightarrow{PQ}| = 5$, a unit vector in the same direction will have components one-fifth of those of $\overrightarrow{PQ}$.

The unit vector parallel to $\overrightarrow{PQ}$ is $\begin{pmatrix} \frac{4}{5} \\ \frac{3}{5} \end{pmatrix} = \frac{4}{5}\mathbf{i} + \frac{3}{5}\mathbf{j}$

Check: its magnitude

$= \sqrt{\left(\frac{4}{5}\right)^2 + \left(\frac{3}{5}\right)^2} = \sqrt{\frac{16+9}{25}} = 1$

Two vectors are **equal** if they have the same magnitude and the same direction.

One vector is the **negative** of another vector if they have the same magnitude but opposite directions.

One vector is a **scalar multiple** of another vector if they have the same direction but different magnitudes.
In general, the vector $k\mathbf{a}$ is parallel to the vector $\mathbf{a}$ and has a magnitude k times that of $\mathbf{a}$.

If k is positive, $k\mathbf{a}$ and $\mathbf{a}$ have the same direction.
If k is negative, $k\mathbf{a}$ and $\mathbf{a}$ have opposite directions.

e.g. Consider $\mathbf{a} = \begin{pmatrix} 2 \\ -1 \end{pmatrix}$, $\mathbf{b} = \begin{pmatrix} 4 \\ -2 \end{pmatrix}$ and $\mathbf{c} = \begin{pmatrix} -2 \\ 1 \end{pmatrix}$

You have $\mathbf{b} = 2\mathbf{a}$, $\mathbf{c} = -\mathbf{a}$ and $\mathbf{b} = -2\mathbf{c}$

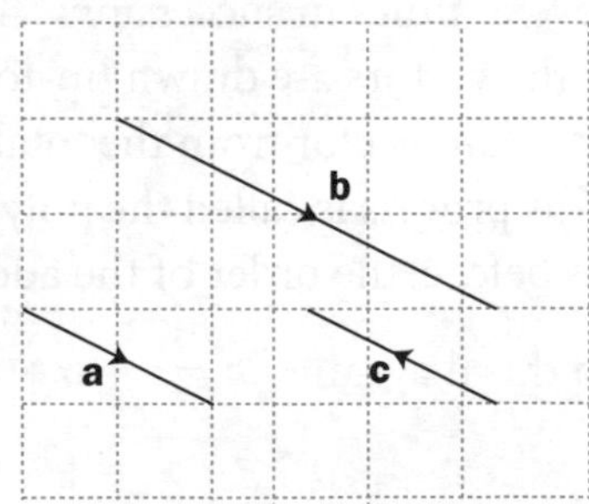

Adding and subtracting vectors

You can represent two successive displacements by two vectors $\overrightarrow{PQ}$ and $\overrightarrow{QR}$.
They are equivalent to one displacement given by the vector $\overrightarrow{PR}$.

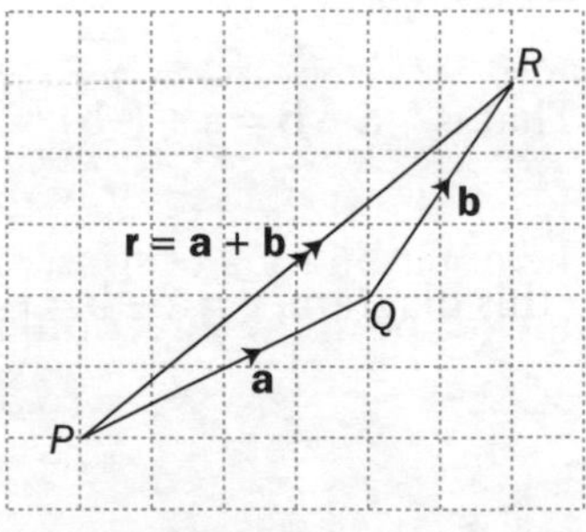

So $\overrightarrow{PR} = \overrightarrow{PQ} + \overrightarrow{QR}$
or $\mathbf{r} = \mathbf{a} + \mathbf{b}$

C4

$\overrightarrow{PR}$ (or **r**) is called the **resultant vector** or, simply, the **resultant**.

Using the components of the vectors, the resultant of the vectors $\overrightarrow{PQ}$ and $\overrightarrow{QR}$ is $\begin{pmatrix} 6 \\ 5 \end{pmatrix}$.

On diagrams, the resultant is marked by a double arrow.

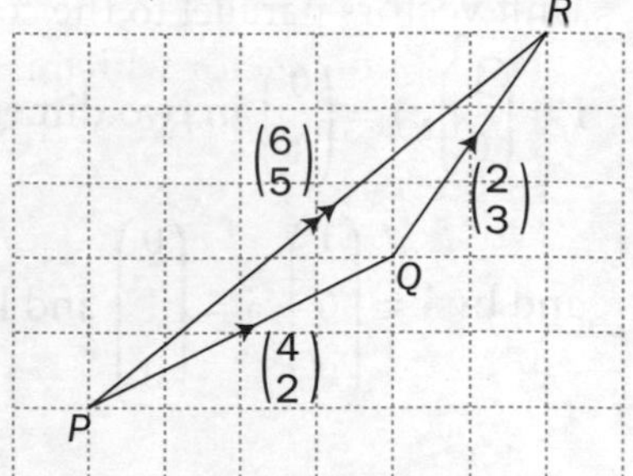

You can change the order of displacements without affecting the result.

So $\mathbf{r} = \mathbf{a} + \mathbf{b} = \mathbf{b} + \mathbf{a}$

When the two vectors are drawn tip-to-tail to make a triangle, the addition process is called the **triangle law** of addition.

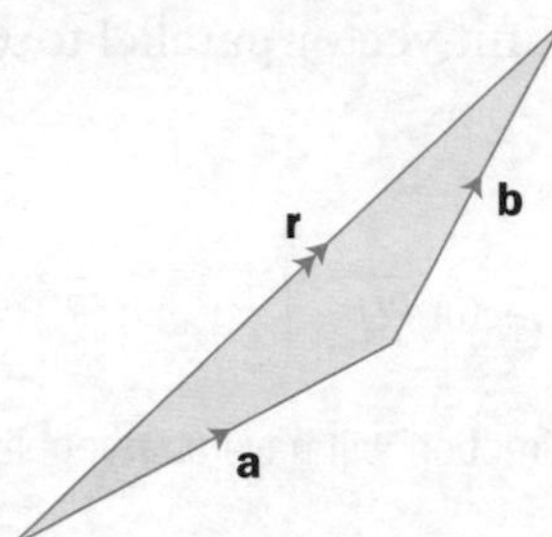

When the two vectors start from the same point to make a parallelogram, the addition process is called the **parallelogram law** of addition.

The triangle law and the parallelogram law give the same resultant.

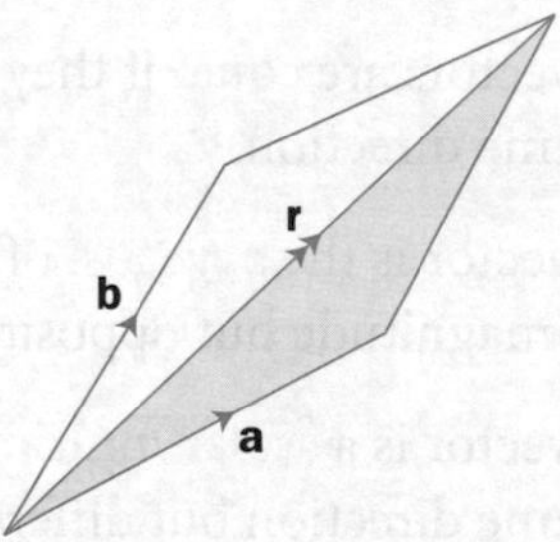

To add several vectors you need to apply the triangle law several times in succession.
If the vectors are drawn tip-to-tail, you can find the resultant vector from the total displacement.
This process is called the **polygon law** of addition.
As before, the order of the addition does not matter.

In this diagram, $\mathbf{r} = \mathbf{a} + \mathbf{b} + \mathbf{c} + \mathbf{d}$

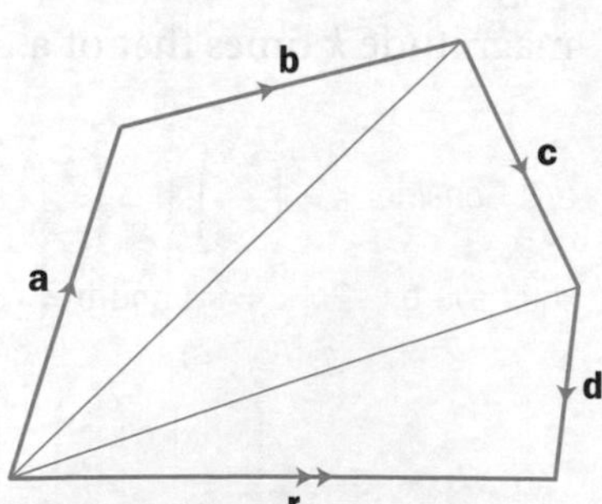

Subtracting a vector is equivalent to adding the negative of that vector.

That is $\mathbf{a} - \mathbf{b} = \mathbf{a} + (-\mathbf{b})$

In this diagram, $\mathbf{r} = \mathbf{a} - \mathbf{b} = \begin{pmatrix} 4 \\ 2 \end{pmatrix} - \begin{pmatrix} 2 \\ 3 \end{pmatrix} = \begin{pmatrix} 4 \\ 2 \end{pmatrix} + \begin{pmatrix} -2 \\ -3 \end{pmatrix} = \begin{pmatrix} 2 \\ -1 \end{pmatrix}$

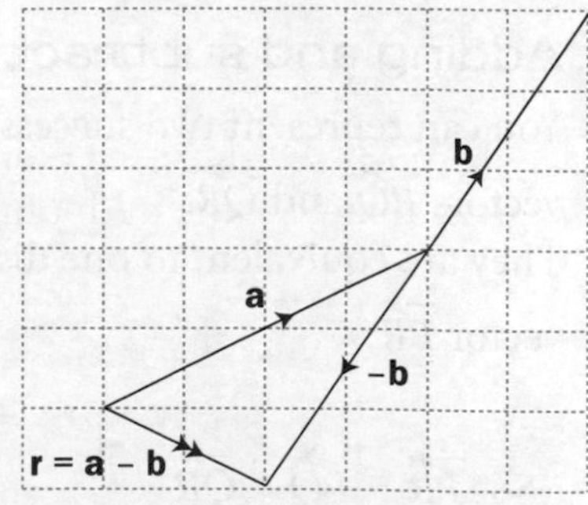

C4

EXAMPLE 1

Find a unit vector which is parallel to $6\mathbf{i} + 8\mathbf{j}$.
Find the angle between this vector and the direction of the x-axis.

Sketch a diagram to help you to visualise the problem.

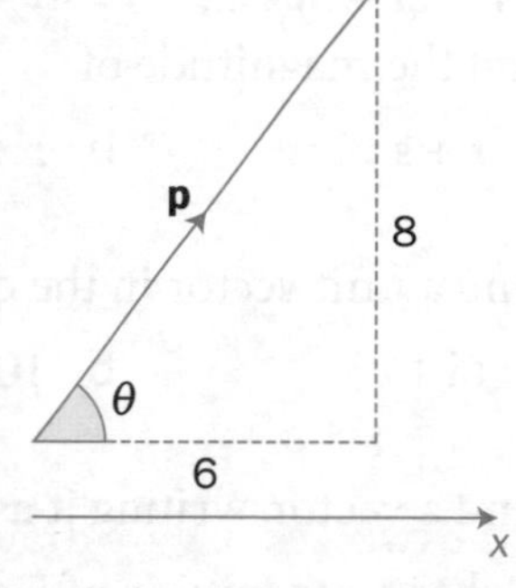

If $\mathbf{p} = 6\mathbf{i} + 8\mathbf{j}$

then the magnitude of $\mathbf{p}$, $|\mathbf{p}| = \sqrt{6^2 + 8^2} = 10$

and the unit vector $\hat{\mathbf{p}} = \frac{\mathbf{p}}{|\mathbf{p}|} = \frac{6\mathbf{i} + 8\mathbf{j}}{10} = \frac{3}{5}\mathbf{i} + \frac{4}{5}\mathbf{j}$

From the diagram, $\tan\theta = \frac{4}{3}$

and the required angle, $\theta = \tan^{-1}\left(\frac{4}{3}\right) = 53.1°$

EXAMPLE 2

Find the values of k and n if $\mathbf{q} = k\mathbf{p}$ where $\mathbf{p} = \begin{pmatrix} 3 \\ 2 \end{pmatrix}$ and $\mathbf{q} = \begin{pmatrix} n \\ 8 \end{pmatrix}$.

You have $\begin{pmatrix} n \\ 8 \end{pmatrix} = k\begin{pmatrix} 3 \\ 2 \end{pmatrix}$

The y-components give $8 = 2k$, so $k = 4$
The x-components give $n = 3k$, so $n = 12$

EXAMPLE 3

If $\mathbf{p} = 2\mathbf{i} + \mathbf{j} - 3\mathbf{k}$ and $\mathbf{q} = 4\mathbf{i} + 2\mathbf{k}$
find **a** $|\mathbf{p} + \mathbf{q}|$ **b** $|2\mathbf{p} - \mathbf{q}|$

a $\mathbf{p} + \mathbf{q} = \begin{pmatrix} 2 \\ 1 \\ -3 \end{pmatrix} + \begin{pmatrix} 4 \\ 0 \\ 2 \end{pmatrix} = \begin{pmatrix} 6 \\ 1 \\ -1 \end{pmatrix}$ and $|\mathbf{p} + \mathbf{q}| = \sqrt{36 + 1 + 1} = \sqrt{38}$

b $2\mathbf{p} - \mathbf{q} = 2\begin{pmatrix} 2 \\ 1 \\ -3 \end{pmatrix} - \begin{pmatrix} 4 \\ 0 \\ 2 \end{pmatrix} = \begin{pmatrix} 0 \\ 2 \\ -8 \end{pmatrix}$ and $|2\mathbf{p} - \mathbf{q}| = \sqrt{0 + 4 + 64}$
$= \sqrt{68} = 2\sqrt{17}$

EXAMPLE 4

Find $\overrightarrow{KL}$ when $\overrightarrow{LN} = 2\mathbf{i} - 3\mathbf{j} + \mathbf{k}$ and $\overrightarrow{KM} = 4\mathbf{i} - 2\mathbf{j} + 3\mathbf{k}$ given that M is the midpoint of KN.

$$\overrightarrow{KL} = \overrightarrow{KN} + \overrightarrow{NL} = 2\overrightarrow{KM} - \overrightarrow{LN}$$
$$= 2(4\mathbf{i} - 2\mathbf{j} + 3\mathbf{k}) - (2\mathbf{i} - 3\mathbf{j} + \mathbf{k})$$
$$= 6\mathbf{i} - \mathbf{j} + 5\mathbf{k}$$

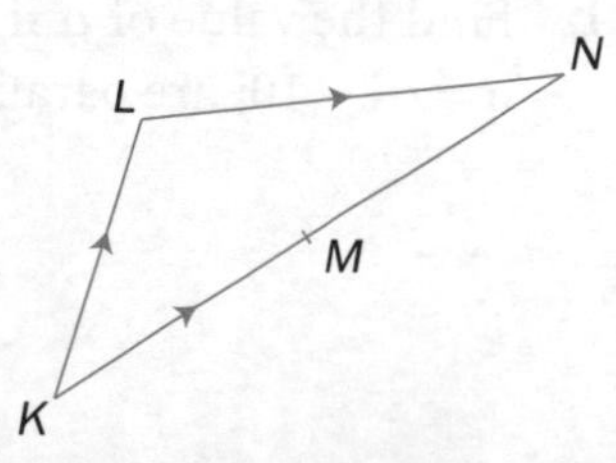

C4

Exercise 12.1

1 Find the magnitude of

a $5\mathbf{i} + 12\mathbf{j}$ b $5\mathbf{i} - 12\mathbf{j}$ c $\mathbf{i} + 2\mathbf{j} + 4\mathbf{k}$ d $2\mathbf{i} - 3\mathbf{j} + \mathbf{k}$

2 If $\mathbf{r} = 2\mathbf{i} - 3\mathbf{j} + \mathbf{k}$, $\mathbf{s} = 4\mathbf{i} - 2\mathbf{j} + 2\mathbf{k}$ and $\mathbf{t} = 2\mathbf{i} + \mathbf{k}$
find the magnitude of

a $\mathbf{r} + \mathbf{s}$ b $\mathbf{r} + \mathbf{s} - \mathbf{t}$ c $2\mathbf{s} + \mathbf{t}$ d $2\mathbf{r} - 3\mathbf{s} + 4\mathbf{t}$

3 Find a unit vector in the direction of

a $6\mathbf{i} + 8\mathbf{j}$ b $10\mathbf{i} - 24\mathbf{j}$ c $\mathbf{i} + \mathbf{j} + \mathbf{k}$ d $2\mathbf{i} - \mathbf{j} + 2\mathbf{k}$

4 Find a vector, writing it as a column vector, which

a has a magnitude of 5 and makes an angle of 30° with the positive direction of the x-axis

b has a magnitude of 2 and makes an angle of 120° with the positive direction of the x-axis.

5 a Find a vector which has a magnitude 20 and is parallel to $4\mathbf{i} + 3\mathbf{j}$.

b Find a vector which has a magnitude of 5 and is parallel to $2\mathbf{i} - 2\mathbf{j} - \mathbf{k}$.

C4

6 Find the value of n, given that

a $|3\mathbf{i} - 3\mathbf{j} + n\mathbf{k}| = \sqrt{22}$ b $|n\mathbf{i} + n\mathbf{j} + 2\mathbf{k}| = 4$

7 The vectors $\mathbf{p}$ and $\mathbf{q}$ are given by $\mathbf{p} = 3\mathbf{i} + 2\mathbf{j}$ and $\mathbf{q} = 2\mathbf{i} + 5\mathbf{j}$
Find

a the angle that $\mathbf{p}$ makes with the positive direction of the x-axis

b the angle between the directions of $\mathbf{p}$ and $\mathbf{q}$

c a unit vector in the direction of $\mathbf{p} - \mathbf{q}$.

8 a If $\overrightarrow{LM} = 5\mathbf{i} + 2\mathbf{j}$ and $\overrightarrow{MN} = 4\mathbf{i} - \mathbf{j}$, find $\overrightarrow{LN}$.

b If $\overrightarrow{PQ} = 3\mathbf{i} + 4\mathbf{j} - 3\mathbf{k}$ and $\overrightarrow{RQ} = \mathbf{i} - 2\mathbf{j} + \mathbf{k}$, find $\overrightarrow{PR}$.

c If $\overrightarrow{GE} = 2\mathbf{i} - \mathbf{j} - \mathbf{k}$ and $\overrightarrow{EF} = 3\mathbf{i} + 2\mathbf{j} + 3\mathbf{k}$, find $\overrightarrow{FG}$.

9 a Find the values of λ and μ if $\mathbf{p} = 3\mathbf{i} + 2\mathbf{j} - 4\mathbf{k}$, $\mathbf{q} = 9\mathbf{i} + \lambda\mathbf{j} + \mu\mathbf{k}$ and $\mathbf{q} = 3\mathbf{p}$

b Find the value of α if the vectors $\mathbf{p} = 2\mathbf{i} + 5\mathbf{j}$ and $\mathbf{q} = \alpha\mathbf{i} - 10\mathbf{j}$ are parallel.

10 If the vectors **r** and **s** are parallel, find λ and μ if

a $\mathbf{u} = 4\mathbf{i} + 2\mathbf{j} - 3\mathbf{k}$, $\mathbf{v} = \lambda\mathbf{i} + 4\mathbf{j} + \mu\mathbf{k}$

b $\mathbf{u} = \lambda\mathbf{i} + \mathbf{j} + 2\mathbf{k}$, $\mathbf{v} = 12\mathbf{i} + \mu\mathbf{j} - 6\mathbf{k}$

11 Use this diagram to write these vectors as column vectors.

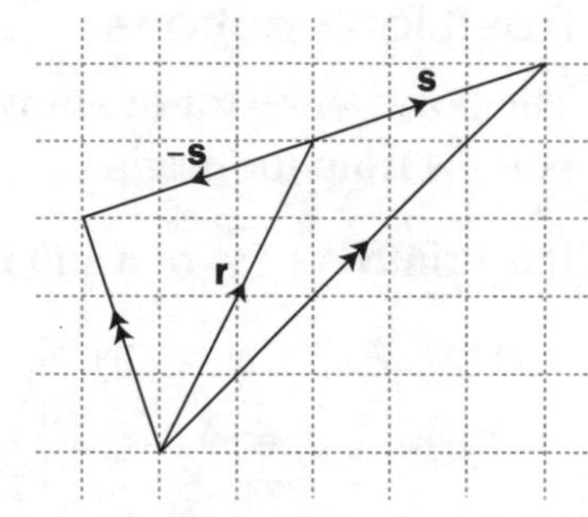

a **r** b **s**

c **r** + **s** d **r** − **s**

12 Using this diagram, write the resultants of these vector additions as column vectors.

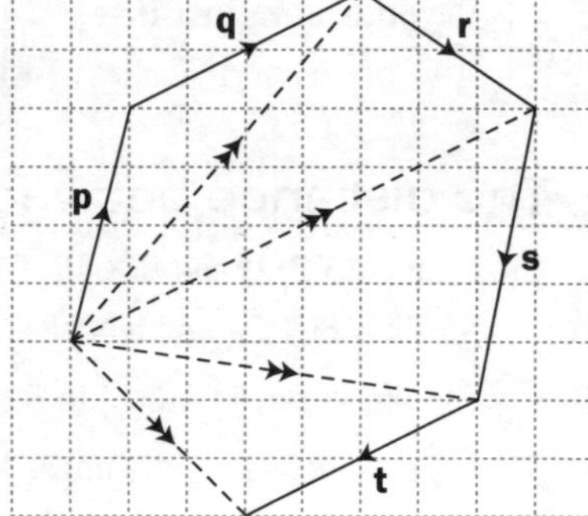

a **p** + **q** b **p** + **q** + **r**

c **p** + **q** + **r** + **s** d **p** + **q** + **r** + **s** + **t**

13 $\overrightarrow{AB} - 5\mathbf{i} + 7\mathbf{j}$, $\overrightarrow{CB} = 11\mathbf{j}$ and $\overrightarrow{CD} = -4\mathbf{i} + 6\mathbf{j}$

Prove that the vectors $\overrightarrow{AC}$ and $\overrightarrow{BD}$ are perpendicular.

14 $\overrightarrow{PQ} = 6\mathbf{i} + 3\mathbf{j} - 4\mathbf{k}$ and $\overrightarrow{PR} = 2\mathbf{i} - \mathbf{j} - 2\mathbf{k}$

M is the midpoint of *QR*. Find the vector $\overrightarrow{PM}$.

INVESTIGATION

15 Points *A* and *B* have position vectors **a** and **b**. *M* is the midpoint of *AB*. *P* is the midpoint of *AM*. *Q* is the midpoint of *MB*. *P*, *M* and *Q* are the points of quadrisection of *AB*.

a Find the position vectors for *M*, *P* and *Q* in terms of **a** and **b**.

b Find the position vectors for the two points of trisection of *AB*.

c Can you write down the position vectors for the four points of quintisection of *AB*?

C4

12.2 Applications in geometry

Position vectors

The **position vector** of point A is the fixed vector to the point A from the origin O.

It is printed as $\overrightarrow{OA}$ or $\mathbf{a}$ and is written as $\overrightarrow{OA}$ or $\underset{\sim}{a}$.

The lower-case notation $\underset{\sim}{a}$ is potentially confusing as it can be used either as a general vector or as the position vector for point A.

E.g. In two dimensions, for the point $A(3,4)$,

the position vector $\overrightarrow{OA} = \begin{pmatrix} 3 \\ 4 \end{pmatrix} = 3\mathbf{i} + 4\mathbf{j}$

In three dimensions, for the point $B(2,1,-3)$,

the position vector $\overrightarrow{OB} = \begin{pmatrix} 2 \\ 1 \\ -3 \end{pmatrix} = 2\mathbf{i} + \mathbf{j} - 3\mathbf{k}$

The distance between two points

E.g. Points $A(5,8,2)$ and $B(8,12,14)$ have position vectors

$\mathbf{a} = 5\mathbf{i} + 8\mathbf{j} + 2\mathbf{k}$ and $\mathbf{b} = 8\mathbf{i} + 12\mathbf{j} + 14\mathbf{k}$

The displacement from A to B can be $\overrightarrow{AB}$ direct

or in two stages via the origin $\overrightarrow{AO} + \overrightarrow{OB}$.

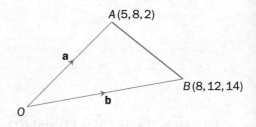

You have
$$\begin{aligned}\overrightarrow{AB} = \overrightarrow{AO} + \overrightarrow{OB} &= -\mathbf{a} + \mathbf{b} \\ &= \mathbf{b} - \mathbf{a} \\ &= (8\mathbf{i} + 12\mathbf{j} + 14\mathbf{k}) - (5\mathbf{i} + 8\mathbf{j} + 2\mathbf{k}) \\ &= 3\mathbf{i} + 4\mathbf{j} + 12\mathbf{k}\end{aligned}$$

The distance from A to $B = AB$
$$\begin{aligned}&= \sqrt{9 + 16 + 144} \\ &= \sqrt{169} \\ &= 13\end{aligned}$$

In general, $\overrightarrow{AB} = \mathbf{b} - \mathbf{a}$

Using coordinates in three dimensions, the distance from $A(x_1, y_1, z_1)$ to $B(x_2, y_2, z_2)$ is

$$AB = \sqrt{(x_2 - x_1)^2 + (y_2 - y_1)^2 + (z_2 - z_1)^2}$$

C4

The midpoint of a line

Points A and B have position vectors **a** and **b**.
The midpoint M of the line AB has the position vector

$$\mathbf{m} = \frac{1}{2}(\mathbf{a} + \mathbf{b})$$

You can prove this result in two ways.

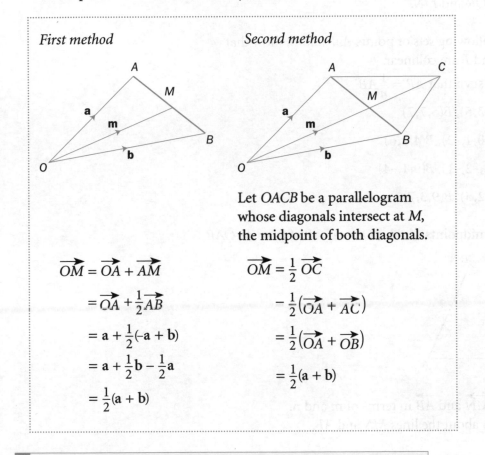

First method

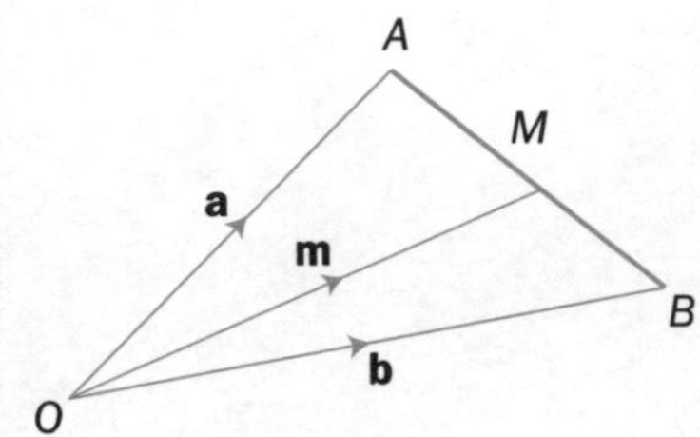

$$\overrightarrow{OM} = \overrightarrow{OA} + \overrightarrow{AM}$$
$$= \overrightarrow{OA} + \frac{1}{2}\overrightarrow{AB}$$
$$= \mathbf{a} + \frac{1}{2}(-\mathbf{a} + \mathbf{b})$$
$$= \mathbf{a} + \frac{1}{2}\mathbf{b} - \frac{1}{2}\mathbf{a}$$
$$= \frac{1}{2}(\mathbf{a} + \mathbf{b})$$

Second method

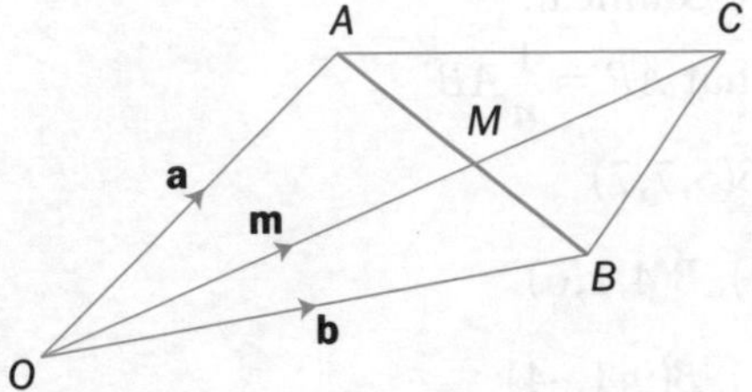

Let $OACB$ be a parallelogram whose diagonals intersect at M, the midpoint of both diagonals.

$$\overrightarrow{OM} = \frac{1}{2}\overrightarrow{OC}$$
$$- \frac{1}{2}(\overrightarrow{OA} + \overrightarrow{AC})$$
$$= \frac{1}{2}(\overrightarrow{OA} + \overrightarrow{OB})$$
$$= \frac{1}{2}(\mathbf{a} + \mathbf{b})$$

C4

EXAMPLE 1

Prove that the points $A(0,-1,2)$, $B(1,1,5)$ and $C(3,5,11)$ are collinear.

Collinear points lie on the same straight line.

$$\overrightarrow{AB} = \mathbf{b} - \mathbf{a} = (\mathbf{i} + \mathbf{j} + 5\mathbf{k}) - (0\mathbf{i} - \mathbf{j} + 2\mathbf{k})$$
$$= \mathbf{i} + 2\mathbf{j} + 3\mathbf{k}$$

From the coordinates of A, B and C, the position vectors are
$\mathbf{a} = -\mathbf{j} + 2\mathbf{k}$
$\mathbf{b} = \mathbf{i} + \mathbf{j} + 5\mathbf{k}$
$\mathbf{c} = 3\mathbf{i} + 5\mathbf{j} + 11\mathbf{k}$

$$\overrightarrow{BC} = \mathbf{c} - \mathbf{b} = (3\mathbf{i} + 5\mathbf{j} + 11\mathbf{k}) - (\mathbf{i} + \mathbf{j} + 5\mathbf{k})$$
$$= 2\mathbf{i} + 4\mathbf{j} + 6\mathbf{k}$$
$$= 2(\mathbf{i} + 2\mathbf{j} + 3\mathbf{k})$$

So, $\overrightarrow{BC} = 2\overrightarrow{AB}$

Hence, $\overrightarrow{BC}$ and $\overrightarrow{AB}$ are parallel and have the point B in common. So the points A, B and C are in the same straight line.

Exercise 12.2

1 a Find the distance between points $A(2,4,1)$ and $B(3,5,3)$.

b Given point $C(4,8,1)$, find the length AC and show that triangle ABC is right-angled.

2 In triangle $P(1,4,5)$, $Q(3,-2,1)$, $R(5,0,3)$, M and N are the midpoints of sides PQ and PR respectively.
Find the lengths QR and MN.

3 For each of the following sets of points A, P and B, show that the points A, P and B are collinear.
Find the fraction such that $AP = \frac{1}{n}AB$

a $A(2,1,4)$, $P(3,3,5)$, $B(5,7,7)$

b $A(-1,1,-4)$, $P(0,1,-2)$, $B(4,1,6)$

c $A(1,-3,0)$, $P(3,-2,-1)$, $B(9,1,-4)$

d $A(3,0,1)$, $P(7,2,5)$, $B(9,3,7)$

C4

4 M and N are the midpoints of sides OA and OB of triangle OAB.

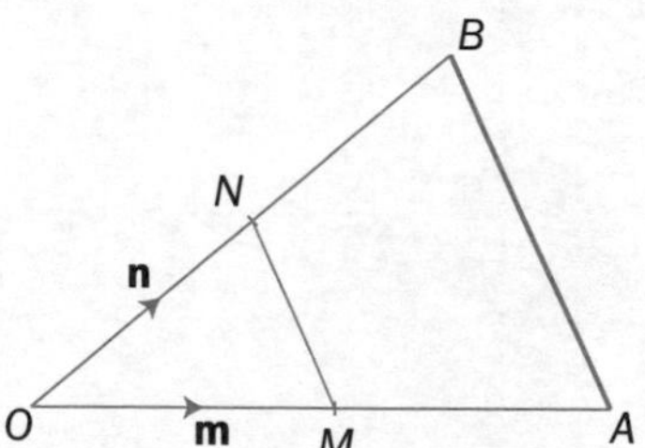

Find the vectors $\overrightarrow{MN}$ and $\overrightarrow{AB}$ in terms of **m** and **n**.
Make a deduction about the lines MN and AB.

5 In triangle OAB, point P lies on AB and points A, B and P have position vectors **a**, **b** and **p**.
Find **p** in terms of **a** and **b** given that $AP:PB$ is

a 1:2 b 1:3 c 2:3

d 3:5 e m:n

6 In triangle OPQ, $\overrightarrow{OP} = \mathbf{p}$ and $\overrightarrow{OQ} = \mathbf{q}$. Point M lies on OP such that $OM:MP = 1:k$ and point N lies on OQ such that $ON:NQ = 1:k$
Find the vector $\overrightarrow{MN}$ in terms of **p** and **q**.
Deduce two facts about MN and PQ.

7 The parallelogram $OACB$ has $\overrightarrow{OA} = \mathbf{a}$ and $\overrightarrow{OB} = \mathbf{b}$
Point R lies on OA and point S on AC such that $OR : OA = 1 : 4$ and $CS : CA = 1 : 4$. Find the vector $\overrightarrow{RS}$ in terms of **a** and **b**.
Deduce two facts relating RS and OC.

8 The parallelogram $OACB$ has $\overrightarrow{OA} = \mathbf{a}$, $\overrightarrow{OB} = \mathbf{b}$ and $\overrightarrow{OC} = \mathbf{c}$
Point P lies on OB such that $OP : PB = 1 : 2$
Point Q lies on AB such that $AQ : QB = 2 : 1$
Prove that OC and PQ are parallel and find the ratio $PQ : OC$.

9 L and M are the points $\left(\frac{1}{3}, -1, 2\right)$ and $(1, 5, 6)$ respectively.
The position vector of point P relative to the origin O is $\mathbf{i} + 3\mathbf{j} + 6\mathbf{k}$. Point R lies on OP such that $OR : RP$ is $k : 1$
Find the value of k if R, L and M are collinear.

10 The skew quadrilateral $OABC$ has $\overrightarrow{OA} = \mathbf{a}$, $\overrightarrow{OB} = \mathbf{b}$ and $\overrightarrow{OC} = \mathbf{c}$
The midpoints of its four sides in order are P, Q, R and S.
Find the vectors $\overrightarrow{PQ}$ and $\overrightarrow{SR}$ in terms of **a**, **b** and **c**.
Prove that $PQRS$ is a parallelogram.

C4

INVESTIGATION

11 **a** The distance between points $A(1, 2, t)$ and $B(t, t, 0)$ is x, where t is a variable.
Find x^2 as a function of t and hence find the minimum distance between these two points.

b Use a similar method to find the minimum distance between $P(2, 0, t)$ and $Q(t, t, 3)$.

12.3 The scalar (dot) product

The **scalar product** (or **dot product**) of two vectors **a** and **b** is

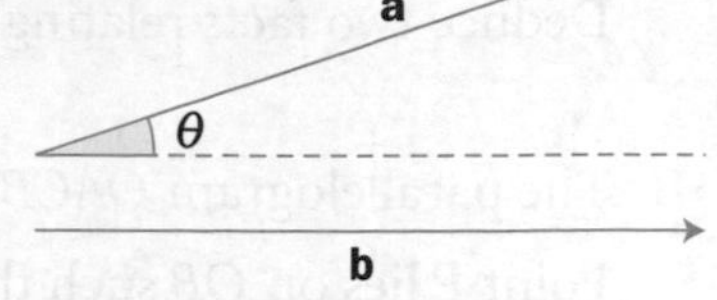

$\mathbf{a} \cdot \mathbf{b} = |\mathbf{a}|\,|\mathbf{b}| \cos\theta$

where θ is the angle between the directions of vectors **a** and **b**.

The scalar product is a scalar quantity.

|a|, |b| and cos θ are all scalar.

Angle θ will lie between 0° and 180°. Therefore, the scalar product is positive when θ is acute, zero when θ is 90°, and negative when θ is obtuse.

Properties of the scalar product

If **a** and **b** are perpendicular, then $\mathbf{a} \cdot \mathbf{b} = 0$

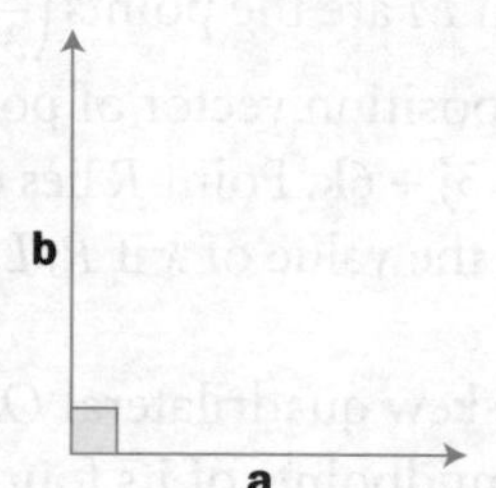

When **a** and **b** are perpendicular,
$\theta = 90°$, $\cos 90° = 0$ and $\mathbf{a} \cdot \mathbf{b} = |\mathbf{a}||\mathbf{b}|\cos 90° = 0$
In particular, $\mathbf{i} \cdot \mathbf{j} = \mathbf{j} \cdot \mathbf{k} = \mathbf{k} \cdot \mathbf{i} = 0$

$\mathbf{a} \cdot \mathbf{a} = a^2$

The angle between two equal vectors is 0° as they have the same direction.
$\mathbf{a} \cdot \mathbf{a} = |\mathbf{a}||\mathbf{a}| \cos 0° = a \times a \times 1 = a^2$
In particular, $\mathbf{i} \cdot \mathbf{i} = \mathbf{j} \cdot \mathbf{j} = \mathbf{k} \cdot \mathbf{k} = 1$

$(\lambda\mathbf{a}) \cdot \mathbf{b} = \mathbf{a} \cdot (\lambda\mathbf{b}) = \lambda(\mathbf{a} \cdot \mathbf{b})$

The length of $\lambda\mathbf{a}$ is λ times the length of **a**.
That is, $|\lambda\mathbf{a}| = \lambda|\mathbf{a}|$.
So, $\lambda(\mathbf{a} \cdot \mathbf{b}) = \lambda|\mathbf{a}||\mathbf{b}| \cos\theta = |\lambda\mathbf{a}||\mathbf{b}| \cos\theta$
$= (\lambda\mathbf{a}) \cdot \mathbf{b}$
Similarly for the lengths of $\lambda\mathbf{b}$ and **b**,
$|\lambda\mathbf{b}| = \lambda|\mathbf{b}|$ and $\lambda(\mathbf{a} \cdot \mathbf{b}) = \mathbf{a} \cdot (\lambda\mathbf{b})$

$\mathbf{a} \cdot \mathbf{b} = \mathbf{b} \cdot \mathbf{a}$

By definition, $\mathbf{a} \cdot \mathbf{b} = |\mathbf{a}||\mathbf{b}| \cos\theta = |\mathbf{b}||\mathbf{a}| \cos\theta = \mathbf{b} \cdot \mathbf{a}$

$\mathbf{a} \cdot (\mathbf{b} + \mathbf{c}) = \mathbf{a} \cdot \mathbf{b} + \mathbf{a} \cdot \mathbf{c}$

By definition, $\mathbf{a} \cdot (\mathbf{b} + \mathbf{c}) = |\mathbf{a}||\mathbf{b} + \mathbf{c}| \cos\theta$
$= OA \times OV \times \cos\theta$ — $OV\cos\theta = OU$
$= OA \times OU$
$= OA \times (OT + BW)$
$= OA \times (OB\cos\alpha + BV\cos\beta)$
$= |\mathbf{a}||\mathbf{b}|\cos\alpha + |\mathbf{a}||\mathbf{c}|\cos\beta$
$= \mathbf{a} \cdot \mathbf{b} + \mathbf{a} \cdot \mathbf{c}$

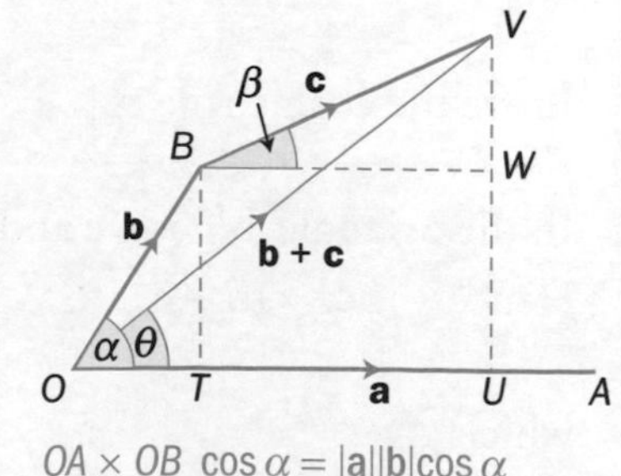

$OA \times OB\ \cos\alpha = |\mathbf{a}||\mathbf{b}|\cos\alpha$
$OA \times BV\ \cos\beta = |\mathbf{a}||\mathbf{c}|\cos\beta$

In **component form** in three dimensions, you have

$$\mathbf{a}\cdot\mathbf{b} = a_1b_1 + a_2b_2 + a_3b_3$$

For $\mathbf{a} = a_1\mathbf{i} + a_2\mathbf{j} + a_3\mathbf{k}$ and $\mathbf{b} = b_1\mathbf{i} + b_2\mathbf{j} + b_3\mathbf{k}$

$$\begin{aligned}\mathbf{a}\cdot\mathbf{b} &= (a_1\mathbf{i} + a_2\mathbf{j} + a_3\mathbf{k})\cdot(b_1\mathbf{i} + b_2\mathbf{j} + b_3\mathbf{k})\\ &= a_1b_1\mathbf{i}\cdot\mathbf{i} + a_1b_2\mathbf{i}\cdot\mathbf{j} + a_1b_3\mathbf{i}\cdot\mathbf{k} + a_2b_1\mathbf{j}\cdot\mathbf{i} + a_2b_2\mathbf{j}\cdot\mathbf{j} + a_2b_3\mathbf{j}\cdot\mathbf{k}\\ &\quad + a_3b_1\mathbf{k}\cdot\mathbf{i} + a_3b_2\mathbf{k}\cdot\mathbf{j} + a_3b_3\mathbf{k}\cdot\mathbf{k}\\ &= a_1b_1 + a_2b_2 + a_3b_3\end{aligned}$$

$\mathbf{i}\cdot\mathbf{j} = \mathbf{i}\cdot\mathbf{k} = \mathbf{j}\cdot\mathbf{k} = 0$
$\mathbf{i}\cdot\mathbf{j} = |\mathbf{i}|\,|\mathbf{j}|\cos 90° = 0$
because $\cos 90° = 0$

$\mathbf{i}\cdot\mathbf{i} = \mathbf{j}\cdot\mathbf{j} = \mathbf{k}\cdot\mathbf{k} = 1$
$\mathbf{i}\cdot\mathbf{i} = |\mathbf{i}|\,|\mathbf{j}|\cos 0° = 1$
because $\cos 0° = 1$

You can calculate the angle between two vectors using

$$\mathbf{a}\cdot\mathbf{b} = |\mathbf{a}||\mathbf{b}|\cos\theta = a_1b_1 + a_2b_2 + a_3b_3$$

EXAMPLE 1

Find the angle between the vectors $\mathbf{a} = 3\mathbf{i} + 2\mathbf{j} + \mathbf{k}$ and $\mathbf{b} = \mathbf{i} + 3\mathbf{j} - 4\mathbf{k}$

$$\mathbf{a}\cdot\mathbf{b} = \begin{pmatrix}3\\2\\1\end{pmatrix}\cdot\begin{pmatrix}1\\3\\-4\end{pmatrix} = (3\times1) + (2\times3) + (1\times-4) = 3 + 6 - 4 = 5$$

$\mathbf{a}\cdot\mathbf{b} = a_1b_1 + a_2b_2 + a_3b_3$

$$\begin{aligned}\mathbf{a}\cdot\mathbf{b} = |\mathbf{a}|\,|\mathbf{b}|\cos\theta &= \sqrt{9+4+1}\times\sqrt{1+9+16}\times\cos\theta\\ &= \sqrt{14}\times\sqrt{26}\times\cos\theta\end{aligned}$$

As $|\mathbf{a}||\mathbf{b}|\cos\theta = a_1b_1 + a_2b_2 + a_3b_3$

$$\cos\theta = \frac{5}{\sqrt{14}\times\sqrt{26}} = \frac{5}{\sqrt{364}} \text{ so } \theta = 74.8° \text{ to nearest } 0.1°$$

EXAMPLE 2

Given the four points $P(2,4,1)$, $Q(3,-2,4)$, $R(2,0,1)$ and $S(-1,2,6)$, prove that the lines PQ and RS are perpendicular.

$\overrightarrow{PQ} = \mathbf{q} - \mathbf{p} = (3\mathbf{i} - 2\mathbf{j} + 4\mathbf{k}) - (2\mathbf{i} + 4\mathbf{j} + \mathbf{k}) = \mathbf{i} - 6\mathbf{j} + 3\mathbf{k}$

$\overrightarrow{RS} = \mathbf{s} - \mathbf{r} = (-\mathbf{i} + 2\mathbf{j} + 6\mathbf{k}) - (2\mathbf{i} + \mathbf{k}) = -3\mathbf{i} + 2\mathbf{j} + 5\mathbf{k}$

So $\overrightarrow{PQ}\cdot\overrightarrow{RS} = (\mathbf{i} - 6\mathbf{j} + 3\mathbf{k})\cdot(-3\mathbf{i} + 2\mathbf{j} + 5\mathbf{k}) = -3 - 12 + 15 = 0$

As the scalar product is zero, PQ and RS are perpendicular.

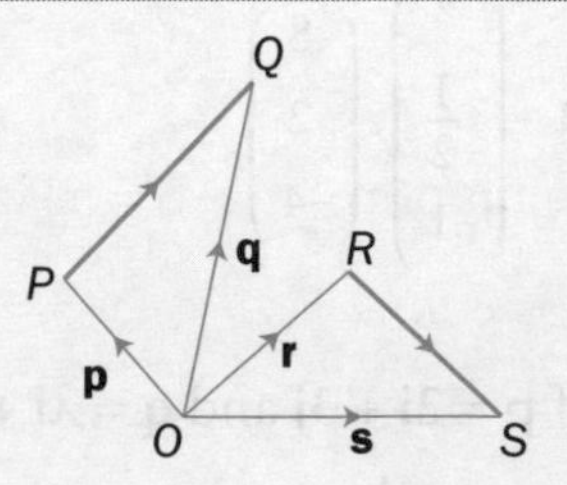

EXAMPLE 3

Prove the cosine rule $c^2 = a^2 + b^2 - 2ab\cos C$ for this triangle.

In this triangle, $\mathbf{b} = \mathbf{a} + \mathbf{c}$ and $\mathbf{c} = \mathbf{b} - \mathbf{a}$.

Hence, $\mathbf{c}\cdot\mathbf{c} = (\mathbf{b} - \mathbf{a})\cdot(\mathbf{b} - \mathbf{a}) = \mathbf{b}\cdot\mathbf{b} - \mathbf{b}\cdot\mathbf{a} - \mathbf{a}\cdot\mathbf{b} + \mathbf{a}\cdot\mathbf{a}$
$= \mathbf{a}\cdot\mathbf{a} + \mathbf{b}\cdot\mathbf{b} - 2\mathbf{a}\cdot\mathbf{b}$

which gives $c^2 = a^2 + b^2 - 2ab\cos C$

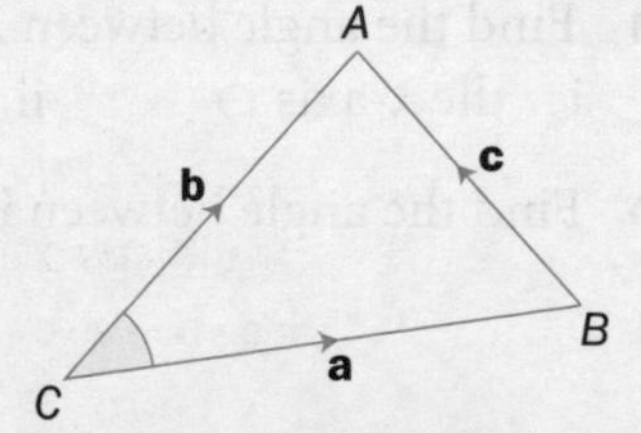

C4

Exercise 12.3

1 Find $\mathbf{p}\cdot\mathbf{q}$ when

a $\mathbf{p} = 3\mathbf{i} + 2\mathbf{j} - 4\mathbf{k}$ $\quad \mathbf{q} = 2\mathbf{i} + 5\mathbf{j} - 2\mathbf{k}$

b $\mathbf{p} = 2\mathbf{i} + \mathbf{j} - 3\mathbf{k}$ $\quad \mathbf{q} = 5\mathbf{i} - 2\mathbf{j} + 2\mathbf{k}$

c $\mathbf{p} = \mathbf{i} - \mathbf{j} - \mathbf{k}$ $\quad \mathbf{q} = \mathbf{i} + 3\mathbf{k}$

2 If $\mathbf{p} = 2\mathbf{i} + 3\mathbf{j} - \mathbf{k}$, $\mathbf{q} = 4\mathbf{i} - \mathbf{j} + 2\mathbf{k}$ and $\mathbf{r} = \mathbf{i} + \mathbf{j} - \mathbf{k}$, find the values of

a $\mathbf{p}\cdot\mathbf{q}$

b $\mathbf{p}\cdot\mathbf{r}$

c $\mathbf{p}\cdot(\mathbf{q}+\mathbf{r})$

d $\mathbf{p}\cdot\mathbf{q}+\mathbf{p}\cdot\mathbf{r}$

e $\mathbf{p}\cdot\mathbf{q}-\mathbf{p}\cdot\mathbf{r}$

f $\mathbf{p}\cdot(\mathbf{q}-\mathbf{r})$

3 Find the acute angle between the vectors $\mathbf{p}$ and $\mathbf{q}$ when

a $\mathbf{p} = 3\mathbf{i} + 4\mathbf{j} - 3\mathbf{k}$ $\quad \mathbf{q} = 2\mathbf{i} - \mathbf{j} - 2\mathbf{k}$

b $\mathbf{p} = 2\mathbf{i} - \mathbf{j} - 2\mathbf{k}$ $\quad \mathbf{q} = 5\mathbf{i} + 3\mathbf{j} + \mathbf{k}$

c $\mathbf{p} = \mathbf{i} - \mathbf{j} - 4\mathbf{k}$ $\quad \mathbf{q} = \mathbf{i} + \mathbf{j} - 2\mathbf{k}$

d $\mathbf{p} = 4\mathbf{i} - 3\mathbf{j} + 3\mathbf{k}$ $\quad \mathbf{q} = \mathbf{i} + 2\mathbf{j} - \mathbf{k}$

e $\mathbf{p} = 2\mathbf{i} - \mathbf{j} - \mathbf{k}$ $\quad \mathbf{q} = \mathbf{i} + 3\mathbf{k}$

f $\mathbf{p} = \mathbf{i} + \mathbf{k}$ $\quad \mathbf{q} = 2\mathbf{j} - \mathbf{k}$

g $\mathbf{p} = \begin{pmatrix} 2 \\ 5 \\ -2 \end{pmatrix}$ $\quad \mathbf{q} = \begin{pmatrix} 0 \\ 1 \\ 2 \end{pmatrix}$

h $\mathbf{p} = \begin{pmatrix} -3 \\ 0 \\ 2 \end{pmatrix}$ $\quad \mathbf{q} = \begin{pmatrix} 4 \\ 5 \\ 6 \end{pmatrix}$

C4

4 These pairs of vectors are perpendicular. Find the values of λ.

a $3\mathbf{i} + \mathbf{j} + 4\mathbf{k}$ and $3\mathbf{i} - 5\mathbf{j} + \lambda\mathbf{k}$

b $4\mathbf{i} - \lambda\mathbf{j} + \lambda\mathbf{k}$ and $\mathbf{i} + 3\mathbf{j} - 5\mathbf{k}$

5 For which values of μ are these vectors perpendicular?

a $\begin{pmatrix} 4 \\ -2 \\ \mu \end{pmatrix}$ and $\begin{pmatrix} 2 \\ 1 \\ 2 \end{pmatrix}$

b $\begin{pmatrix} 3 \\ \mu \\ -1 \end{pmatrix}$ and $\begin{pmatrix} -3 \\ \mu \\ 7 \end{pmatrix}$

6 State whether these pairs of vectors are parallel, perpendicular or neither.

a $\begin{pmatrix} 2 \\ \frac{1}{2} \\ -1 \end{pmatrix}, \begin{pmatrix} 8 \\ 2 \\ -4 \end{pmatrix}$

b $\begin{pmatrix} 6 \\ 2 \\ \frac{1}{2} \end{pmatrix}, \begin{pmatrix} 3 \\ -4 \\ 1 \end{pmatrix}$

c $\begin{pmatrix} 2 \\ -4 \\ 5 \end{pmatrix}, \begin{pmatrix} 6 \\ 3 \\ 0 \end{pmatrix}$

7 If $\mathbf{p} = 2\mathbf{i} + 3\mathbf{j}$ and $\mathbf{q} = \lambda\mathbf{i} + 2\mathbf{j}$, find the value of λ such that

a $\mathbf{p}$ and $\mathbf{q}$ are perpendicular

b $\mathbf{p}$ and $\mathbf{q}$ are parallel

c the angle between $\mathbf{p}$ and $\mathbf{q}$ is $\frac{\pi}{4}$.

8 a Find the angle between $2\mathbf{i} + 3\mathbf{j} + 12\mathbf{k}$ and

i the x-axis

ii the y-axis

iii the z-axis.

b Find the angle between $\mathbf{i} + \mathbf{j} + \mathbf{k}$ and each of the three coordinate axes.

9 Find the acute angle between LM and LN when L, M and N are the points $L(2,1,4)$, $M(4,-1,2)$ and $N(3,0,1)$.

10 Find the size of angle EFG, given the three points $E(1,-1,3)$, $F(2,1,3)$ and $G(-1,0,4)$.

11 Show that triangle PQR is right-angled when its vertices are

a $P(3,-3,7)$, $Q(4,0,2)$, $R(1,-2,3)$ b $P(1,0,-1)$, $Q(2,1,0)$, $R(3,-1,1)$

12 Triangle PQR has vertices $P(2,5,4)$, $Q(1,6,-2)$ and $R(4,6,3)$. Find

a angles P, Q and R to the nearest 0.1° b the lengths of the three sides

c the area of triangle PQR.

Do not round during your working.

13 a Show that the vectors $\mathbf{u} = 3\mathbf{i} + \mathbf{j} - \mathbf{k}$ and $\mathbf{v} = \mathbf{i} - 2\mathbf{k} + \mathbf{j}$ are perpendicular.

b Find a vector $\mathbf{r}$ that is perpendicular to both $\mathbf{u}$ and $\mathbf{v}$.

Use $a\mathbf{i} + b\mathbf{j} + c\mathbf{k}$

14 Simplify

a $(\mathbf{p}+\mathbf{q})\cdot\mathbf{r} + (\mathbf{p}-\mathbf{q})\cdot\mathbf{r}$ b $\mathbf{p}\cdot(\mathbf{q}+\mathbf{r}) - \mathbf{p}\cdot(\mathbf{q}-\mathbf{r})$

c $(\mathbf{p}+\mathbf{r})\cdot(\mathbf{p}-\mathbf{r})$ d $\mathbf{p}\cdot(\mathbf{p}-\mathbf{r}) - (\mathbf{p}-\mathbf{r})\cdot\mathbf{r}$

15 A semicircle, centre O, contains the vectors $\mathbf{a}$ and $\mathbf{b}$ as shown.

Prove that the angle XYZ is a right angle.

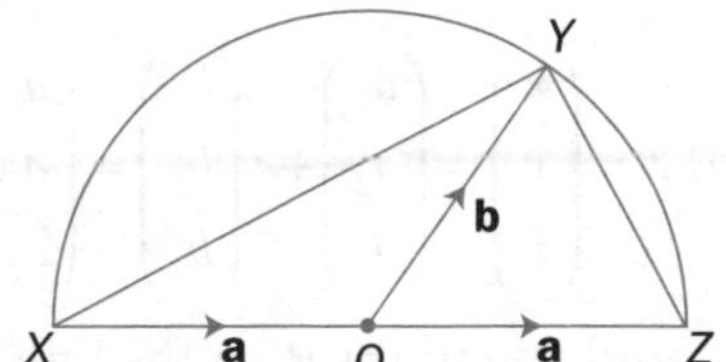

16 The rhombus $PQRS$ has $\overrightarrow{PQ} = \mathbf{a}$ and $\overrightarrow{PS} = \mathbf{b}$

Prove that the diagonals of the rhombus are perpendicular.

17 Find a vector which is perpendicular to both $\mathbf{i} + 2\mathbf{j} + 3\mathbf{k}$ and $2\mathbf{i} - 3\mathbf{j} - 8\mathbf{k}$

18 Find the angle between vectors $\mathbf{a}$ and $\mathbf{b}$ when $|\mathbf{a}| = 1$, $|\mathbf{b}| = 2$ and $|\mathbf{a} - \mathbf{b}| = 3$

19 Triangle OAB has vertices $A(1,2,-2)$, $B(6,8,0)$ and the origin O. Find the value of cos AOB and prove that the area of triangle $OAB = 2\sqrt{26}$

INVESTIGATION

20 The tetrahedron $ABCD$ has two pairs of opposite edges perpendicular. Prove that the third pair of opposite edges is also perpendicular.

C4

12.4 The vector equation of a straight line

Consider a straight line through point A parallel to vector **b** as in this diagram.

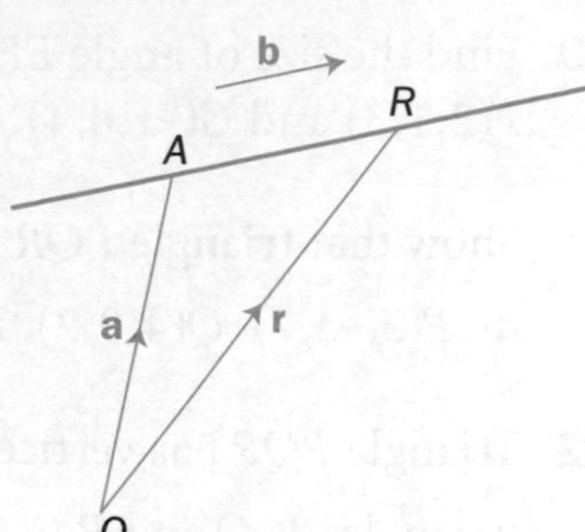

Point A has a position vector **a** with respect to the origin O.

Any other point R on the straight line has a position vector **r**.

Since the line AR is parallel to **b**, then the vector $\overrightarrow{AR}$ is a multiple of **b**.

That is, $\overrightarrow{AR} = t\mathbf{b}$, where t is a scalar.

The vector equation of the straight line through point A in the direction **b** is

$$\mathbf{r} = \overrightarrow{OA} + \overrightarrow{AR} = \mathbf{a} + t\mathbf{b}$$

A can be any point on the line and **b** can be any parallel direction, so the equation $\mathbf{r} = \mathbf{a} + t\mathbf{b}$ is not unique. One line can have many vector equations.

Compare this equation with the cartesian equation of a straight line, $y = mx + c$.

Given $A(a_1, a_2, a_3)$, $R(x, y, z)$ and $\mathbf{b} = b_1\mathbf{i} + b_2\mathbf{j} + b_3\mathbf{k}$, then the vector equation of the line in component form is

$$\mathbf{r} = \begin{pmatrix} x \\ y \\ z \end{pmatrix} = \begin{pmatrix} a_1 \\ a_2 \\ a_3 \end{pmatrix} + t\begin{pmatrix} b_1 \\ b_2 \\ b_3 \end{pmatrix} = \begin{pmatrix} a_1 + tb_1 \\ a_2 + tb_2 \\ a_3 + tb_3 \end{pmatrix}$$

and any point R on the line has coordinates $(a_1 + tb_1, a_2 + tb_2, a_3 + tb_3)$.

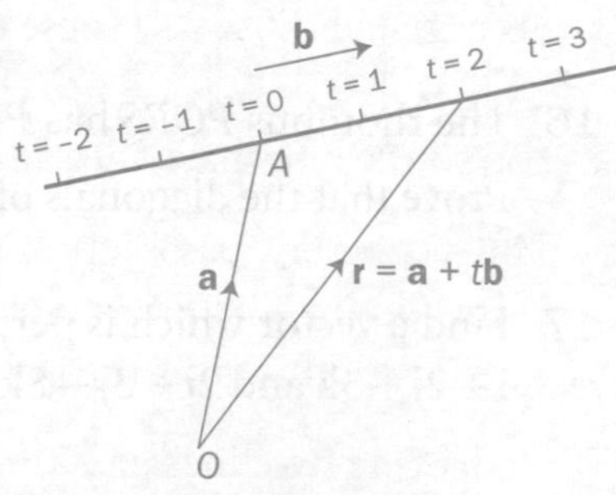

As t varies, R moves along the line.

When $t = 0$, R and A coincide.

When t is positive, R is on one side of A and when t is negative, R is on the other side of A.

EXAMPLE 1

Find the vector equation of the straight line through the point $A(3, 2, -1)$ in the direction of $4\mathbf{i} - \mathbf{j} + 2\mathbf{k}$.

The vector equation is $\mathbf{r} = 3\mathbf{i} + 2\mathbf{j} - \mathbf{k} + t(4\mathbf{i} - \mathbf{j} + 2\mathbf{k})$

$\mathbf{r} = \mathbf{a} + t\mathbf{b}$

An alternative form of the equation is
$\mathbf{r} = (3 + 4t)\mathbf{i} + (2 - t)\mathbf{j} + (-1 + 2t)\mathbf{k}$

Using column vectors with $R(x, y, z)$, you can also write the equation as $\begin{pmatrix} x \\ y \\ z \end{pmatrix} = \begin{pmatrix} 3 \\ 2 \\ -1 \end{pmatrix} + t\begin{pmatrix} 4 \\ -1 \\ 2 \end{pmatrix}$

C4

EXAMPLE 2

Find the vector equation of the line through the points $P(2,3,0)$ and $Q(3,5,1)$.

The direction of the line is

$\overrightarrow{PQ} = \mathbf{q} - \mathbf{p} = (3\mathbf{i} + 5\mathbf{j} + \mathbf{k}) - (2\mathbf{i} + 3\mathbf{j}) = \mathbf{i} + 2\mathbf{j} + \mathbf{k}$

The line passes through point P, so the vector equation of the line is

$$\mathbf{r} = \mathbf{p} + t(\mathbf{q} - \mathbf{p}) = (2\mathbf{i} + 3\mathbf{j}) + t(\mathbf{i} + 2\mathbf{j} + \mathbf{k})$$

Alternatively, as the line passes through Q, the vector equation could also be

$$\mathbf{r} = \mathbf{q} + s(\mathbf{q} - \mathbf{p}) = (3\mathbf{i} + 5\mathbf{j} + \mathbf{k}) + s(\mathbf{i} + 2\mathbf{j} + \mathbf{k})$$

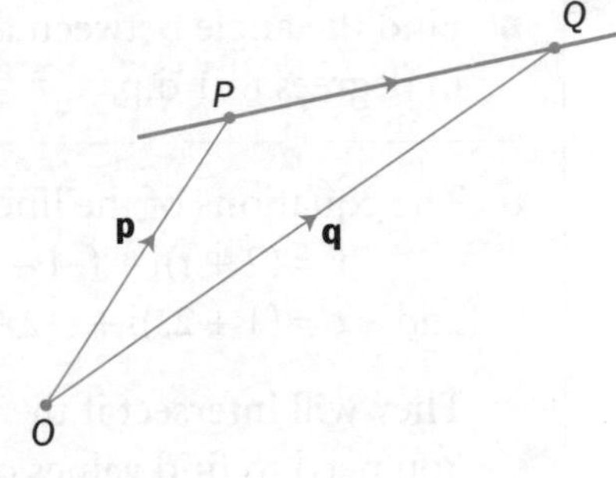

In general, the vector equation of the line through points P and Q is $\quad \mathbf{r} = \mathbf{p} + t(\mathbf{q} - \mathbf{p})$

EXAMPLE 3

Show that the two vector equations $\mathbf{r} = \mathbf{i} + 2\mathbf{j} + 3\mathbf{k} + t(2\mathbf{i} - \mathbf{j} + 2\mathbf{k})$ and $\mathbf{r} = 7\mathbf{i} - \mathbf{j} + 9\mathbf{k} + s(4\mathbf{i} - 2\mathbf{j} + 4\mathbf{k})$ both describe the same straight line.

The two equations give lines with directions $2\mathbf{i} - \mathbf{j} + 2\mathbf{k}$ and $4\mathbf{i} - 2\mathbf{j} + 4\mathbf{k}$.

Because $\begin{pmatrix} 4 \\ -2 \\ 4 \end{pmatrix} = 2\begin{pmatrix} 2 \\ -1 \\ 2 \end{pmatrix}$ these two vectors are parallel

and so the lines have the same direction.

The second line passes through the point $(7,-1,9)$ when $s = 0$. Points on the first line are given by $(1 + 2t, 2 - t, 3 + 2t)$ and, when $t = 3$, this gives the point $(7,-1,9)$.

Hence, the two lines pass through the same point in the same direction. The two equations represent the same line.

Intersection of straight lines

In two dimensions, two straight lines can coincide (that is, they are the same line), they can be parallel to each other, or they can intersect each other.

In three dimensions, two straight lines can coincide, be parallel, intersect or be **skew**.

When two lines are skew, one passes above the other without intersecting it. There is still an angle between them.

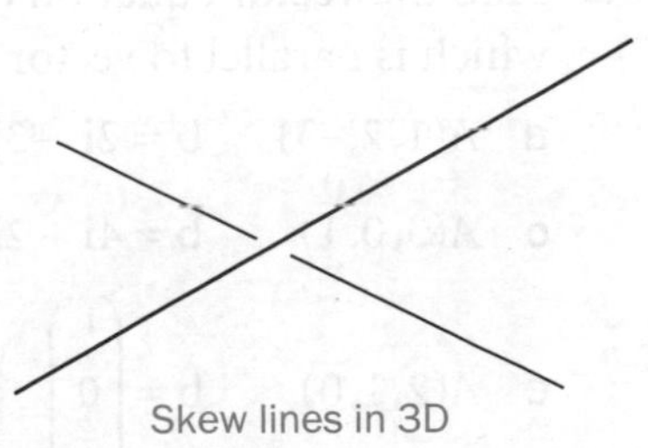

Skew lines in 3D

C4

EXAMPLE 4

a Determine whether the lines given by
$\mathbf{r} = 2\mathbf{i} - \mathbf{j} + 4\mathbf{k} + t(\mathbf{i} + \mathbf{j} - \mathbf{k})$ and $\mathbf{r} = \mathbf{i} - 2\mathbf{j} + 3\mathbf{k} + s(2\mathbf{i} + 2\mathbf{j} - \mathbf{k})$
intersect or are skew.
If they intersect, find the point of intersection.

b Find the angle between the two lines, giving your answer in degrees to 1 d.p.

---

a The equations of the lines can be written as

$$\mathbf{r} = (2 + t)\mathbf{i} + (-1 + t)\mathbf{j} + (4 - t)\mathbf{k}$$

and $\mathbf{r} = (1 + 2s)\mathbf{i} + (-2 + 2s)\mathbf{j} + (3 - s)\mathbf{k}$

They will intersect if the coefficients of **i**, **j** and **k** are equal.
You need to find values of s and t to make

$$2 + t = 1 + 2s \quad (1)$$
$$-1 + t = -2 + 2s \quad (2)$$
$$4 - t = 3 - s \quad (3)$$

Add (2) and (3): $3 = 1 + s$ so $s = 2$

Substitute into (2): $-1 + t = -2 + 2 \times 2$ so $t = 3$

Substitute in (1): LHS $= 2 + 3 = 5$ and RHS $= 1 + 2 \times 2 = 5$

Substitute into one of the vector equations:
$\mathbf{r} = 2\mathbf{i} - \mathbf{j} + 4\mathbf{k} + 3(\mathbf{i} + \mathbf{j} - \mathbf{k}) = 5\mathbf{i} + 2\mathbf{j} + \mathbf{k}$
so the point of intersection is $(5, 2, 1)$.

The values $s = 2$ and $t = 3$ satisfy all three equations. The two lines do intersect. If all three equations were not satisfied, then the lines would not intersect.

b The directions of the two lines are given by
$\mathbf{b}_1 = \mathbf{i} + \mathbf{j} - \mathbf{k}$ and $\mathbf{b}_2 = 2\mathbf{i} + 2\mathbf{j} - \mathbf{k}$

The angle θ between the lines is found from $\mathbf{b}_1 \cdot \mathbf{b}_2 = |\mathbf{b}_1||\mathbf{b}_2| \cos\theta$

$$1 \times 2 + 1 \times 2 + (-1) \times (-1) = \sqrt{1 + 1 + 1} \times \sqrt{4 + 4 + 1} \times \cos\theta$$

$$5 = \sqrt{3} \times \sqrt{9} \times \cos\theta$$

$$\cos\theta = \frac{5}{3\sqrt{3}}$$

$|\mathbf{b}_1| = \sqrt{1^2 + 1^2 + (-1)^2} = \sqrt{3}$

$|\mathbf{b}_2| = \sqrt{2^2 + 2^2 + (-1)^2} = \sqrt{9}$

The acute angle between the two lines is 15.8°

Exercise 12.4

1 Find the vector equation of the line through point A which is parallel to vector **b**, when

a $A(1, 2, -3)$ $\quad \mathbf{b} = 2\mathbf{i} - 3\mathbf{j} + \mathbf{k}$ **b** $A(4, -1, 3)$ $\quad \mathbf{b} = -2\mathbf{i} + \mathbf{j} + 2\mathbf{k}$

c $A(3, 0, 1)$ $\quad \mathbf{b} = 4\mathbf{i} + 2\mathbf{k}$ **d** $A(1, -1, 0)$ $\quad \mathbf{b} = \mathbf{i} + \mathbf{j} - \mathbf{k}$

e $A(2, 1, 0)$ $\quad \mathbf{b} = \begin{pmatrix} 1 \\ 0 \\ 3 \end{pmatrix}$ **f** $A(0, 0, 1)$ $\quad \mathbf{b} = \begin{pmatrix} 2 \\ 0 \\ 0 \end{pmatrix}$

C4

2 Find a vector equation of the line through the points P and Q when

a $P(2,3,1)$ $Q(3,0,4)$ b $P(2,-1,1)$ $Q(5,0,1)$

c $P(1,0,0)$ $Q(4,5,-2)$ d $P(0,0,0)$ $Q(1,2,3)$

e $P(1,-2,-3)$ $Q(2,-1,-2)$ f $P(2,-3,-1)$ $Q(5,0,-1)$

3 Find the vector equation of the line through the point $A(2,1,-2)$ which is

a parallel to the x-axis b perpendicular to the xy-plane.

4 Show that the given point lies on the given line and find the corresponding value of t.

a $(4,1,-1)$ and $\mathbf{r} = 2\mathbf{i} - 3\mathbf{j} + \mathbf{k} + t(\mathbf{i} + 2\mathbf{j} - \mathbf{k})$

b $(9,-4,1)$ and $\mathbf{r} = -\mathbf{i} + \mathbf{j} + 6\mathbf{k} + t(-2\mathbf{i} + \mathbf{j} + \mathbf{k})$

c $(3,2,0)$ and $\mathbf{r} = (3 - t)\mathbf{i} + (2 + 3t)\mathbf{j} - t\mathbf{k}$

5 Determine whether these pairs of lines are parallel, intersect or are skew. Find the acute angle between the lines. If they intersect, find their point of intersection.

a $\mathbf{r} = 3\mathbf{i} + 5\mathbf{j} + 7\mathbf{k} + t(\mathbf{i} + 2\mathbf{j} - 3\mathbf{k})$ $\mathbf{r} = 4\mathbf{i} + \mathbf{j} + 2\mathbf{k} + s(3\mathbf{j} + \mathbf{k})$

b $\mathbf{r} = 6\mathbf{i} - 2\mathbf{j} + 8\mathbf{k} + t(\mathbf{i} - 5\mathbf{j} + 7\mathbf{k})$ $\mathbf{r} = \mathbf{i} + 3\mathbf{j} + 2\mathbf{k} + s(3\mathbf{i} + 5\mathbf{j} - 8\mathbf{k})$

c $\mathbf{r} = 2\mathbf{i} - 4\mathbf{j} + \mathbf{k} + t(2\mathbf{i} - \mathbf{j} - \mathbf{k})$ $\mathbf{r} = 3\mathbf{i} + \mathbf{j} - 6\mathbf{k} + s(4\mathbf{i} - 2\mathbf{j} - 2\mathbf{k})$

d $\mathbf{r} = 2\mathbf{j} - \mathbf{k} + t(4\mathbf{i} - 2\mathbf{j} + 3\mathbf{k})$ $\mathbf{r} = 4\mathbf{i} + \mathbf{j} - 2\mathbf{k} + s(\mathbf{i} + 4\mathbf{j} + 4\mathbf{k})$

e $\mathbf{r} = 2\mathbf{i} - \mathbf{k} + t(6\mathbf{i} + \mathbf{j} + 8\mathbf{k})$ $\mathbf{r} = 12\mathbf{i} - 15\mathbf{j} - 5\mathbf{k} + s(9\mathbf{i} + 3\mathbf{j} + 4\mathbf{k})$

6 a Find the vector equations of the two lines PQ and RS where the four points are $P(-1,1,3)$, $Q(8,7,6)$, $R(0,5,2)$ and $S(-2,7,0)$.

b Show that the two lines intersect and find the point of intersection.

c Find angle PRS to the nearest 0.1°

7 a Find the vector equation of the line through the points $A(1,2,3)$ and $B(2,-1,-1)$. Show that the point $R(1 + t, 2 - 3t, 3 - 4t)$ lies on this line for all t.

b C is the point $(5,-2,-6)$. Find the value of t that makes the vectors $\overrightarrow{CR}$ and $\overrightarrow{AB}$ perpendicular.

c Find the shortest distance from point C to line AB.

INVESTIGATION

8 Using the parameter λ, find the general point R on the line joining points $A(1,2,3)$ and $B(0,1,2)$.

Repeat using the parameter μ for the general point S on the line joining $C(0,-2,2)$ and $D(1,0,-1)$.

Find λ and μ so that the line RS is perpendicular to both AB and CD. Find the shortest distance between lines AB and CD.

C4

Review 12

1 Find **a** the magnitude of the vector $\mathbf{a} = 3\mathbf{i} + 12\mathbf{j} + 4\mathbf{k}$

b the unit vector parallel to vector $\mathbf{a}$

c a vector of magnitude 5 which is parallel to vector $\mathbf{a}$

d the angle between $\mathbf{a}$ and the direction of the x-axis.

2 The points A, B and C have coordinates $(1,4,3)$, $(3,6,6)$ and $(7,10,12)$ respectively. Find the vector $\overrightarrow{AB}$ and show that the points A, B and C are collinear.

3 A triangle has vertices $L(3,1,0)$, $M(1,-2,5)$ and $N(2,5,2)$.

a Prove that the triangle is right-angled and find its area.

b Find point K such that the quadrilateral $KLMN$ is a parallelogram.

c Show that the midpoint of KM coincides with the midpoint of LN.

4 a Show that the vectors $\begin{pmatrix} 1 \\ 2 \\ -2 \end{pmatrix}$ and $\begin{pmatrix} 4 \\ -1 \\ 1 \end{pmatrix}$ are perpendicular.

b Find the acute angle between the two vectors $\mathbf{i} + 2\mathbf{j} - 2\mathbf{k}$ and $3\mathbf{i} - \mathbf{j} - \mathbf{k}$.

5 a Given points $A(-1,0,2)$, $B(2,4,1)$ and $C(0,5,-3)$, find the vectors $\overrightarrow{AB}$ and $\overrightarrow{BC}$. Hence, use the scalar product to find angle ABC to the nearest degree.

b D is the point $(1,1,k)$. Find the value of k such that the lines AB and BD are perpendicular.

6 Find a unit vector $a\mathbf{i} + b\mathbf{j} + c\mathbf{k}$ which is perpendicular to the two vectors $2\mathbf{i} + 2\mathbf{j} - \mathbf{k}$ and $4\mathbf{i} + 2\mathbf{k}$.

7 Relative to a fixed origin O, the point A has position vector $3\mathbf{i} + 2\mathbf{j} - \mathbf{k}$, the point B has position vector $5\mathbf{i} + \mathbf{j} + \mathbf{k}$, and the point C has position vector $7\mathbf{i} - \mathbf{j}$.

a Find the cosine of angle ABC.

b Find the exact value of the area of triangle ABC.

c The point D has position vector $7\mathbf{i} + 3\mathbf{k}$.
Show that AC is perpendicular to CD.

d Find the ratio $AD : DB$.

8 a Write down the vector equation of the line L_1 which passes through point $P(1,-2,4)$ and is parallel to the vector $2\mathbf{i}+3\mathbf{j}-\mathbf{k}$.

b Find the vector equation of the line L_2 which contains point P and also point $Q(0,1,3)$.

c Find the angle between the two lines L_1 and L_2.
Hence, or otherwise, find the shortest distance from Q to L_1.

9 a Show that the point $A(7,8,1)$ lies on the line L with the vector equation
$\mathbf{r}=\begin{pmatrix}3\\0\\-1\end{pmatrix}+t\begin{pmatrix}2\\4\\1\end{pmatrix}$, but that the point $B(1,-4,0)$ does not lie on L.

b Find the acute angle between L and the line through A and B.

c Find the shortest distance from B to L.

10 a Show that the two lines with these vector equations intersect.

$$\mathbf{r}=\begin{pmatrix}3\\-1\\2\end{pmatrix}+\lambda\begin{pmatrix}2\\1\\-1\end{pmatrix}\text{ and }\mathbf{r}=\begin{pmatrix}1\\4\\-7\end{pmatrix}+\mu\begin{pmatrix}1\\-1\\2\end{pmatrix}$$

b Find their point of intersection.

c Find the acute angle between the two lines.

11 Two straight lines have the vector equations $\mathbf{r}=\mathbf{i}-2\mathbf{k}+t(-2\mathbf{i}+\mathbf{j}+2\mathbf{k})$ and $\mathbf{r}=3\mathbf{i}-4\mathbf{j}+\mathbf{k}+s(\mathbf{i}-\mathbf{j}+\mathbf{k})$

a Find the acute angle between the directions of the two lines.

b Do the two lines intersect or are they skew?

12 a Find the values of t and s which show that the point $P(-2,6,-1)$ lies on the line $\mathbf{r}=\mathbf{i}-4\mathbf{k}+t(-\mathbf{i}+2\mathbf{j}+\mathbf{k})$ and that point $Q(1,7,0)$ lies on the line $\mathbf{r}=8\mathbf{j}+2\mathbf{k}+s(-\mathbf{i}+\mathbf{j}+2\mathbf{k})$

b Prove that PQ is perpendicular to both lines.

c Find the shortest distance between the two lines.

d Find the angle between the two lines.

12 Exit

Summary

Refer to

- Equal vectors have the same **magnitude** and direction. 12.1
- The vector $k\mathbf{a}$ is parallel to vector $\mathbf{a}$ and has a magnitude k times that of $\mathbf{a}$. 12.1
- The vector $\mathbf{p} = x\mathbf{i} + y\mathbf{j} + z\mathbf{k}$ has a magnitude (or modulus) $|\mathbf{p}| = \sqrt{x^2 + y^2 + z^2}$ 12.1
- The **unit vector** parallel to $\mathbf{p}$ is $\hat{\mathbf{p}} = \dfrac{\mathbf{p}}{|\mathbf{p}|} = \dfrac{x\mathbf{i} + y\mathbf{j} + z\mathbf{k}}{\sqrt{x^2 + y^2 + z^2}}$ 12.1
- The **position vector** of point $A(a_1, a_2, a_3)$ relative to the origin O is given by $\overrightarrow{OA} = \mathbf{a} = a_1\mathbf{i} + a_2\mathbf{j} + a_3\mathbf{k}$ 12.2
- If points A and B have position vectors $\mathbf{a}$ and $\mathbf{b}$, then $\overrightarrow{AB} = \mathbf{b} - \mathbf{a}$ 12.2
- The **midpoint** M of AB has a position vector $\mathbf{m} = \frac{1}{2}(\mathbf{a} + \mathbf{b})$ 12.2
- The distance between points $A(a_1, a_2, a_3)$ and $B(b_1, b_2, b_3)$ is given by $AB = \sqrt{(a_1 - b_1)^2 + (a_2 - b_2)^2 + (a_3 - b_3)^2}$ 12.2
- The **scalar** (or **dot**) **product** $\mathbf{a} \cdot \mathbf{b} = |\mathbf{a}||\mathbf{b}| \cos\theta = a_1b_1 + a_2b_2 + a_3b_3$ where θ is the angle between $\mathbf{a}$ and $\mathbf{b}$. 12.3
- The **vector equation** of the line through point A and parallel to vector $\mathbf{b}$ is $\mathbf{r} = \mathbf{a} + t\mathbf{b}$ 12.4
- The vector equation of the line through points P and Q is $\mathbf{r} = \mathbf{p} + t(\mathbf{q} - \mathbf{p})$ 12.4

Links

To ensure safety during aircraft landing, the aviation industry uses vectors to guide the aircraft to the runway.

As an aircraft approaches an airport, the Instrument Landing System (ILS) uses a series of radars, which are sent up from the perimeter of the runway, to guide it to a safe landing.

Vectors and the dot product are used to determine if the aircraft has intercepted the area covered by the beams.

Suppose that the aircraft has position vector $\mathbf{s}$ and moves with velocity $\mathbf{v}$, and that $\mathbf{p}$ is the position vector of a point P in the area covered by the radar beams. Then, if the dot product $\mathbf{v} \cdot (\mathbf{s} - \mathbf{p})$ is positive, the aircraft is moving towards P. When this product is 0 the aircraft is passing P and when it is negative it is moving away from P.

Revision 4

1 Express these as partial fractions.

a $\frac{6}{x(x+2)}$ b $\frac{3x-1}{(x+1)^2(x-3)}$ c $\frac{x^3}{x^2-4}$

2 $f(x) = \frac{11+2x+x^2}{(1-2x)(3+x)^2} = \frac{A}{1-2x} + \frac{B}{3+x} + \frac{C}{(3+x)^2}$ $|x| < \frac{1}{2}$

a Find the values of A and C. Show that $B = 0$

b Hence, find an expansion of f(x), in ascending powers of x up to and including the term in x^3. Simplify your answer as far as possible.

3

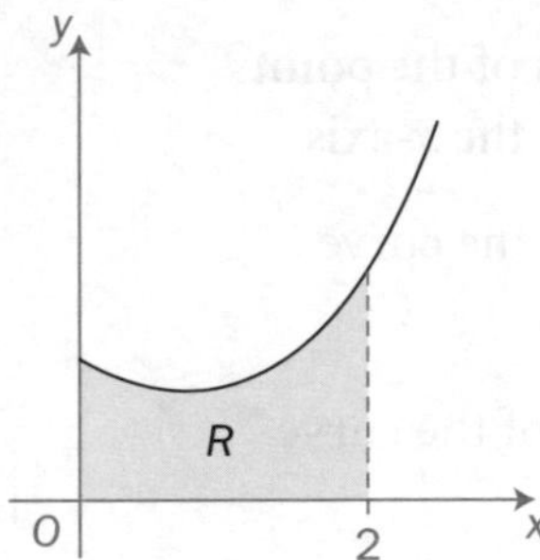

The diagram shows part of the curve with equation $y = f(x)$, where

$$f(x) = \frac{x^2+1}{(1+x)(3-x)}, \quad 0 \leqslant x < 3$$

a Given that $f(x) = A + \frac{B}{1+x} + \frac{C}{3-x}$, find the values of the constants A, B and C.

b The finite region R, shown in the diagram, is bounded by the curve with equation $y = f(x)$, the x-axis, the y-axis and the line $x = 2$
Find the area of R, giving your answer in the form $p + q\ln r$, where p, q and r are rational constants to be found. [(c) Edexcel Limited 2002]

4 A curve has parametric equations $x = \tan^2 t$ $y = \sin t$, $0 < t < \frac{\pi}{2}$

a Find an equation for $\frac{dy}{dx}$ in terms of t. You do not need to simplify your answer.

b Find an equation of the tangent to the curve at the point where $t = \frac{\pi}{4}$
Give your answer in the form $y = ax + b$, where a and b are constants to be determined.

c Find the Cartesian equation of the curve in the form $y^2 = f(x)$ [(c) Edexcel Limited 2007]

C4

5

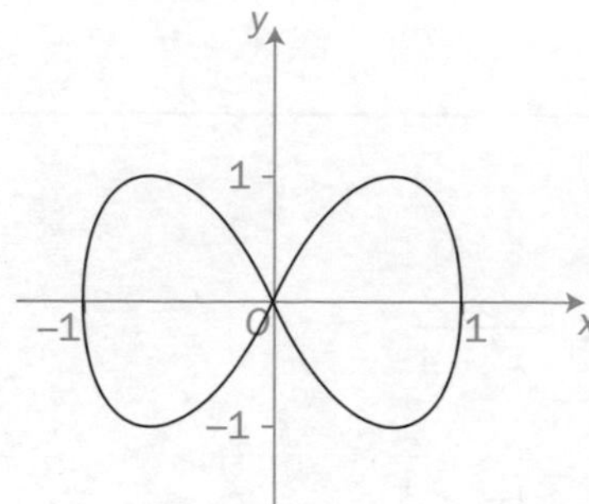

The curve shown in the diagram has parametric equations

$x = \cos t \quad y = \sin 2t \quad 0 \leqslant t < 2\pi$

a Find an expression for $\frac{dy}{dx}$ in terms of the parameter t.

b Find the values of the parameter t at the points where $\frac{dy}{dx} = 0$

c Hence, give the exact values of the coordinates of the points on the curve where the tangents are parallel to the x-axis.

d Show that a Cartesian equation for the part of the curve where $0 \leqslant t < \pi$ is $y = 2x\sqrt{1 - x^2}$

e Write down a Cartesian equation for the part of the curve where $\pi \leqslant t < 2\pi$

[(c) Edexcel Limited 2003]

6

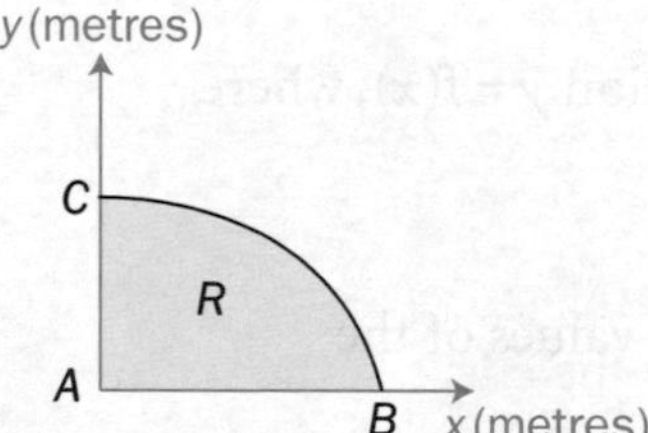

This diagram shows a cross-section R of a dam. The line AC is the vertical face of the dam, AB is the horizontal base and the curve BC is the profile.
Taking x and y to be the horizontal and vertical axes, then A, B and C have coordinates $(0, 0)$, $(3\pi^2, 0)$ and $(0, 30)$ respectively.
The area of the cross-section is to be calculated.
Initially the profile BC is approximated by a straight line.

a Find an estimate for the area of the cross-section R using this approximation.

The profile BC is actually described by the parametric equations

$$x = 16t^2 - \pi^2, \quad y = 30\sin 2t, \quad \frac{\pi}{4} \leqslant t \leqslant \frac{\pi}{2}$$

b Find the exact area of the cross-section R.

c Calculate the percentage error in the estimate of the area of the cross-section R that you found in part **a**.

[(c) Edexcel Limited 2004]

7

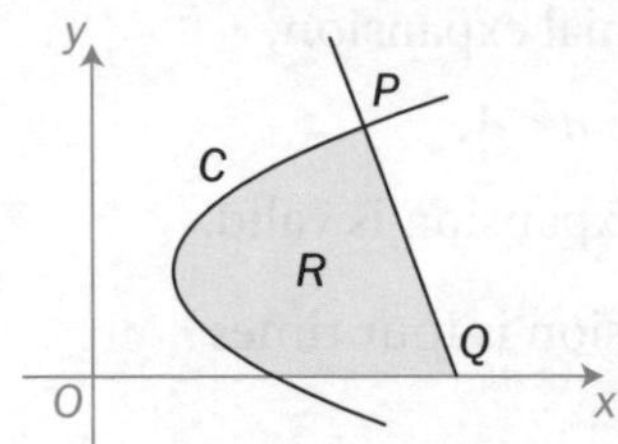

The diagram shows a sketch of part of the curve C with parametric equations
$x = t^2 + 1, \quad y = 3(1 + t)$

The normal to C at the point $P(5,9)$ cuts the x-axis at the point Q, as shown in the diagram.

a Find the x-coordinate of Q.

b Find the area of the finite region R bounded by C, the line PQ and the x-axis. [(c) Edexcel Limited 2005]

8

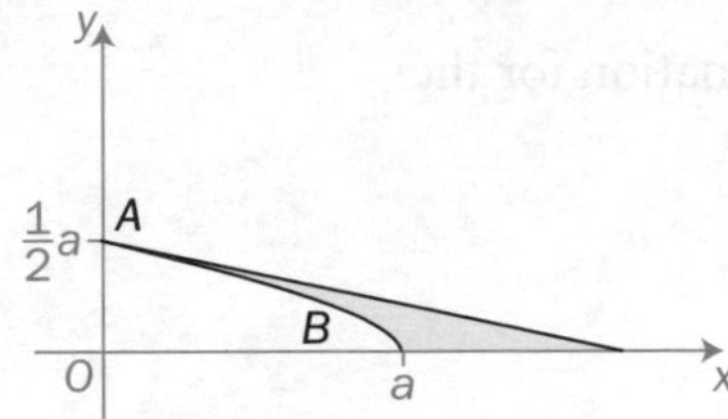

The curve shown in the diagram has parametric equations

$x = a\cos 3t, \quad y = a\sin t, \quad 0 \leqslant t \leqslant \frac{\pi}{6}$

The curve meets the axes at points A and B as shown.
The straight line shown is part of the tangent to the curve at the point A.
Find, in terms of a,

a an equation of the tangent at A

b an exact value for the area of the finite region between the curve, the tangent at A and the x-axis, shown shaded in the diagram. [(c) Edexcel Limited 2006]

9 Expand as a series of ascending powers of x up to and including x^3.
State the range of values for which each expansion is valid.

a $\frac{1}{1 - 3x}$ **b** $\sqrt[3]{1 + \frac{1}{2}x}$ **c** $(1 - x)\sqrt{1 + x}$

d $\frac{2 + x}{\sqrt{4 - x}}$ **e** $\frac{3}{(1 - x)\left(1 + \frac{1}{2}x\right)}$

C4

10 a Write down the first four terms of the binomial expansion, in ascending powers of x, of $\left(1-\frac{1}{3}x\right)^n$, where $n < 1$.
State the range of values of x for which the expansion is valid.

b Given that the coefficient of x^3 in this expansion is four times the coefficient of x^2, find
i the value of n
ii the coefficient of x^4 in the expansion.

11 a Prove that, when $x = \frac{1}{12}$, the value of $\sqrt{1-3x}$ is exactly equal to $\cos 30°$

b i Expand $\sqrt{1-3x}$, $|x| < \frac{1}{3}$, in ascending powers of x up to and including the term in x^3, writing your answer in as simple a form as possible.
ii Use your expansion to find an approximation for $\cos 30°$

c Find the percentage error in your approximation for the value of $\cos 30°$

12 Find $\frac{dy}{dx}$, in terms of x and y, when

a $y^2 = xy + 2$

b $x^3 - 3x^2y + y^3 = 1$

c $\frac{1}{x^2} + \frac{1}{xy} + \frac{1}{y^2} = 1$

d $x \ln y = y \ln x$

13 A set of curves is given by the equation $\sin x + \cos y = 0.5$

a Use implicit differentiation to find an expression for $\frac{dy}{dx}$

b For $-\pi < x < \pi$ and $-\pi < y < \pi$, find the coordinates of the points where $\frac{dy}{dx} = 0$ [(c) Edexcel Limited 2007]

14 a Given that $y = 2^x$, and using the result $2^x = e^{x\ln 2}$, or otherwise, show that $\frac{dy}{dx} = 2^x \ln 2$

b Find the gradient of the curve with the equation $y = 2^{(x^2)}$ at the point with the coordinates $(2, 16)$. [(c) Edexcel Limited 2007]

15 The volume of a spherical balloon of radius r cm is V cm^3, where $V = \frac{4}{3}\pi r^3$

a Find $\frac{dV}{dr}$

The volume of the balloon increases with time t seconds according to the formula $\frac{dV}{dt} = \frac{1000}{(2t+1)^2}$, $t \geqslant 0$

b Using the chain rule, or otherwise, find an expression in terms of r and t for $\frac{dr}{dt}$

c Given that $V=0$ when $t=0$, solve the differential equation $\frac{dV}{dt} = \frac{1000}{(2t+1)^2}$, to obtain V in terms of t.

d Hence, at time of $t=5$

i find the radius of the balloon, giving your answer to 3 significant figures

ii show that the rate of increase of the radius of the balloon is approximately 2.90×10^{-2} cm s^{-1} [(c) Edexcel Limited 2006]

16 a Copy and complete this table by finding the three missing values of y, given that $y = \sec x$

x	0	$\frac{\pi}{16}$	$\frac{\pi}{8}$	$\frac{3\pi}{16}$	$\frac{\pi}{4}$
y		1.01959	1.08239		

b By using the trapezium rule with all five y-values in the table, find an estimate for $\int_0^{\frac{\pi}{4}} \sec x \, dx$

Show all your working, giving your answer to 4 decimal places.

c Use a standard integral to show that the exact value of

$\int_0^{\frac{\pi}{4}} \sec x \, dx$ is $\ln(1+\sqrt{2})$

d Find the percentage error in the estimate that you obtained using the trapezium rule.

17 Evaluate using an appropriate method.

a $\int e^x\sqrt{e^x - 2}\, dx$ **b** $\int (x-1)(x+2)^7\, dx$ **c** $\int 2\cos^2 x\, dx$

d $\int \cos x \sin^4 x\, dx$ **e** $\int \frac{9x^2}{(x-1)(x+2)^2}\, dx$ **f** $\int (x+3)e^x\, dx$

g $\int e^x \cos 2x\, dx$ **h** $\int \frac{x^2}{x^3+3}\, dx$ **i** $\int \frac{1}{2}\tan x \sin 2x\, dx$

C4

18 Use the substitution $u = 2^x$ to find the exact value of $\int_0^1 \frac{2^x}{(2^x+1)^2}\,dx$ [(c) Edexcel Limited 2007]

19 a By using the formulae for $\sin(A \pm B)$, with $A = 5x$ and $B = 2x$, show that $2\sin 2x\cos 5x$ can be written as $\sin \lambda x - \sin \mu x$, where λ and μ are positive integers. State the values of λ and μ.

b Hence, or otherwise, find $\int \sin 2x \cos 5x\,dx$

c Hence find the exact value of $\int_{\frac{\pi}{4}}^{\frac{3\pi}{4}} \sin 2x \cos 5x\,dx$

20 a Use the identity for $\cos(A + B)$ to prove that $\cos 2A = 2\cos^2 A - 1$

b Use the substitution $x = 2\sqrt{2}\sin\theta$ to prove that

$$\int_2^{\sqrt{6}} \sqrt{(8 - x^2)}\,dx = \frac{1}{3}(\pi + 3\sqrt{3} - 6)$$

c A curve is given by the parametric equations

$x = \sec\theta,\quad y = \ln(1 + \cos 2\theta),\quad 0 < \theta < \frac{\pi}{2}$

Find an equation of the tangent to the curve at the point where $\theta = \frac{\pi}{3}$ [(c) Edexcel Limited 2003]

21 a Show, by using the substitution $x = \sin\theta$, that, for $|x| < 1$,

$$\int \frac{1}{(1-x^2)^{\frac{3}{2}}}\,dx = \frac{x}{(1-x^2)^{\frac{1}{2}}} + c,$$ where c is an arbitrary constant.

b Use integration by parts to show that the exact value of $\int_2^4 x^2 \ln x\,dx$ can be written as $\frac{8}{9}(p\ln 2 - q)$, where p and q are integers.
Find the values of p and q.

22 a Use integration by parts to show that

$$\int x\operatorname{cosec}^2\left(x + \frac{\pi}{6}\right)dx = -x\cot\left(x + \frac{\pi}{6}\right) + \ln\left[\sin\left(x + \frac{\pi}{6}\right)\right] + c, \quad -\frac{\pi}{6} < x < \frac{\pi}{3}$$

b Solve the differential equation

$$\sin^2\left(x + \frac{\pi}{6}\right)\frac{dy}{dx} = 2xy(y+1)$$

to show that $\frac{1}{2}\ln\left|\frac{y}{y+1}\right| = -x\cot\left(x + \frac{\pi}{6}\right) + \ln\left[\sin\left(x + \frac{\pi}{6}\right)\right] + c$

c Given that $y = 1$ when $x = 0$, find the exact value of y when $x = \frac{\pi}{12}$ [(c) Edexcel Limited 2005]

C4

23

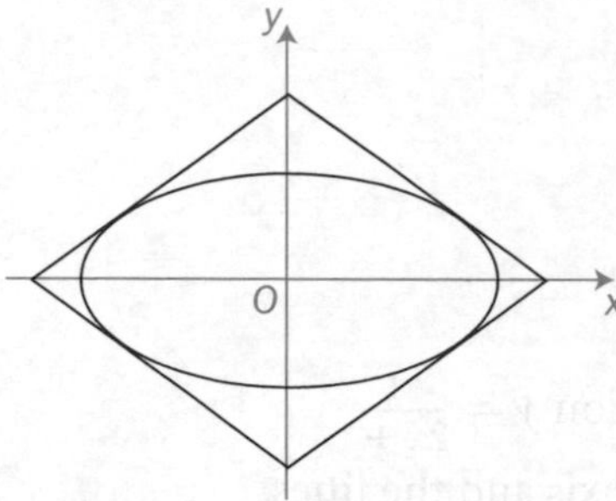

A table top, in the shape of a parallelogram, is made from two types of wood. The design is shown in the diagram. The area inside the ellipse is made from one type of wood, and the surrounding area is made from a second type of wood.

The ellipse has parametric equations

$$x = 5\cos\theta, \quad y = 4\sin\theta, \quad 0 \leqslant \theta < 2\pi$$

The parallelogram consists of four line segments, which are tangents to the ellipse at the points where $\theta = \alpha$, $\theta = -\alpha$, $\theta = \pi - \alpha$, $\theta = -\pi + \alpha$

a Find an equation of the tangent to the ellipse at $(5\cos\alpha, 4\sin\alpha)$, and show that it can be written in the form $5y\sin\alpha + 4x\cos\alpha = 20$

b Find by integration the area enclosed by the ellipse.

c Hence show that the area enclosed between the ellipse and the parallelogram is $\dfrac{80}{\sin 2\alpha} - 20\pi$

d Given that $0 < \alpha < \dfrac{\pi}{4}$, find the value of α for which the areas of the two types of wood are equal. [(c) Edexcel Limited 2002]

24

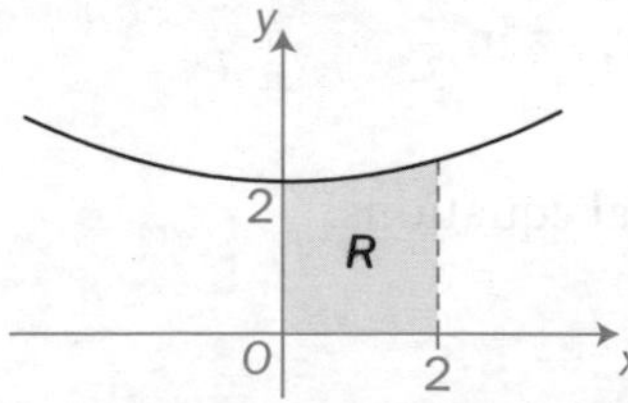

This diagram shows part of the curve with equation $y = x^2 + 2$

The finite region R is bounded by the curve, the x-axis and the lines $x = 0$ and $x = 2$.

a Use the trapezium rule with four strips of equal width to estimate the area of R.

b State, with a reason, whether your answer in part **a** is an under-estimate or over-estimate of the area of R.

c Using integration, find the volume of the solid generated when R is rotated through 360° about the x-axis, giving your answer in terms of π. [(c) Edexcel Limited 2002]

25

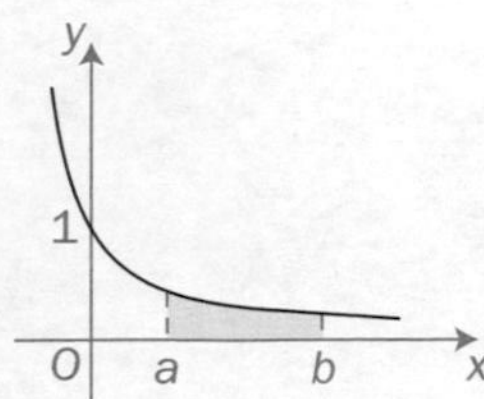

The curve shown in the diagram has the equation $y = \dfrac{1}{2x+1}$

The finite region bounded by the curve, the x-axis and the lines $x = a$ and $x = b$ is shown shaded. This region is rotated through 360° about the x-axis to generate a solid of revolution.

Find the volume of the solid generated. Express your answer as a single simplified fraction, in terms of a and b.

26 a The curve C_1 in Figure 1 has parametric equations

$x = \dfrac{2}{t},\ y = 3t^2$

The area bounded by the curve, the x-axis and the ordinates $x = 2$ and $x = 4$ is rotated through an angle of 360° about the x-axis.

Find **i** the values of t when $x = 2$ and $x = 4$

ii the volume of the solid of revolution.

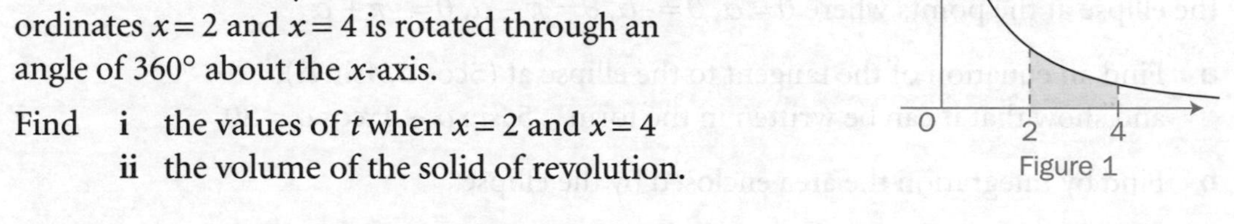

Figure 1

b The curve C_2 in Figure 2 is defined parametrically by

$x = t^2 + 4t,\ y = \dfrac{1}{2+t}$

The region between the x-axis and that part of C_2 from the point $P\left(5, \dfrac{1}{3}\right)$ to the point $Q\left(12, \dfrac{1}{4}\right)$ is rotated a half-turn about the x-axis.

Show that the volume of the solid generated is $\pi \ln \dfrac{4}{3}$

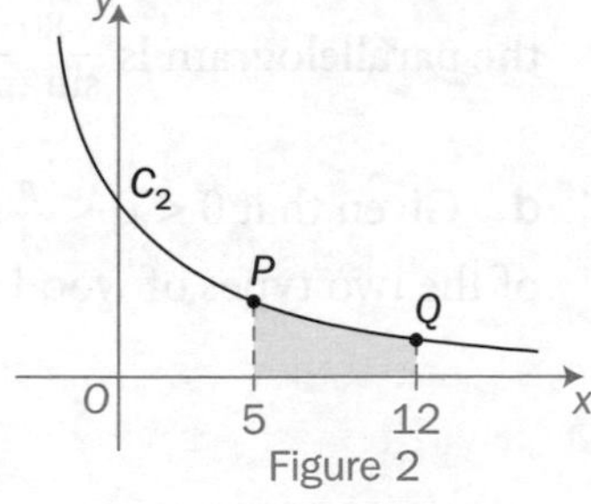

Figure 2

27 a Find the general solutions of these differential equations.

i $\dfrac{dy}{dx} = e^{2y}$ **ii** $(x+1)\dfrac{dy}{dx} = y$

iii $\dfrac{dy}{dx} + xy = x^2y$ **iv** $\cos y\dfrac{dy}{dx} = \sec^2 x \sin y$

b Find the particular solutions of these differential equations.

i $\left(\dfrac{dy}{dx}\right)^2 = yx^2$, given that $y = 4$ when $x = 2$

ii $(x^2+1)\dfrac{dy}{dx} - xy = 0$, given that $y = 10$ when $x = 1$

28 In an experiment a scientist considered the loss of mass of a collection of picked leaves. The mass, M grams, of a single leaf was measured at times t days after the leaf was picked.
The scientist attempted to find a relationship between M and t.
In a preliminary model she assumed that the rate of loss of mass was proportional to the mass M grams of the leaf.

a Write down a differential equation for the rate of change of mass of the leaf, using this model.

b Show, by differentiation, that $M = 10(0.98)^t$ satisfies this differential equation.

Further studies implied that the mass, M grams, of a certain leaf satisfied a modified differential equation

$$10\frac{dM}{dt} = -k(10M - 1) \qquad (1)$$

where k is a positive constant and $t \geqslant 0$

c Given that the mass of this leaf at time $t = 0$ is 10 grams, and that its mass at time $t = 10$ is 8.5 grams, solve the modified differential equation (1) to find the mass of this leaf at time $t = 15$. [(c) Edexcel Limited 2003]

29 Fluid flows out of a cylindrical tank with constant cross-section. At time t minutes, $t \geqslant 0$, the volume of fluid remaining in the tank is $V\,\text{m}^3$. The rate at which the fluid flows, in $\text{m}^3\,\text{min}^{-1}$, is proportional to the square root of V.

a Show that the depth, h metres, of fluid in the tank satisfies the differential equation $\frac{dh}{dt} = -k\sqrt{h}$, where k is a positive constant.

b Show that the general solution of the differential equation may be written as $h = (A - Bt)^2$, where A and B are constants.

c Given that, at time $t = 0$, the depth of fluid in the tank is 1 m, and that 5 minutes later the depth of fluid has reduced to 0.5 m, find the time, T minutes, which it takes for the tank to empty.

d Find the depth of water in the tank at time $0.5T$ minutes. [(c) Edexcel Limited 2003]

30 Points $P(2, 3, -1)$, $Q(4, -1, 4)$ and $R(2, -1, 3)$ are the vertices of a triangle.

a Find the vectors $\overrightarrow{PR}$ and $\overrightarrow{PQ}$ and the angle between them, to the nearest tenth of a degree.

b Find the lengths PR and PQ, giving your answers in surd form.

c Find the area of triangle PQR, correct to 3 significant figures.

C4

31 Relative to a fixed origin O, the point A has position vector $4\mathbf{i} + 8\mathbf{j} - \mathbf{k}$ and the point B has position vector $7\mathbf{i} + 14\mathbf{j} + 5\mathbf{k}$.

a Find the vector $\overrightarrow{AB}$.

b Calculate the cosine of $\angle OAB$.

c Show that, for all values of λ, the point P with position vector $\lambda\mathbf{i} + 2\lambda\mathbf{j} + (2\lambda - 9)\mathbf{k}$ lies on the line through A and B.

d Find the value of λ for which OP is perpendicular to AB.

e Hence find the coordinates of the foot of the perpendicular from O to AB. [(c) Edexcel Limited 2002]

32 Referred to an origin O, the points A, B and C have position vectors $(9\mathbf{i} - 2\mathbf{j} + \mathbf{k})$, $(6\mathbf{i} + 2\mathbf{j} + 6\mathbf{k})$ and $(3\mathbf{i} + p\mathbf{j} + q\mathbf{k})$ respectively, where p and q are constants.

a Find, in vector form, an equation of the line l which passes through A and B.

Given that C lies on l,

b find the value of p and the value of q

c calculate, in degrees, the acute angle between OC and AB.

The point D lies on AB and is such that OD is perpendicular to AB.

d Find the position vector of D. [(c) Edexcel Limited 2002]

33 The points A and B have position vectors $2\mathbf{i} + 6\mathbf{j} - \mathbf{k}$ and $3\mathbf{i} + 4\mathbf{j} + \mathbf{k}$. The line L_1 passes through the points A and B.

a Find the vector $\overrightarrow{AB}$.

b Find a vector equation for the line L_1.

A second line L_2 passes through the origin and is parallel to the vector $\mathbf{i} + \mathbf{k}$. The line L_1 meets the line L_2 at the point C.

c Find the acute angle between L_1 and L_2.

d Find the position vector of the point C. [(c) Edexcel Limited 2008]

34 Lines L_1 and L_2 are given by the equations

L_1: $\mathbf{r} = \mathbf{i} - 2\mathbf{k} + \lambda(2\mathbf{i} + 4\mathbf{j} - \mathbf{k})$

L_2: $\mathbf{r} = 8\mathbf{i} + 5\mathbf{j} - 10\mathbf{k} + \mu(-\mathbf{i} + \mathbf{j} + 2\mathbf{k})$, where λ and μ are parameters.

a Show that L_1 and L_2 intersect and find the coordinates of the point of intersection.

b Show that L_1 and L_2 are mutually perpendicular.

c Show that the point P (3, 4, –3) lies on L_1.

d P is reflected in L_2. Find the coordinates of the image of P.

C4

Answers

Before you start Answers

Chapter 1

1 a $1\frac{1}{24}$ b $\frac{1}{4}$

2 a

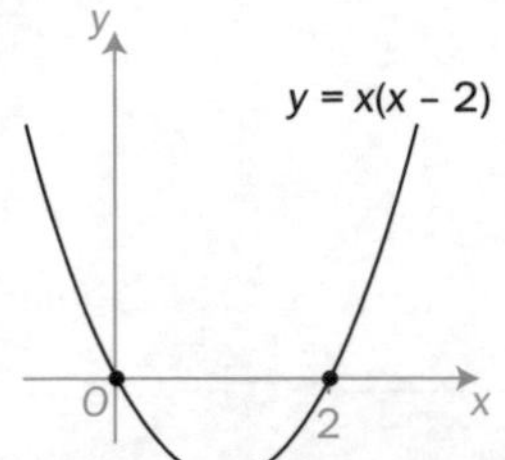

b

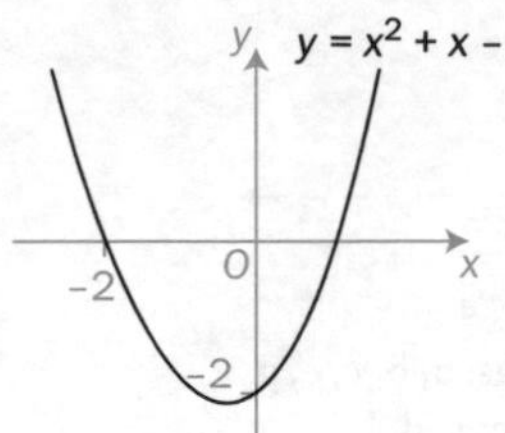

c

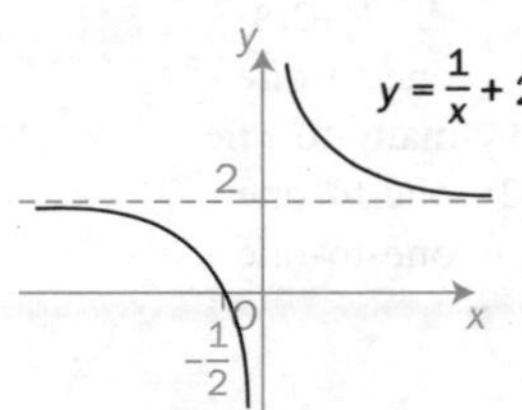

3 a $y = -x^2$ b $y = x^2 + \sqrt{x}$
c $y = x^2 + 11x + 30$ d $y = 4x^2$

Exercise 1.1

1 a $\frac{2x}{z}$ b $\frac{6a^2}{bc}$ c $6pr$
d $\frac{1}{(x-1)x}$ e $\frac{a}{a-1}$ f $\frac{1}{x^2-4}$
g $\frac{1}{x-3}$ h $\frac{1}{x}$

2 a $\frac{xz+y^2}{yz}$ b $\frac{a^2+b^2}{ab}$ c $\frac{b-a}{abx}$
d $\frac{x+y}{x^2y^2}$ e $\frac{2x+1}{x(x+1)}$ f $\frac{2}{a^2-1}$
g $\frac{a+2}{(a+1)^2}$ h $\frac{2x+1}{x}$ i $\frac{3x+1}{x+1}$
j $\frac{2y+8}{3(y-2)}$ k $\frac{2x+3}{x^2-1}$ l $\frac{3x+5}{x(x+1)}$
m $\frac{z-4}{z^2-z-2}$ n $\frac{1}{y+1}$
o $\frac{3x+7}{(x+1)(x+2)(x+3)}$ p $\frac{2(2y^2-2y-1)}{(y-1)(y-2)(y+2)}$
q $\frac{z}{(z+1)(z+3)}$ r $\frac{2x^3-8x-2}{x(x^2-4)}$

3 a $-x$ b $\frac{y}{y-1}$
c $\frac{x^2-x+1}{x-1}$ d $\frac{x^2+x+1}{x}$

Exercise 1.2

1 a x^2-6x+8 b $x^2-2x-24$
c $x^2-10x+25$ d $x^2-8x+15$
e x^2+2x-3 f $3x^2+x+2$
g n^2-3n+2 h n^2-3n-2
i $3y^2-2y-4$ j $2a^3+a^2+3a-1$
k $4a^3-16a^2+24a-6$ l $4z-5$
m $3x-4$ n x^2-x+1

2 a $1+\frac{4}{x+2}$ b $1-\frac{1}{x+2}$ c $1-\frac{4}{x+2}$
d $3+\frac{1}{x+2}$ e $2+\frac{7}{x-3}$ f $2+\frac{1}{2x+1}$
g $1-\frac{2}{x^2+1}$ h $3+\frac{4}{x^2-1}$ i $-1+\frac{3}{x+1}$
j $1+\frac{x+1}{x^2-2}$

3 a $3-\frac{5}{x^2}$ b $3+\frac{11}{x^2-3}$
c $2+\frac{3}{2x^2-1}$ d $7-\frac{2}{2x^2+3}$

4 a x^2+6x+1 rem 2 b x^2+5x-2 rem 2
c $2x^2-7x-4$ rem −2 d x^2-3x-5 rem 0
e n^2+2n-3 rem 3 f $3n^2+4n-2$ rem −2
g n^2-4n+2 rem −1 h x^2+2x+3 rem 2
i $2x^2+x-2$ rem 2 j x rem $x-2$

5 a x^2-3x-5 b x^2-8x-2
c $2x^2-x+10$ d x^2-3x+1

6 a $x^2-x+1, 2$ b $x^2+9x+15, 14$
c $2x^2+x-1, 1$ d $x^2+x+2, 6x-2$

7 a $(x-2)(x-3)(x+1)$
b $(x-4)(x-3)(x-1)$
c $(2x+3)(2x+1)(x-2)$
d $(x+2)(x-2)(x+3)(x-3)$
e $(x+2)(x^2-2x+4)$
f $(x-2)(x^2+2x+4)$

8 $x^4-1=(x-1)(x^3+x^2+x+1)$
$x^5-1=(x-1)(x^4+x^3+x^2+x+1)$
$x^6-1=(x-1)(x^5+x^4+x^3+x^2+x+1)$
$x^n-1=(x-1)(x^{n-1}+x^{n-2}+\cdots+x+1)$

Exercise 1.3

1 a

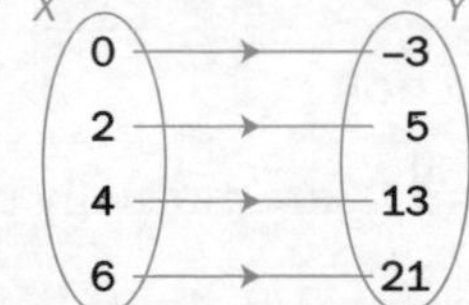

Range $y = \{-3, 5, 13, 21\}$

C3

b

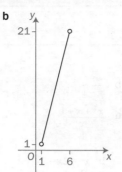

Range $y \in \mathbb{R}, 1 < y < 21$

c

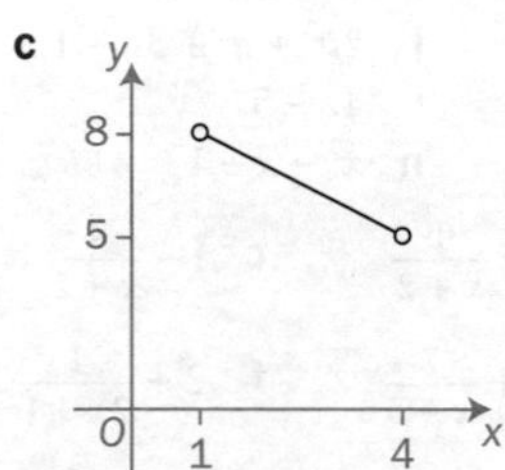

Range $y \in \mathbb{R}, 5 < y < 8$

d

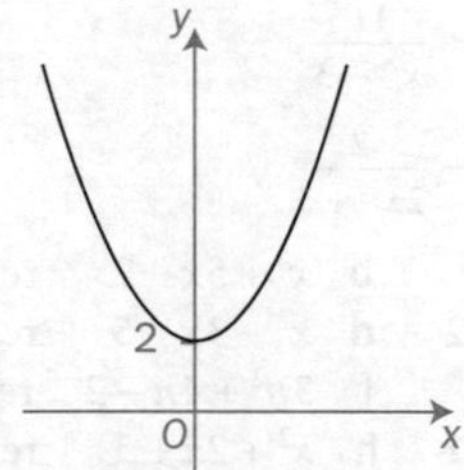

Range $y \in \mathbb{R}, y \geqslant 2$

e

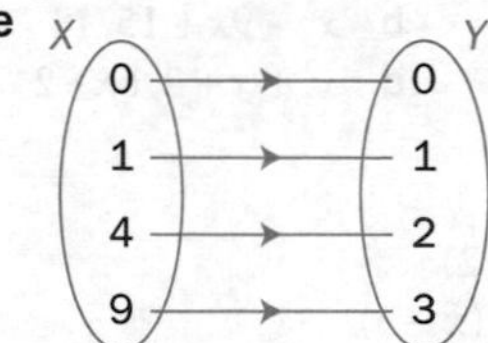

Range $y = \{0, 1, 2, 3\}$

f

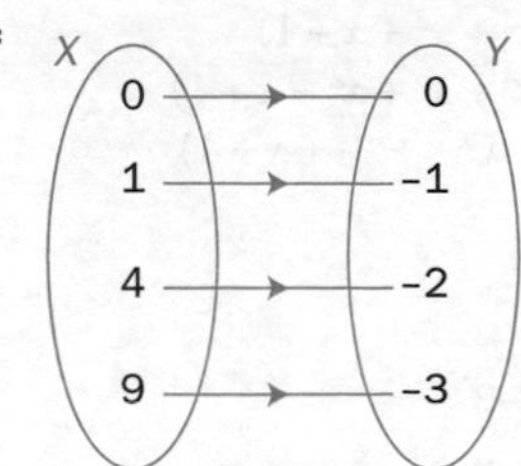

Range $y = \{0, -1, -2, -3\}$

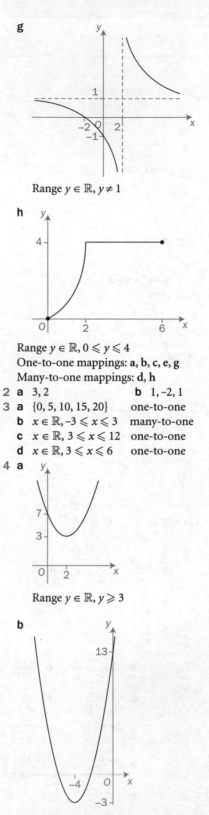

g

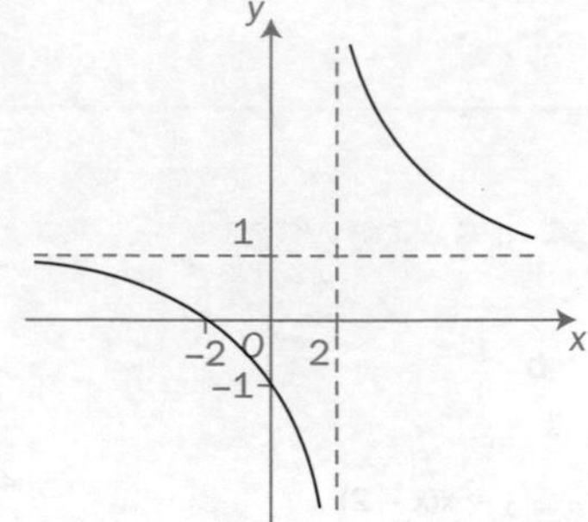

Range $y \in \mathbb{R}, y \neq 1$

h

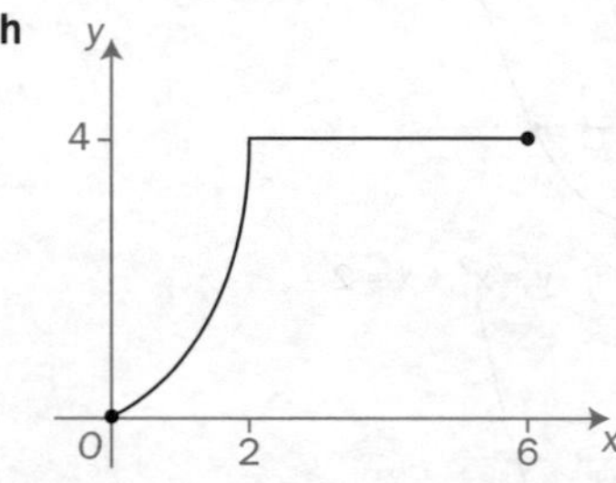

Range $y \in \mathbb{R}, 0 \leqslant y \leqslant 4$

One-to-one mappings: a, b, c, e, g

Many-to-one mappings: d, h

2 **a** 3, 2 **b** 1, –2, 1

3 **a** $\{0, 5, 10, 15, 20\}$ one-to-one

b $x \in \mathbb{R}, -3 \leqslant x \leqslant 3$ many-to-one

c $x \in \mathbb{R}, 3 \leqslant x \leqslant 12$ one-to-one

d $x \in \mathbb{R}, 3 \leqslant x \leqslant 6$ one-to-one

4 **a**

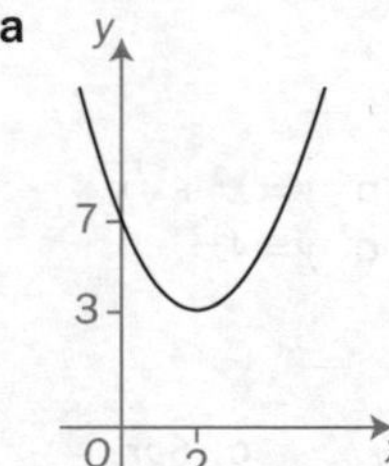

Range $y \in \mathbb{R}, y \geqslant 3$

b

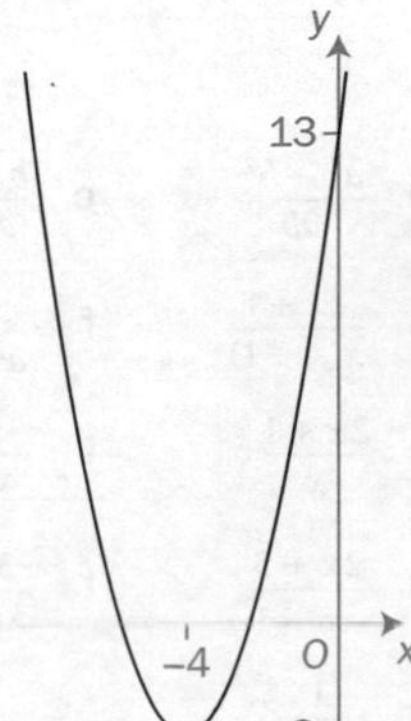

Range $y \in \mathbb{R}, y \geqslant -3$

c

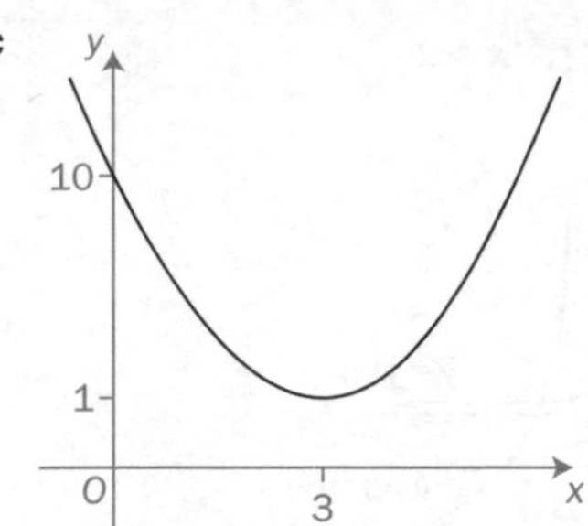

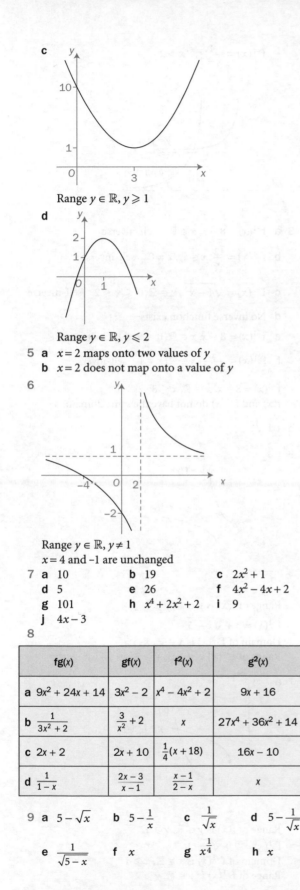

Range $y \in \mathbb{R}, y \geqslant 1$

d

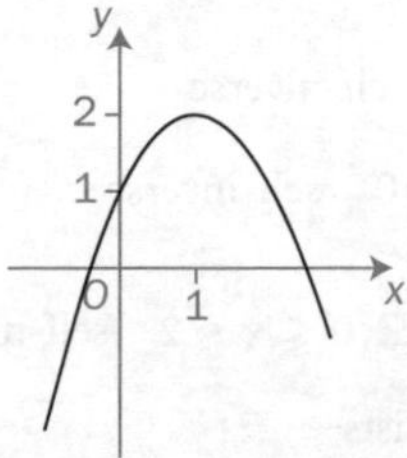

Range $y \in \mathbb{R}, y \leqslant 2$

5 a $x = 2$ maps onto two values of y
 b $x = 2$ does not map onto a value of y

6

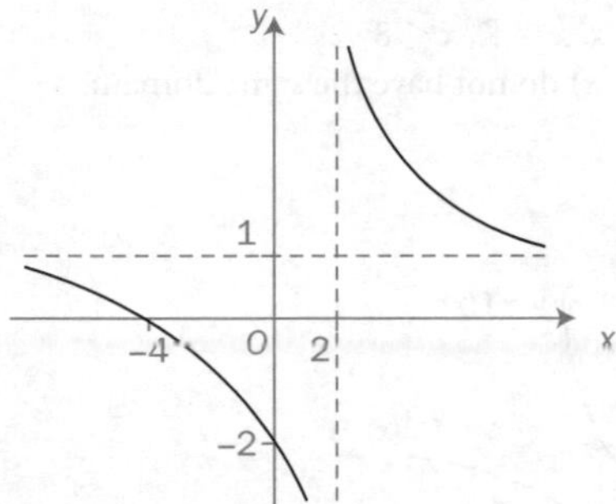

Range $y \in \mathbb{R}, y \neq 1$
$x = 4$ and -1 are unchanged

7 a 10 b 19 c $2x^2 + 1$
 d 5 e 26 f $4x^2 - 4x + 2$
 g 101 h $x^4 + 2x^2 + 2$ i 9
 j $4x - 3$

8

	fg(x)	gf(x)	f²(x)	g²(x)
a	$9x^2 + 24x + 14$	$3x^2 - 2$	$x^4 - 4x^2 + 2$	$9x + 16$
b	$\frac{1}{3x^2 + 2}$	$\frac{3}{x^2} + 2$	x	$27x^4 + 36x^2 + 14$
c	$2x + 2$	$2x + 10$	$\frac{1}{4}(x + 18)$	$16x - 10$
d	$\frac{1}{1 - x}$	$\frac{2x - 3}{x - 1}$	$\frac{x - 1}{2 - x}$	x

9 a $5 - \sqrt{x}$ b $5 - \frac{1}{x}$ c $\frac{1}{\sqrt{x}}$ d $5 - \frac{1}{\sqrt{x}}$
 e $\frac{1}{\sqrt{5 - x}}$ f x g $x^{\frac{1}{4}}$ h x

10 a $\alpha = 1$ b i ± 5 ii $2, -1$ iii 4
 c $-\frac{1}{2}, 1$

11 a 2 b -1

12 $p = \pm 1, \quad q = \pm 2$

13 a i $4x + 9$ ii $8x + 21$ iii $16x + 45$
 b $2^n x + 3(2^n - 1)$

14 The range of $g(x)$ is not a subset of the domain of $f(x)$. Change domain of $g(x)$ to $0 \leqslant x \leqslant \sqrt{5}$

Exercise 1.4

1 a $\frac{x + 2}{3}$ b $\frac{x - 1}{2}$ c $\frac{x}{2} - 1$
 d $2x - 3$ e $2(x - 3)$ f $\sqrt{x} - 1$
 g $\sqrt{x - 1}$ h $6 - x$ i $x^2 + 3$

2 a $f^{-1}(x) = \frac{x - 1}{2}$

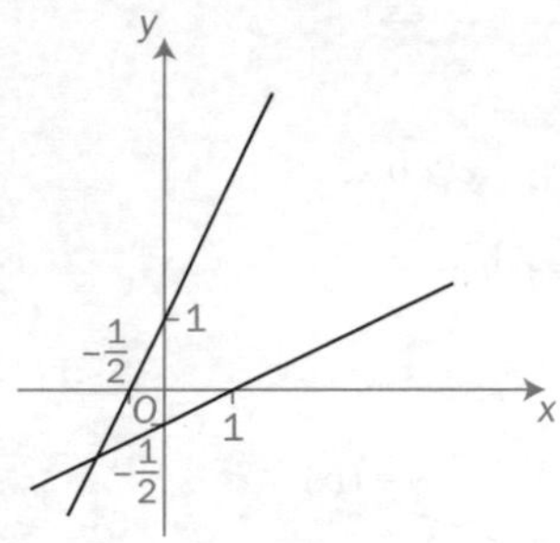

b $f^{-1}(x) = 5 - \frac{1}{2}x$

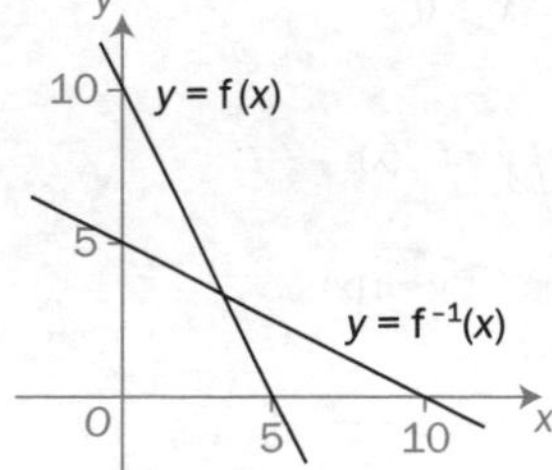

c $f^{-1}(x) = 2x - 4$

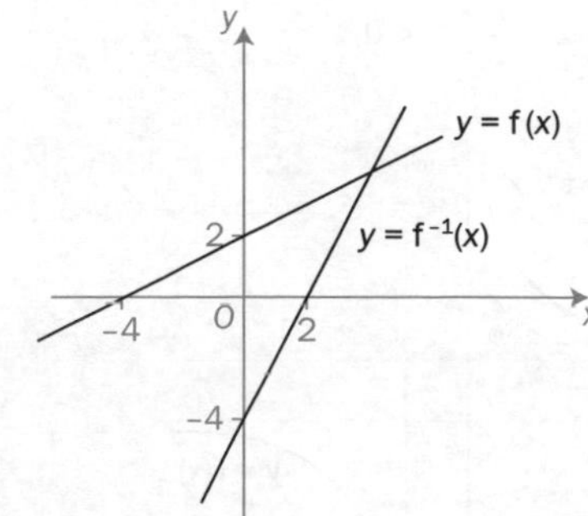

C3

d $f^{-1}(x) = \sqrt{x-2}, \quad x \geqslant 2$

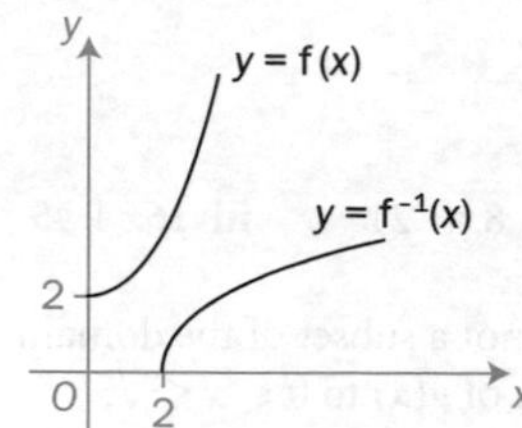

e $f^{-1}(x) = \sqrt{x} + 2, \quad x \geqslant 0$

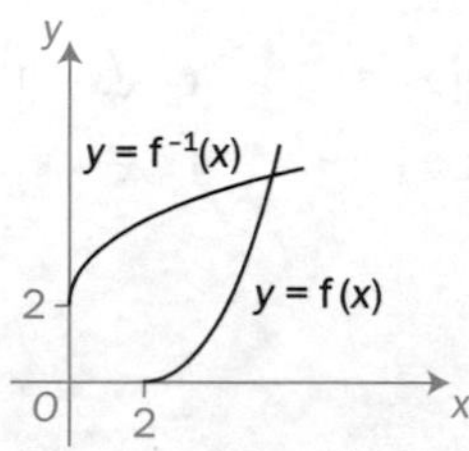

f $f^{-1}(x) = x^2 + 4, \quad x \geqslant 0$

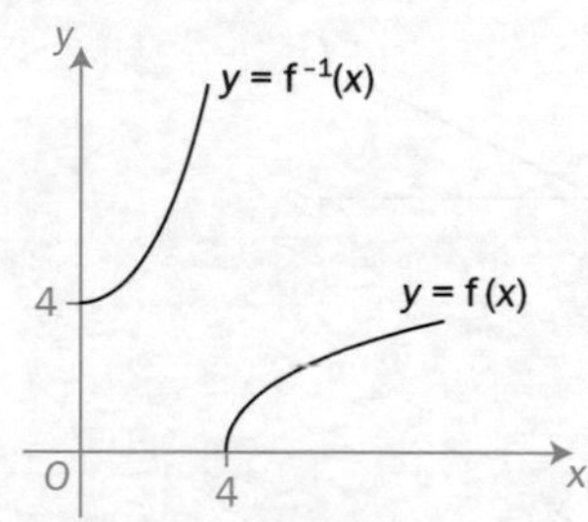

g $f^{-1}(x) = x^2 - 3, \quad x \geqslant 0$

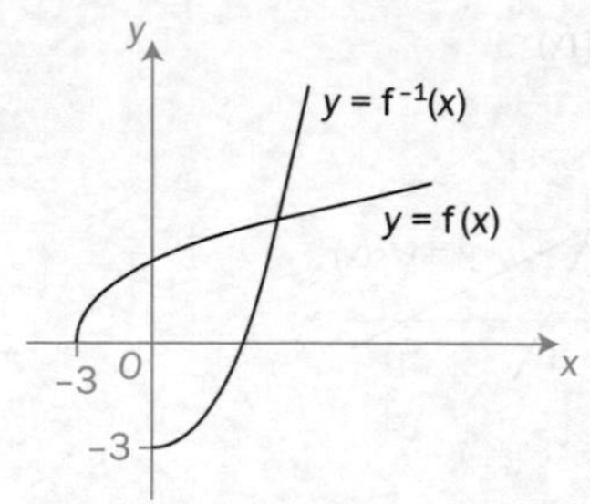

h $f^{-1}(x) = 2 - \frac{1}{x}, \quad x < 0$

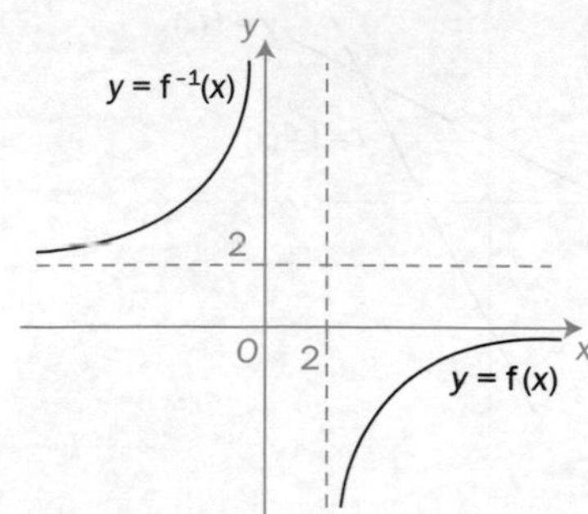

i $f^{-1}(x) = \frac{1}{x-2}, \quad x > 2$

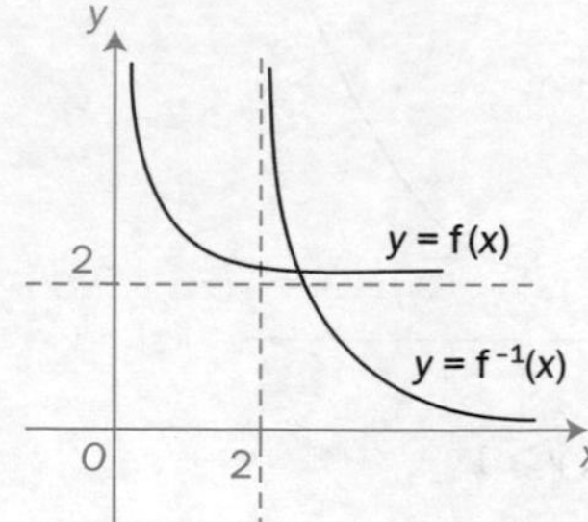

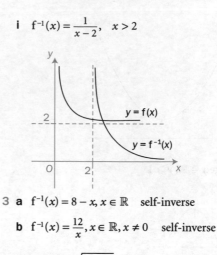

3 **a** $f^{-1}(x) = 8 - x, x \in \mathbb{R}$ self-inverse

b $f^{-1}(x) = \frac{12}{x}, x \in \mathbb{R}, x \neq 0$ self-inverse

c $f^{-1}(x) = \sqrt{4 - x^2}, x \in \mathbb{R}, 0 \leqslant x \leqslant 2$ self-inverse

d No inverse function exists

e $f^{-1}(x) = 8 - x, x \in \mathbb{R}, 0 \leqslant x \leqslant 8$ self-inverse

f $f^{-1}(x) = \frac{x}{x-1}, x \in \mathbb{R}, x \neq 1$ self-inverse

4 $f^{-1}(x) = 8 - x, x \in \mathbb{R}, x \leqslant 8$

$f(x)$ and $f^{-1}(x)$ do not have the same domain.

5 $1 \pm \sqrt{2}$

6

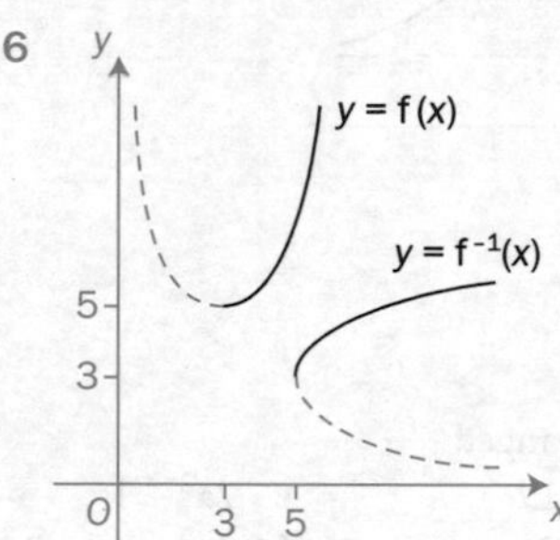

Range of $f(x)$ is $y \in \mathbb{R}, y \geqslant 5$

$f^{-1}(x) = 3 + \sqrt{x-5}$

Domain of $f^{-1}(x)$ is $x \in \mathbb{R}, x \geqslant 5$

Range of $f^{-1}(x)$ is $y \in \mathbb{R}, y \geqslant 3$

7 **a**

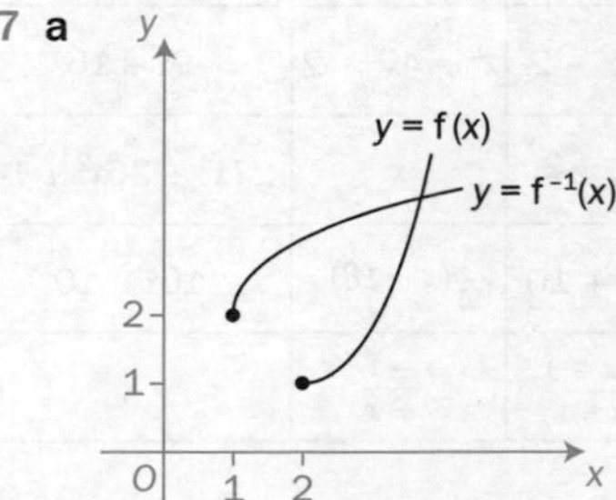

Range of $f(x)$ is $y \in \mathbb{R}, y \geqslant 1$

$f^{-1}(x) = 2 + \sqrt{x-1}$

Domain of $f^{-1}(x)$ is $x \in \mathbb{R}, x \geqslant 1$

Range of $f^{-1}(x)$ is $y \in \mathbb{R}, y \geqslant 2$

C3

b

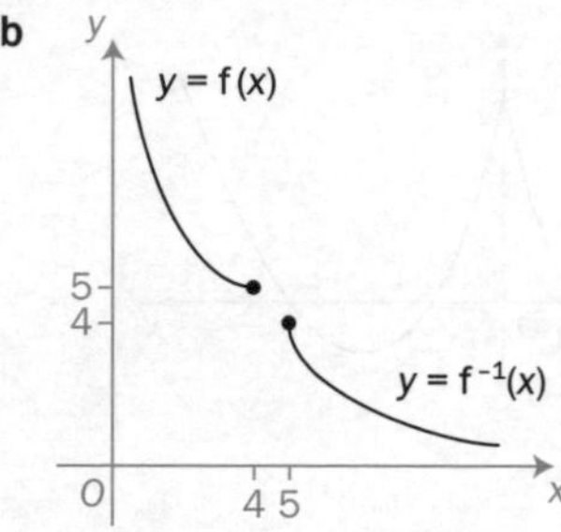

Range of f(x) is $y \in \mathbb{R}, y \geqslant 5$
$f^{-1}(x) = 4 - \sqrt{x - 5}$
Domain of $f^{-1}(x)$ is $x \in \mathbb{R}, x \geqslant 5$
Range of $f^{-1}(x)$ is $y \in \mathbb{R}, y \leqslant 4$

c

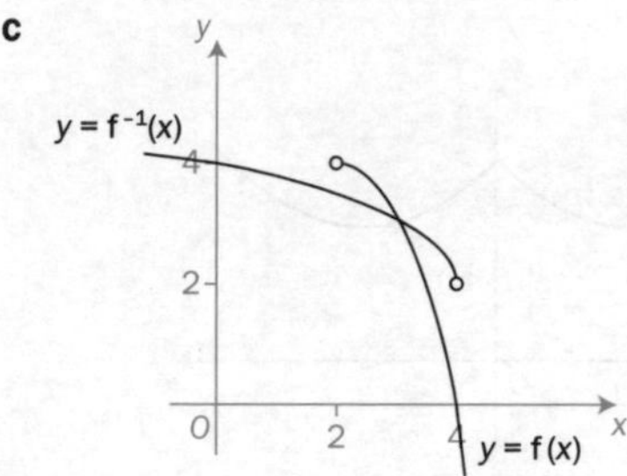

Range of f(x) is $y \in \mathbb{R}, y < 4$
$f^{-1}(x) = 2 + \sqrt{4 - x}$
Domain of $f^{-1}(x)$ is $x \in \mathbb{R}, x < 4$
Range of $f^{-1}(x)$ is $y \in \mathbb{R}, y > 2$

d

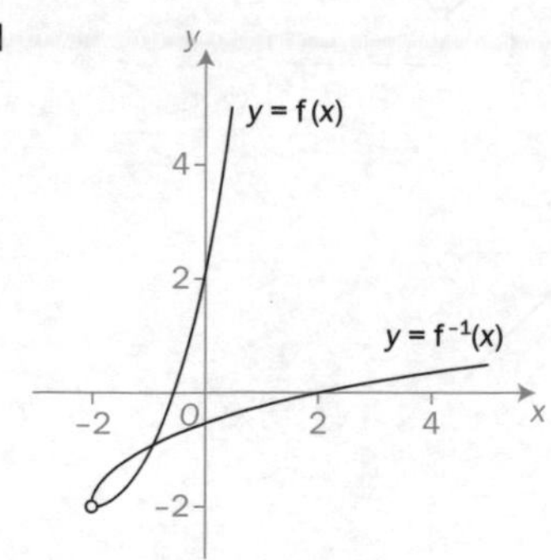

Range of f(x) is $y \in \mathbb{R}, y > -2$
$f^{-1}(x) = \sqrt{x + 2} - 2$
Domain of $f^{-1}(x)$ is $x \in \mathbb{R}, x > -2$
Range of $f^{-1}(x)$ is $y \in \mathbb{R}, y > -2$

8 **a** $f^{-1}(x) = 2x - 8, x \in \mathbb{R}, \quad x = 8 \quad (8, 8)$
b $f^{-1}(x) = \sqrt{x}, x \in \mathbb{R}, x \geqslant 0 \quad x = 0, 1 \quad (0, 0), (1, 1)$
c $f^{-1}(x) = 2 + \sqrt{x}, x \in \mathbb{R}, x \geqslant 0 \quad x = 4 \quad (4, 4)$
d $f^{-1}(x) = \sqrt{x + 4} - 4, x \in \mathbb{R}, x \geqslant -4$
$x = -3, (-3, -3)$

9 **a** $a = -2, b = 0, c = 1$
b i 4 ii 48 iii 3

10 $g^{-1}(x) = 2 + \frac{2}{x}, x \in \mathbb{R}, x \neq 0$ solutions are $x = 1 \pm \sqrt{3}$

11 $x = 3 \pm \sqrt{10}$

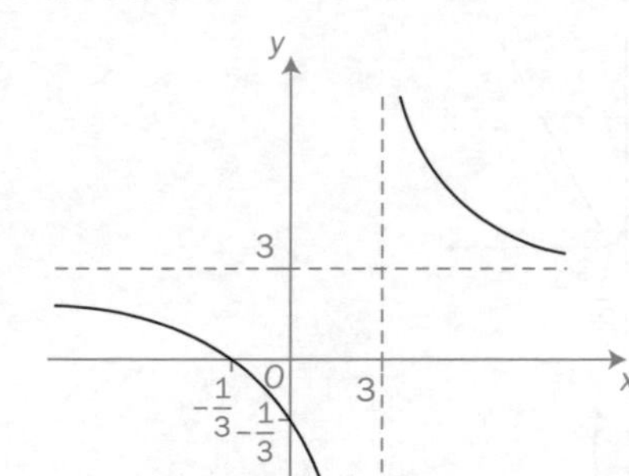

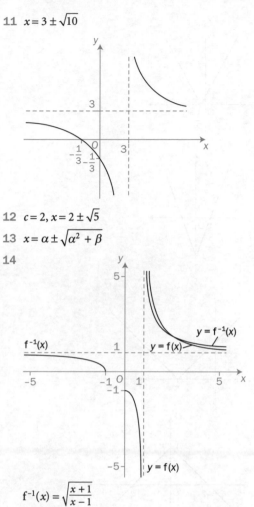

12 $c = 2, x = 2 \pm \sqrt{5}$

13 $x = \alpha \pm \sqrt{\alpha^2 + \beta}$

14

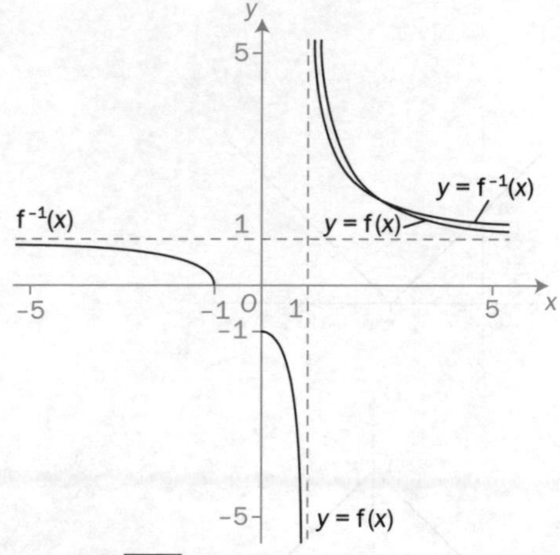

$f^{-1}(x) = \sqrt{\frac{x + 1}{x - 1}}$
Domain $x \in \mathbb{R}, x \leqslant -1, x > 1$
Range $y \in \mathbb{R}, y \geqslant 0, y \neq 1$

15 f_3 has an inverse.
All the others are many-to-one functions.

Exercise 1.5

1 **a**

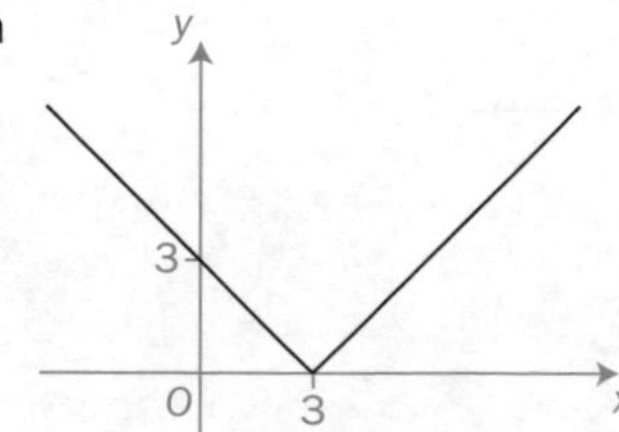

b

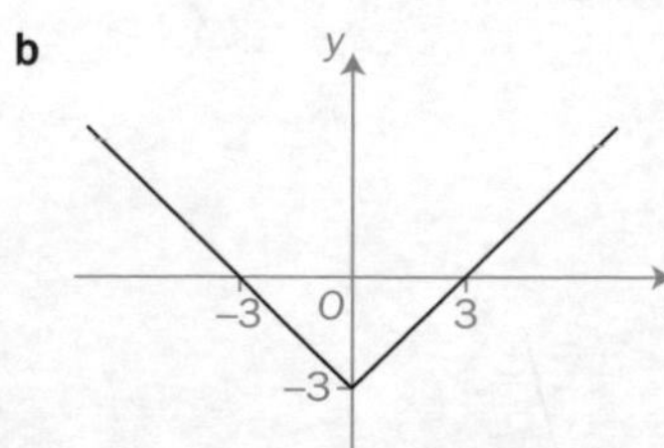

c

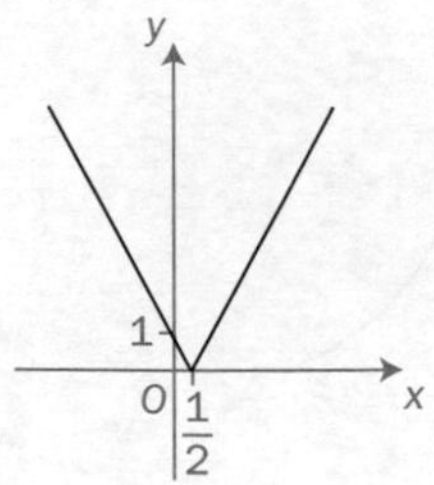

d

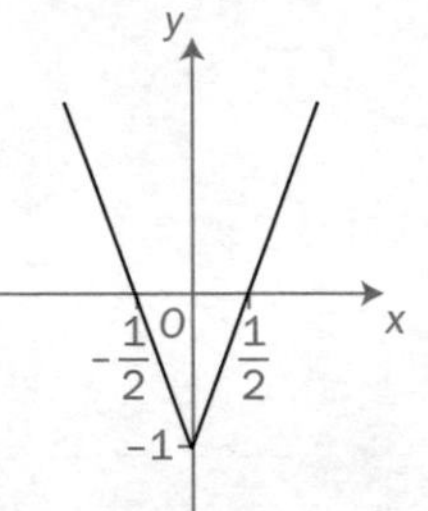

e

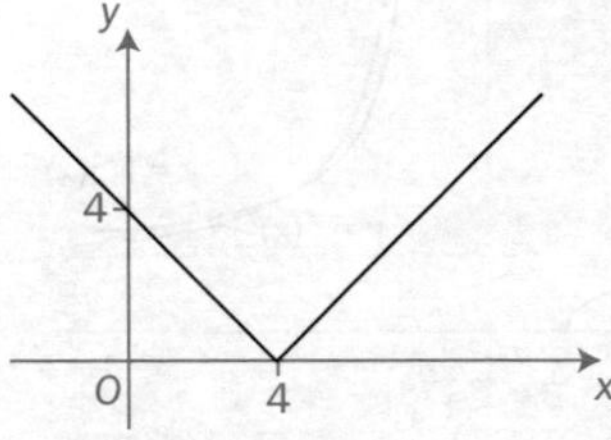

f

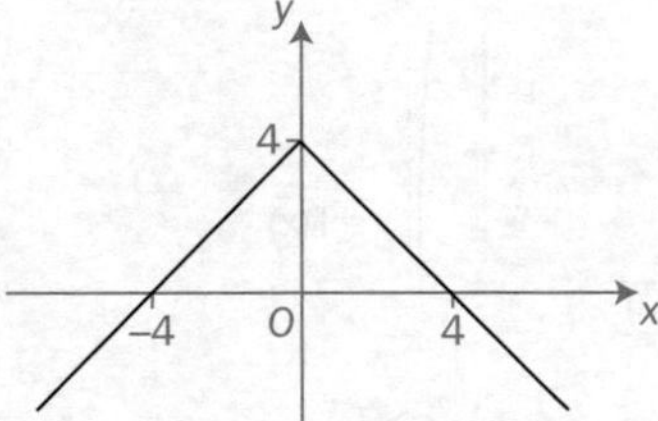

g

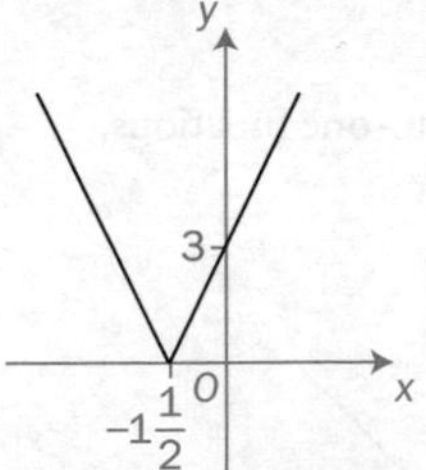

h

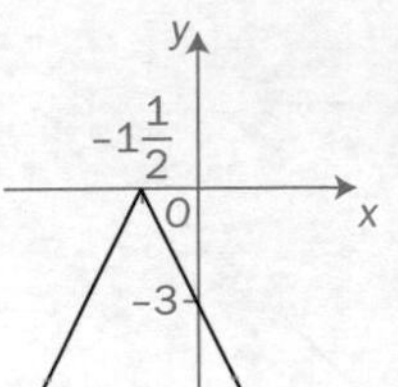

i

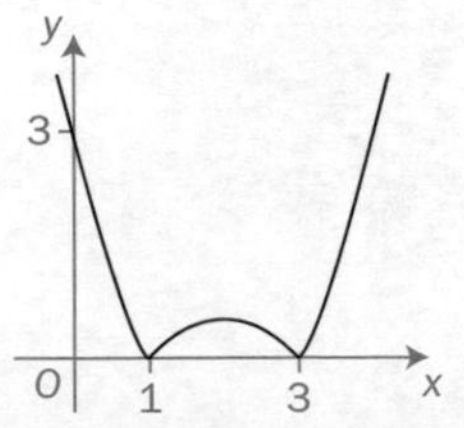

j

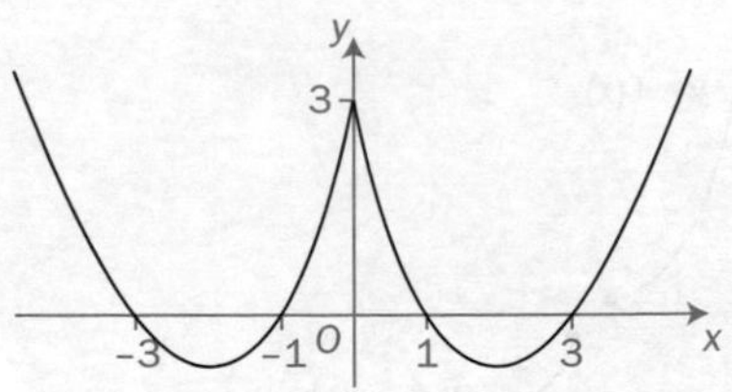

k

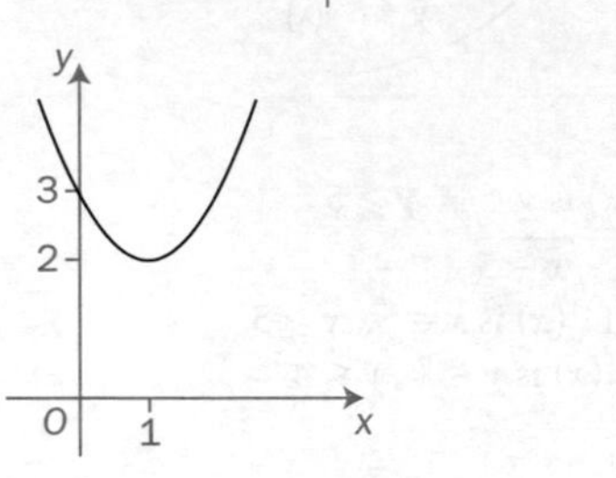

l

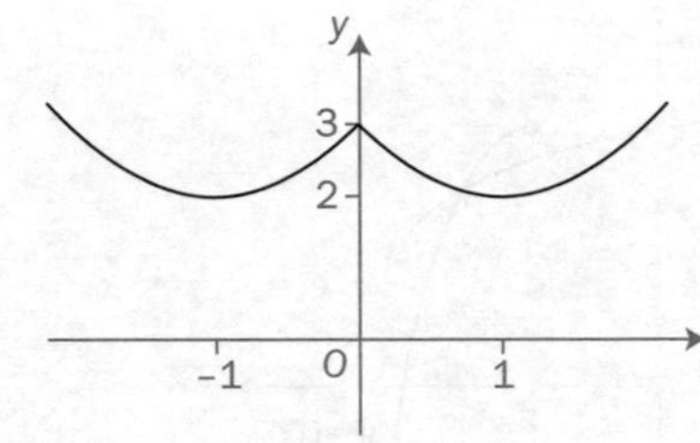

m

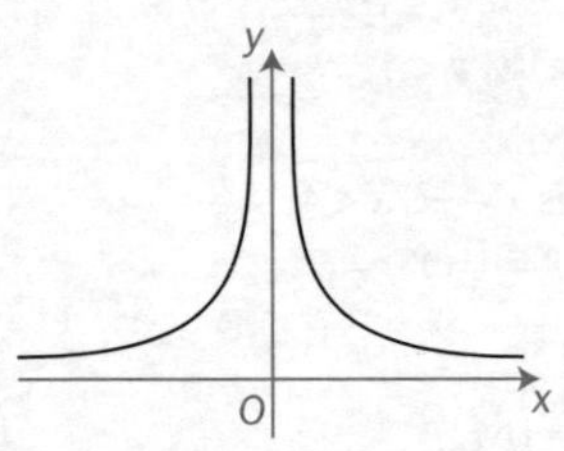

n

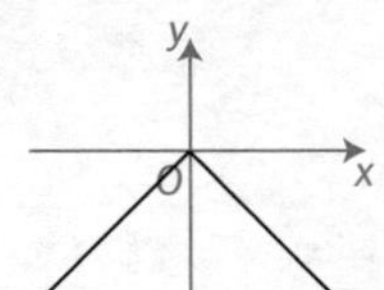

o

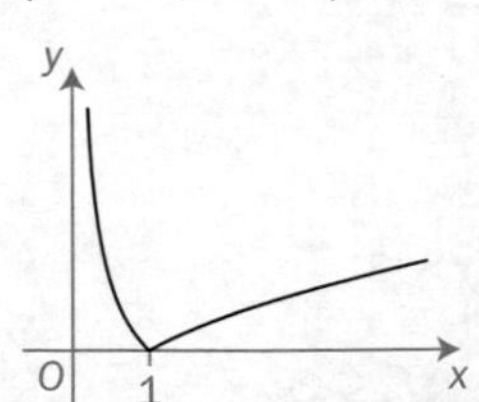

2 **a**

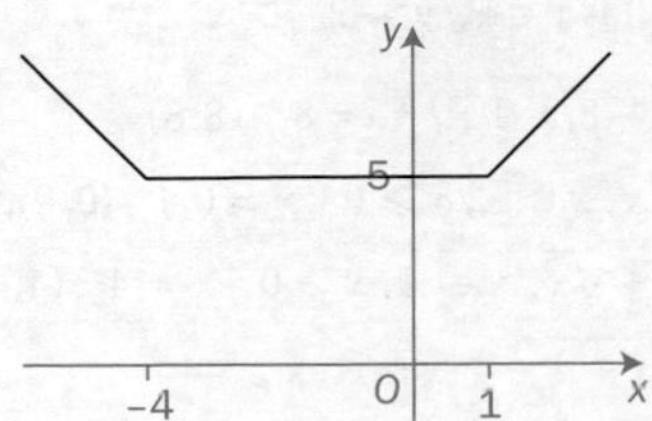

b

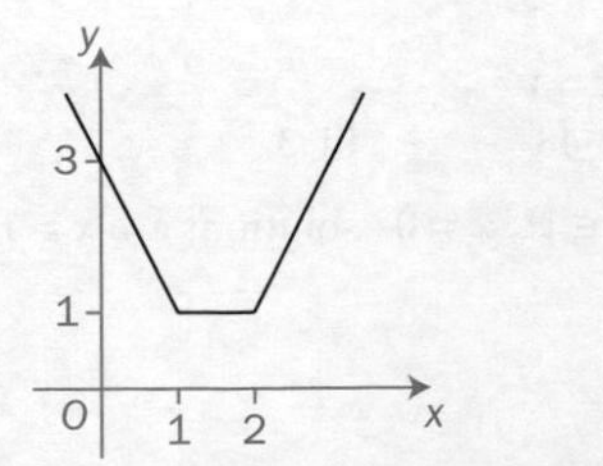

C3

3 a

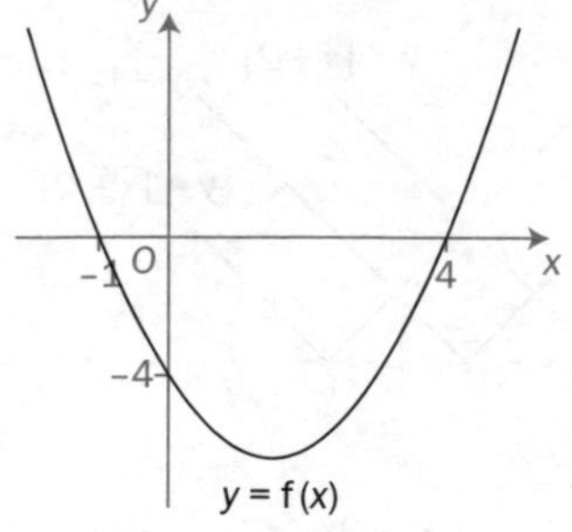

$y = f(x)$

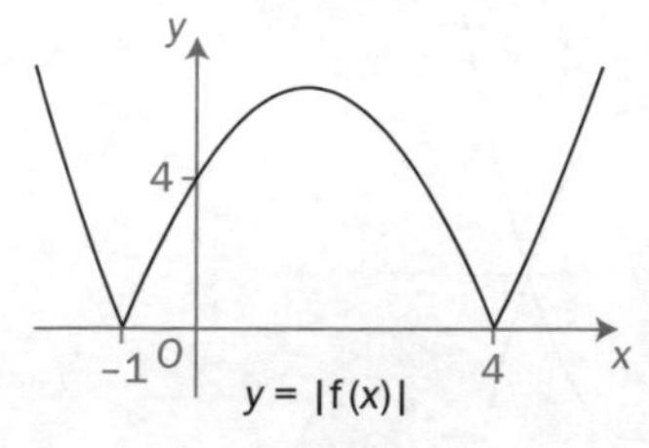

$y = |f(x)|$

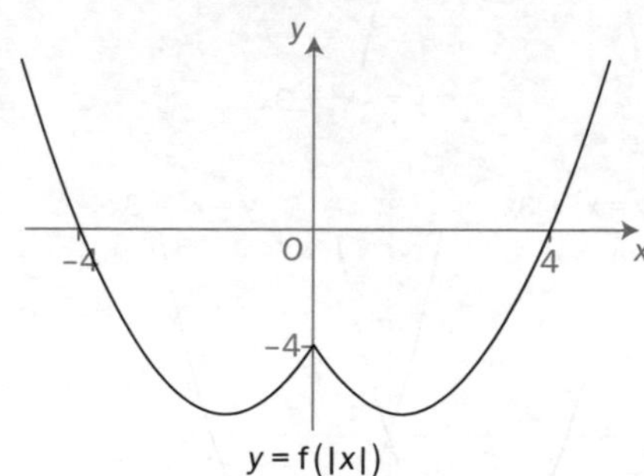

$y = f(|x|)$

b

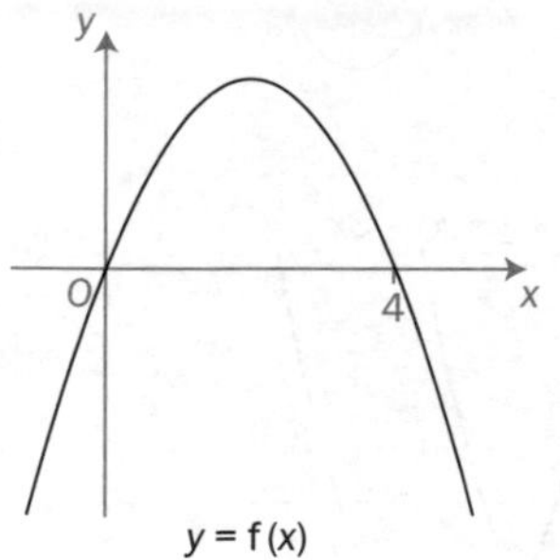

$y = f(x)$

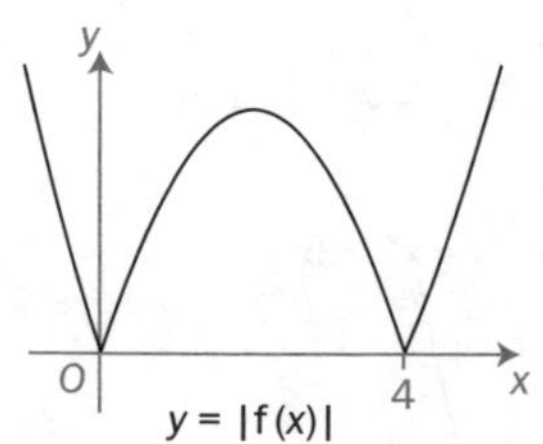

$y = |f(x)|$

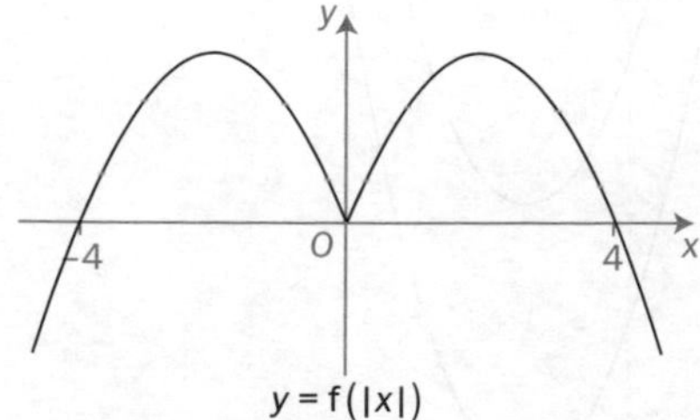

$y = f(|x|)$

4

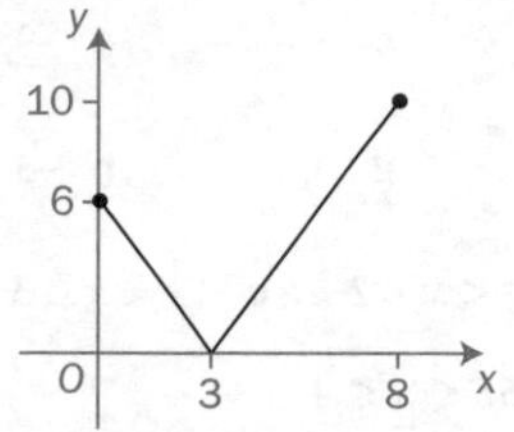

Range $0 \leqslant y \leqslant 10$
Solutions $x = 1$ or 5

5

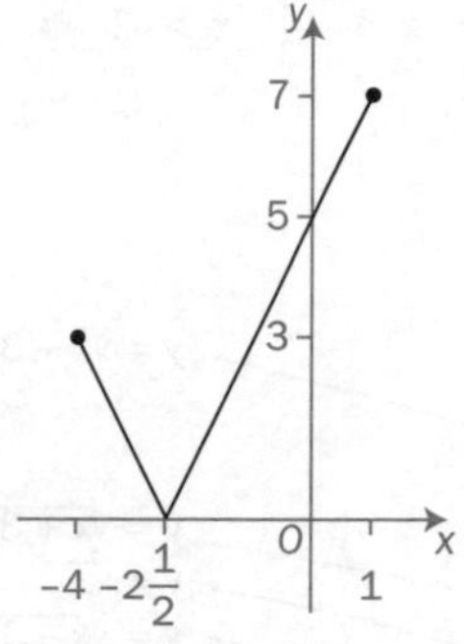

Range $0 \leqslant y \leqslant 7$
Solutions $x = -1$ or -4

6 a

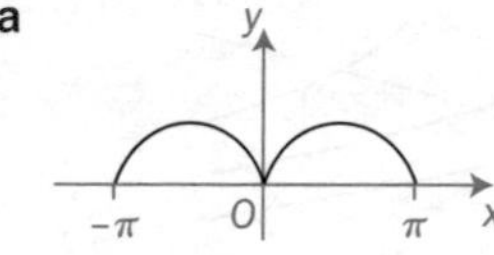

b

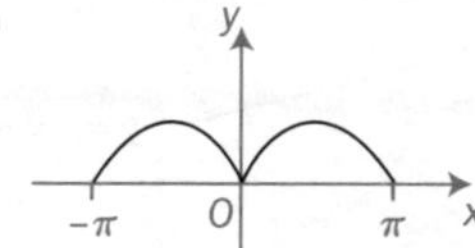

c

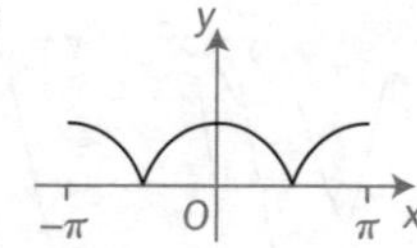

d

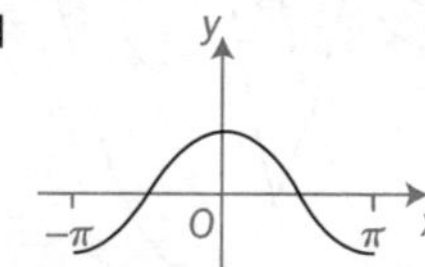

7 (2, 3)
8 (−2, 0), (6, 8)
9 4 solutions; (−6, 16), (0, 4), (2, 0), (4, 4)

Exercise 1.6

1 a i $-1, 9$ ii $-1 < x < 9$
b i 1 ii $x < 1$
2 a i $-7, 3$ ii $-7 < x < 3$
b i $-5, 2$ ii $x < -5, x > 2$
c i $-2\frac{1}{3}, 3$ ii $x \leqslant 2\frac{1}{3}, x \geqslant 3$
d i $2, 4$ ii $2 \leqslant x \leqslant 4$
e i 1 ii $x > 1$
f i $\frac{1}{4}, 1\frac{1}{2}$ ii $\frac{1}{4} < x < 1\frac{1}{2}$

C3

3 **a** $-2\frac{1}{2}$ **b** $-1\frac{1}{2}$ **c** 5 **d** $-9, 1$

e $1\frac{1}{2}$ **f** $6, \frac{2}{3}$ **g** $3, 6$ **h** $-3, 0, 5$

4 **a** $x > 2, x < 0$ **b** $-\frac{2}{3} < x < 2$ **c** $1\frac{1}{2} < x < 3$

d $x \in \mathbb{R}$ **e** $2.56 \leqslant x \leqslant 4$

f $-3 < x < -1$ and $1 < x < 3$

5 **a** $-2\sqrt{2}, 1-\sqrt{5}, 2, 4$ **b** $x < -4\frac{2}{3},\ x > -2$

6 $b^2 \leqslant 4c$

Exercise 1.7

1

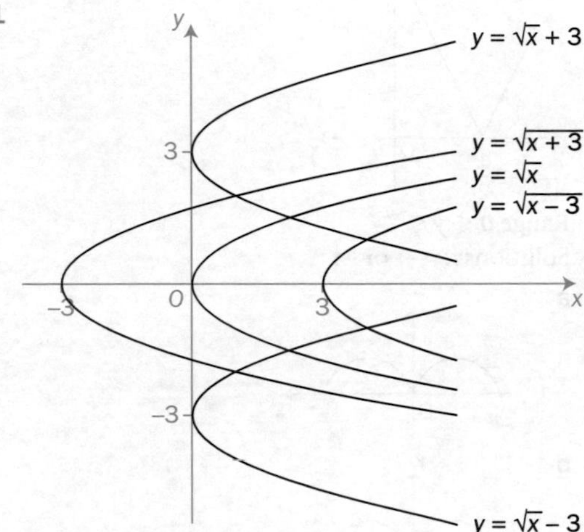

2

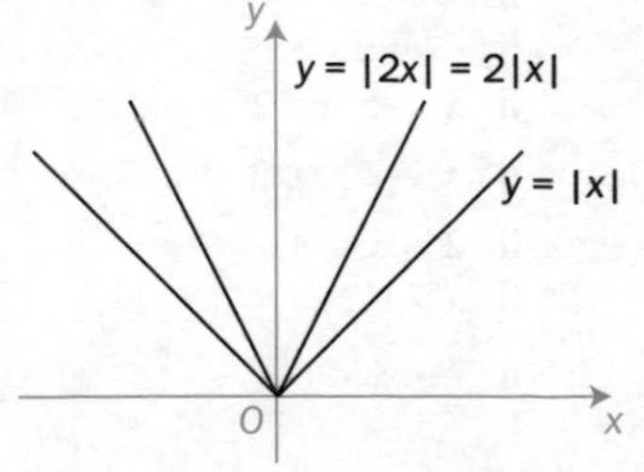

3 **a**

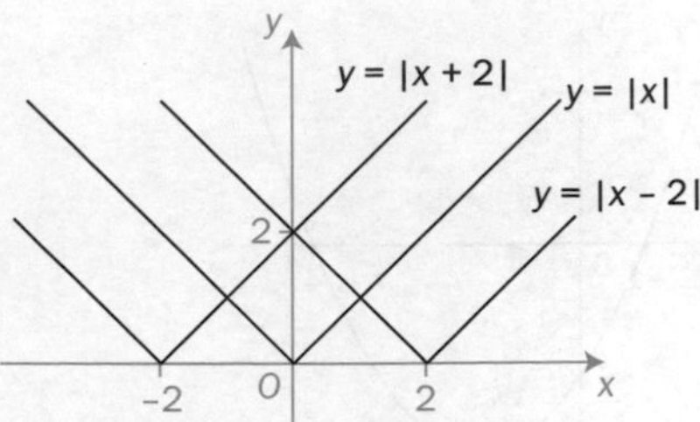

b

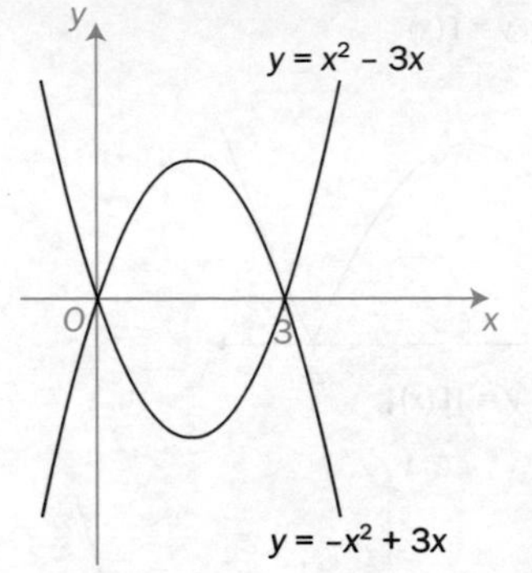

4 **a**

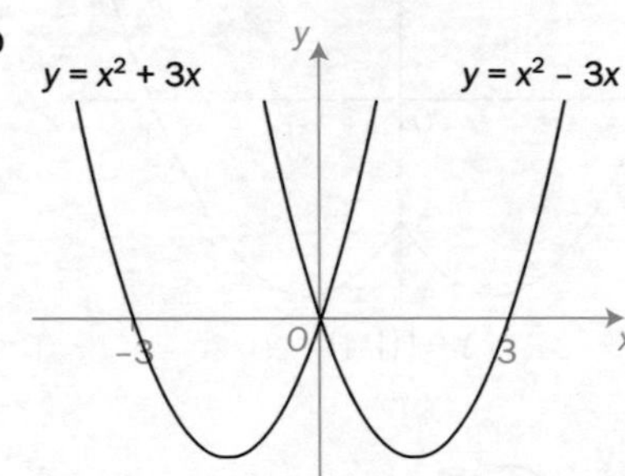

b

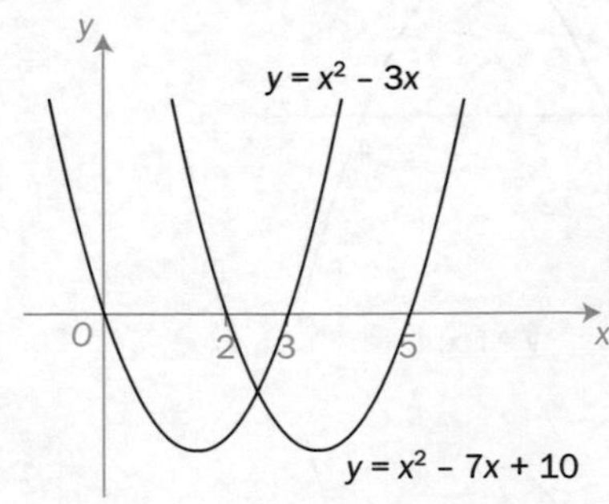

c

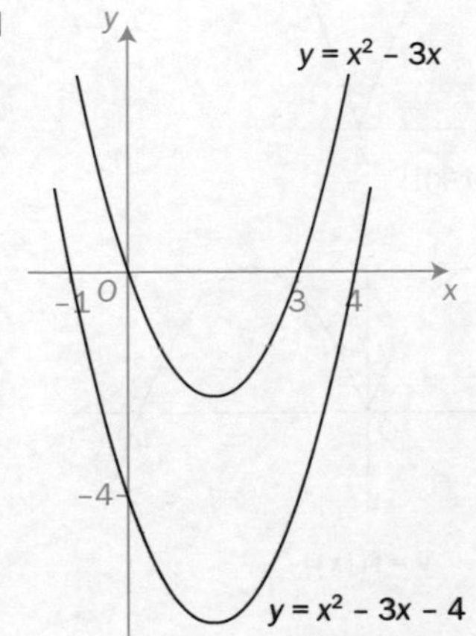

d

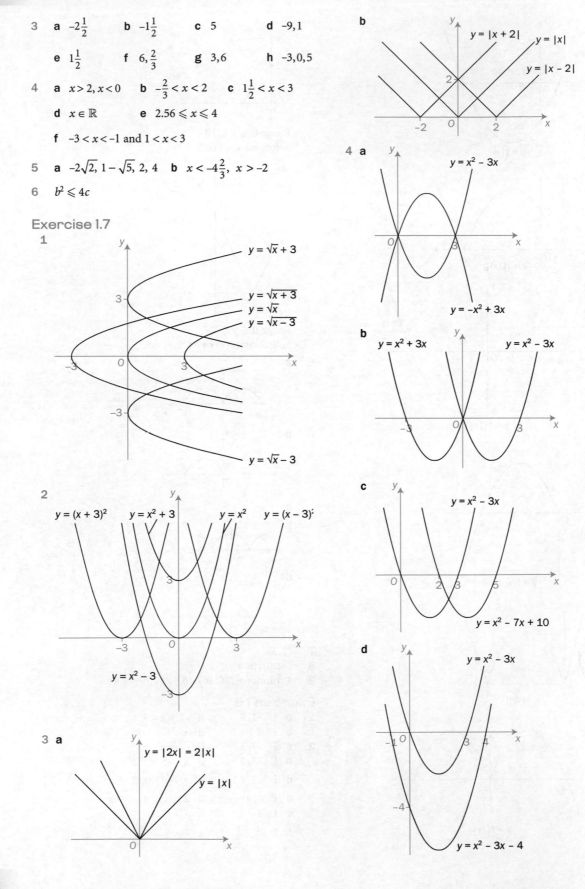

5

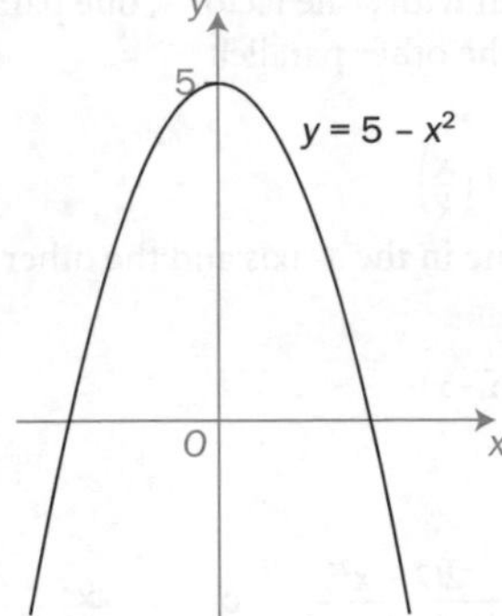

6 a

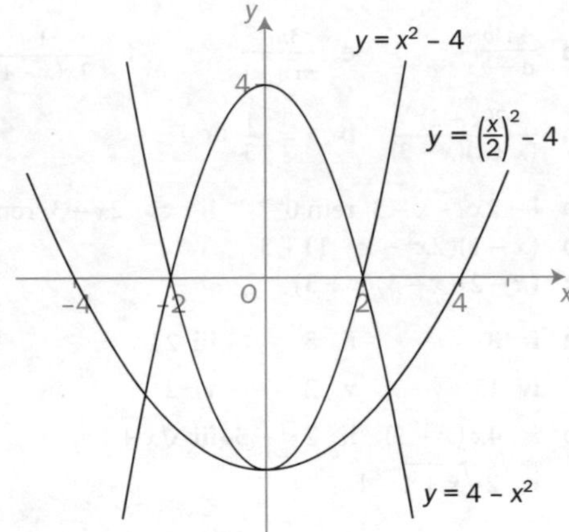

b

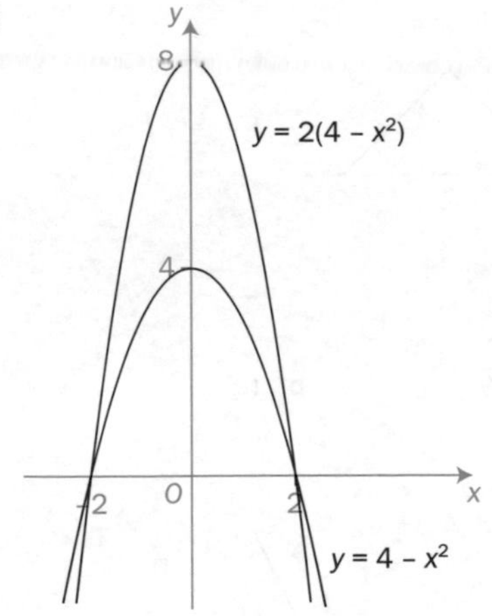

7 a

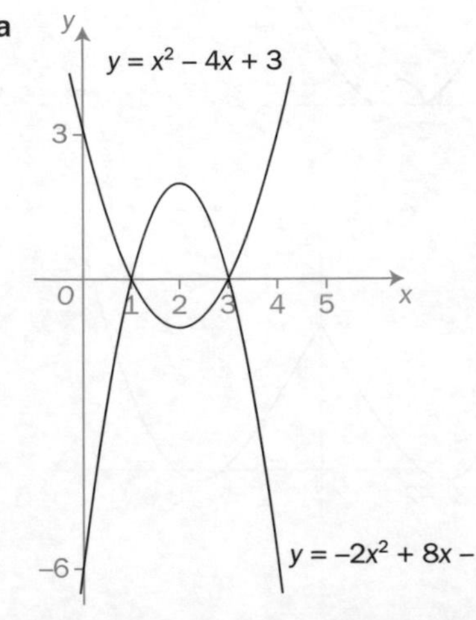

b

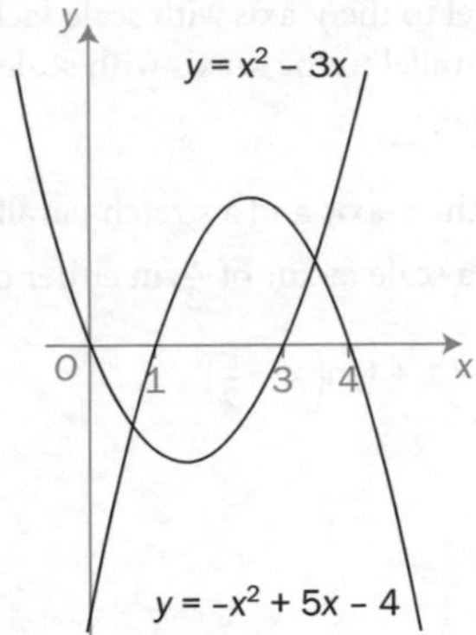

c

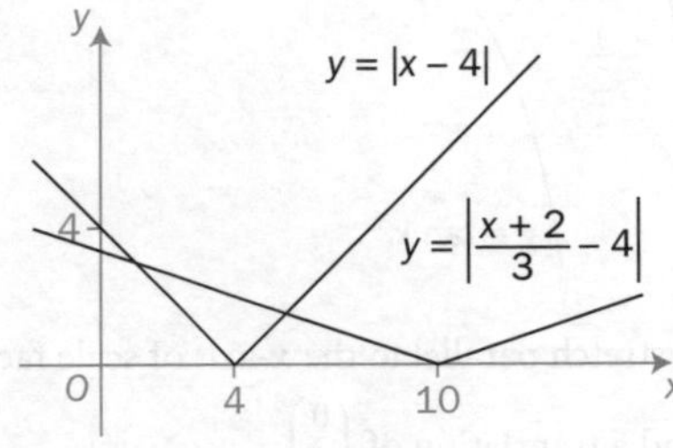

d

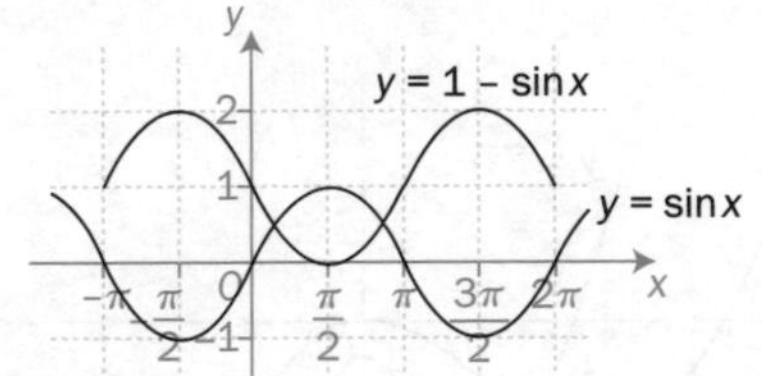

e

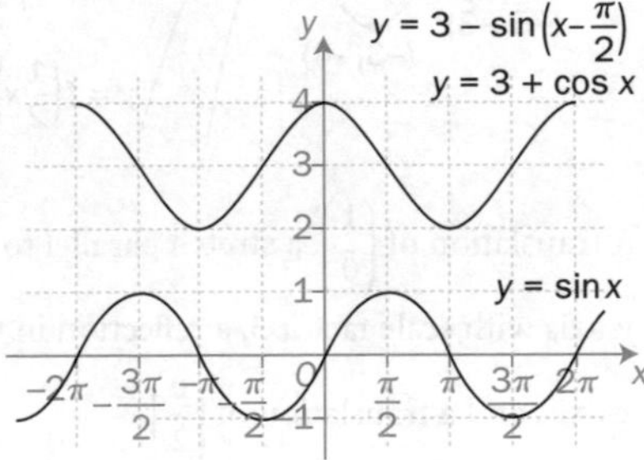

8 a First: Stretch parallel to y-axis with scale factor of 2

Second: Translation of $\begin{pmatrix} 0 \\ 1 \end{pmatrix}$

b Translation of $\begin{pmatrix} \frac{-\pi}{4} \\ 0 \end{pmatrix}$ and translation of $\begin{pmatrix} 0 \\ 3 \end{pmatrix}$ in either order

c First: A stretch parallel to x-axis with scale factor of $\frac{1}{2}$

Second: A reflection in the x-axis

Third: A translation of $\begin{pmatrix} 0 \\ 3 \end{pmatrix}$

d A stretch parallel to the y-axis with scale factor of 4 and a translation of $\begin{pmatrix} \frac{\pi}{2} \\ 0 \end{pmatrix}$, in either order.

e A stretch parallel to the y-axis with scale factor 2 and a stretch parallel to the x-axis with scale factor $\frac{1}{3}$, in either order.

f A reflection in the y-axis and a stretch parallel to the y-axis with a scale factor of $\frac{1}{2}$, in either order.

9

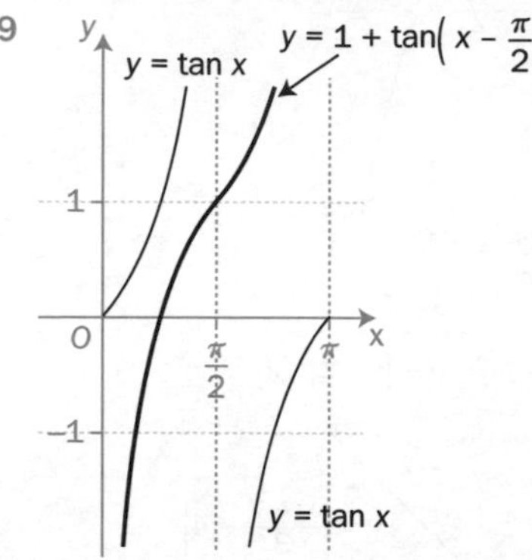

10 a i A stretch parallel to the x-axis of scale factor 2 and a translation of $\begin{pmatrix}0\\1\end{pmatrix}$.

ii

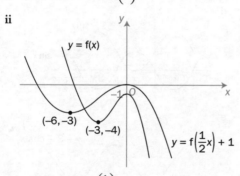

b i A translation of $\begin{pmatrix}1\\0\end{pmatrix}$, a stretch parallel to the y-axis with scale factor 3, a reflection in the x-axis, and a translation of $\begin{pmatrix}0\\2\end{pmatrix}$.

$y = -3x^3 + 18x^2 - 33x + 20$
image point $(0, 20)$

ii

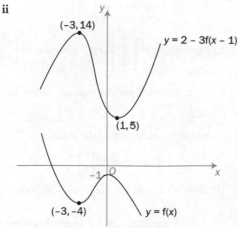

11 a Two stretches, both with scale factor k, one parallel to the x-axis and the other parallel to the y-axis

$y = \mathrm{f}(x) \longrightarrow y = k\mathrm{f}\left(\frac{x}{k}\right)$

b Two reflections, one in the x-axis and the other in the y-axis

$y = \mathrm{f}(x) \longrightarrow y = -\mathrm{f}(-x)$

Review 1

1 a $\frac{4x+1}{(x-1)(2x+3)}$ **b** $\frac{2(2-x^2)}{(x+1)(x+2)}$ **c** $\frac{6x}{(x-4)(x^2-1)}$

d $\frac{4a^3b}{a-b}$ **e** $\frac{3m^3}{m^2-1}$ **f** $\frac{1}{2x(x-1)}$

2 a $\frac{2(x+2)}{(x+1)(x-3)}$ **b** $x = -\frac{1}{3}$ or 1

3 a i $2x^2 + x - 3$ rem 0 **ii** $x^2 + 2x - 3$ rem 2
b $(x-1)(2x^2 - x - 1) + 3$
c $(x-2)(x-3)(x+3)$

4 a i 8 **ii** 5 **iii** 24
iv 15 **v** 2 **vi** 2
b i $4x(x-1)$ **ii** $2x^2 - 3$ **iii** $\sqrt{x+1}$
iv $2\sqrt{x+1} - 1$

5 a

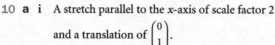

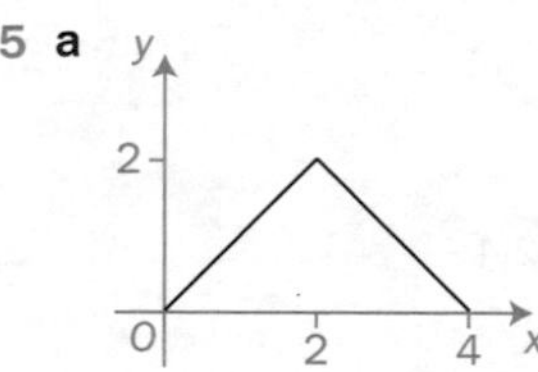

b $1\frac{1}{3}$, $2\frac{2}{5}$

6 a $\frac{1}{2}\sqrt{x-9}$ **b** 1.5

8 a

b

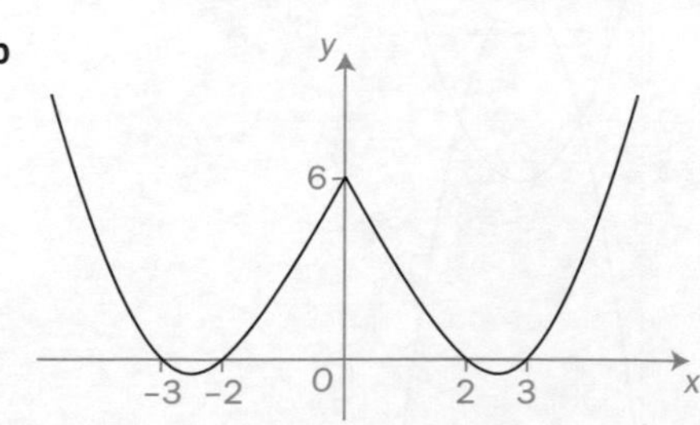

9 a

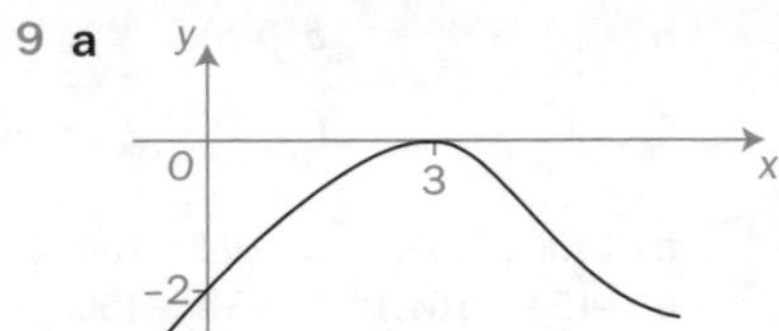

Turning point (3, 0)

b

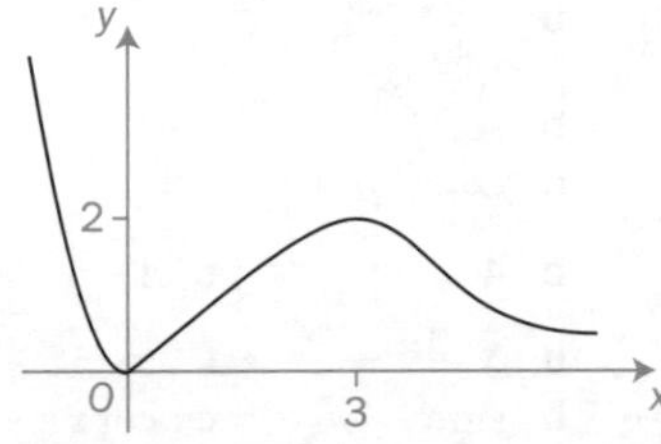

Turning point (3, 2)

c

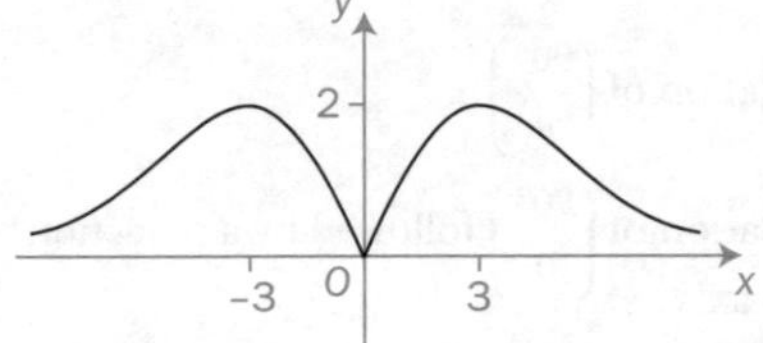

Turning points (±3, 2)

10

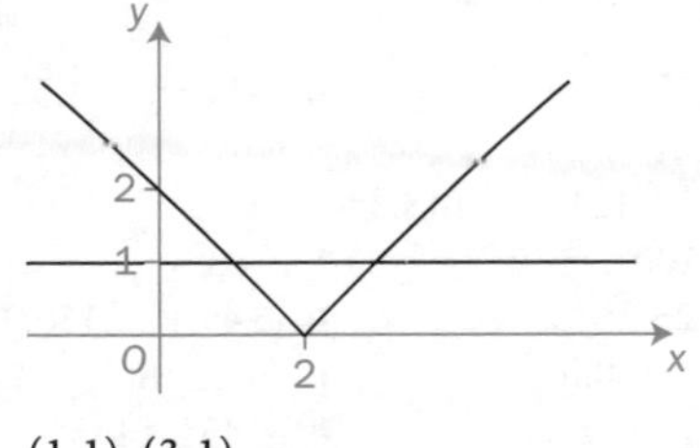

(1, 1), (3, 1)

11 a

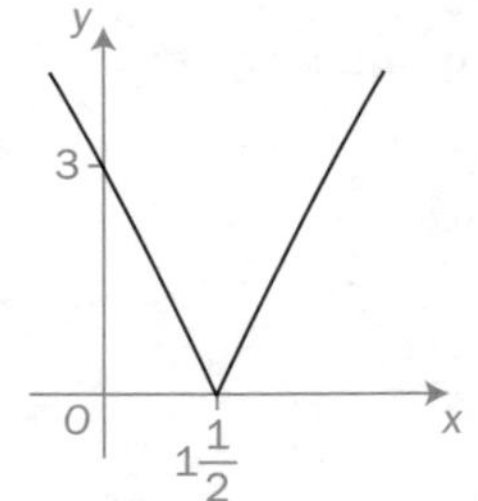

b

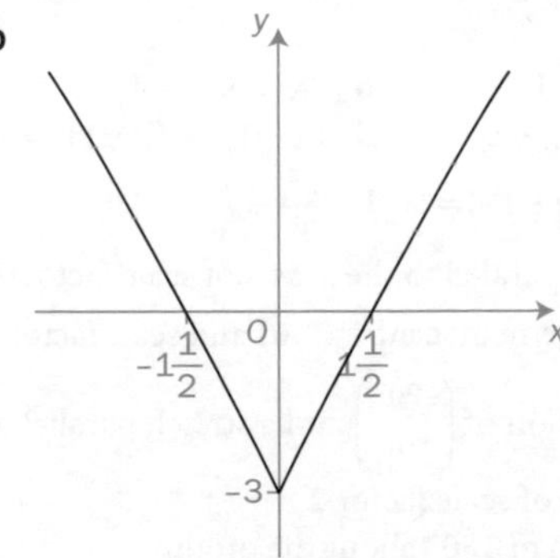

c

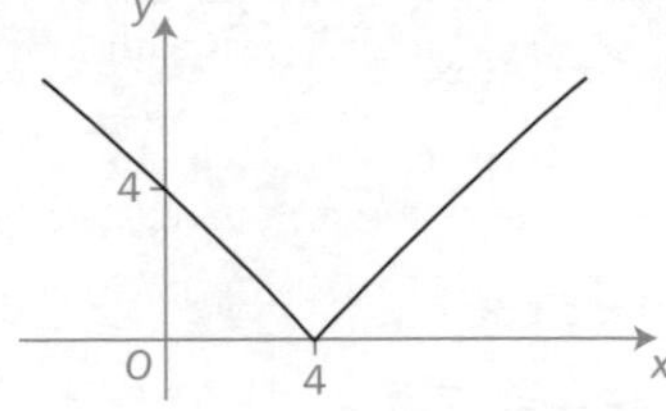

d

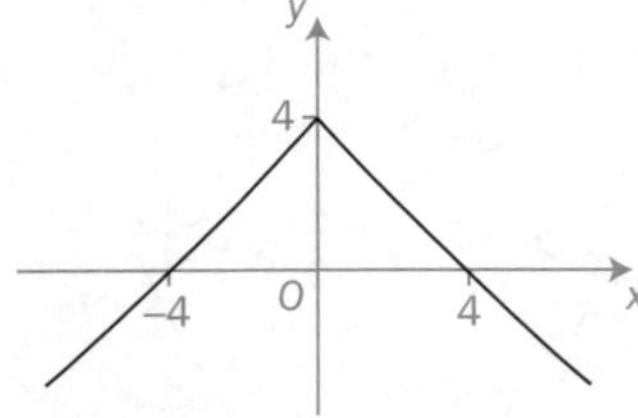

e

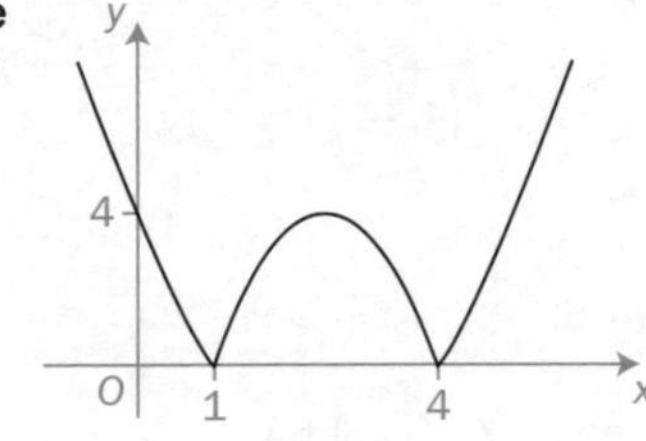

f

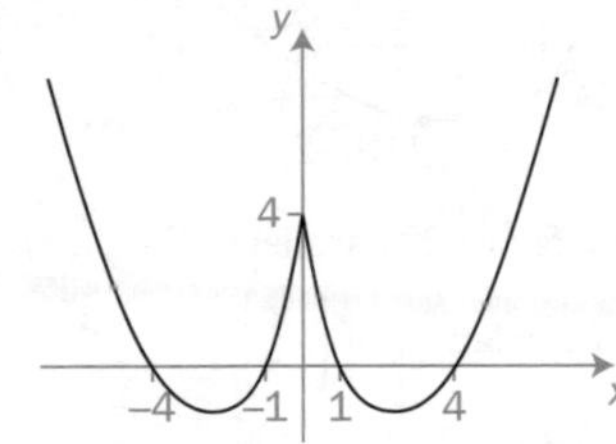

g

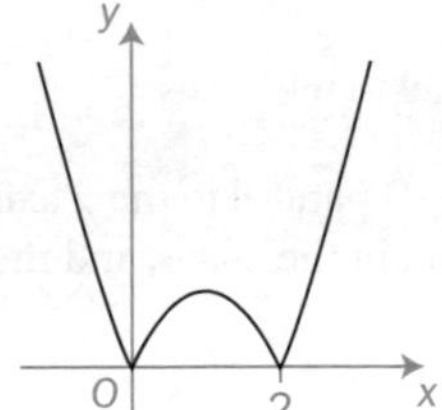

h

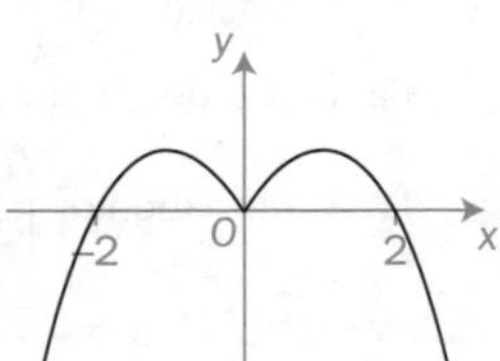

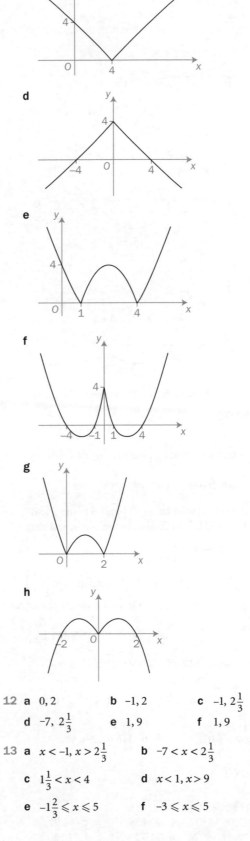

12 **a** 0, 2 **b** −1, 2 **c** $-1,\ 2\frac{1}{3}$

d $-7,\ 2\frac{1}{3}$ **e** 1, 9 **f** 1, 9

13 **a** $x < -1,\ x > 2\frac{1}{3}$ **b** $-7 < x < 2\frac{1}{3}$

c $1\frac{1}{3} < x < 4$ **d** $x < 1,\ x > 9$

e $-1\frac{2}{3} \leqslant x \leqslant 5$ **f** $-3 \leqslant x \leqslant 5$

14 **a** $-2, 1, 3$ **b** $x \leqslant 1, x \geqslant 3$
c $x \leqslant -3, -1 \leqslant x \leqslant 1, x \geqslant 3$

15 **a** $y \leqslant 3$ **b** **i** 6 **ii** 16
c

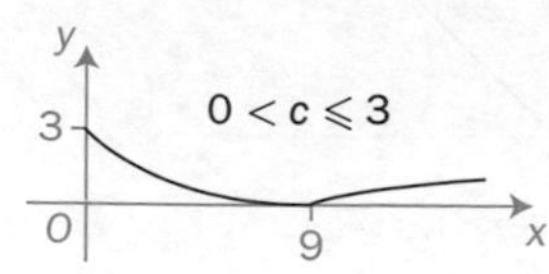

16 **a**

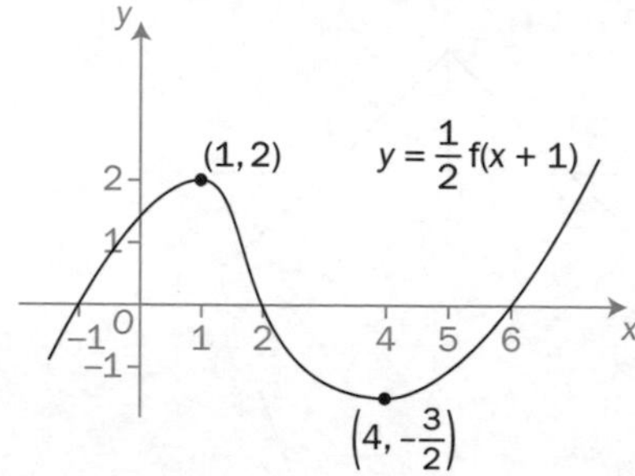

b

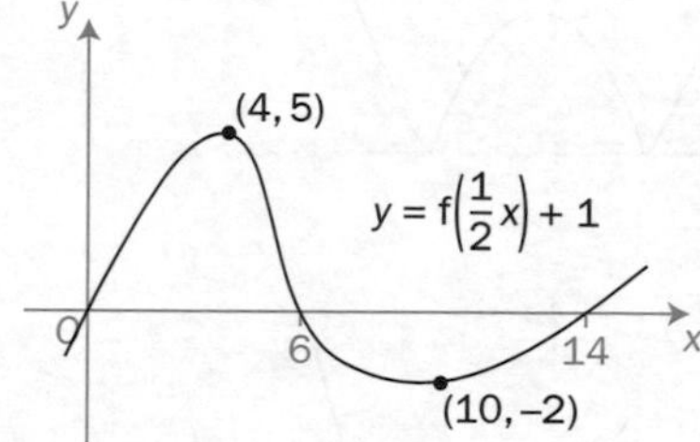

17 **a** A stretch (scale factor 2) parallel to the y-axis followed by a translation of $\begin{pmatrix} 0 \\ -5 \end{pmatrix}$

b A translation of $\begin{pmatrix} 2 \\ 0 \end{pmatrix}$ followed by a stretch $\left(\text{scale factor } \frac{1}{2}\right)$ parallel to the y-axis

c A stretch (scale factor 2) parallel to the y-axis followed by a reflection in the x-axis, and then a translation of $\begin{pmatrix} 0 \\ 1 \end{pmatrix}$

18 **a** $\dfrac{2x+3}{x+2}$
c T_1: A translation of $\begin{pmatrix} -2 \\ 0 \end{pmatrix}$
T_2: A reflection in the x-axis
T_3: A translation of $\begin{pmatrix} 0 \\ 2 \end{pmatrix}$

Before you start Answers

Chapter 2

1 **a** $\sqrt{3}$ **b** $\sqrt{3}+1$ **c** 1 **d** 0

2 **a** $\frac{15}{17}$ **b** $\frac{8}{15}$

3 **a** 17.5°, 162.5° **b** 116.6°, 296.6°
c 75.5°, 284.5°

4 **a** $\frac{-\sqrt{21}}{5}$

Exercise 2.1

1 **a** −1.06 **b** −0.839 **c** −2.92
d −3.24 **e** −1.73 **f** 1.56

2 **a** −1 **b** −2 **c** −2
d $\frac{1}{\sqrt{3}}$ **e** $\sqrt{2}$ **f** −1

3 **a** ±36.9° **b** 21.8°, −158.2° **c** 19.5°, 160.5°
d ±143.3° **e** −15.9°, 164.1° **f** −30°, −150°

4 **a** $-\frac{17}{15}$ **b** $-\frac{15}{8}$

5 **a** $-\frac{41}{40}$ **b** $-\frac{9}{40}$

6 **a** $\cot^2\beta$ **b** $\sec^3\beta$
c $\operatorname{cosec}\beta$ **d** $\operatorname{cosec}^5\beta$

7 **a** $\frac{2}{3}$ **b** 4 **c** 1
d ±3 **e** 3 **f** $\pm\sqrt{3}$

8 **a** 1 **b** $\sin x$ **c** $\cot x$
d $\tan x$ **e** 1 **f** 1
g 1 **h** 1 **i** 1

9 A translation of $\begin{pmatrix} 90° \\ 0 \end{pmatrix}$
A translation of $\begin{pmatrix} 90° \\ 0 \end{pmatrix}$ followed by a reflection in the y-axis
$\sec(x - 90°) = \operatorname{cosec} x$
$\cot(90° - x) = \tan x$

Exercise 2.2

1 **a** 80.5°, −60.5° **b** −34.5°, 174.5°
c −56.6°, 123.4°
d 31.7°, −58.3°, 121.7°, −148.3°
e 40°, −80°, 100°, −140° **f** −63.6°
g ±54.7°, ±125.3° **h** ±49.1°, ±130.9°
i ±90°, 19.5°, 160.5° **j** 0°, ±180°, ±41.4°
k ±60° **l** ±45°, ±135°

2 **a** $\pm\frac{\pi}{4}, \pm\frac{3\pi}{4}$ **b** $\pm\frac{\pi}{6}, \pm\frac{5\pi}{6}$
c $\pm\frac{\pi}{3}, \pm\frac{2\pi}{3}$ **d** $0, \pm\pi, \pm\frac{\pi}{3}, \pm\frac{2\pi}{3}$

3 **a** 45°, 225°, 71.6°, 251.6°
b 45°, 225°, 116.6°, 296.6°
c 90°, 221.8°, 318.2°
d 60°, 300°
e 0°, 360°, 138.6°, 221.4°
f 26.6°, 206.6°
g 0°, 360°
h 45°, 225°
i 30°, 150°

4 **a** $x^2 = 4y^2 + 16$ **b** $4x^2 = 9y^2 + 36$
c $x^2y^2 = 144 - 9x^2$ **d** $(x-1)^2 + (y-1)^2 = 1$
e $x^2(y^2 \quad 8y + 17) = 9$ **f** $\frac{x^2}{a^2} + \frac{b^2}{y^2} = 1$

5 **a** A stretch parallel to the x-axis of scale factor 2
b An enlargement, centre (0,0) and scale factor $\frac{1}{2}$
c A translation of $\begin{pmatrix} -90° \\ 0 \end{pmatrix}$ and a stretch parallel to the y-axis of scale factor 2
d A rotation of 180° about the origin

6 **a** $(\pm180°, 0), \left(\pm60°, \frac{1}{2}\right), \left(\pm300°, \frac{3}{2}\right)$

b $(30°, 2.2), (150°, -0.15), (-210°, -0.15), (-330°, 2.2)$
c $(\pm 52°, 1.6), (\pm 310°, 1.6)$
d $(0, 0), (\pm 180°, 0), (\pm 360°, 0), \left(45°, \frac{1}{2}\right), \left(225°, \frac{1}{2}\right), \left(-135°, \frac{1}{2}\right), \left(-315°, \frac{1}{2}\right)$

Exercise 2.3

1 a $\frac{\pi}{4}$ b $\frac{\pi}{3}$ c 0 d $-\frac{\pi}{4}$
e $\frac{\pi}{6}$ f 0 g $\frac{3\pi}{4}$ h $-\frac{\pi}{6}$

2 a $\frac{2}{3}\sqrt{2}$ b $\frac{\sqrt{7}}{3}$ c 0 d $\frac{5}{8}$
e $\frac{1}{\sqrt{2}}$ f $\frac{\sqrt{3}}{2}$ g $-\frac{\sqrt{5}}{2}$

3 a 0 b 90° or $\frac{\pi}{2}$

4 a i $\sqrt{1-x^2}$ ii $\frac{x}{\sqrt{1-x^2}}$ b $\frac{1}{x}$

5 a $\frac{1+x}{\sqrt{1+x^2}}$ b $\frac{\pi}{2}-\theta$

Exercise 2.4

1 a $\frac{1}{\sqrt{2}}$ b $\frac{\sqrt{3}}{2}$ c $\frac{\sqrt{3}}{2}$
d $-\frac{1}{2}$ e 1 f -1

2 a $\sin 3A$ b $\cos 5\alpha$ c $\tan 3x$ d $\cot 2x$

3 a $\sin(45° - x)$ or $\cos(45° + x)$
b $\sin(60° + x)$ or $\cos(30° - x)$
c $\tan(60° + x)$ d $\tan(45° + x)$
e 1 f $\cos 2x$

5 a $\frac{\sqrt{3}-1}{2\sqrt{2}}$ b $\frac{\sqrt{3}-1}{2\sqrt{2}}$ c $\frac{\sqrt{3}+1}{\sqrt{3}-1}$
d $\frac{\sqrt{3}-1}{\sqrt{3}+1}$ e $-\frac{\sqrt{3}+1}{\sqrt{3}-1}$ f $\frac{2\sqrt{2}}{\sqrt{3}-1}$

6 a $\frac{63}{65}$ b $-\frac{63}{16}$ c $-\frac{65}{16}$

7 a $-\frac{13}{85}$ b $-\frac{84}{85}$ c $-\frac{13}{84}$

8 a $12+5\sqrt{5}$ b $\frac{2+\sqrt{5}}{\sqrt{30}}$

9 a $\frac{1}{13}$ b $\frac{1}{2}$ c $\frac{5+\tan\beta}{1-5\tan\beta}$

11 a 17.1°, 197.1° b 113.8°, 293.8°
c 160.9°, 340.9° d 77.9°, 147.1°, 257.9°, 327.1°
e 106.1°, 286.1° f 38.2°, 141.8°
g 52.5°, 142.5°, 232.5°, 322.5°
h 114.3°, 335.7°

13 a i +1, 50° ii −1, 230°
b i +1, 340° ii −1, 160°

16 a $\frac{1}{3}$ b $\frac{33}{65}$ c $\frac{\pi}{2}$

Exercise 2.5

1 a sin 46° b cos 84° c tan 140°
d cos 100° e $\sin 6\theta$ f $\cos 8\theta$
g $\cos^2 20°$ h $2\cos^2\theta$ i $\frac{1}{2}\sin 2\theta$
j $\tan 6\theta$ k $\cos\frac{2\pi}{5}$ l $\cot 8\theta$
m $2\cos^2\frac{\theta}{2}$ n $2\operatorname{cosec} 2\theta$ o $2\cot 2\theta$

2 a $\frac{1}{2}$ b $\frac{1}{\sqrt{2}}$ c $\frac{1}{\sqrt{2}}$
d 2 e $\frac{2-\sqrt{3}}{4}$ f $\frac{1}{4}\sqrt{2}$

3 a $\frac{24}{25}, -\frac{7}{25}, -\frac{24}{7}$ b $\pm\frac{4\sqrt{2}}{9}, \frac{7}{9}, \pm\frac{4\sqrt{2}}{7}$
c $\pm\frac{120}{169}, \frac{119}{169}, \pm\frac{120}{119}$

4 a $\frac{1}{8}, \pm\frac{3\sqrt{7}}{8}$ b $-\frac{7}{9}, \pm\frac{4\sqrt{2}}{9}$

5 a $\frac{2}{5}, \frac{\sqrt{21}}{5}, \frac{2}{\sqrt{21}}$ b $\frac{2}{3}, \frac{\sqrt{5}}{3}, \frac{2}{\sqrt{5}}$ c $\frac{1}{\sqrt{10}}, \frac{3}{\sqrt{10}}, \frac{1}{3}$

6 a $\frac{2}{3}, \frac{\sqrt{5}}{3}, \frac{2}{\sqrt{5}}$ b $\frac{1}{\sqrt{10}}, \frac{3}{\sqrt{10}}, \frac{1}{3}$ c $\frac{1}{\sqrt{5}}, \frac{2}{\sqrt{5}}, \frac{1}{2}$

7 a $1-2x^2$ b $\frac{6x}{9-x^2}$ c $\frac{18-x^2}{x^2}$

8 a 22.5°, 67.5°, 202.5°, 247.5°
b 15°, 165°, 195°, 345°
c 45°, 135°
d 30°, 150°, 270°
e 210°, 330°
f 30°, 150°, 90°, 270°
g 60°, 90°, 270°, 300°
h 0°, 30°, 150°, 210°, 330°, 360°
i 60°, 109.5°, 250.5°, 300°
j 90°, 270°, 45°, 255°
k 0°, 180°, 60°, 300°
l 0°, 165.6°, 360°
m 0°, 78.5°, 281.5°, 360°
n 0°, 120°
o 90°, 180°, 270°
p No solutions in range
q 19.1°, 70.9°, 199.1°, 250.9°
r 112.8°, 247.2°
s 35.3°, 144.7°, 215.3°, 324.7°
t 23.6°, 156.4°, 90°, 270°
u 90°, 323.1°

11 $\tan 3A = \frac{3\tan A - \tan^3 A}{1 - 3\tan^2 A}$

Exercise 2.6

1 a 5, 36.9° b 13, 67.4° c $\sqrt{5}$, 63.4°
d $2\sqrt{5}$, 26.6° e 5, −53.1° f 17, −61.9°

2 a 4.9°, 129.9° b 17.6°, 229.8°
c 82.1°, 334.1° d 102.3°, 195.7°

5 a $\frac{\pi}{3}$ b $\frac{7\pi}{12}, -\frac{\pi}{12}$
c $\frac{\pi}{2}$, −0.927 d $\frac{\pi}{2}$, 0.643

6 a 63.4°, −116.6°, 0°, 180°, 360°
b −11.3°, −131.3°, 108.7°, −36.5°, 83.5°, −156.5°
c −38.8° d 180°

7 a $\sqrt{5}$, 50.8° b $\sqrt{5}$, 50.8° c $\frac{1}{\sqrt{5}}$, 50.8°

8 a 10, 36.9° b −10, 71.6° c 10, 161.6°

9 **a** 13, 67.4° **b** Max 13, 157.4°; Min −13, 337.4°
c 33, 157.4° **d** 10, 53.1°
e −2, 233.1° **f** 1, 157.4°

10 **a** $\sqrt{13}\sin(\theta - 33.7°)$

b

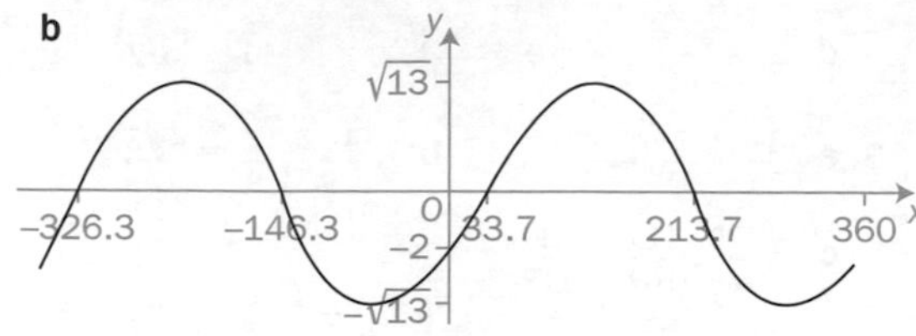

c A stretch parallel to the y-axis of scale factor $\sqrt{13}$ and a translation of $\begin{pmatrix} 33.7° \\ 0 \end{pmatrix}$

11 $2\sqrt{5}$, $t = 1.29$

12 $r_1 = r_2 = r_3 = r_4 = 5$
$\alpha_1 = 36.9°$ $\alpha_2 = -36.9$ $\alpha_3 = 233.1°$ $\alpha_4 = 126.9°$

Review 2

1 **a** i $\frac{1}{\sqrt{2}}$ ii $\sqrt{5}$ iii $\sqrt{\frac{5}{2}}$

b i $\frac{13}{5}$ ii $-\frac{12}{5}$ iii $-\frac{13}{12}$

2 **a** 14.5°, 165.5° **b** 174° **c** 24.2°, 114.2°
d 3.2°, 50.2°, 123.2°, 170.2°
e 22.5°, 67.5°, 112.5°, 157.5° **f** 120°

4 **a** 45°, 63.4°, −116.6°, −135°
b 30°, 150°, −19.5°, −160.5°
c ±90°, 19.5°, 160.5°
d ±75.5°, ±120°
e ±90°, 56.3°, −123.7°
f 45°, −135°, 63.4°, −116.6°

5 **a** i A translation of $\begin{pmatrix} -\frac{\pi}{2} \\ 0 \end{pmatrix}$ and a stretch (scale factor 2) parallel to the y-axis

ii A stretch $\left(\text{scale factor } \frac{1}{2}\right)$ parallel to the x-axis, followed by a reflection in the x-axis, and a translation of $\begin{pmatrix} 0 \\ 3 \end{pmatrix}$

iii A translation of $\begin{pmatrix} \frac{\pi}{4} \\ 0 \end{pmatrix}$, followed by a stretch (scale factor 2) parallel to the y-axis, and a translation of $\begin{pmatrix} 0 \\ 1 \end{pmatrix}$

b

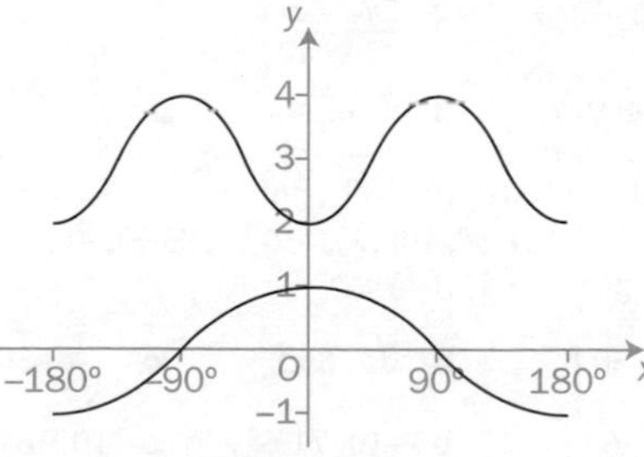

(−90°, 0), (90°, 0)
(0, 1), (0, 2)

6 **a** i $\frac{\pi}{6}$ ii $\frac{\pi}{3}$ iii π

b i $\frac{1}{2}$ ii $\frac{1}{\sqrt{2}}$ iii $\sqrt{3}$

7 **b** 0°, 228.2°, 131.8°

8

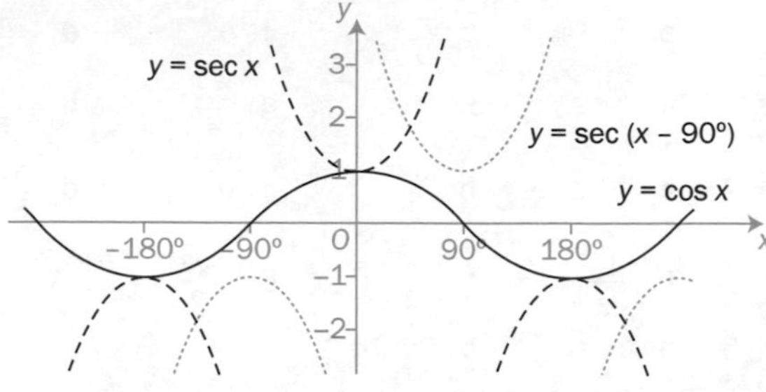

A translation of $\begin{pmatrix} 90° \\ 0 \end{pmatrix}$

9 **a** $\frac{84}{85}$ **b** $-\frac{13}{85}$ **c** $\frac{36}{77}$

10 **a** $\frac{1}{4}\sqrt{2}(\sqrt{3} - 1)$ **b** $\frac{17 - 7\sqrt{3}}{26\sqrt{2}}$

11 **a** 19.1°, 199.1° **b** 90°, 270°, 120°, 240°
c 30°, 150°, 270° **d** 60°, 300°
e 60°, 300° **f** 0°, 180°, 30°, 150°, 210°, 330°, 360°
g 0°, 360°, 124.8°, 235.2° **h** 90°, 199.5°, 340.5°
i 0, 90°, 360°

13 **a** 0° **b** 0°, ±180°, 76.0°, −104°
c ±18.4°, ±161.6°

14 **a** $y = 1 - 2x^2$ **b** $2x^2(y + 1) = 1$ **c** $x(1 - y^2) = 2y$

15 **a** $3\frac{4}{7}$

16 **a** $\frac{\pi}{3}$ **b** $\frac{2\pi}{3}$ **c** $\frac{2}{\sqrt{13}}$ **d** $\frac{2}{\sqrt{5}}$

17 $\frac{3 + 8\sqrt{2}}{15}$

18 60°

19 **b** 24° **c** 99.7°, 170.3°, 279.7°, 350.3°

20 **b** 0°, 180°, 360°, 26.6°, 206.6°

22 **a** 13, 22.6° **b** 17, 61.9°
c $\sqrt{5}$, 26.6° **d** $\sqrt{2}$, 45°

23 **a** 5, 53.1° **b** 5 **c** 103.3°, 330.5°

24 **a** i $\pm\sqrt{10}$, 71.6°, 251.6° ii $\pm\sqrt{5}$, 296.6°, 116.6°
b i 20.8°, 122.4° ii 0°, 233.1°

25 **a** $4\sqrt{10}$, 18.4° **b** 38.0°, 285.2°
c i $-4\sqrt{10}$ ii 161.57°

Revision 1

1 $\frac{x + 4}{(1 + x)(2 + x)^2}$

2 **a** $\frac{x(x + 2)}{x + 1}$ **b** $\frac{x^2 + 4x + 2}{(x + 1)(x + 2)}$

3 $Q(x) = x^2 + 3x - 2$, $R(x) = -3$

4 $Q(x) = x^2 - x + 4$, $R(x) = x + 1$

5 a $Q(x) = 2x + 5$, $R(x) = 2x - 4$ **b** $\lambda = -11$, $\mu = 10$

6 a $5\sin(\theta + 36.9°)$ **b** 113°, 353°

7 a $R = \sqrt{17}$, $\alpha = 76°$

b 119° (2.08 rad), 33° (0.576 rad)

8 a $R = 13$ **b** $\theta = 157°$

9 a $f(x) \leqslant 9$ **b** Does not exist **c** −40

10 a $(x - 2)^2 - 3$

b i $f(x) \geqslant -3$ **ii** $f^{-1}(x) = 2 + \sqrt{x + 3}, x \geqslant -3$

c

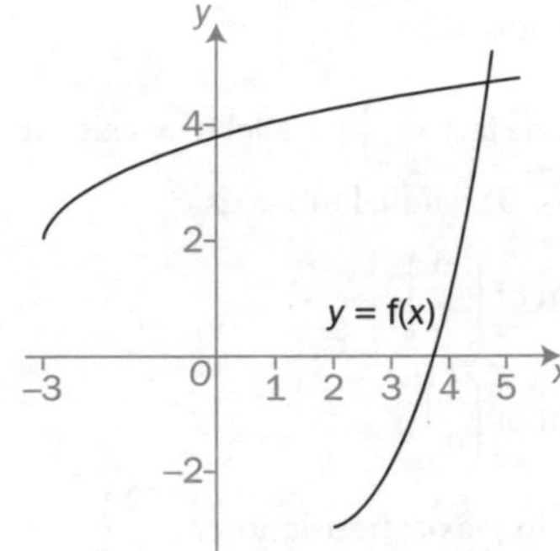

11 a Stretch parallel to the y-axis by scale factor 2, reflection in the x-axis then translation by 1 unit parallel to the y-axis

b $y = x^2 + 4x + 8$

12 a $-2.25 \leqslant f(x) \leqslant 4$ **b** $k = 1$

13 a $f^{-1}(x) = \frac{x}{5}, x \in \mathbb{R}$

b $gf^{-1}(x) = \frac{3}{25}x^2 - 2$, $gf^{-1}(x) \geqslant -2$

14 a i

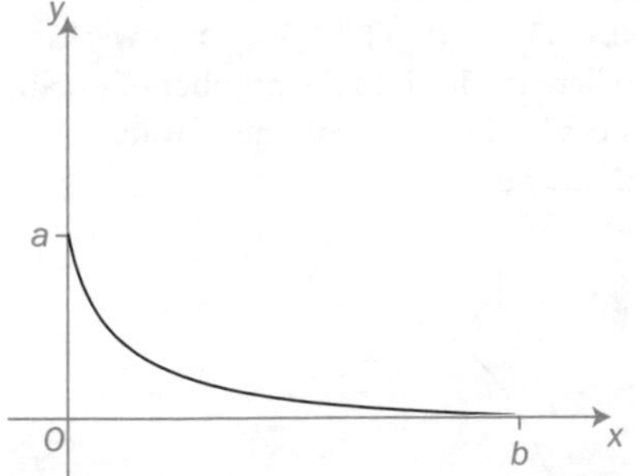

ii

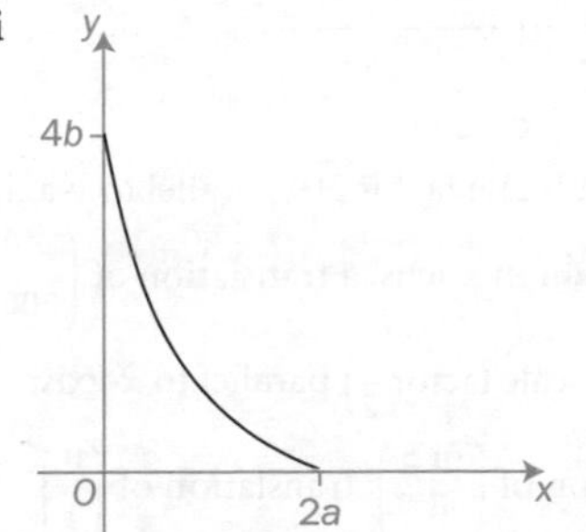

b $a = 3, b = 9, 0 \leqslant f(x) \leqslant 9$

c $gf(x) = 4 - x$

15 a

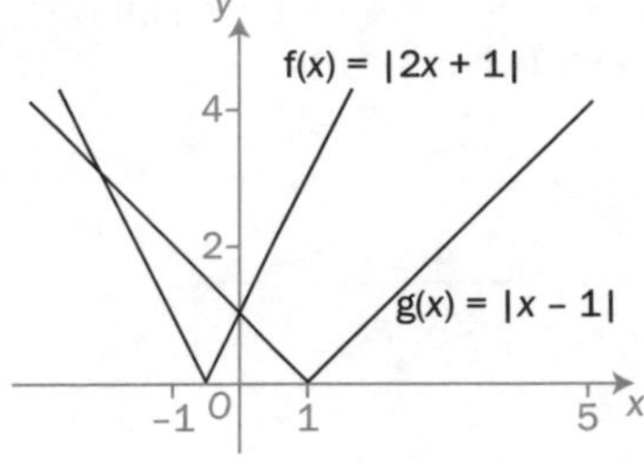

b $x = 0, -2$

16 a i $3, -\frac{7}{3}$ **ii** 3 **iii** 3

b i $x < -\frac{7}{3}, x > 3$ **ii** $x \in \mathbb{R}$ **iii** $x > 0$

17 a

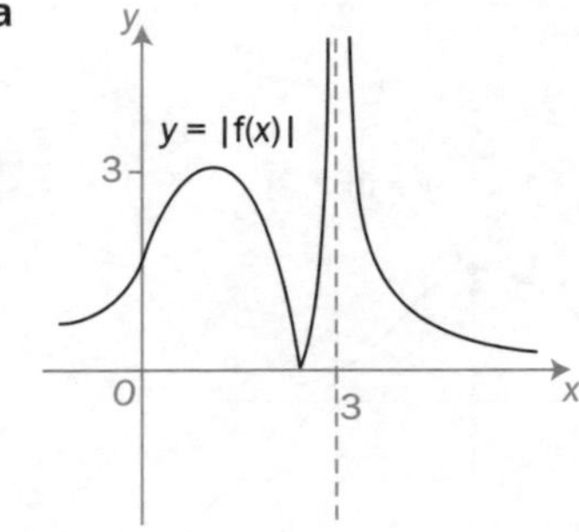

b

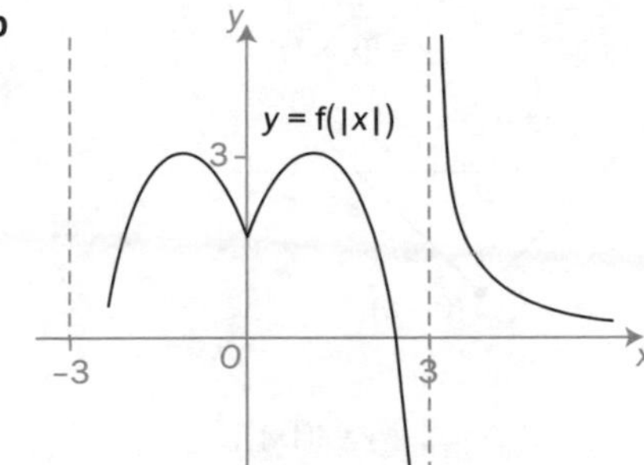

c

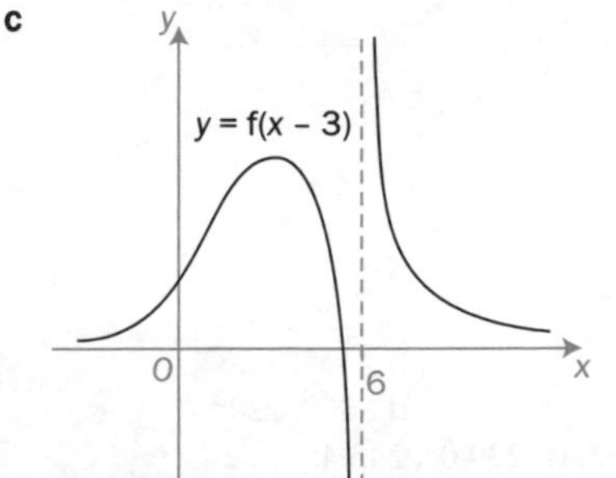

18 a

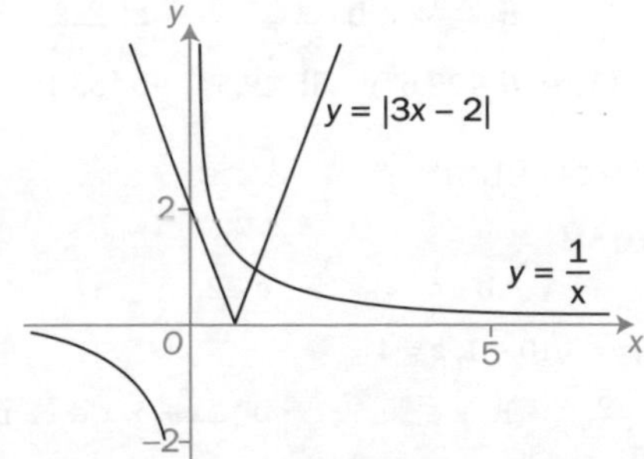

b There is only one intersection.

c $x = 1$

19 a

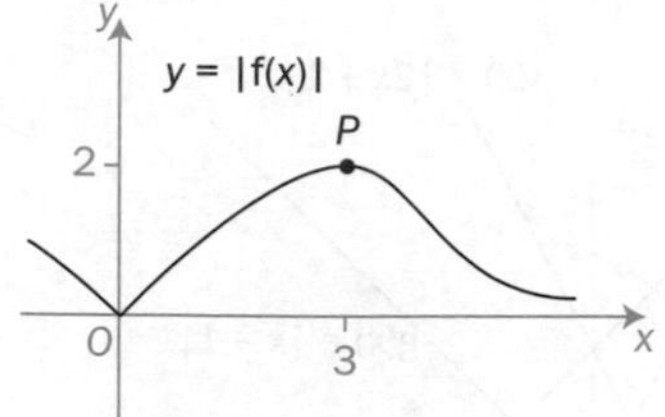

b

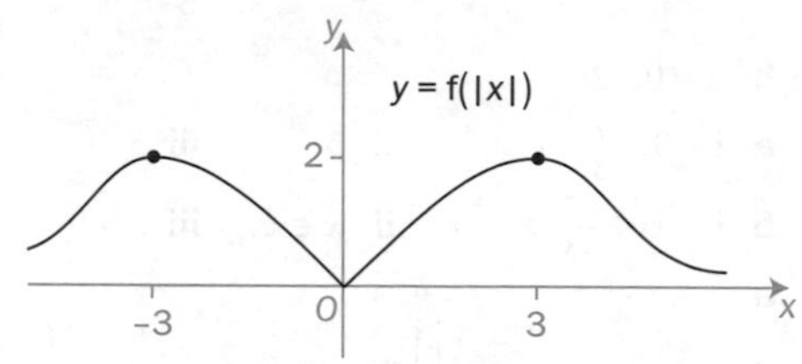

c

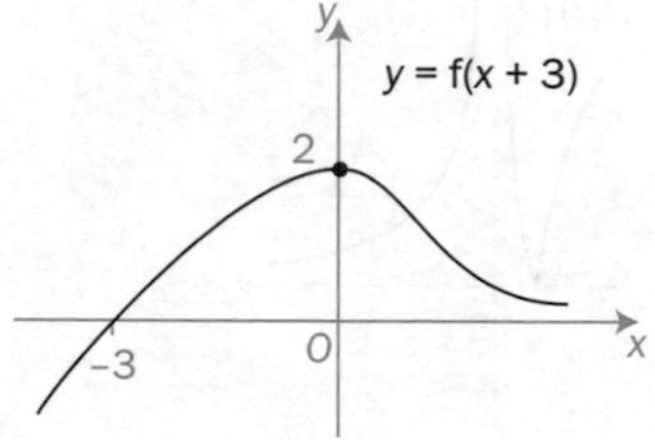

20 a

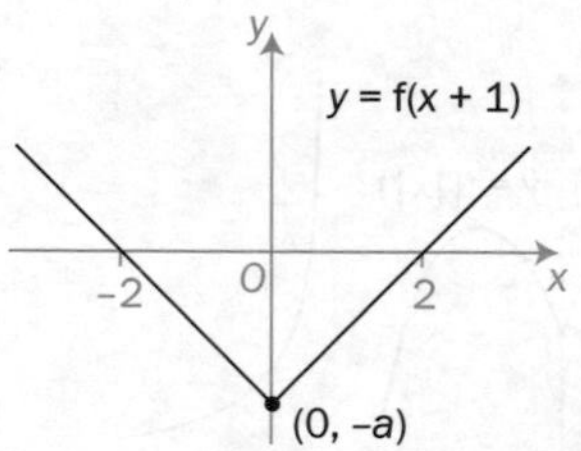

b

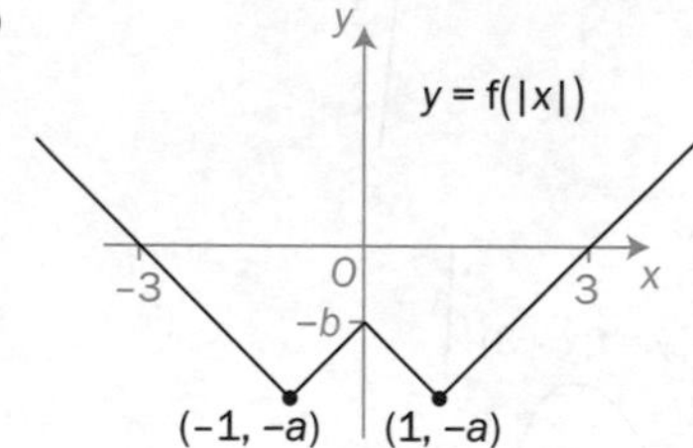

c i $a = 2, b = 1$ ii $-\frac{1}{6}$

22 a 0°, 60° **b** 45°, 71.6°
c 70.9° **d** 90°, 120°

23 b 59.0°, 63.4°, 239.0°, 243.4°

24 b 22.5°, 67.5°

25 a i $\frac{\pi}{4}$ ii $\frac{\sqrt{3}}{2}$ **b** $-\frac{1}{\sqrt{2}}$ **c** $-\frac{36}{85}$

26 a i $13\cos(\theta + 22.6°)$ ii 49.5° **b** 53.1°

Before you start Answers

Chapter 3

1 a $\frac{1}{9}$ **b** $\frac{1}{3}$ **c** 2

2 $x = 9, y = \sqrt{10} - 1, z = 4$

3 a $\frac{x-5}{3}, x \in \mathbb{R}, y \in \mathbb{R}$ **b** $2x + 3, x \in \mathbb{R}, y \in \mathbb{R}$
c $\frac{x^2+1}{2}, x \geqslant 0, y \geqslant \frac{1}{2}$

4 a $3x^2 + 6$ **b** $\frac{x^2}{4} - 3$

Exercise 3.1

1 a Reflection in y-axis; translation of $\begin{pmatrix}0\\1\end{pmatrix}$
b Reflection in y-axis; reflection in x-axis; translation of $\begin{pmatrix}0\\1\end{pmatrix}$
c Stretch (scale factor 3) parallel to y-axis; translation of $\begin{pmatrix}0\\2\end{pmatrix}$
d Stretch (scale factor 3) parallel to y-axis; reflection in x-axis; translation of $\begin{pmatrix}0\\2\end{pmatrix}$
e Stretch $\left(\text{scale factor } \frac{1}{2}\right)$ parallel to x-axis; stretch (scale factor 3) parallel to y-axis
f Translation of $\begin{pmatrix}-1\\0\end{pmatrix}$
g Translation of $\begin{pmatrix}2\\0\end{pmatrix}$
h Reflection in y-axis; translation of $\begin{pmatrix}2\\0\end{pmatrix}$

2 $A = 100$

t	0	5	10	15	20
P	100	128	165	212	272

13.9 weeks

3 $M_0 = 6$

t	0	5	10	15	20
M	6	3.64	2.21	1.34	0.81

Half-life = 6.93 sec

4 a 0.1733 **b** 113 **c** 7 weeks
d There is a limit to the number of organisms possible in a unit volume of water.

5 b 332.2 hours

6 a

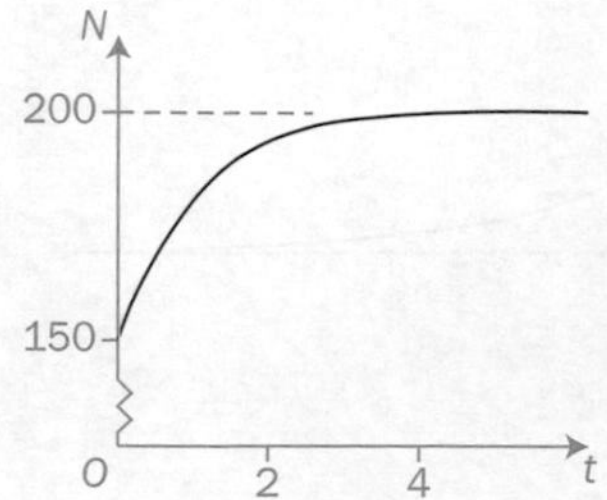

b 150 **c** 200
d A stretch (scale factor 50) parallel to y-axis; a reflection in x-axis; a translation of $\begin{pmatrix}0\\200\end{pmatrix}$

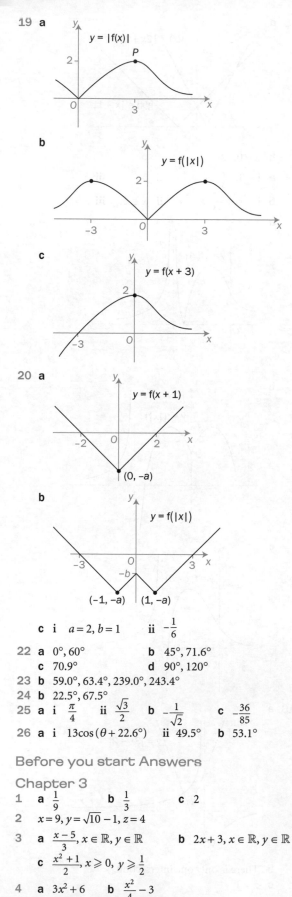

7 a Stretch $\left(\text{scale factor } \frac{1}{2}\right)$ parallel to x-axis; translation of $\begin{pmatrix}-1.5\\0\end{pmatrix}$; translation of $\begin{pmatrix}0\\4\end{pmatrix}$
b Stretch $\left(\text{scale factor } \frac{1}{2}\right)$ parallel to x-axis; translation of $\begin{pmatrix}0.5\\0\end{pmatrix}$; translation of $\begin{pmatrix}0\\-4\end{pmatrix}$

c Reflection in y-axis; translation of $\begin{pmatrix}2\\0\end{pmatrix}$; translation of $\begin{pmatrix}0\\-3\end{pmatrix}$

Exercise 3.2

1 **a** i 4.48 ii 0.223 iii 0.223
iv 0.405 v −0.405 vi 2.47

b i 5 ii x iii 5
iv x v $5+\ln 2$ vi 25

2 **a** Translation of $\begin{pmatrix}0\\2\end{pmatrix}$; $x>0, x\in\mathbb{R}$

b Reflection in x-axis; translation of $\begin{pmatrix}0\\3\end{pmatrix}$; $x>0, x\in\mathbb{R}$

c Translation of $\begin{pmatrix}-2\\0\end{pmatrix}$; $x>-2, y\in\mathbb{R}$

d Translation of $\begin{pmatrix}2\\0\end{pmatrix}$; $x>2, y\in\mathbb{R}$

e Stretch (scale factor 3) parallel to y-axis; translation of $\begin{pmatrix}0\\1\end{pmatrix}$; $x>0, y\in\mathbb{R}$

f Stretch $\left(\text{scale factor } \frac{1}{2}\right)$ parallel to x-axis; translation of $\begin{pmatrix}0\\1\end{pmatrix}$; $x>0, y\in\mathbb{R}$

g Reflection in x-axis; translation of $\begin{pmatrix}0\\1\end{pmatrix}$; $x>0, y\in\mathbb{R}$

h Reflection in y-axis; translation of $\begin{pmatrix}1\\0\end{pmatrix}$; stretch (scale factor 2) parallel to y-axis; $x<1, y\in\mathbb{R}$

3 **a** Translation of $\begin{pmatrix}-3\\0\end{pmatrix}$; stretch (scale factor 3) parallel to y-axis; translation of $\begin{pmatrix}0\\1\end{pmatrix}$

b $(-2.4, 0), (0, 3.2)$

4 **a**

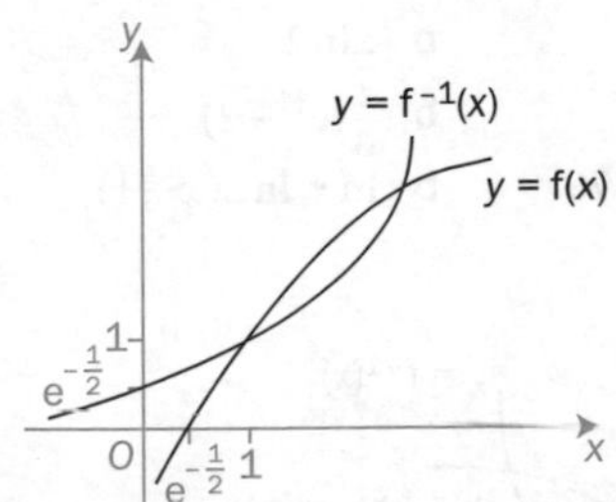

$f^{-1}(x)=e^{\frac{x-1}{2}}$
Domain $x\in\mathbb{R}$
Range $y>0$

b

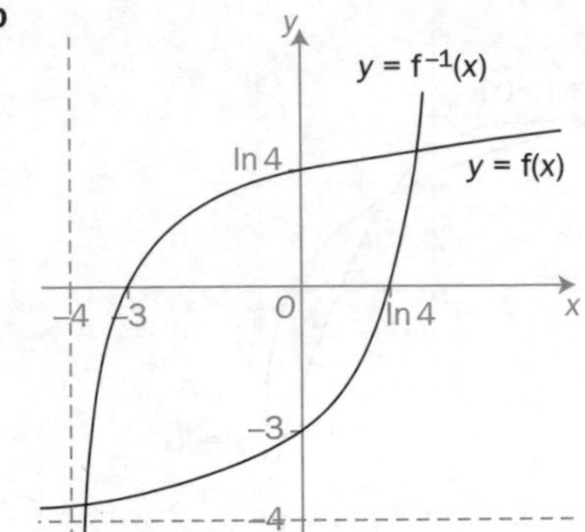

$f^{-1}(x)=e^{x}-4$
Domain $x\in\mathbb{R}$
Range $y>-4$

c

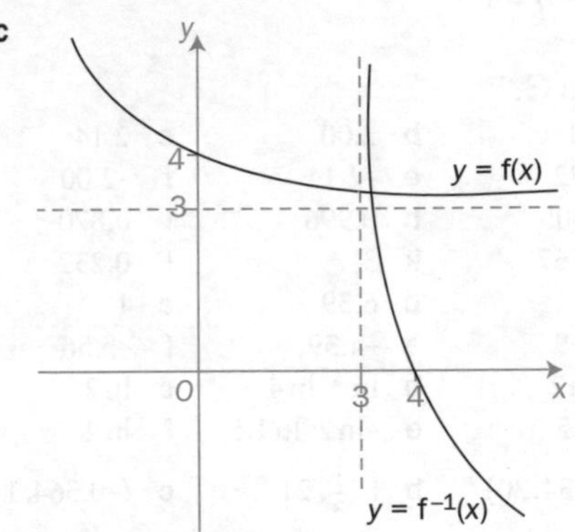

$f^{-1}(x)=-\ln(x-3)$
Domain $x>3$
Range $y\in\mathbb{R}$

d

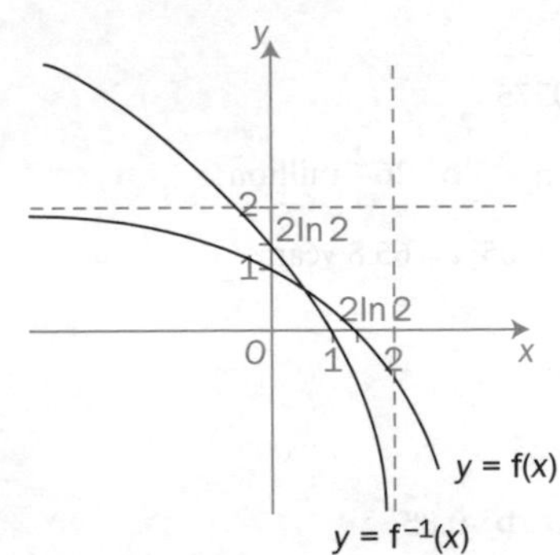

$f^{-1}(x)=2\ln(2-x)$
Domain $x<2$
Range $y\in\mathbb{R}$

5 **a** $f^{-1}(x)=3-\frac{1}{2}e^{x}$
Domain $x\in\mathbb{R}$
Range $y<3$

$g^{-1}(x)=2\ln x$
Domain $x>0$
Range $y\in\mathbb{R}$

C3

b

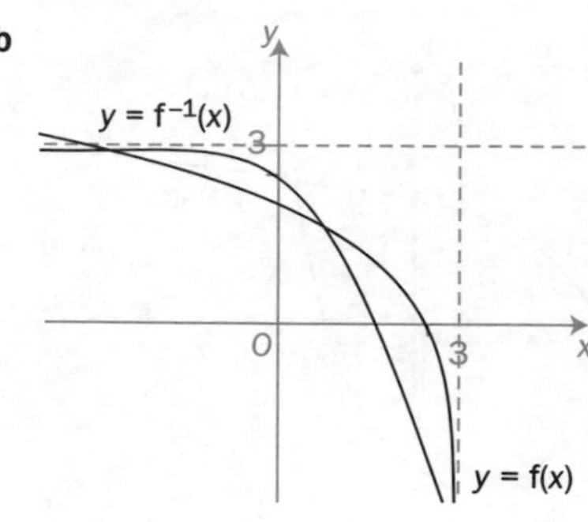

f(x) passes through $\left(2\frac{1}{2}, 0\right)$ and $(0, \ln 6)$

$f^{-1}(x)$ passes through $(\ln 6, 0)$ and $\left(0, 2\frac{1}{2}\right)$

c $gf(x) = \sqrt{6 - 2x}$

$gf(-5) = 4$

Exercise 3.3

1	**a** 2.20	**b** 2.00	**c** 2.14
	d 0.882	**e** −2.14	**f** −2.00
	g −0.405	**h** 0.996	**i** 0.870
	j 0.0767	**k** 3	**l** 0.232
2	**a** 403	**b** 6.39	**c** 1
	d 0.859	**e** −4.39	**f** −5.56
3	**a** $0, \ln 2$	**b** $\ln 3, \ln 4$	**c** $\ln 2$
	d $\ln 2.5$	**e** $-\ln 2, \ln 1.5$	**f** $\ln 4$
4	**a** $(0.434, 20)$	**b** $\left(-\frac{1}{3}, 2\right)$	**c** $(-0.564, 1)$

5 $\left(-\ln 2, -\frac{1}{2}\right)$ and $(0, 1)$

6 **a** 2.5 **b** 2.68

7 **a** $y = 5$ **b** $(0, 8), \left(\ln\frac{5}{3}, \frac{40}{3}\right)$

8 $(\ln 2, 3)$

9 $\frac{1}{5}\ln\frac{4}{3} = 0.0575$

a 15 million **b** $26\frac{2}{3}$ million

10 $\frac{1}{10}\ln\frac{10}{9} = 0.0105, t = 65.8$ years

11 193 years

12 36.4°C

Review 3

1 **a** 1.2 **b** 0.85

c 11 **d** 0.95

2

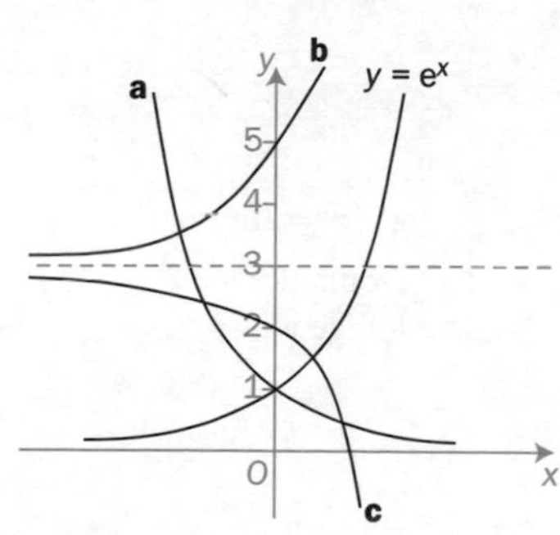

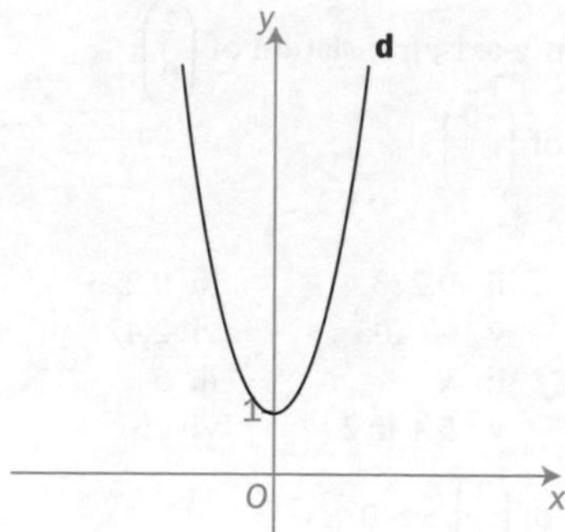

3

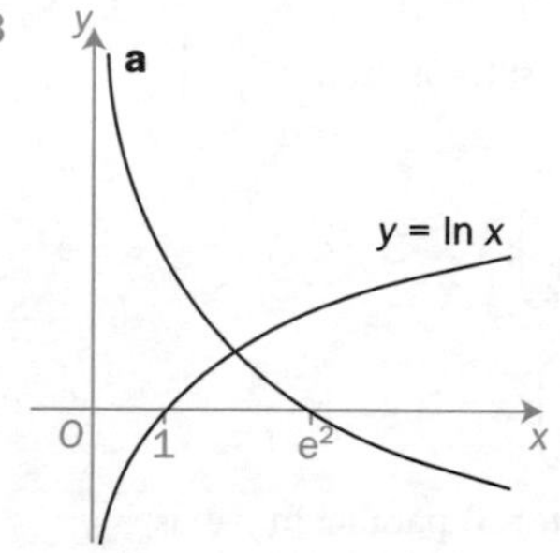

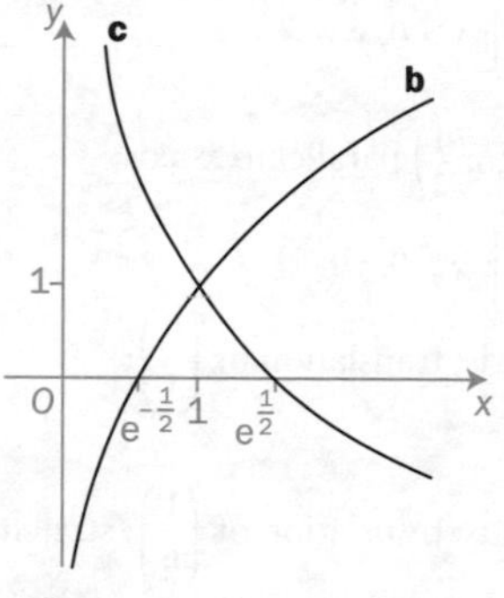

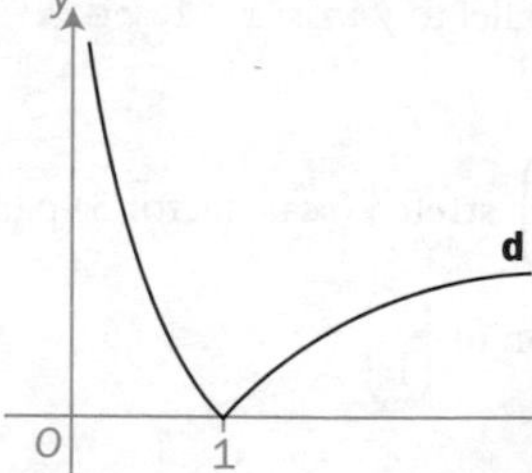

4 **a** $\ln 2$ **b** $\pm\ln 2$

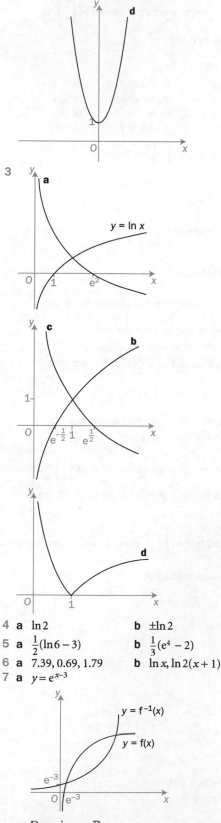

5 **a** $\frac{1}{2}(\ln 6 - 3)$ **b** $\frac{1}{3}(e^4 - 2)$

6 **a** 7.39, 0.69, 1.79 **b** $\ln x, \ln 2(x + 1)$

7 **a** $y = e^{x-3}$

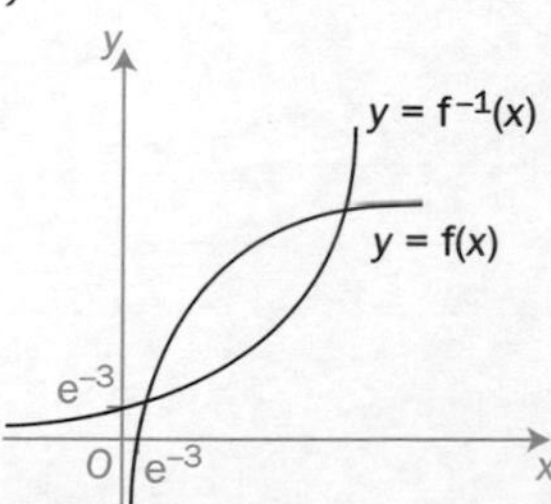

Domain $x \in \mathbb{R}$

Range $y > 0$

b $y = e^x + 2$

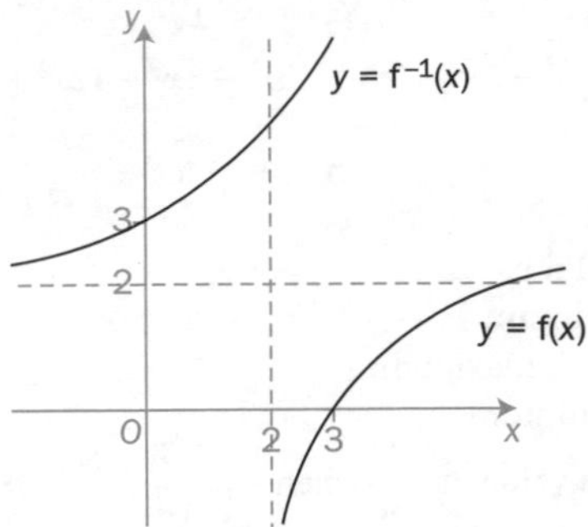

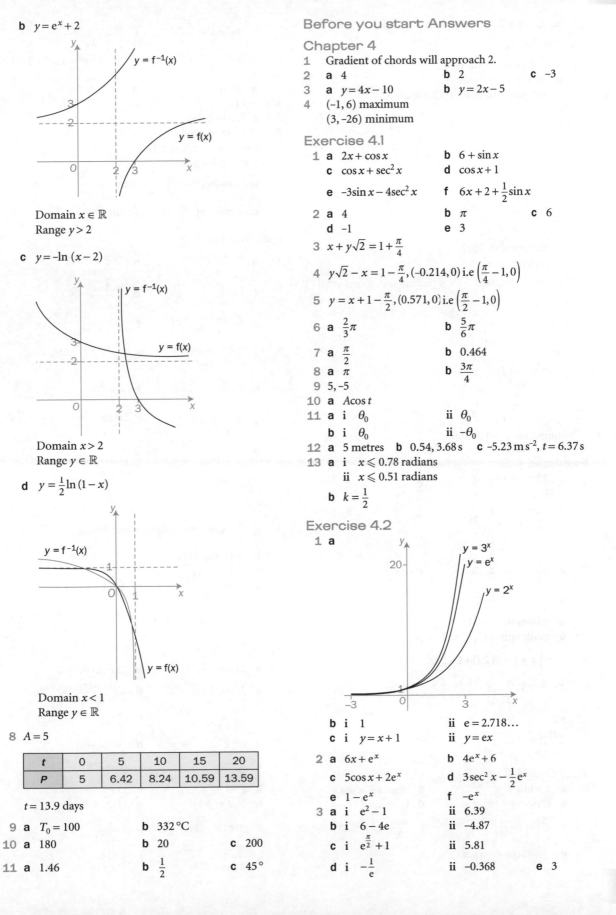

Domain $x \in \mathbb{R}$
Range $y > 2$

c $y = -\ln(x-2)$

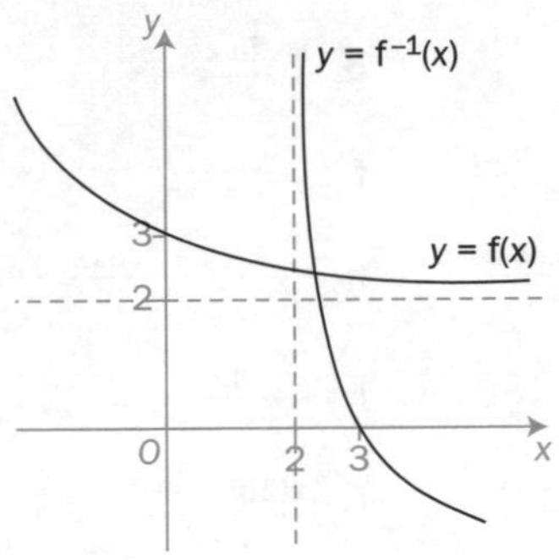

Domain $x > 2$
Range $y \in \mathbb{R}$

d $y = \frac{1}{2}\ln(1-x)$

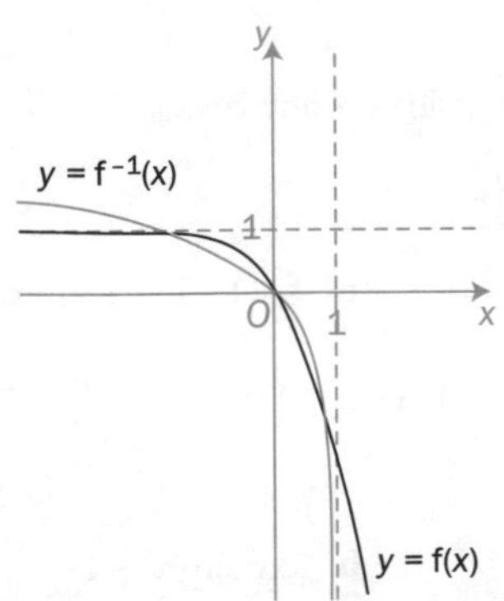

Domain $x < 1$
Range $y \in \mathbb{R}$

8 $A = 5$

t	0	5	10	15	20
P	5	6.42	8.24	10.59	13.59

$t = 13.9$ days

9 **a** $T_0 = 100$ **b** 332 °C

10 **a** 180 **b** 20 **c** 200

11 **a** 1.46 **b** $\frac{1}{2}$ **c** 45°

Before you start Answers

Chapter 4

1 Gradient of chords will approach 2.

2 **a** 4 **b** 2 **c** −3

3 **a** $y = 4x - 10$ **b** $y = 2x - 5$

4 (−1, 6) maximum
(3, −26) minimum

Exercise 4.1

1 **a** $2x + \cos x$ **b** $6 + \sin x$
c $\cos x + \sec^2 x$ **d** $\cos x + 1$
e $-3\sin x - 4\sec^2 x$ **f** $6x + 2 + \frac{1}{2}\sin x$

2 **a** 4 **b** π **c** 6
d −1 **e** 3

3 $x + y\sqrt{2} = 1 + \frac{\pi}{4}$

4 $y\sqrt{2} - x = 1 - \frac{\pi}{4}$, (−0.214, 0) i.e $\left(\frac{\pi}{4} - 1, 0\right)$

5 $y = x + 1 - \frac{\pi}{2}$, (0.571, 0) i.e $\left(\frac{\pi}{2} - 1, 0\right)$

6 **a** $\frac{2}{3}\pi$ **b** $\frac{5}{6}\pi$

7 **a** $\frac{\pi}{2}$ **b** 0.464

8 **a** π **b** $\frac{3\pi}{4}$

9 5, −5

10 **a** $A\cos t$

11 **a** i θ_0 ii θ_0
b i θ_0 ii $-\theta_0$

12 **a** 5 metres **b** 0.54, 3.68 s **c** $-5.23\ \text{m s}^{-2}$, $t = 6.37$ s

13 **a** i $x \leqslant 0.78$ radians
ii $x \leqslant 0.51$ radians
b $k = \frac{1}{2}$

Exercise 4.2

1 **a**

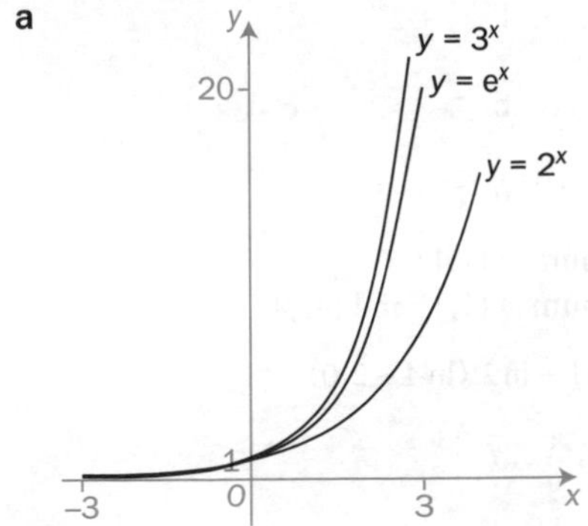

b i 1 ii $e = 2.718\ldots$
c i $y = x + 1$ ii $y = ex$

2 **a** $6x + e^x$ **b** $4e^x + 6$
c $5\cos x + 2e^x$ **d** $3\sec^2 x - \frac{1}{2}e^x$
e $1 - e^x$ **f** $-e^x$

3 **a** i $e^2 - 1$ ii 6.39
b i $6 - 4e$ ii −4.87
c i $e^{\frac{\pi}{2}} + 1$ ii 5.81
d i $-\frac{1}{e}$ ii −0.368 **e** 3

C3

4 a $y = 5x + 2$ b $y = 2x + 8$

5 $y = -2x + \frac{1}{2}$

8 a minimum at point $(0, 2)$
b maximum at point $(0, 0)$
c maximum at point $(\ln 2, 4\ln 2 - 3)$

9

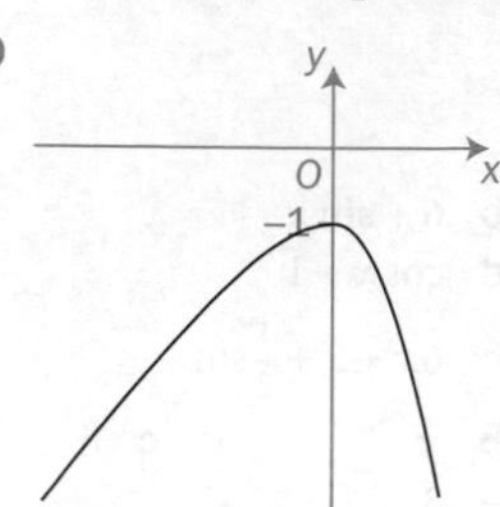

maximum at $(0, -1)$
asymptote $y = x$

10 a $(0,0), (\ln 2, 0)$ b minimum when $x = \ln\left(\frac{3}{2}\right)$

11

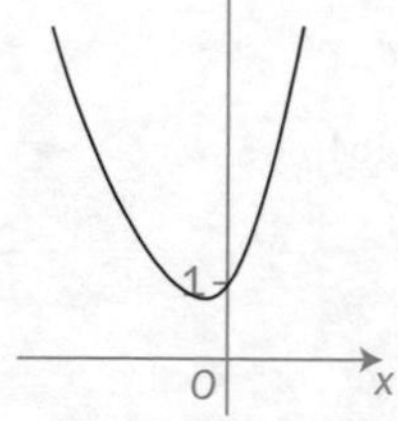

minimum at $(-0.35, 0.83)$

Exercise 4.3

1 a $\frac{2}{x}$ b $\frac{2}{x}$ c $\frac{3}{x}$ d $\frac{1}{x}$
e $-\frac{1}{x}$ f $\frac{3}{2x}$ g $-\frac{1}{2x}$ h 2
i $\frac{1}{x} + 1$ j $\frac{2}{x} + 2$

2 a $2\frac{1}{2}$ b 1 c $e + \frac{1}{2}$
d $\frac{2}{3}$ e 2

3 a minimum at $(1, 1)$
b minimum at $(1, 1)$ and $(-1, 1)$

4 $y = \frac{1}{2}x + 1 - \ln 2$; $(\ln 4 - 2, 0)$

5 $x + 6y = 1$, $\frac{1}{6}\sqrt{37}$

6 $\frac{1}{4(2e+1)}$

7 $\frac{d(\log_{10} x)}{dx} = \frac{1}{x \ln 10}$

Exercise 4.4

1 a $1 + \ln x$ b $2x\sin x + x^2\cos x$
c $e^x(\sec^2 x + \tan x)$ d $(x^2 - x - 2)e^x$
e $e^x\left(\ln x + \frac{1}{x}\right)$ f $\frac{\sin x}{x} + \cos x \ln x$
g $-\frac{1}{x^3}(2\cos x + x\sin x)$ h $\frac{e^x(x-1)}{x^2}$
i $3x^2\ln x + \frac{x^3 - 1}{x}$ j $\frac{\tan x + 2x\sec^2 x}{2\sqrt{x}}$
k $5x^4 - 4x^3 - 3x^2 + 4x$ l $5x^4 - 8x^3 + 12x^2 + 2x - 2$

2 π

3 a $y = 0$ b $y = -27e^{-3}$

4 $x = 1$

5 b 0.6 to 1 d.p.

6 $(0, 0)$ minimum
$(-0.2, 0.007)$ maximum
$(-1, 0)$ minimum

7 minimum value of $-\frac{1}{3e}$ when $x = \sqrt[3]{\frac{1}{e}}$

8 maximum at $\left(\frac{\pi}{4}, \frac{1}{2}\right)$; minimum at $\left(\frac{3\pi}{4}, -\frac{1}{2}\right)$

Exercise 4.5

1 a $\frac{\sin x - x\cos x}{\sin^2 x}$ b $\frac{2x\tan x - x^2\sec^2 x}{\tan^2 x}$
c $1 - \frac{1}{x^2}$ d $\frac{1 - \ln x}{x^2}$
e $\frac{e^x(\sin x - \cos x)}{\sin^2 x}$ f $-\frac{x}{e^x}$
g $\frac{\sin x - 2x\cos x}{2\sqrt{x}\sin^2 x}$ h $-\frac{x\sin x + 2\cos x}{x^3}$
i $\frac{-2}{(1+x)^2}$ j $-\frac{4x}{(x^2-1)^2}$
k $\frac{6x(x^3 - x + 1)}{(3x^2 - 1)^2}$ l $\frac{x(2\ln x - 1)}{(\ln x)^2}$

2 a 7 b 1

3 $(3, 0)$

4 $x + 2y = 0$

5 $\frac{2}{x^3}$

7 a $\frac{1}{x^2}(x\cos^2 x - x\sin^2 x - \sin x\cos x)$
b $\tan x + x\sec^2 x$
c $\frac{e^x[(1+x)\ln x - 1]}{(\ln x)^2}$ d $\frac{e^x}{x^2}(1 - \ln x + x\ln x)$
e $\frac{x}{e^x}(1 + 2\ln x - x\ln x)$ f $\frac{(x\cos x + \sin x)\ln x - \sin x}{(\ln x)^2}$

Exercise 4.6

1 a $2x\cos(x^2 + 1)$ b $-3x^2\sin(x^3 - 1)$
c $2x\sec^2(x^2)$ d $\frac{2(x+1)}{x^2 + 2x + 3}$
e $\frac{3x^2}{x^3 + 1}$ f $\frac{3}{x}$
g $\cot x$ h $\frac{1}{x}\cos(\ln x)$
i $\frac{1}{x\ln x}$ j $\cos x e^{\sin x}$
k e^{x+4} l $2xe^{x^2}$
m $10x(x^2 + 1)^4$ n $14(2x + 6)^6$
o $-6x^2(x^3 - 1)^{-3}$ p $\frac{x}{\sqrt{x^2 - 1}}$
q $\frac{3x^2}{2\sqrt{x^3 + 1}}$ r $-\frac{1}{(x-1)^2}$

s $-\frac{2(2x+3)}{(x^2+3x-1)^2}$ **t** $-3(2x+1)^{-\frac{3}{2}}$

u $-x(x^2+1)^{-\frac{3}{2}}$ **v** $-\frac{1}{x(\ln x)^2}$

w $-\frac{2e^x}{(e^x+1)^2}$ **x** $\frac{2}{3}(2x-1)^{-\frac{2}{3}}$

2 a $2\theta\cos(\theta^2)$ **b** $2\sin\theta\cos\theta$

c $\frac{\cos\theta}{2\sqrt{\sin\theta}}$ **d** $3\theta^2\sec^2(\theta^3)$

e $3\tan^2\theta\sec^2\theta$ **f** $-\frac{\sec^2\theta}{\sqrt{\tan^3\theta}}$

3 a $-\frac{\cos x}{\sin^2 x}$ **b** $e^x \times e^{e^x}$

c $-\frac{8(x+2)}{(x^2+4x-1)^2}$ **d** $-\frac{4e^{\frac{1}{x}}}{x^2}$

e $\frac{1}{2(x-1)}$ **f** $-\text{cosec}\,x$

4 a $k\cos kx$ **b** $-k\sin kx$ **c** $k\sec^2 kx$

d $\frac{1}{x}$ **e** ke^{kx}

5 $y=24x-164$

6 $\left(\frac{3}{4}, \frac{1}{16}\right)$

7 $\frac{1}{2}(\sqrt{2}-1)$

8 $\sqrt{y}=3x+16$

9 a $\frac{b^2x}{\sqrt{a^2+b^2x^2}}$ **b** $\frac{-ax}{\sqrt{(ax^2+b)^3}}$

c $\frac{4}{3}a^2x(a^2x^2-b^2)^{-\frac{1}{3}}$

10 $x=0$

11 b $\alpha=4$

13 minima at $(n\pi, 0)$, maxima at $\left(\left(n+\frac{1}{2}\right)\pi, 1\right)$, $n=0, \pm1, \pm2, \ldots$

14 minimum at $(0,0)$

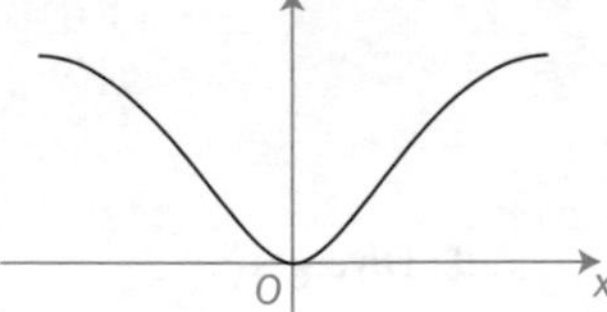

Exercise 4.7

1 a $(x^2-1)^2\,[6x\sin x+(x^2-1)\cos x]$

b $(x^3+1)\,[6x^2\tan x+(x^3+1)\sec^2 x]$

c $e^x(3x+2)^3(3x+14)$

d $\frac{e^x(2x+3)}{2\sqrt{x+1}}$ **e** $\frac{2}{x^3}(x\cos 2x-\sin 2x)$

f $\frac{x^2(1+\ln x)-1}{x\sqrt{x^2-1}}$ **g** $\sin(x^2)+2x^2\cos(x^2)$

h $\sin^2 x+2x\sin x\cos x$ **i** $\cos 3x-3x\sin 3x$

j $2x(\tan 2x+x\sec^2 2x)$ **k** $(2x+1)e^{2x+1}$

l $2x\ln(x^2-1)+\frac{2x^3}{x^2-1}$

m $\cos x\cos 2x-2\sin x\sin 2x$

n $4\cos 4x\cos x-\sin 4x\sin x$ **o** $\cos x(1-3\sin^2 x)$

p $e^x(3\cos 3x+\sin 3x)$ **q** $-e^{-2x}(\sin x+2\cos x)$

r $\frac{e^{3x}(3x-1)}{x^2}$ **s** $\frac{e^{\sin x}(x\cos x-1)}{x^2}$

t $e^{-x}(2\cos 2x-\sin 2x)$

2 a $2x\cot(x^2)$ **b** $2x\sec(x^2)\cot(x^2)$

c $6\sin 3x\cos 3x$ **d** $-12\cos^2 4x\sin 4x$

e $4\tan 2x\sec^2 2x$ **f** $3\tan^2(x+4)\sec^2(x+4)$

g $\sin 2x e^{\sin^2 x}$ **h** $2x$

3 a $4^x\ln 4$ **b** $5^x\ln 5$ **c** $c^x\ln c$

d $2^{2x}\ln 4$ **e** $5^{2x}\ln 25$ **f** $x^x(1+\ln x)$

4 $-\frac{1}{2}(1+\sqrt{5})$

5 4

6 -1

7 $\frac{\ln 2}{\ln 5}=0.431$

8 a $\frac{1}{6}\sqrt{\frac{3}{x+4}}$ **b** $\frac{x}{\sqrt{x^2-3}}$ **c** $\frac{1}{2\sqrt{x}}$

9 a $\frac{1}{12y^2+1}$ **b** $\frac{1}{12\sin^3 y\cos y}$

c $\frac{1}{30}\cot^2(2y)\cos^2(2y)$ **d** $\frac{1}{1+\ln y}$

e $\frac{e^y}{\cos y-\sin y}$ **f** $\frac{1}{12\cos 4y-8\sin 2y}$

g $\frac{2y(y-2)}{5y-8}$

10 $6y=x+4$

11 $8y=\pm(x+4)$

13 maxima at (1.02, 2.47) and (4.16, 57.2)
minima at (2.59, –11.9) and (5.73, –275.3)

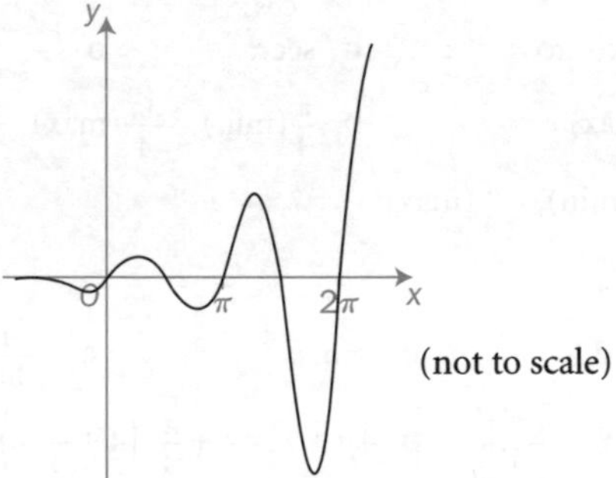

14 a (1,2), (–1,–2)

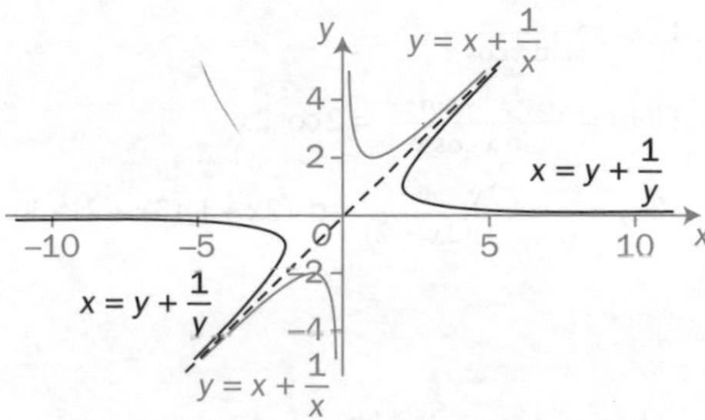

c $x=y+\frac{1}{y}$ is the inverse of $y=x+\frac{1}{x}$
Their graphs are reflections in the line $y=x$

C3

Review 4

1 a $x^3\sec^2 x + 3x^2\tan x$ b $3\tan^2 x\sec^2 x$
c $3x^2\sec^2(x^3)$ d $\dfrac{x\sec^2 x - 3\tan x}{x^4}$
e $e^x(\sin x + \cos x)$ f $\cos x e^{\sin x}$
g $-\dfrac{\cos x}{e^{\sin x}}$ h $\dfrac{1}{x}\cos x - \sin x \ln x$
i $3\cos 3x$ j $6\sec^2 6x$
k $\dfrac{2}{2x+3}$ l $3e^{3x-1}$
m $\dfrac{\cos x - \sin x}{e^x}$ n $e^x(x^2 - x + 1)$
o $\dfrac{1-3x^2}{2\sqrt{x}\,(x^2+1)^2}$

2 a ae^{ax} b ae^{ax+b}
c $f'(x)e^{f(x)}$ d $a\cos(ax)$
e $a\cos(ax+b)$ f $f'(x)\cos[f(x)]$
g $a\sec^2 ax$ h $a\sec^2(ax+b)$
i $f'(x)\sec^2[f(x)]$ j $\dfrac{1}{x}$
k $\dfrac{a}{ax+b}$ l $\dfrac{f'(x)}{f(x)}$

3 a $e^x(\cos 3x - 3\sin 3x)$ b $e^{3x}(3\tan x + \sec^2 x)$
c $e^{3x}(3\cos 2x - 2\sin 2x)$ d $2\sec^2\left(\dfrac{x}{2}\right)$
e $\dfrac{3\sin x + x\cos x}{x\sin x}$ f $\dfrac{1}{x} + \dfrac{1}{2(x+1)}$
g $-\dfrac{6(3x+1)(x+2)}{(2x-1)^4}$ h 1 i $5\sec^2 5x$
j $2x\left[\sin\left(2x+\dfrac{\pi}{2}\right) + x\cos\left(2x+\dfrac{\pi}{2}\right)\right]$
k $3e^{3\tan x}\sec^2 x$ l $-\dfrac{2}{\cos x}$
m $\cot x - x\operatorname{cosec}^2 x$ n $\sec x$ o $\dfrac{1-x\tan x}{\sec x}$

4 a 1(max) b $\dfrac{\pi}{4}$(min), $-\dfrac{3\pi}{4}$(max)
c $\dfrac{3\pi}{4}$(min), $-\dfrac{\pi}{4}$(max)

5 $y = -\theta + \dfrac{\pi}{2}$

7 a $2^x\ln 2$ b $a^x\ln a$ c $\dfrac{1}{x\ln 10}$

8 a $y + x = \dfrac{6+\pi}{4}$ b $4y + x = 4 + \dfrac{\pi}{4}, \left(4+\dfrac{\pi}{4}, 0\right)$

9 a $y = \sin\left(\dfrac{\pi x}{180}\right)$ b $y = \dfrac{\pi}{180}\cos\left(\dfrac{\pi x}{180}\right) = \dfrac{\pi}{180}\cos(x°)$

10 a $f'(x) = \dfrac{1}{\sin x\cos x}$
b $f'(x) = \dfrac{\cos^2 x - \sin^2 x}{\sin x\cos x} = 2\cot 2x$
c $f'(x) = \dfrac{30x-1}{(2x-1)(3x+2)}$ d $(2x-1)(3x+2)^2(30x-1)$

12 4

14 $\dfrac{4}{(1-x)^3}$

15 a $\dfrac{1}{2\sqrt{x}}$ b $\dfrac{1}{2\sqrt{x-3}}$

16 a $3x^2e^{3x}(1+x)$ b $\dfrac{2(\cos x + x\sin x)}{\cos^2 x}$
c $2\tan x\sec^2 x$ d $-\dfrac{1}{2y\sin y^2}$

17 a i $xe^{3x+2}(2+3x)$ ii $-\dfrac{6x^3\sin 2x^3 + \cos 2x^3}{3x^2}$
b $\dfrac{1}{2\sqrt{(16-x^2)}}$

19 (0,–729) min, (3,0) and (–3,0) points of inflexion

Before you start Answers

Chapter 5

1 a $x^3 - 1$ b $2x^4 - x^2 + 1$
2 a –1.2, 3.2 b (1,0)
3 –2.1, 0.25, 1.9
4 1

Exercise 5.1

1 a 2 b 1 c 2 d 2
e 3 f 1 g 3 h 2
2 a 1 b 2 c 1
d 1 e 2 f 0
3 b There is only one point of intersection.
c [1,2] d 1.9
4 c 0.703
5 b 3 c 2.13
7 3.7
8 $a = 1, x = 1.28$
9 c [–2,–1], [1,2], [3,4] d 3.93
10 b [–1,0], [0,1], [4,5] c –0.88
11 a [2,3] b 2.19
12 b 2.67
13 a 2 b 1.32
14 b 3.29
15 a 1 c 6.846
16 3, 2.478
17 0, $(2n+1)\pi, n \in \mathbb{Z}$
18 1.9 radians

Exercise 5.2

1 c 4.56
2 c 2.46 d Diverges
3 b 1.466
4 0.35
5 2.659
6 a $\lambda = 5, \mu = 7$ c 5.25
7 a –0.856, +0.473 b 5, 6, 2 c 1.304
8 $x^3 - 6x + 3$, 2.145
9 a $x^3 - 5x + 7 = 0$ b $x^3 - 5x - 1 = 0$
c $x^3 - 8x + 2 = 0$ d $3x^4 = 100$
e $x^3 - 2x - 1 = 0$ f $x^2 + 2e^{-x} - 6 = 0$
10 a 30, 20 b The second
c 2.11474
11 c –1.3734 d –53.9°
12 c 2.303 d 1.2

Review 5

1 **a**

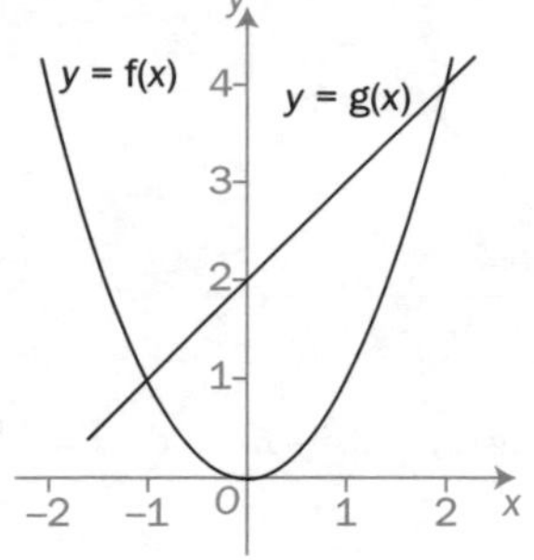

Two roots

b

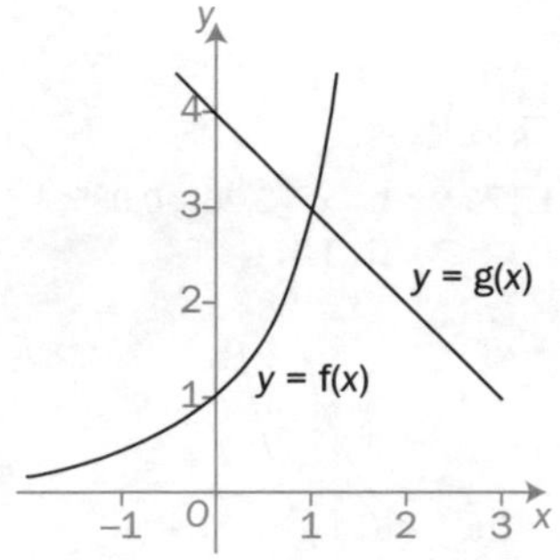

One root

2 **a**

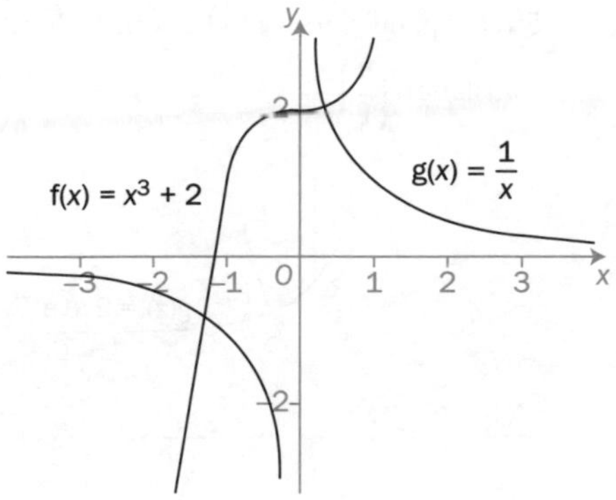

Two roots

b

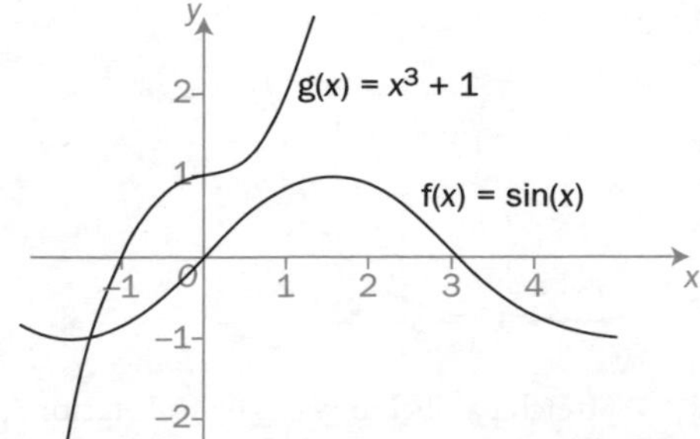

One root

c

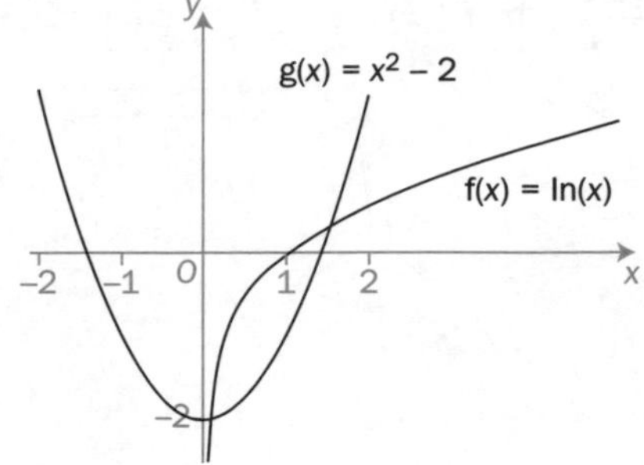

Two roots

d

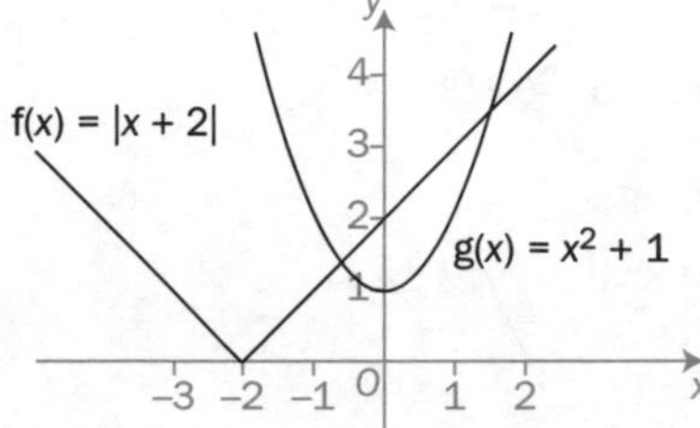

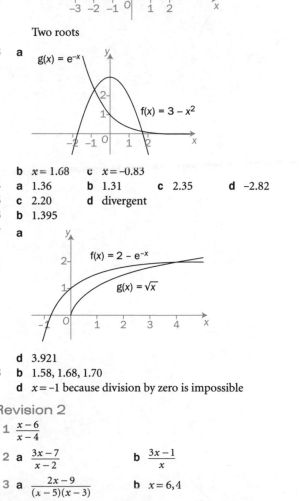

Two roots

3 **a**

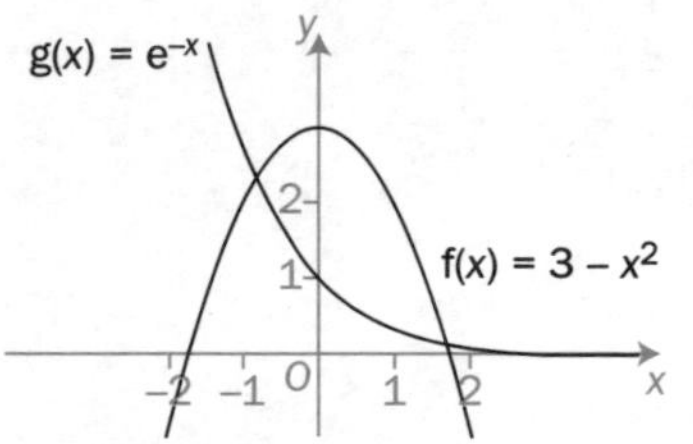

b $x = 1.68$ **c** $x = -0.83$

4 **a** 1.36 **b** 1.31 **c** 2.35 **d** −2.82

5 **c** 2.20 **d** divergent

6 **b** 1.395

7 **a**

$f(x) = 2 - e^{-x}$

$g(x) = \sqrt{x}$

d 3.921

8 **b** 1.58, 1.68, 1.70

d $x = -1$ because division by zero is impossible

Revision 2

1 $\dfrac{x-6}{x-4}$

2 **a** $\dfrac{3x-7}{x-2}$ **b** $\dfrac{3x-1}{x}$

3 **a** $\dfrac{2x-9}{(x-5)(x-3)}$ **b** $x = 6, 4$

4 $a = 2, b = 0, c = -1, d = 1, e = 0$

C3

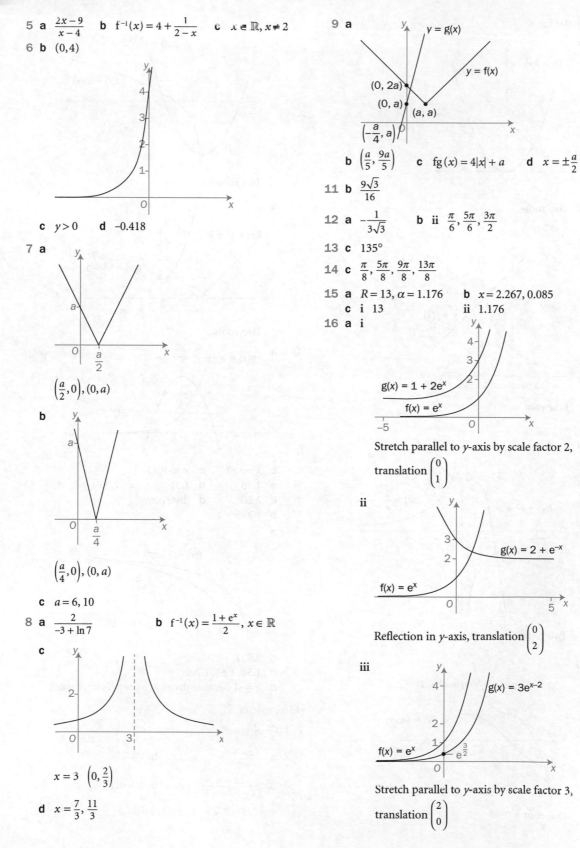

5 a $\frac{2x-9}{x-4}$ **b** $f^{-1}(x) = 4 + \frac{1}{2-x}$ **c** $x \in \mathbb{R}, x \neq 2$

6 b $(0, 4)$

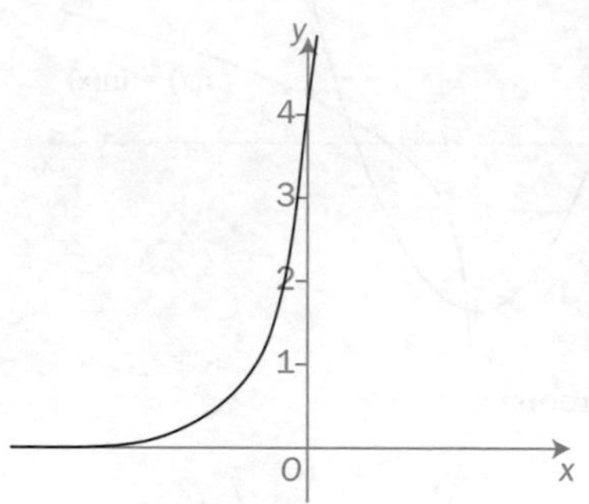

c $y > 0$ **d** -0.418

7 a

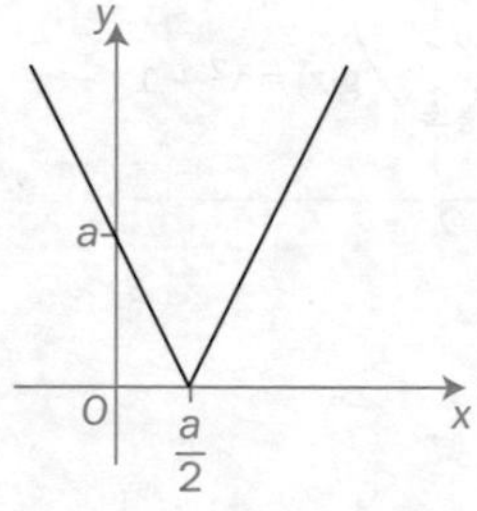

$\left(\frac{a}{2}, 0\right), (0, a)$

b

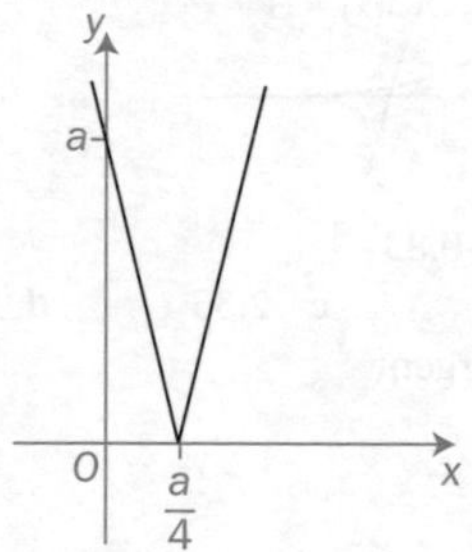

$\left(\frac{a}{4}, 0\right), (0, a)$

c $a = 6, 10$

8 a $\frac{2}{-3 + \ln 7}$ **b** $f^{-1}(x) = \frac{1 + e^x}{2}, x \in \mathbb{R}$

c

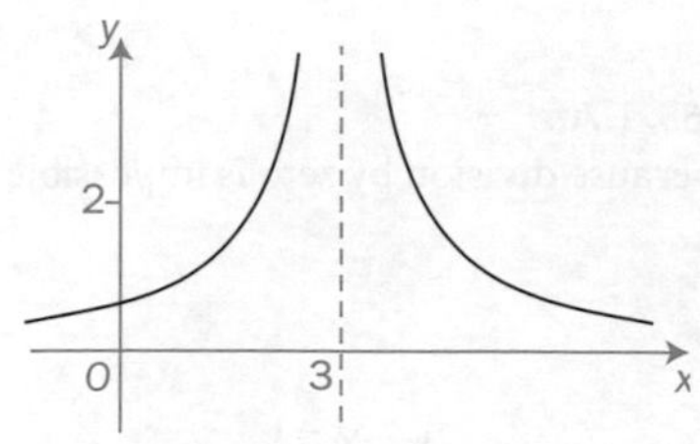

$x = 3 \quad \left(0, \frac{2}{3}\right)$

d $x = \frac{7}{3}, \frac{11}{3}$

9 a

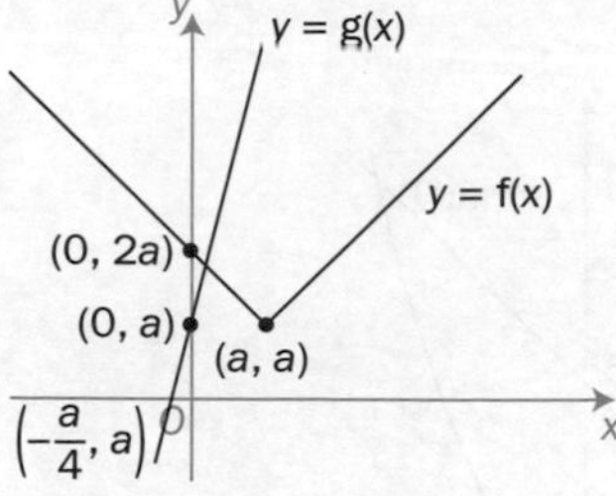

b $\left(\frac{a}{5}, \frac{9a}{5}\right)$ **c** $fg(x) = 4|x| + a$ **d** $x = \pm\frac{a}{2}$

11 b $\frac{9\sqrt{3}}{16}$

12 a $-\frac{1}{3\sqrt{3}}$ **b ii** $\frac{\pi}{6}, \frac{5\pi}{6}, \frac{3\pi}{2}$

13 c $135°$

14 c $\frac{\pi}{8}, \frac{5\pi}{8}, \frac{9\pi}{8}, \frac{13\pi}{8}$

15 a $R = 13, \alpha = 1.176$ **b** $x = 2.267, 0.085$

c i 13 **ii** 1.176

16 a i

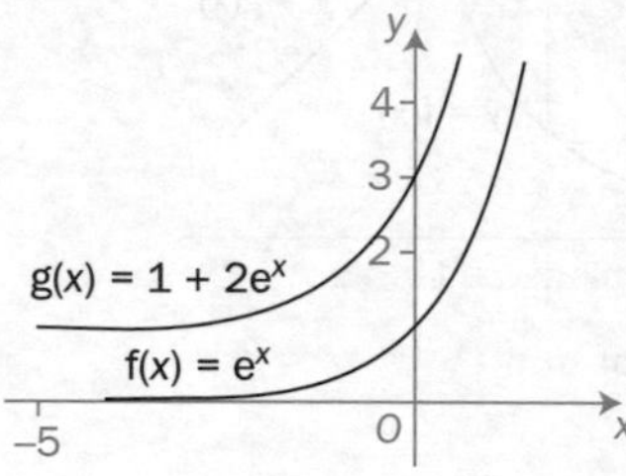

Stretch parallel to y-axis by scale factor 2, translation $\begin{pmatrix} 0 \\ 1 \end{pmatrix}$

ii

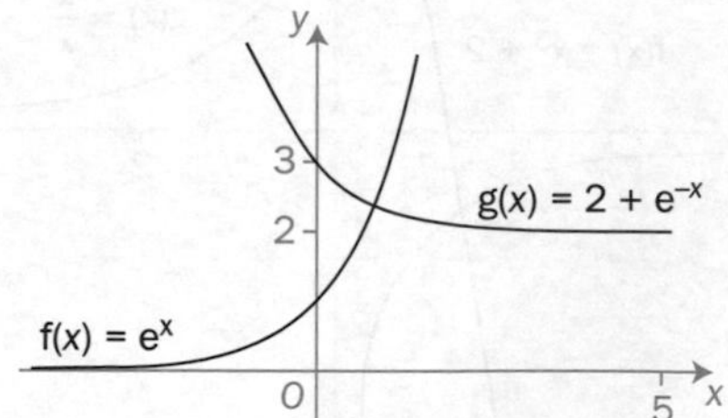

Reflection in y-axis, translation $\begin{pmatrix} 0 \\ 2 \end{pmatrix}$

iii

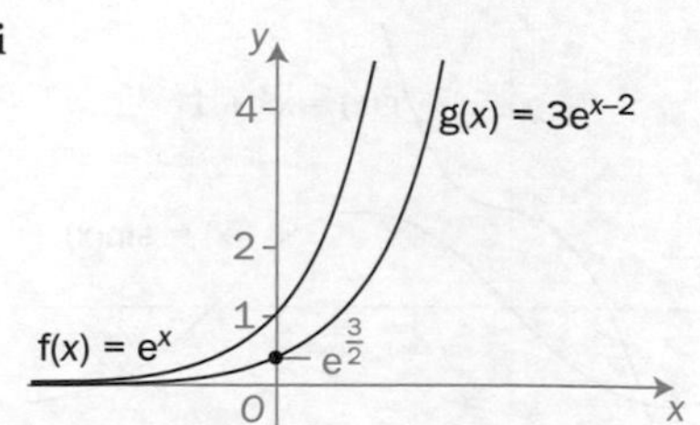

Stretch parallel to y-axis by scale factor 3, translation $\begin{pmatrix} 2 \\ 0 \end{pmatrix}$

b i

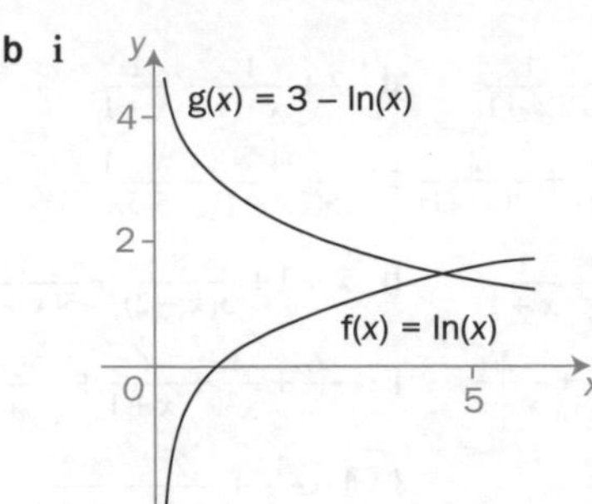

Reflection in x-axis translation $\begin{pmatrix}0\\3\end{pmatrix}$

ii

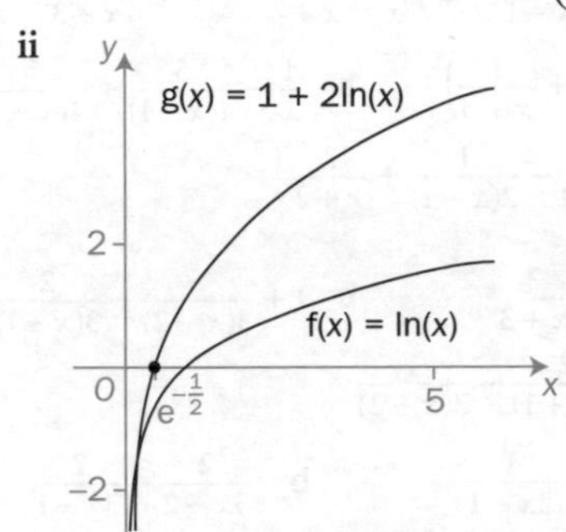

Stretch parallel to y-axis by scale factor 2, translation $\begin{pmatrix}0\\1\end{pmatrix}$

iii

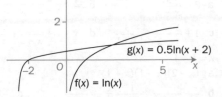

y
2
g(x) = 0.5ln(x + 2)
−2
0
5
x
f(x) = ln(x)

Stretch parallel to y-axis by scale factor 0.5, translation $\begin{pmatrix}-2\\0\end{pmatrix}$

17 **a** 7 days **b** i 105 cells ii 80 cells
18 **b** £670 **c** 15 years **d** 7.2%
19 **b** 14 years
20 **a** 425 °C **b** 7.49 mins **c** 1.64 °C/min
21 **a** 0.405 **b** 4.39 **c** 0, 0.693
22 **a** $e^x(\sin x + \cos x)$ **b** $x^2(1 + 3\ln x)$
c $e^{-x}(3 + 2x - x^2)$ **d** $\dfrac{e^x(\sin x - \cos x)}{\sin^2 x}$
e $\dfrac{x^2 - 1 - 2x^2\ln x}{x(x^2 - 1)^2}$ **f** $\dfrac{6x^2}{(x^3 + 1)^2}$
g $\dfrac{1}{\sin x\cos x}$ or $\dfrac{2}{\sin 2x}$ **h** $3x^2e^{x^3}$
i $\dfrac{2x - 2}{3(x^2 - 2x + 5)^{\frac{2}{3}}}$ **j** $\dfrac{-9x^2}{2(x^2 - 1)^{\frac{3}{2}}}$
k $\dfrac{-e^x}{(e^x - 1)^2}$ **l** $\cos^3 x - 2\cos x\sin^2 x$
23 **a** i $(2 + 3x)xe^{3x+2}$ ii $-\dfrac{6x^2\sin(2x^3) + \cos(2x^3)}{3x^2}$
b $\dfrac{1}{2\sqrt{16 - x^2}}$

24 **a** i $e^{3x}(\sin x + 7\cos x)$ ii $\dfrac{5x^3}{5x + 2} + 3x^2\ln(5x + 2)$
c $\dfrac{-60}{(x + 1)^4}$; 1, −3
25 **a** $\left(3, \frac{1}{6}\right), \left(-3, -\frac{1}{6}\right)$ **b** 18
26 **a** (1, e) min **b** $(-1, e^{-1})$ min
c $\left(1, \frac{1}{2}\right)$ max, $\left(-1, -\frac{1}{2}\right)$ min **d** (0.464, 0.177) max (−2.678, −94.74) min
27 **a** $xe^x(x + 2)$ **b** (0, 0), $(-2, 4e^{-2})$
c $e^x(x^2 + 4x + 2)$
d (0,0) minimum, $(-2, 4e^{-2})$ maximum
28 **a** $\dfrac{1 + x^2}{(x^2 - 1)^2}$ **b** $9y + 5x = 16$, $15y - 27x = 44$
c (−0.49, 2.05)
29 **b** $y = x$
30 **a** i $x = a^y$ **c** $y = \dfrac{1}{10\ln 10}(x - 10 + 10\ln 10)$
d $B(10(1 - \ln 10), 0)$
31 **a**

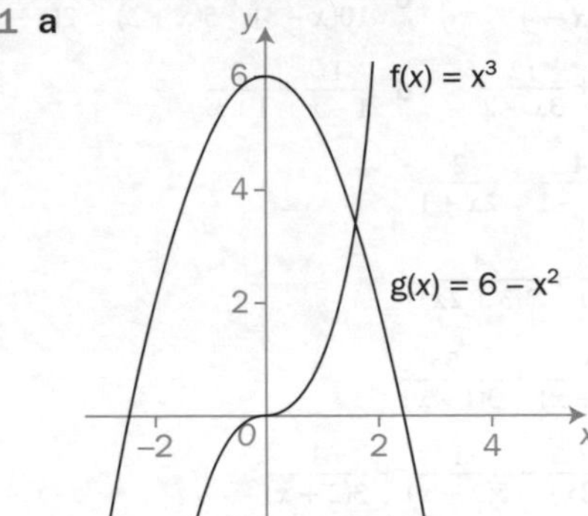

c [1, 2] **d** 1.5
32 **a** [1, 2] **b** 1.32
33 **c** 2.754
34 **a** $x^3 + x - 3 = 0$ **b** $x^3 - 5x + 10 = 0$
c $x^3 - 3x - 2 = 0$
35 **a** $\alpha = 12, \beta = 2$ **c** 2.73
36 **b** 1.395
37 **a** $2y = x + 1$ **c** 2.1530
38 **b** $-2\ln 2$
e $x_1 = 4.9192, x_2 = 4.9111, x_3 = 4.9103$
39 **a** $3e^x - \dfrac{1}{2x}$
c $x_1 = 0.0613, x_2 = 0.1568, x_3 = 0.1425, x_4 = 0.1445$

Before you start Answers

Chapter 6

1 **a** 24 **b** 28
2 **a** $x(2x - 3)(2x + 3)$ **b** $(x - 1)(x + 1)(x^2 + 1)$
3 **a** $A = 5, B = 9$ **b** $A = 4, B = 1$
4 **a** $\dfrac{3x + 7}{(x + 1)^2}$ **b** $\dfrac{(2x + 1)(3x - 1)}{x(x^2 - 1)}$
5 **a** $x^2 - 4x + 3$ **b** $x + 8 + \dfrac{15}{x - 2}$

Exercise 6.1

1 **a** $\dfrac{3}{x + 2} + \dfrac{1}{x + 1}$ **b** $\dfrac{4}{x - 3} - \dfrac{3}{x + 4}$
c $\dfrac{1}{4(x - 3)} - \dfrac{1}{4(x + 5)}$ **d** $\dfrac{3}{x} + \dfrac{1}{x - 1}$

e $\frac{3}{x+5}+\frac{1}{2x-1}$ **f** $\frac{3}{2(x-4)}+\frac{1}{2(x-2)}$

2 **a** $\frac{5}{x-3}-\frac{1}{x-2}$ **b** $\frac{2}{x+1}+\frac{3}{x+4}$

c $\frac{4}{x}-\frac{3}{x+1}$ **d** $\frac{2}{x-1}-\frac{1}{x-6}$

e $\frac{3}{2(x-1)}-\frac{6}{x-2}+\frac{9}{2(x-3)}$

f $\frac{1}{2(x-1)}-\frac{1}{2(x+1)}+\frac{1}{2-x}$

3 **a** $\frac{3}{x-2}-\frac{3}{x+1}$ **b** $\frac{2}{x-1}+\frac{1}{x+4}$

c $\frac{1}{2(x-5)}-\frac{1}{2(x-3)}$ **d** $\frac{3}{2x}-\frac{1}{2(x-2)}$

e $\frac{2}{x}-\frac{5}{x+1}+\frac{3}{x+2}$ **f** $-\frac{1}{x}+\frac{1}{2x+1}+\frac{1}{2x-1}$

4 **a** $\frac{-2}{x-2}+\frac{4}{x-4}$ **b** $\frac{3}{10(x-3)}+\frac{1}{5(x+2)}-\frac{1}{2(x-1)}$

c $-\frac{6}{2x-1}+\frac{12}{3x-2}$ **d** $\frac{1}{1-x}+\frac{2}{1+x}$

e $-\frac{3}{x}+\frac{4}{2x-1}+\frac{2}{2x+1}$

f $\frac{1}{3(3-2x)}+\frac{4}{3(3+2x)}$

g $\frac{2}{x}-\frac{4}{3(2+x)}+\frac{5}{3(1-x)}$

h $-\frac{4}{15(1+2x)}+\frac{1}{5(2-x)}+\frac{1}{3(2+x)}$

i $-\frac{1}{x-1}+\frac{1}{x+1}+\frac{1}{2(x-2)}-\frac{1}{2(x+2)}$

Exercise 6.2

1 **a** $\frac{2}{x-1}-\frac{2}{x-2}+\frac{3}{(x-2)^2}$ **b** $\frac{1}{x+1}-\frac{1}{x-3}+\frac{4}{(x-3)^2}$

c $\frac{2}{x+1}+\frac{2}{(x+1)^2}-\frac{3}{2x-1}$ **d** $\frac{1}{x-4}+\frac{4}{(x-4)^2}$

e $\frac{2}{x}+\frac{5}{x^2}-\frac{2}{x-2}$ **f** $\frac{1}{2x^2}+\frac{1}{4x}+\frac{3}{4(x-2)}$

g $\frac{1}{x}-\frac{3}{3x-1}+\frac{3}{(3x-1)^2}$

h $\frac{1}{2(x+2)}+\frac{1}{(x+2)^2}+\frac{3}{(x+2)^3}-\frac{1}{2x}$

2 **a** $1-\frac{2}{x+2}$ **b** $1+\frac{1}{2(x-1)}-\frac{1}{2(x+1)}$

c $1-\frac{1}{4(x-1)}-\frac{7}{4(x+3)}$ **d** $1+\frac{1}{x}-\frac{2}{x+1}$

e $x+1+\frac{1}{x-1}$ **f** $x+\frac{1}{2(x-1)}+\frac{1}{2(x+1)}$

g $x-4+\frac{5}{x+2}+\frac{1}{x-1}$ **h** $-1+\frac{3}{x+3}+\frac{3}{x-3}$

i $\frac{1}{4}x+\frac{1}{16(2x-1)}+\frac{1}{16(2x+1)}$

3 $A=1, B=\frac{1}{2}, C=\frac{7}{2}, D=5$

4 **a** $\frac{8}{5(2x-3)}-\frac{4}{5(x+1)}$ **b** $\frac{2}{3(x-2)}-\frac{2}{3(x+1)}-\frac{2}{(x+1)^2}$

c $1-\frac{1}{x+1}+\frac{1}{2x-1}$ **d** $2+\frac{1}{x-1}-\frac{1}{x+1}$

e $1-\frac{7}{2(x+3)}+\frac{3}{2(x-1)}$ **f** $\frac{4}{5(2x-1)}-\frac{1}{5(3x+1)}$

g $x-1+\frac{2}{x}-\frac{1}{x+1}$ **h** $x+1+\frac{8}{3(x-2)}+\frac{1}{3(x+1)}$

i $1-\frac{1}{x}-\frac{1}{x^2}+\frac{2}{x-1}$ **j** $-\frac{6}{x}+\frac{3}{x^2}+\frac{6}{x+1}+\frac{3}{(x+1)^2}$

k $1+\frac{1}{x}-\frac{3}{x+1}$ **l** $1-\frac{1}{x^2}+\frac{2}{x-1}-\frac{2}{x+1}$

Review 6

1 **a** $\frac{5}{x-2}-\frac{4}{x-1}$ **b** $\frac{2}{x}-\frac{3}{x+4}$ **c** $\frac{1}{x-3}-\frac{1}{x+3}$

2 **a** $\frac{1}{2}\left(\frac{1}{x-1}+\frac{1}{x+3}\right)$ **b** $\frac{1}{x}-\frac{5}{4(x-1)}+\frac{5}{4(x-5)}$

c $-\frac{1}{2(x+1)}-\frac{1}{2(x-1)}+\frac{2}{x+2}$

3 **a** $\frac{2}{x-1}-\frac{2}{x+3}$ **b** $1+\frac{5}{3(x-2)}-\frac{2}{3(x+1)}$

c $\frac{2}{x}-\frac{2}{3(x+1)}-\frac{1}{3(x-2)}$

4 **a** $\frac{1}{2x+1}+\frac{1}{2x-1}$ **b** $\frac{2}{3x-2}-\frac{2}{3x-1}$

c $\frac{3}{x-2}-\frac{3}{x-1}-\frac{2}{(x-1)^2}$ **d** $-\frac{1}{x}-\frac{2}{x^2}+\frac{1}{x-3}$

e $\frac{1}{x}-\frac{2}{2x-3}+\frac{6}{(2x-3)^2}$ **f** $-\frac{1}{x+1}+\frac{2}{x-1}+\frac{2}{(x-1)^2}$

5 $A=1, B=2, C=-2$

6 **a** $3-\frac{3}{x+2}$ **b** $2+\frac{2}{x+2}-\frac{3}{x-3}$

c $1+\frac{2}{x}-\frac{2}{x^2}-\frac{3}{x+1}$ **d** $x+2+\frac{3}{x-2}+\frac{1}{x+2}$

7 $a=k+1$

8 **a** $(x-2)(x-1)(x+1)$

b $\frac{1}{3(x-2)}-\frac{1}{2(x-1)}+\frac{1}{6(x+1)}$ **c** $-1, 1$ or 2

9 **a** $\frac{2}{x+4}+\frac{1}{x-1}$ **b** $f'(x)=-\left[\frac{2}{(x+4)^2}+\frac{1}{(x-1)^2}\right]$

10 **a** $\frac{1}{x+2}+\frac{2}{x-1}$

11 **a** $-\frac{1}{x-2}-\frac{2}{x+3}$ **b** $\frac{9}{8}$

c $y'=\frac{1}{(x-2)^2}+\frac{2}{(x+3)^2}>0$ for all $x\in\mathbb{R}$

12 **a** $\frac{1}{r}-\frac{1}{r+1}$ **c** 1

13 **a** $\frac{1}{r+1}-\frac{1}{r+2}$ **c** 1

Before you start Answers

Chapter 7

1 **a** $\frac{1}{2}(4x-1)$ **b** $17x^2+4x+12$

2 **a** $x=2, y=-1$ **b** $x=\frac{2\pm\sqrt{14}}{5}$

3 **a** $2x-\frac{2}{x^3}; \frac{x^3}{3}+x-\frac{1}{x}+c$

b $2+2x; x+x^2+\frac{x^3}{3}+c$ **c** $-\frac{2}{x^3}; x-\frac{1}{x}+c$

C4

Exercise 7.1

1

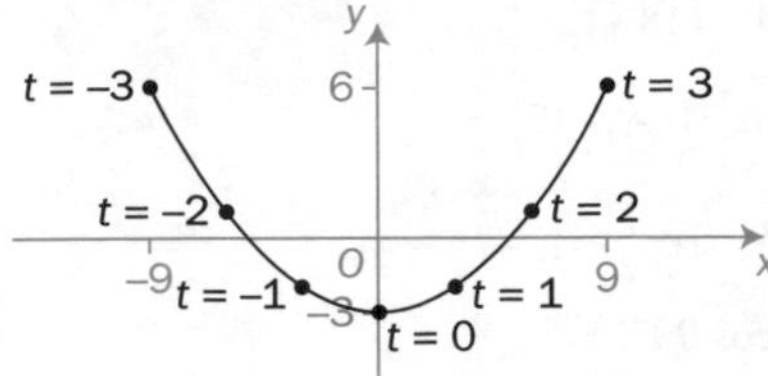

2

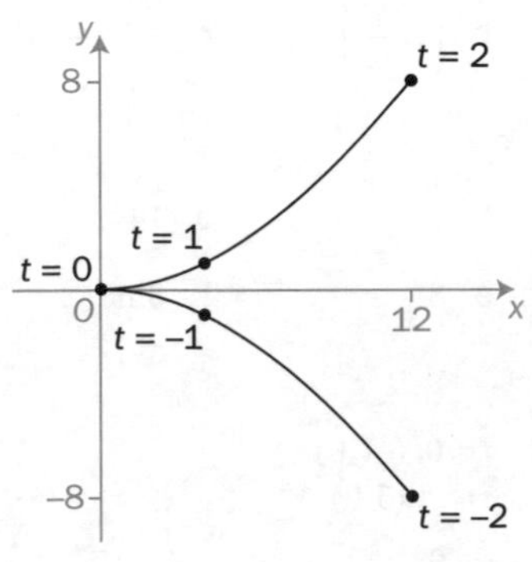

3 a

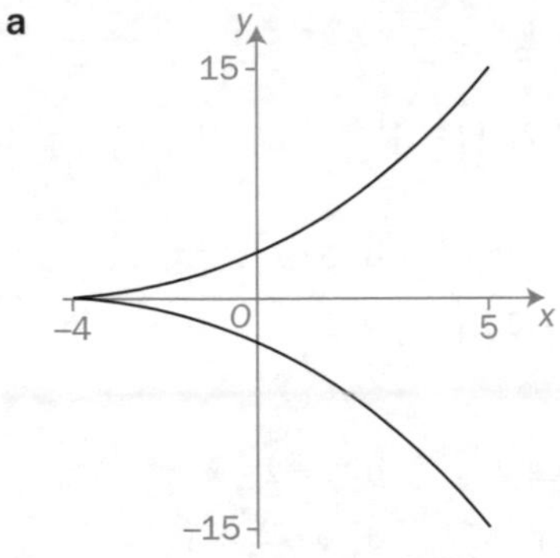

b

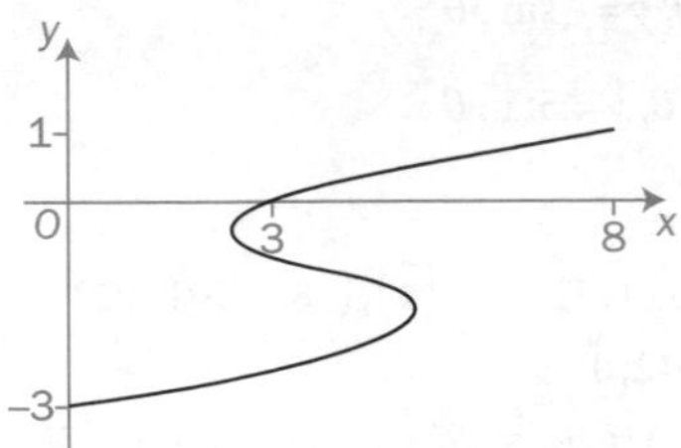

c

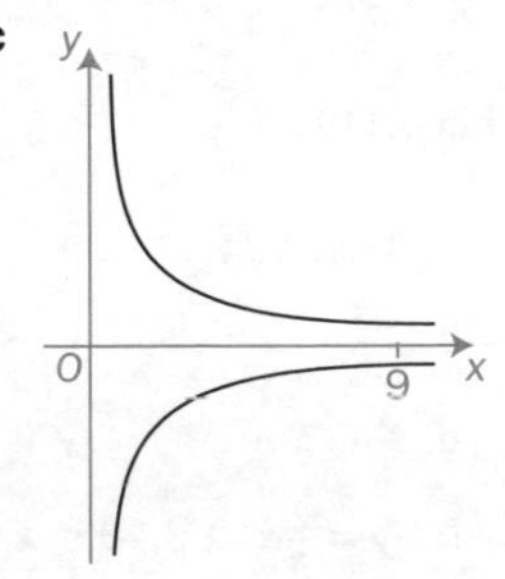

d

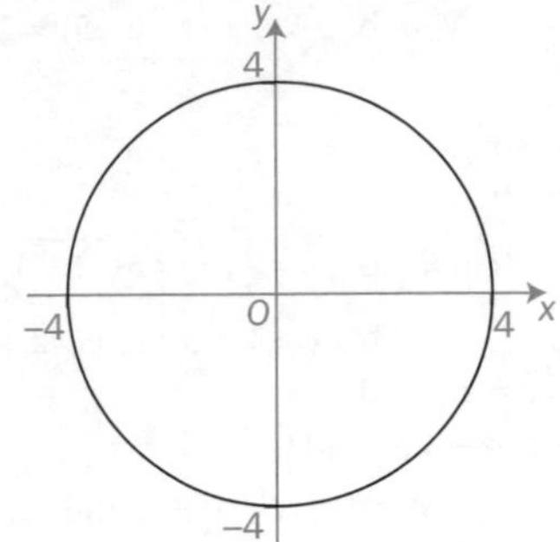

e

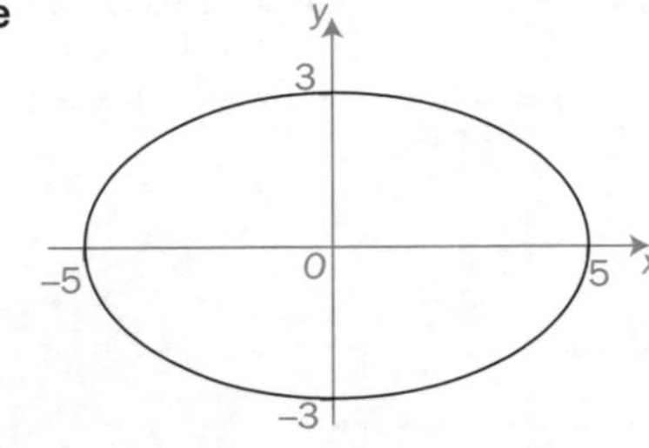

f

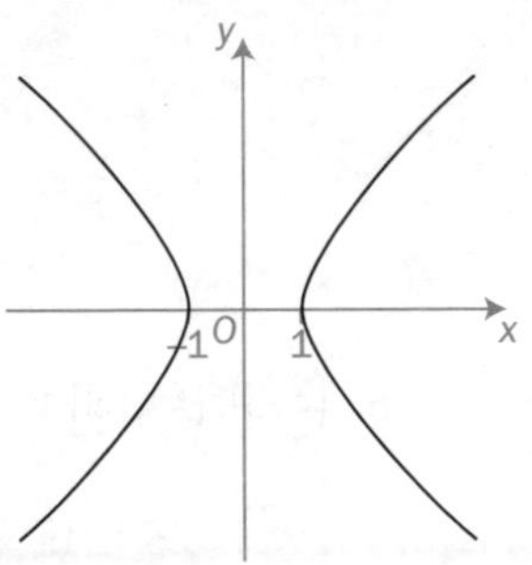

4 **a** $y + 2x = 9$ **b** $xy = 12$
c $y = x^2 - 2x - 1$ **d** $y = x^{\frac{3}{2}}$
e $y = (x+1)^{\frac{3}{2}} + 2$ **f** $xy^2 = 4$
g $y = x + 2$ **h** $\frac{x^2}{9} + \frac{y^2}{16} = 1$
i $y = 1 - 2x^2$ **j** $y = \frac{10}{9}x^2 - 5$
k $\frac{x^2}{4} - \frac{y^2}{9} = 1$ **l** $y = \frac{3x+1}{2}$

5 6

6 $\frac{3}{4}$

7 2, –4

8 ±2, $y = \frac{1}{4}x^2 - 1$

9 **a** (10,0) **b** (9,0)
c (13,0), (–7,0) **d** (±3,0)

10 **a** (0,23) **b** (0,5), (0,6)
c (0,0), (0,1) **d** (0,±1)

11 5

12 5

13 3

14 a $\sqrt{29}$ b $\frac{2}{5}$ c $5y = 2x + 33$

15 $y = 2x + 1$

16 $x = 2\cos\theta, y = 2\sin 2\theta$

17 $x = \tan\theta, y = \sin\theta$

18 a $x = \sin\theta, y = 3\cot\theta$ b $x = \frac{1}{t}, y = 3\sqrt{t^2 - 1}$

19 a $\alpha = 3, \beta = 2$ b $x = 3 + \cos\theta, y = 2 + \sin\theta$

20 a $a = 1, b = 2, c = 2, d = 3$

b $x = 1 + 2\sec\theta, y = 2 + 3\tan\theta$

21 a $\frac{\pi}{3}, \frac{5\pi}{3}$ b $\beta = 1.9$; $(0, -1)$ and $(0, 1.6)$

22 a $x = \frac{1}{t^3}, y = \frac{1}{t^2}$ b $x = t + 2, y = t(t + 2)$

c $x = \frac{1}{1 - t^3}, y = \frac{t}{1 - t^3}$ d $x = \frac{1}{t} - 1, y = 1 - t$

23 c

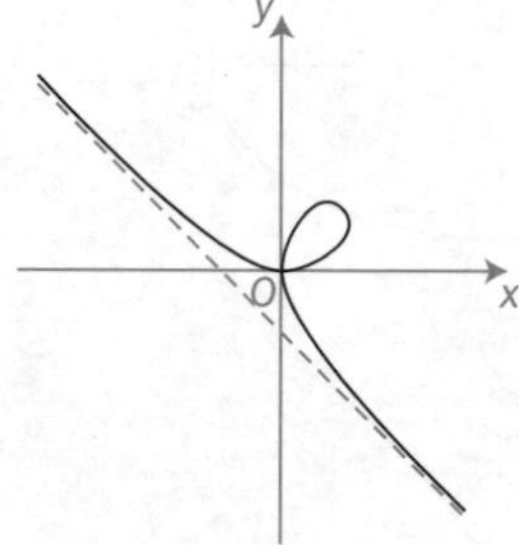

Asymptote $x + y = -1$

Exercise 7.2

1 a $(9,-6), (1,2)$ b $\left(\frac{9}{4},3\right), \left(\frac{25}{4},-5\right)$

2 a $(-1,1), (3,5)$ b $(1,3), (-1,-1), \left(\frac{1}{8},\frac{5}{4}\right)$

c $(-1,1), (3,3)$

3 $(2,3)$ and $(2,-3)$

4 $(3,0), \left(\frac{81}{25},-\frac{12}{5}\right)$

5 a $(0,-3), (3,0)$ b $(0,-5), (1,0), (-5,0)$

c $(10,0)$ d $(0,-3), (7,0), (-9,0)$

e $\left(0,-\frac{2}{3}\right), (2,0)$ f $(4\pi n, 0), n \in \mathbb{Z}$

6 $(1,-2), (4,4)$

7 $Q\left(\frac{1}{2}t^2, 2t\right), y^2 = 8x$

8 a $Q(2t^2,0), M(2t^2,3t)$ b $2y^2 = 9x$

9 $(-4,2), (16,-18)$

10 $\frac{1}{4}$

Exercise 7.3

1 a 12 b 6 c 8 d $-\frac{5}{3}$

e $\frac{1}{16}$ f $\frac{1}{2}$

2 a $y = 2x + 1, 2y + x = 4\frac{1}{2}$

b $2y = 3x - 6, 3y + 2x = 4$

c $y = \sqrt{2}x - 2, y = -\frac{x}{\sqrt{2}} + 1$

d $3y = 4x - 5, 4y + 3x = 0$

3 a $\left(-\frac{1}{\sqrt{3}},\frac{2}{3\sqrt{3}}\right), \left(\frac{1}{\sqrt{3}},-\frac{2}{3\sqrt{3}}\right)$

b $(1,2), (1,-2)$

c $\left(\frac{\pi}{2},1\right), \left(\frac{3\pi}{2},-1\right)$ d $(2,8), (2,2)$

4 $y + 2x = 24$ $(18,-12)$

5 $y = 4x - 45$ $\left(-\frac{3}{4},-48\right)$

6 $y + 2x = 3$ $Q\left(-\frac{1}{2},4\right)$

7 a $AP = 3\cos\theta + 5$

$BP = 5 - 3\cos\theta$

8 a $\frac{1}{t^2}, y = \frac{x}{t^2}$ b $\left(1,\frac{1}{t^2}\right)$ c $y(2x - 1)^2 = x$

Exercise 7.4

1 a $19\frac{1}{2}$ b $25\frac{1}{3}$ c 144

d 28 e $8\frac{2}{3}$ f $8\ln 2$

2 $19\frac{1}{5}, 12\frac{4}{5}$

3 a $t = -1, (0,-6)$ $t = 0, (-1,0)$

$t = 1, (0,2)$ $t = 2, (3,0)$

b $3\frac{2}{3}$ c 9

4 $t = 0, \pm 2$ $17\frac{1}{15}$

5 a $\left(4\frac{1}{4},1\right), (2,2), \left(4\frac{1}{4},4\right)$ b $4\frac{1}{2}$

6 $(0,5), (0,17), 16$

7 a $(2,2)$ b 5.67

8 a $\pm 2\sqrt{2}, \left(\frac{1}{\sqrt{2}},2\right), \left(-\frac{1}{\sqrt{2}},-2\right)$ b $\frac{3}{4} - \ln 2$

Review 7

1 a $y = x^2 + 2x + 2$ b $y = 2 - x$

c $\frac{x^2}{16} + \frac{y^2}{9} = 1$ d $y = \frac{x^2}{2} - 1$

2 a $x = \sin\theta, y = \frac{3}{2}\sin 2\theta$

b $x = 2\sin\theta, y = 5\tan\theta$

3 $-1, \frac{1}{3}$

4 a $\left(-8 + 3\sqrt{14}, 22 - 6\sqrt{14}\right), \left(-8 - 3\sqrt{14}, 22 + 6\sqrt{14}\right)$

b $(4,0), (-12,0)$

c $\left(\frac{\sqrt{3}}{2},\frac{\sqrt{5}}{2}\right), \left(\frac{\sqrt{3}}{2},-\frac{\sqrt{5}}{2}\right), \left(-\frac{\sqrt{3}}{2},\frac{\sqrt{5}}{2}\right), \left(-\frac{\sqrt{3}}{2},-\frac{\sqrt{5}}{2}\right)$

5 a $2y - 3x = 1, 3y + 2x = 8$

b $12y = 5x + 49, 5y + 12x = 119$

c $x - y = 4, y + x = 2$

d $2\sqrt{3}y = 2x + 1, 2y + 2\sqrt{3}x = 3\sqrt{3}$

6 a 19.5 b $\frac{28}{3}$

c 0.574 (3 s.f.) d $\frac{3}{2}$

7 a $4\cos\theta$ b $y = \frac{2}{3}x^2 + 3, -3 \leqslant x \leqslant 3$

8 b $\frac{2}{s + r}$ d $\left(\frac{121}{9},-\frac{22}{3}\right)$

9 a $y = 2 + x - \frac{x^2}{20}$ b $t = 1 + \sqrt{\frac{7}{5}}$ or 2.18 s (3 s.f.)

c 21.8 m (3 s.f.)

10 **a** $a = 1 + \frac{6}{\pi}$ **c** 3

11 **a** $-2\sin^3\theta\cos\theta$ **b** $2y + x = 4$

c $y = \frac{8}{x^2 + 4}, x \geqslant 0$

Before you start Answers

Chapter 8

1 **a** $(1+x)^{\frac{5}{2}}$ **b** $(1+x)^{-\frac{3}{2}}$

2 **a** $\frac{\sqrt{121}}{10}$ **b** $\frac{\sqrt{112}}{3}$ **c** $\frac{\sqrt{30}}{6}$

3 **a** $8 + 12x + 6x^2 + x^3$

b $56; 256 + 1024x + 1792x^2 + 1792x^3 + 1120x^4 + 448x^5 + 112x^6 + 16x^7 + x^8$

4 **a** $-\frac{5}{2}x$ **b** $-\frac{25}{4}x^2$

5 **a** $\frac{2}{1-2x} + \frac{1}{1+x}$ **b** $\frac{4}{3(1-2x)} + \frac{2}{3(1+x)} + \frac{1}{(1+x)^2}$

Exercise 8.1

1 **a** $1 - 2x + 4x^2 - 8x^3, \quad |x| < \frac{1}{2}$

b $1 + \frac{1}{2}x - \frac{1}{8}x^2 + \frac{1}{16}x^3, \quad |x| < 1$

c $1 + x - \frac{1}{2}x^2 + \frac{1}{2}x^3, \quad |x| < \frac{1}{2}$

d $1 + 3x + 6x^2 + 10x^3, \quad |x| < 1$

e $1 + x - x^2 + \frac{5}{3}x^3, \quad |x| < \frac{1}{3}$

f $1 + \frac{x}{2} + \frac{3}{8}x^2 + \frac{5}{16}x^3, \quad |x| < 1$

g $1 + 6x + 27x^2 + 108x^3, \quad |x| < \frac{1}{3}$

h $1 + \frac{x}{2} - \frac{3}{8}x^2 + \frac{7}{16}x^3, \quad |x| < \frac{1}{2}$

i $1 + x + \frac{3}{2}x^2 + \frac{5}{2}x^3, \quad |x| < \frac{1}{2}$

j $1 + x + \frac{3}{4}x^2 + \frac{1}{2}x^3, \quad |x| < 2$

k $1 + 3x + \frac{3}{2}x^2 - \frac{1}{2}x^3, \quad |x| < \frac{1}{2}$

l $1 - \frac{1}{3}x - \frac{1}{36}x^2 - \frac{1}{162}x^3, \quad |x| < 2$

2 **a** **i** $1 - x + x^2 - x^3 + x^4$

ii $1 + x + x^2 + x^3 + x^4$

b **i** $\frac{1}{1+x}$ **ii** $\frac{1}{1-x}$

3 **a** $2 + \frac{1}{4}x - \frac{1}{64}x^2, \quad |x| < 4$

b $\frac{1}{2} - \frac{x}{4} + \frac{x^2}{8}, \quad |x| < 2$

c $\frac{1}{4} + \frac{3}{4}x + \frac{27}{16}x^2, \quad |x| < \frac{2}{3}$

d $\frac{1}{3}\left(1 + \frac{x}{18} + \frac{x^2}{216}\right), \quad |x| < 9$

4 **a** $1 + \frac{1}{2}x - \frac{5}{8}x^2, \quad |x| < 1$

b $1 - \frac{5}{2}x + \frac{11}{8}x^2, \quad |x| < 1$

c $2 + 3x + \frac{5}{2}x^2, \quad |x| < 2$

d $-3 - 4x - 4x^2, \quad |x| < 1$

e $1 + x + \frac{1}{2}x^2, \quad |x| < 1$

f $3 + 3x - \frac{8}{3}x^2, \quad |x| < \frac{1}{2}$

g $4 - 5x^2, \quad |x| < \frac{1}{2}$

5 $\frac{3}{8}$

6 $1 + \frac{1}{x} - \frac{1}{2x^2} + \frac{1}{2x^3}, \quad |x| > 2$

7 $a = \pm 4, 1 + 2x - 2x^2 + 4x^3, 1 - 2x - 2x^2 - 4x^3$

8 $4, \frac{1}{2}, 4$

9 *Either* $k = 3, n = \frac{1}{3}, -x^2, |x| < \frac{1}{3}$

or $k = -\frac{3}{2}, n = -\frac{2}{3}, \frac{5}{4}x^2, |x| < \frac{2}{3}$

10 $1 + \frac{1}{2x} - \frac{1}{8x^2} + \frac{1}{16x^3} + \dots,$

$1 + \frac{1}{2}x^{\frac{1}{2}} - \frac{1}{8}x^{\frac{3}{2}} + \frac{1}{16}x^{\frac{5}{2}} + \dots$

Exercise 8.2

1 **a** $3 + 9x + 21x^2 + 45x^3, \quad |x| < \frac{1}{2}$

b $4 - 8x + 28x^2 - 80x^3, \quad |x| < \frac{1}{3}$

c $5 - \frac{15}{2}x + \frac{65}{4}x^2 - \frac{255}{8}x^3, \quad |x| < \frac{1}{2}$

d $\frac{3}{2} - \frac{3}{4}x + \frac{9}{8}x^2 - \frac{15}{16}x^3, \quad |x| < 1$

e $1 - \frac{5}{6}x + \frac{19}{36}x^2 - \frac{65}{216}x^3, \quad |x| < 2$

f $\frac{2}{3} + \frac{16}{9}x + \frac{104}{27}x^2 + \frac{640}{81}x^3, \quad |x| < \frac{1}{2}$

g $4 + 4x + 8x^2 + 8x^3, \quad |x| < 1$

2 **a** $A = 2, B = 11$

b $\frac{3}{16} + \frac{7x}{32} + \frac{25x^2}{256} + \frac{9x^3}{256} + \frac{47x^4}{4096}, \quad |x| < 4$

c It converges more slowly to the correct answer as $|x|$ increases.

3 **a** $\frac{5}{1-x} - \frac{4}{1+2x}$ **b** $\ln\left(\frac{125}{64}\right)$

c $1 + 13x - 11x^2 + 37x^3$

Exercise 8.3

1 0.73742

2 $729 + 1458x + 1215x^2 + 540x^3 + 135^4 + 18x^5 + x^6$

3 **a** 2.23607 **b** 1.73205

4 3.16228

5 **a** $1 + x - x^2 + \frac{5}{3}x^3$ **b** 10.009990

6 $1 - 6x + 24x^2 - 80x^3$; 0.99402392

7 **a** $1 - \frac{1}{2x} - \frac{1}{8x^2} - \frac{1}{16x^3}, \quad |x| > 1$ **b** 9.949874

c $x = -100$, 10.049876

9 $1 + \frac{3}{4}x + \frac{3}{32}x^2 - \frac{1}{128}x^3$

10 **a** $1 + 3x + \frac{7}{2}x^2$ **b** $\frac{5}{2} + \frac{15}{4}x + \frac{65}{8}x^2 + \frac{255}{16}x^3$

11 2, 4, 3

12 $\frac{1}{(1+x)^3} = 1 - 3x + 6x^2 - 10x^3 \ldots$ for $|x| < 1$

Review 8

1 a $1 + 3x + 9x^2 + 27x^3, \quad |x| < \frac{1}{3}$

b $1 + x - \frac{1}{2}x^2 + \frac{1}{2}x^3, \quad |x| < \frac{1}{2}$

c $1 + \frac{1}{3}x + \frac{5}{36}x^2 + \frac{5}{81}x^3, \quad |x| < 2$

d $\frac{1}{3}\left(3 + \frac{1}{3}x - \frac{x^2}{54} + \frac{x^3}{486}\right), \quad |x| < \frac{9}{2}$

e $\frac{1}{2}\left(1 + \frac{1}{2}x + \frac{1}{4}x^2 + \frac{1}{8}x^3\right), \quad |x| < 2$

f $\frac{1}{2}\left(1 + \frac{1}{8}x + \frac{3}{128}x^2 + \frac{5}{1024}x^3\right), \quad |x| < 4$

g $1 - \frac{1}{2}x - \frac{3}{8}x^2 - \frac{1}{16}x^3, \quad |x| < 1$

h $-(3 + 4x + 4x^2 + 4x^3), \quad |x| < 1$

2 $A = 0.5, B = 0, C = -\frac{1}{16}, \quad |x| < 1$

3 $-\frac{5}{8}$

4 $1 - \frac{1}{x} - \frac{1}{2x^2} - \frac{1}{2x^3}, \quad |x| > 2$

5 a $a = 3, n = -2$ b -108 c $|x| < \frac{1}{3}$

6 a $\frac{1}{1-x} + \frac{2}{1+2x}, 3 - 3x + 9x^2 - 15x^3, |x| < \frac{1}{2}$

b $\frac{1}{1-x} + \frac{1}{1+x} - \frac{1}{(1+x)^2}, 1 + 2x - x^2 + 4x^3, |x| < 1$

7 a $A = 1, B = 2$ b $3 - x + 11x^2$

c No as $x = \frac{1}{2}$ is outside $|x| < \frac{1}{3}$

9 3.162 28 (5 d.p.)

11 a $p = -13.5, q = 67.5$ b 1.424 6875

12 a $A = 3, C = 4, B = 0$ b $4 + 8x + \frac{111}{4}x^2 + \frac{161}{2}x^3$

Before you start Answers

Chapter 9

1 a $y = 6x - 5$ b $y = \sqrt{2}(x - 4)$

2 a $3\tan^2 x\sec^2 x$ b $\frac{3}{2}x^2(x^3 + 1)^{-\frac{1}{2}}$

c $e^x(\sin x + \cos x)$ d $x^2(1 + 3\ln x)$

3 a $\frac{1}{\ln 1.5}$ b $\frac{1}{2}(3 + \ln 20)$ c $50(1 - \ln 1.5)$

Exercise 9.1

1 a $\frac{x}{3y}$ b $\frac{x^2}{y^2}$ c $\frac{3-2x}{2y}$

d $\frac{1}{3y(y+2)}$ e $-\frac{y}{x}$ f $-\frac{1+y}{1+x}$

g $-\frac{2x+y}{x+2y}$ h $-\frac{x(1+y^2)}{y(1+x^2)}$ i $\frac{x(3-y^2)}{y(x^2-2)}$

j -1 k $\frac{y^2}{x^2}$ l $-\frac{3y^3}{4x^3}$

m $\tan x \tan y$ n $\cot^2 y$

2 a $\frac{1}{2xy}$ b $-\frac{y^2}{x(1 - y\ln x)}$ c $-\sqrt{\frac{y}{x}}$

d $\frac{\cos x}{\cos(x+y)} - 1$ e $\frac{3}{2}\tan 3x \tan 2y$ f $e^{-(x+y)} - 1$

g $-\frac{ye^x + e^y}{xe^y + e^x}$ h $\frac{y(1 - e^x \ln y)}{e^x}$

3 a $\frac{4}{3}$ b -3 c $-\frac{3}{7}$

d 0 e 2 f -2

g $y = x + 3$ h $x = 3$

i $7y = 33 - 13x$ j $y = 1 - \frac{x}{2}$

4 a $10y = 7x - 54, 7y = -10x - 8$

b $y = x, y = \frac{\pi}{2} - x$ c $x = 2, y = \frac{1}{2}$

d $y = -x, y = x$

5 $y = x + 1$

6 $26\frac{2}{3}$

7 a $5y + 8x = 21$ b $4\frac{1}{5}$

8 (2,3)

9 (6,14)

10 $\left(1\frac{1}{2}, 0\right)$, 63.4°

11 (0,0), (4,2), 90°, 31°

12 $y = \pm 3$

13 $\frac{1}{3}, 1$

14 a $\left(4, \frac{1}{3}\right)$, (4, 1) b $\left(1, -\frac{1}{2}\right)$, (1, −2)

15 a (−5,−2) min; (−5,−3) max

b $\left(-3, \frac{3}{2}\right)$ min; (−3,−3) max

16 (1,1) and (1,−1)

17 a maximum of 3, minimum of −1

b maximum of −1, minimum of 3

18 a $\sqrt[3]{2}$ b $\sqrt[3]{4}$, max

19 $\frac{dy}{dx} = \frac{x^2 + 8x - 4}{(x+4)^2}$

Exercise 9.2

1 $x = \pm 1$

2 $y = x + 2\sqrt{2},\ y = x - 2\sqrt{2},$

$y = -x + 2\sqrt{2},\ y = -x - 2\sqrt{2}$, area = 16

3 (−0.618, ±0.300)

5 $\left(\frac{1}{5}, -\frac{1}{10}\right)$

Exercise 9.3

1 a $\ln 3 \times 3^x$ b $2\ln 3 \times 3^{2x-1}$

c $20\ln 3 \times 3^{5x+2}$ d $-2\ln 3 \times 3^{-2x}$

e $2\ln 10 \times 10^{2x+5}$ f $-6\ln 5 \times 5^{1-2x}$

2 $\frac{1}{2}\ln 2\left(2^{\frac{1}{2}x+2} + 6 \times 2^x\right)$, $16\ln 2$

3 a $y = 6.59x - 0.59$

$y = -0.15x + 6.15$

b $y = 4.16x + 6$

$y = -0.24x + 6$

4 a 40, 83.2 b 5.28, 1.46

c 13.4, 20.1 d 0.55, 0.22

5 a 3.21, −0.71 b 5.80, −2.55

c 1.98, −0.020 d 54.6, −109

6 a 5, 37 b 3.19 c 2.72

7 a 126 000, 0.0314

b $3956e^{0.0314t}$, 5415 people/year

c 323 200, 6.7%

8 a 150, 100 b 200, 0.03 cells/hour

9 a 20, 0.0575

b i 12 h ii 40 h

c $1.15\,\text{g h}^{-1}$, $0.863\,\text{g h}^{-1}$

10 a 70, 0.154

b 10.8° C/min, 4.99° C/min

c $T = 10 + 70e^{-0.154t}$ d 5.5 min

11 Both models give the same estimate of 3 360 000, because $P_0a^t = P_0e^{kt}$ if k is defined by $a = e^k$

Exercise 9.4

1 $0.2\,\text{m}^2$ per min

2 a $47.1\,\text{cm}^2$ per sec b 3.14 cm per sec

3 a $32.4\,\text{mm}^3$ per min b $21.6\,\text{mm}^2$ per min

4 $2\,\text{cm}^3$ per hour

5 a $40.2\,\text{mm}^3$/sec b $15.1\,\text{mm}^2$/sec

6 $15\,\text{cm}^3$/sec

7 a 0.00 111 cm/sec b $0.333\,\text{cm}^2$/sec

8 0.0265 cm/min

9 $126\,\text{mm}^2$ per sec

10 0.305 cm/sec

11 9 m/sec

12 $-0.1\,\text{N m}^{-2}$/sec

13 $\frac{-k}{\pi r^2 L^2}\frac{dL}{dt}$

where k is Boyle's constant. Assumes temperature remains constant.

Review 9

1 a $-\frac{3x}{4y}$ b $-\frac{2x+3y}{3x+2y}$ c $\frac{2y^3}{3x^3}$

d $\frac{2}{3}\tan 2x \tan 3y$ e $\frac{1}{3xy^2}$ f $\frac{y^2e^x - 2xe^y}{x^2e^y - 2ye^x}$

2 $\frac{3y^2 + 4xy - 3x^2}{3y^2 - 6xy - 2x^2}$ $-\frac{17}{7}$

3 a $3y - 2x = 2$ b $7y - 5x = 8$ c $y + 6x = 3$

5 $x - 2y + 2 = 0$

6 b (2,3), (2,5) c $(2 \pm 4\sqrt{2}, -1)$

7 $4y + \sqrt{3}x = 8\sqrt{3}$

8 a $y = 3x - 7$ b $3y = 2x - 7$

9 $y = \frac{b}{a}x\sqrt{2} - b$

10 $\left(\frac{1}{2}, \frac{3\sqrt{3}}{2}\right), \left(\frac{1}{2}, \frac{-3\sqrt{3}}{2}\right)$

11 a $54y + x = 27$ b $t = -6$

12 a $4y = x + 15$

13 a (0,2), (0,–2) b $x = 0$

14 a i $4^x \ln 4$ ii $(2^{4x} + 2^{2x+2})\ln 2$

b i $y = 5 + 2x\ln 2$ ii $y = 4 + 5x\ln 2$

15 a 13, 50 b 12.3 per sec

c 5.8 per sec d $t = 1.5$

16 a £1975 b –£801 per year

c rate of depreciation

Revision 3

1 a $\frac{2}{x-2} - \frac{1}{x+1}$ b $-\frac{3}{x} + \frac{2}{x-1} + \frac{1}{x+1}$

c $\frac{6}{x-3} - \frac{4}{x-2}$ d $\frac{2}{x} - \frac{1}{x+1} - \frac{3}{(x+1)^2}$

e $-\frac{4}{x} - \frac{2}{x^2} + \frac{8}{2x-1}$

f $\frac{2}{5(1-2x)} + \frac{1}{5(x-3)} - \frac{1}{(x-3)^2}$

2 $A = 1, B = -3, C = 2, D = 1$

3 a $1 - \frac{3}{2(x+1)} + \frac{3}{2(x-1)}$ b $1 - \frac{1}{3(x-1)} - \frac{2}{3(x+2)}$

c $1 + \frac{2}{x} - \frac{2}{x-1} + \frac{1}{(x-1)^2}$ d $x + 2 - \frac{1}{2x} + \frac{9}{2(x-2)}$

e $2 - \frac{2}{x} - \frac{1}{x^2} + \frac{6}{2x-1}$

f $1 - \frac{4}{3(x+1)} + \frac{1}{(x+1)^2} - \frac{1}{3(2x+1)}$

4 $p = 1, q = 0, r = 4, s = 3, t = 9$

5 a $x = 21$ b $k = -\frac{2}{3}, -\frac{4}{3}$

6 a 3 b $y = \frac{x^2}{3} - 1$

7 a $\left(\frac{1}{3}, 0\right)$ b (0,–1) c (3,2), (0,–1)

8 a $2y + x = 11$ b (–1,6)

9 a (7,3), (12,4) b 17.7

10 a $t = 0, 3, -3$ b 64.8, 129.6

11 a $P(0, 4.25), Q(2,2)$ b $5\frac{2}{3}$ c $4\frac{1}{2}$

12 a $t = \frac{\pi}{3}$ d $\frac{64}{3} - 8\sqrt{3}$

13 a $1 - 3x + 9x^2 - 27x^3$, $|x| < \frac{1}{3}$

b $1 + \frac{5}{2}x - \frac{25}{8}x^2 + \frac{125}{16}x^3$, $|x| < \frac{1}{5}$

c $3 + \frac{1}{3}x - \frac{1}{54}x^2 + \frac{1}{486}x^3$, $|x| < \frac{1}{2}$

d $x + \frac{1}{8}x^2 + \frac{3}{128}x^3$, $|x| < 4$

e $\frac{1}{4} - \frac{3}{16}x + \frac{9}{64}x^2 - \frac{27}{256}x^3$, $|x| < \frac{4}{3}$

f $\frac{1}{2} + \frac{1}{16}x + \frac{3}{256}x^2 + \frac{5}{2048}x^3$, $|x| < 4$

14 a $1 - \frac{3}{2}x - \frac{9}{8}x^2$, $|x| < 1$

b $1 - \frac{1}{2}x + \frac{15}{8}x^2$, $|x| < \frac{1}{3}$

15 $1 - \frac{3}{2x} - \frac{9}{8x^2} - \frac{27}{16x^3}$, $|x| > 3$

16 $a = \pm 8$, $2 + 2x - x^2$, $2 - 2x - x^2$

17 a $\frac{1}{1-2x} + \frac{1}{1+x}$, $2 + x + 5x^2 + 7x^3$, $|x| < \frac{1}{2}$

b $\frac{1}{2+x} + \frac{1}{8-3x}$, $\frac{3}{2} + \frac{11}{4}x + \frac{73}{8}x^2 + \frac{431}{16}x^3$, $|x| < \frac{1}{3}$

18 a $A = -\frac{3}{2}, B = \frac{1}{2}$ b $-1 - x + 4x^3$

19 a $\frac{5y - 2x}{2y - 5x}$ b $\frac{e^y}{3y^2 - xe^y}$ c $\frac{y - \sin y}{x(\cos y - 1)}$

d $\cot x \cot y$ e $\sqrt{\frac{y}{x}}$ f $\frac{2xy - \sec^2(x+y)}{-x^2 + \sec^2(x+y)}$

20 a $y=\frac{1}{2}x-1$ b $y+x=2$

22 a $-\frac{2}{3}\tan\theta$ c $6\operatorname{cosec}2\alpha$ d $\frac{\pi}{4}, 6$

23 21.4

24 a 60 664 b 485 per year c 87 years

25 a 0.0637 m/min b $0.8\,\text{m}^2/\text{min}$

26 a 0.002 55 cm/s, b $0.48\,\text{cm}^3/\text{s}$

Before you start Answers

Chapter 10

1 a $2x\ln x+x$ b $e^x(3x^2+x^3)$
c $e^x(\tan x+\sec^2 x)$ d $\frac{2x^2+1}{\sqrt{x^2+1}}$
e $\frac{1-\ln x}{x^2}$ f $\cot x$

3 a $\frac{1}{2(x-5)}+\frac{1}{2(x+5)}$ b $-\frac{3}{x}+\frac{2}{x-1}+\frac{1}{x+2}$
c $\frac{1}{x}-\frac{1}{x-1}+\frac{1}{(x-1)^2}$ d $1-\frac{3}{2x}+\frac{7}{2(x-2)}$

(Arbitrary constants are omitted from answers.)

Exercise 10.1

1 a 1.45 b 8.72 c 1.17
d 0.137 e 1.57 f 2.12

2 a 0.9943 b 1 c 0.57%
d convex graph

3 a 0.9185 b 0.9116

4 55.1, $\pi\approx 3.06$

5 $I=2.30$, 18.4

6 a 50.5 b 51.2 c 1.4%

7 Area $=\pi ab$ where a, b are the lengths of the semi-axes. For a circle, $a=b=r$

Exercise 10.2

1 a 2 b $\frac{5}{2}\sqrt{2}$ c $\frac{e^2-1}{e}$
d $\ln 2$ e $-\sqrt{2}$ f $3+\ln 4$
g $\frac{27}{2}+\ln 4$ h $\frac{\pi}{3}+\sqrt{3}$

2 a $\frac{1}{2}\tan\theta+c$ b $\frac{1}{2}x^2-2\ln x+c$
c $5\sin x+3\cos x+c$

3 a e^2-1 b $\ln 3$

4 a $\left(\frac{\pi}{4},\frac{1}{\sqrt{2}}\right)$ b $2-\sqrt{2}$ c $\sqrt{2}-1$

5 $e^2-e-\ln 2\approx 3.98$

6 e^4

7 $\frac{10}{\pi}$

Exercise 10.3

1 a $\frac{1}{5}\sin 5x+c$ b $-\frac{1}{4}\cos 4x+c$
c $\frac{1}{3}\tan 3x+c$ d $2\sin\frac{1}{2}x+c$
e $-\frac{1}{4}\cot 4x+c$ f $\frac{1}{4}e^{4x-3}+c$
g $\frac{1}{15}(3x+2)^5+c$ h $\frac{1}{3}\ln|\sec 3x|+c$
i $\frac{1}{3}\ln|3x-1|+c$ j $-\frac{1}{3(3x-1)}+c$
k $\frac{5}{2}\ln\left|\sin\frac{2}{5}x\right|+c$ l $\frac{1}{2}\sin(2x+3)+c$
m $-\frac{1}{4}\tan(4x+1)+c$ n $\frac{1}{4}\sec 4x+c$
o $\frac{1}{4}\ln|\sec 4x+\tan 4x|+c$
p $-\frac{1}{4}\ln|\operatorname{cosec}4x+\cot 4x|+c$
q $-\frac{1}{2}e^{-2x}+c$ r $\frac{1}{3}\sin 3x-3\cos\frac{1}{3}x+c$
s $-\frac{1}{2}\operatorname{cosec}2x+c$ t $\frac{1}{2}e^{2x}-2x-\frac{1}{2}e^{-2x}+c$

2 a 0 b $\frac{1}{2}$ c $\sqrt{2}$
d $\frac{2\sqrt{2}}{3}\left(\sqrt{2}-1\right)$ e $\frac{e^4-1}{2e}$ f $\frac{1}{3}\ln 7$

3 a $-\frac{1}{x}+\tan x+c$ b $\ln x-\cot x+c$
c $\frac{1}{2}\left(e^{2x-1}-e^{1-2x}\right)+c$ d $\frac{1}{3}\sec 3x+c$
e $-\frac{1}{3}\operatorname{cosec}3x$

4 $\frac{10}{\pi}$

Exercise 10.4

1 a $\ln|x^3-1|+c$ b $\frac{1}{6}\left(x^3-1\right)^6+c$
c $\ln|x^2+3x-1|+c$ d $\frac{1}{5}\left(x^2+3x-1\right)^5+c$
e $\frac{1}{2}\ln|x^2-4x+1|+c$ f $\frac{1}{8}(x^2-4x+1)^4+c$
g $\ln|\sin x+1|+c$ h $\frac{1}{2}\sin(x^2+1)+c$
i $\frac{1}{3}(x^2-1)^{\frac{3}{2}}+c$ j $\sqrt{x^2-1}+c$
k $\frac{1}{2}\ln|x^2-1|+c$ l $\frac{1}{2}e^{x^2}+c$
m $-\frac{1}{2}e^{-x^2}+c$ n $\frac{1}{3}(x^2+2x+3)^{\frac{3}{2}}+c$
o $\frac{1}{2}\ln|x^2+2x+3|+c$ p $e^{\sin x}+c$
q $\frac{1}{3}e^{x^3}+c$ r $\ln|\ln x|+c$

3 $\frac{1}{2}\ln 2$

4 $x+\ln|x-1|+c$
$\frac{1}{n}x^n+\frac{1}{n-1}x^{n-1}+\cdots+x+\ln|x-1|+c$

Exercise 10.5

1 a $\frac{1}{15}(1+x^3)^5+c$ b $\frac{2}{9}(1+x^3)^{\frac{3}{2}}+c$
c $\frac{1}{5}\sin^5 x+c$ d $\frac{1}{2}\tan^2 x+\tan x+c$
e $-\frac{1}{9(3x-1)^3}+c$ f $\ln(x+5)+\frac{5}{x+5}+c$
g $2\ln(x-1)+(x-1)-\frac{1}{x-1}+c$
h $\frac{-1}{4(e^x+2)^4}+\frac{2}{5(e^x+2)^5}+c$

C4

2 **a** $\frac{2}{3}(1+e^x)^{\frac{3}{2}}+c$ **b** $\frac{2}{3}(1-\cos x)^{\frac{3}{2}}+c$

c $\frac{2}{3}(x+2)\sqrt{x-1}+c$ **d** $\frac{1}{6}(1+x^4)^{\frac{3}{2}}+c$

e $\frac{1}{3}(\ln x)^3+c$ **f** $2\sqrt{x}-4\ln(\sqrt{x}+2)+c$

g $2\sqrt{x+1}-\ln(1+\sqrt{x+1})+c$

h $-\sqrt{1-x^2}+c$ **i** $\sin^{-1}(x)+c$

j $\frac{1}{3}\sec^3 x+c$ **k** $\frac{1}{2}\ln\left(\frac{e^x-1}{e^x+1}\right)+c$

l $-\frac{\sqrt{4-x^2}}{4x}+c$ **m** $2\sqrt{x+1}+4\ln\left(\frac{\sqrt{x+1}-1}{\sqrt{x+1}+1}\right)+c$

3 **a** $2(e-1)$ **b** $\frac{9}{20}$ **c** $\sqrt{2}-1$

d 23.2 **e** 2 **f** $\frac{31}{162}$

g $\frac{e-1}{2e}$ **h** $\frac{352}{15}$

4 **a** (2,0) **b** $\frac{1}{30}$

5 **a** $(0,0), (4,0), \frac{128}{15}$ **b** $(0,0), (2,0), \frac{4}{3}$

c $(0,0), \left(\frac{\pi}{2},0\right), (\pi,0); \frac{5}{2}$

6 **a, d, e, g, h on sight**

a $\frac{1}{5}(x^3+1)^5+c$ **b** $\frac{(x^2+1)^4}{4}-\frac{(x^2+1)^2}{2}+c$

c $\frac{2}{15}(3x-2)(x+1)^{\frac{3}{2}}+c$ **d** $\frac{2}{3}(1+\tan x)^{\frac{3}{2}}+c$

e $(x^2+1)^{\frac{1}{2}}+c$ **f** $\frac{2}{15}\sqrt{x+1}(3x^2-4x+8)+c$

g $\frac{1}{2}e^{x^2+1}+c$ **h** $\frac{2}{3}(e^x-1)^{\frac{3}{2}}+c$

Exercise 10.6

1 **a** $\frac{1}{3}\sin 3x+c$ **b** $-\frac{1}{4}\cos 4x+c$

c $2\sin\left(\frac{1}{2}x\right)+c$ **d** $-\frac{2}{3}\cos\left(\frac{3}{2}x\right)+c$

e $\frac{1}{2}\sin(2x+1)+c$ **f** $-\frac{1}{3}\cos(3x-2)+c$

g $\frac{1}{4}\tan(4x)+c$ **h** $\frac{1}{2}\tan(2x-3)+c$

2 **a** $\frac{1}{3}\tan 3x+c$ **b** $-\frac{1}{3}\cos 3x+c$ **c** $-\frac{1}{4}\cot 4x+c$

d $\tan x+c$ **e** $-\frac{1}{2}\ln|\operatorname{cosec} x+\cot x|+c$

f $\sec x+c$ **g** $\ln|\sec x+\tan x|+c$

h $x+\tan x+2\ln|\sec x+\tan x|+c$

i $\frac{1}{3}(\sin 3x-\cos 3x)+c$

j $-\cot x-4\ln|\operatorname{cosec} x+\cot x|+4x+c$

k $-2\cot x-2\operatorname{cosec} x-x+c$ **l** $-\frac{1}{4}\cot 2x+c$

3 **a** $\frac{1}{2}x+\frac{1}{4}\sin 2x+c$ **b** $\frac{1}{2}x-\frac{1}{12}\sin 6x+c$

c $\frac{1}{2}(x+\sin x)+c$ **d** $\frac{1}{2}x-\frac{1}{12}\sin(6x+2)+c$

e $\tan x-2\ln|\sec x|+c$ **f** $\frac{3}{2}x-2\cos x-\frac{1}{4}\sin 2x+c$

g $\frac{1}{5}\tan^5 x+c$ **h** $\sin x-\frac{1}{3}\sin^3 x+c$

i $\frac{1}{3}\cos^3 x-\cos x+c$ **j** $\frac{3}{8}x-\frac{1}{4}\sin 2x+\frac{1}{32}\sin 4x+c$

k $\sin x-\sin^3 x+\frac{3}{5}\sin^5 x-\frac{1}{7}\sin^7 x+c$

l $-\ln|\sec x|+\frac{1}{2}\tan^2 x+c$

4 **a** $-\frac{1}{4}\cos 2x+c$ **b** $-\frac{1}{8}\cos 4x+c$

c $-\frac{1}{2}\cos x+c$ **d** $-\frac{1}{12}\cos 6x+c$

e $-x+\frac{1}{3}\tan 3x+c$ **f** $-x-\frac{1}{3}\cot 3x+c$

g $-x-\frac{1}{2}\cot 2x+c$ **h** $2\tan\left(\frac{1}{2}x\right)+c$

5 $-\frac{1}{16}\cos 8x-\frac{1}{8}\cos 4x+c$

6 **a** $-\frac{1}{10}\cos 5x-\frac{1}{6}\cos 3x+c$ **b** $\frac{1}{2}\sin x+\frac{1}{18}\sin 9x+c$

c $\frac{1}{2}\sin x-\frac{1}{10}\sin 5x+c$

7 **a** $\frac{3}{5}$ **b** $\frac{1}{5}$ **c** $\frac{\sqrt{2}}{5}$

8 **a** $\frac{3\pi}{2}$ **b** $\ln\left(\frac{\sqrt{2}+1}{\sqrt{2}-1}\right)$ **c** 0

9 $\frac{1}{3}$

11 **a** $\frac{1}{2}$ **b** $\frac{8}{3}$

12 $\frac{\pi}{2}+\frac{1}{6}$

13 **a** odd function **b** even function

Exercise 10.7

1 **a** $2\ln\left|\frac{x}{x+2}\right|+c$ **b** $2\ln|x-3|-\ln|x+1|+c$

c $\ln|(2x+1)(x+1)|+c$ **d** $\ln\left|\sqrt{(x-1)(x+3)^3}\right|+c$

e $\ln\left|\sqrt{\frac{2x-1}{2x+1}}\right|+c$ **f** $4\ln\left|\frac{x^2-1}{x^2}\right|+c$

g $\frac{1}{x}+\ln\left|\frac{x-1}{x}\right|+c$ **h** $\ln\left|\frac{x}{x-1}\right|-\frac{1}{x-1}+c$

i i $\ln\left|\frac{x-2}{x+1}\right|+\frac{3}{x+1}+c$

2 **a** $\ln\left(\frac{8}{5}\right)$ **b** $\ln 2+\frac{3}{2}$ **c** $\ln\left(\frac{3}{7}\right)$

4 $\frac{5}{2}\ln 2-\frac{3}{2}\ln 3$

5 **a** $x+\frac{3}{2}\ln\left|\frac{x-3}{x+3}\right|+c$ **b** $x+\ln\left|\frac{x-1}{x+1}\right|+c$

c $x+2\ln\left|\frac{x+1}{x}\right|+c$

d $\frac{1}{2}x^2+2x+\frac{37}{5}\ln|x-3|+\frac{3}{5}\ln|x+2|+c$

6 $\ln|\sqrt{x^2-1}|$

C4

Exercise 10.8

1 a $\sin x - x\cos x + c$ b $xe^x - e^x + c$

c $\frac{1}{2}x^2 \ln|x| - \frac{1}{4}x^4 + c$ d $\frac{1}{4}e^{2x}(2x-1) + c$

e $x\tan x - \ln|\sec x| + c$ f $-e^{-x}(1+x) + c$

g $\frac{1}{4}\sin 2x - \frac{1}{2}x\cos 2x + c$

h $-\frac{1}{2x^2}\ln|x| - \frac{1}{4x^2} + c$ i $\frac{1}{4}e^{2x+1}(2x-1) + c$

j $\frac{1}{9}x^3(3\ln|x| - 1) + c$ k $\frac{2}{3}x^{\frac{3}{2}}\left(\ln|x| - \frac{2}{3}\right) + c$

l $x\sin\left(x - \frac{\pi}{4}\right) + \cos\left(x - \frac{\pi}{4}\right) + c$

m $\sin\left(x + \frac{\pi}{6}\right) - x\cos\left(x + \frac{\pi}{6}\right) + c$

n $\frac{1}{2}x^2 + x\tan x - \ln|\sec x| + c$

o $x(\ln|x|)^2 - 2x\ln|x| + 2x + c$

2 a $\frac{1}{n^2}(nx\sin(nx) + \cos(nx)) + c$

b $\frac{1}{n^2}e^{nx}(nx-1) + c$

c $\frac{x^{n+1}}{(n+1)^2}((n+1)\ln x - 1) + c$

d $-\frac{\cos nx}{n}(\ln|\sec nx| + 1) + c$

3 a $\frac{\pi}{2} - 1$ b $4\ln 2 - \frac{15}{16}$ c $\ln\frac{27}{4} - 1$

d $\log_{10} 4 - \frac{3\log_{10} e}{4}$ e $3\log_{10}4 - 2\log_{10}e$

f $\ln\sqrt{2} + \frac{\pi}{4} - 3\pi^2$ g $\frac{1}{2}$ h $\frac{1}{2}\left(e^{\frac{\pi}{2}} + 1\right)$

4 $\pi, 3\pi$

5 $\left(1, \frac{1}{e}\right), 1 - \frac{6}{e^5}$

6 a $-\frac{1}{e}$ b 1

7 a $\frac{1}{30}(5x-1)(1+x)^5 + c$

b $e^x(1+x^2) + c$

c $\frac{2}{15}(3x+2)(x-1)^{\frac{3}{2}} + c$

8 a $e^x(x^2 - 2x + 2) + c$

b $-x^2\cos x + 2x\sin x + 2\cos x + c$

c $\frac{1}{2}e^{2x}\left(x^2 - x + \frac{1}{2}\right) + c$

d $-\frac{1}{27}e^{-3x}(9x^2 + 6x + 2) + c$

e $\frac{1}{4}e^{2x}(\cos 2x + \sin 2x) + c$

f $\frac{1}{3}x^2\sin 3x + \frac{2}{9}x\cos 3x - \frac{2}{27}\sin 3x + c$

g $\frac{1}{13}e^{3x}(3\sin 2x - 2\cos 2x) + c$

9 a $\frac{1}{5}(e^{\pi} - 2)$ b $\frac{e^2 - 5}{2e^2}$ c $2(\ln 2)^2 - 2\ln 2 + \frac{3}{4}$

10 a $\frac{1}{e^2}$ b $\frac{1}{4} - \frac{13}{4e^4}$

11 $-\frac{1}{3}\cos x\sin^2 x - \frac{2}{3}\cos x$

$-\frac{1}{4}\cos x\sin^3 x - \frac{3}{8}\cos x\sin x + \frac{3}{8}x$

$\frac{5\pi}{16}$

Exercise 10.9

1 $\frac{1}{4}x^4 - x^3 + c$ 2 $\frac{1}{4}x^4 - 2x^3 + \frac{9}{2}x^2 + c$

3 $\frac{1}{7}(x-6)^7 + c$ 4 $\frac{1}{8}(x^3-2)^8 + c$

5 $\frac{2}{9}(x^3+1)^{\frac{3}{2}} + c$ 6 $\frac{1}{3}\ln|x^3+1| + c$

7 $\frac{1}{3}\sin(2x^3) + c$ 8 $\frac{3}{2}x + \frac{3}{4}\sin 2x + c$

9 $x - \frac{1}{6}\sin 6x + c$ 10 $4\tan x + c$

11 $\frac{5}{4}\sin 4x + c$ 12 $\frac{1}{4}\sin 2x - \frac{1}{2}x\cos 2x + c$

13 $\frac{1}{9}x^3(3\ln(2x) - 1) + c$ 14 $\frac{3}{2}\ln(x^2+7) + c$

15 $\frac{1}{30}(5x-1)(x+1)^5 + c$ 16 $\frac{1}{12}(x^2+2)^6 + c$

17 $\frac{1}{4}e^{2x^2} + c$ 18 $\frac{1}{4}e^{2x}(2x-1) + c$

19 $-2x + \tan 2x + c$ 20 $\frac{1}{2}\sin 2x + c$

21 $2\ln\left|\sin\left(\frac{1}{2}x\right)\right| + c$ 22 $x\ln|2x| - x + c$

23 $2x(\ln|x| - 1) + c$ 24 $x + \ln|x+1| + c$

25 $\frac{1}{4}x^4 + x^3 + \frac{1}{2}x^2 + 3x + c$ 26 $\frac{1}{2}\sin\left(2x + \frac{\pi}{2}\right) + c$

27 $\frac{1}{2}\ln|\sec(2x - \pi)| + c$ 28 $\frac{1}{4}e^{2x}(\sin 2x - \cos 2x) + c$

29 $-\frac{1}{8}\cos 4x + c$ 30 $\frac{1}{4}\cos 2x - \frac{1}{12}\cos 6x + c$

31 $\ln\left|\frac{x-4}{x+1}\right| + c$ 32 $\frac{1}{2}\ln\left|\frac{x-2}{x}\right| + c$

33 $\frac{2}{3}\sqrt{x-2}(x+4) + c$ 34 $\frac{2}{3}(e^x+1)^{\frac{3}{2}} + c$

35 $\frac{1}{2}\ln(e^{2x}+1) + c$ 36 $-\frac{1}{3}\sin^{-3}x + c$

37 $-\frac{1}{14}\cos 7x - \frac{1}{6}\cos 3x + c$ 38 $\frac{1}{8}\sin^8 x + c$

39 $\frac{1}{6}\ln\left|\frac{x-3}{x+3}\right| + c$ 40 $-\frac{2}{3}\sqrt{\cos^3 x} + c$

41 a $\frac{\pi}{8}$ b $\ln\frac{3}{2}$

c $\ln(\sqrt{2}+1)$ d $\frac{1}{4}$

42 a $-\frac{1}{2\tan^2 x} + c$ b $\frac{1}{3}\tan^3 x - \tan x + x + c$

c $\frac{1}{3}e^{x^3} + c$ d $\sin x - \frac{2}{3}\sin^3 x + \frac{1}{5}\sin^5 x + c$

e $\frac{1}{4}\left(\frac{3}{2}x - \sin 2x + \frac{1}{8}\sin 4x\right) + c$

f $2\sqrt{\ln|x|} + c$ g $\ln|3 + \sec x| + c$ h $\frac{3^x}{\ln 3} + c$

43 a negative b positive c zero

Exercise 10.10

1 a $\frac{279}{5}\pi$ b 4π c $\frac{32}{5}\pi$

2 a $\frac{38}{15}\pi$ b $\pi\ln\left(\frac{3}{2}\right)$ c $\frac{8}{15}\pi$

d $\frac{\pi^2}{4}$ e π f $\frac{\pi}{4}e^2(3e^2-1)$

g $\frac{\pi}{6}(7-12\ln 1.5)$ h $\frac{\pi}{4}(18\ln 3-8\ln 2-5)$

3 $m=\frac{r}{h}$

4 $\frac{64}{15}\pi$

5 $\frac{1}{8}\pi(4-\pi)$

6 a $\frac{243\pi}{5}$ b $(-4,16)$, $(3,9)$, $868\frac{14}{15}\pi$

7 a $(-2,-4)$, $(2,4)$ b $\frac{80}{3}\pi$

8 $\frac{5\pi}{14}$

9 $\frac{\pi}{2}$

10 a $(0,4)$, $(1,4)$ b 1.3π

11 $\frac{\pi}{2}(3(\ln 3)^2-6\ln 3+4)$

12 $\frac{1946}{3}\pi$

13 a $\frac{381}{7}\pi$ b $\frac{197}{2}\pi$ c 16π

14 $\frac{21\pi}{8}$

15 $\frac{\pi}{2}\left(1-\frac{3}{e^2}\right)$

16 $2\pi\left(\ln\frac{3}{2}-\frac{1}{6}\right)$

17 $\frac{4}{3}\pi ab^2$

18 $\frac{8\pi}{15}$

19 $\pi\left(\sqrt{3}-1-\frac{\pi}{12}\right)$

20 Torus (doughnut), $4\pi^2$

Review 10

1 a 7.40 b 1.47 c 4.04 d 0.477

2 a 1.008 b 1 c 0.8% d concave curve

3 a $0.5\ln 1.5$, $1.5\ln 2.5$ b i 1.792 ii 1.684

4 a $\frac{1}{3}\sin 3x$ b $\frac{1}{5}\tan 5x$ c $-4\cos\left(\frac{x}{4}\right)$

d $\frac{1}{30}(6x+1)^5$ e $0.5\ln|\sin 2x|$ f $0.25\ln|4x+3|$

g $-\frac{1}{4(4x+3)}$ h $0.25e^{4x}$

i $0.5e^{2x}+2e^x+x$ j $-2\cot(0.5x+1)$

k $0.5\ln|\sec 2x+\tan 2x|$ l $\frac{1}{2}\sec 2x$

5 a $\frac{1}{6}(x^2+3)^6$ b $0.5\ln|x^2+3|$ c $0.5\sin(x^2+1)$

d $\frac{1}{7}\sin^7 x$ e $0.5\ln|x^2-2x-1|$

f $\ln|1+\tan x|$ g $-0.5e^{-x^2}$ h $\frac{2}{9}(1+x^3)^{\frac{3}{2}}$

6 a $-\frac{1}{(x+3)}$ b $\frac{1}{5}(x-4)^5+\frac{3}{4}(x-4)^4$

c $\frac{2}{3}(e^x+3)^{\frac{3}{2}}$ d $\frac{2}{3}(x+1)^{\frac{3}{2}}-2(x+1)^{\frac{1}{2}}$

7 0.183

8 $\frac{1}{\sqrt{3}}$

9 a $\ln|\sin x|$ b $\tan x$ c $\ln|\sec x+\tan x|$

d $0.5\left(x+\frac{1}{6}\sin(6x)\right)$ e $0.5\left(x-\frac{1}{6}\sin 6x\right)$

f $3x-4\ln|\cos x|+\tan x$

g $\frac{1}{8}(3x+4\sin x+0.5\sin 2x)$

h $2\sin\left(\frac{x}{2}\right)-\frac{2}{3}\sin^3\left(\frac{x}{2}\right)$ i $-\cos x+\frac{2}{3}\cos^3 x-\frac{1}{5}\cos^5 x$

j $\frac{1}{8}(3x-\sin 4x+\frac{1}{8}\sin 8x)$ k $x-2\tan x$

l $0.5\left(x+2\sin\left(\frac{x}{2}\right)\right)$

10 a $\frac{1}{4}\sin 2x+\frac{1}{20}\sin 10x$ b 0.3

11 a $\ln\left|\frac{(x-3)^3}{(x-1)^2}\right|$ b $\ln\left((x+2)^2\sqrt{2x-1}\right)$

c $\frac{7}{16}\ln\left|\frac{x+1}{x-3}\right|-\frac{7}{4(x-3)}$

12 a 0.182 b 2.886

13 a $A=3, B=2$ b $4\ln 2-\frac{5}{4}\ln 3$

14 a $x\sin x+\cos x$ b $\frac{x}{3}\sin 3x+\frac{1}{9}\cos 3x$

c $\frac{1}{9}e^{3x}(3x-1)$ d $-\frac{x}{n}\cos nx+\frac{1}{n^2}\sin nx$

e $\frac{1}{4}e^{2x}(2x^2-2x+1)$ f $\frac{1}{5}e^{2x}(2\sin x-\cos x)$

g $\frac{x^4}{16}(4\ln|x|-1)$

15 a 0.1517 b 0.7183 c 1.446 d 3.196

16 a $0.5x\sin 2x+0.25\cos 2x$

b $\frac{1}{8}(2x^2+2x\sin 2x+\cos 2x)$

17 a $\frac{1}{6}x^6-\frac{1}{4}x^4$ b $\frac{1}{4}x^4-\frac{2}{3}x^3-\frac{1}{2}x^2+2x$

c $\frac{1}{7}(x-5)^7$ d $\frac{1}{6}(x^2-2)^6$

e $\left(\frac{2}{5}x+\frac{8}{15}\right)(x-2)^{\frac{3}{2}}$ f $0.5\ln|x^2-1|$

g $0.5\ln\left|\frac{x-1}{x+1}\right|$ h $\ln\left|\frac{x-1}{x}\right|$ i $e^{\frac{x^2}{2}}$

j $2e^{\frac{x}{2}}(x-2)$ k $0.25x^2(2\ln|3x|-1)$

l $\frac{1}{9}x^3(3\ln|3x|-1)$

18 a $\ln|\sin x|$ b $\ln|\sec x+\tan x|$

c $\frac{1}{4}\sin 4x-\frac{1}{12}\sin^3 4x$

d $\frac{1}{16}(6x-8\sin x+\sin 2x)$ e $\frac{3}{2}x+\frac{1}{4}\sin 2x$

f $\frac{1}{8}\sin 4x-\frac{1}{20}\sin 10x$ g $\frac{1}{10}e^x(\sin 3x-3\cos 3x)$

h $\frac{1}{32}(8x^2\cos 4x+4x\cos 4x-\sin 4x)$

19 a 0.3096 b 1.071 c 0.125

20 a $\frac{53}{15}\pi$ b $\frac{\pi^2 + 2\pi}{8}$

21 $V = 10\frac{8}{15}\pi$

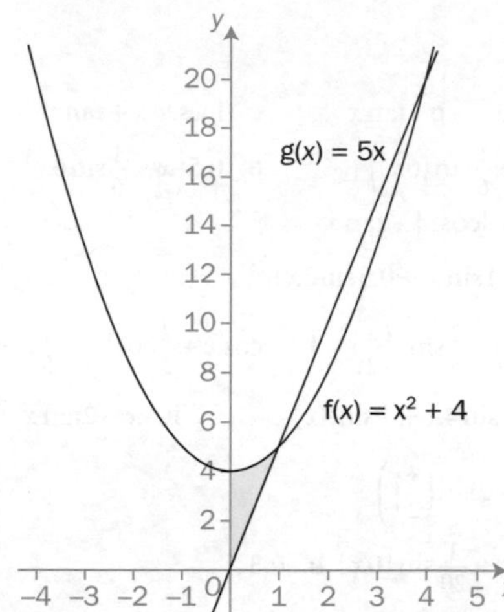

22 a $63\frac{3}{4}\pi$ b $\pi(2e - 1)$

Before you start Answers

Chapter 11

1 a $y = \frac{1}{3}x^3 - 5x + 2$ b $y = \frac{(x+1)^3}{3} - 4$

2 a $\frac{(2x+1)^4}{8}$ b $\frac{1}{2}\sin 2x + \frac{1}{3}\tan 3x$

c $\frac{1}{4}\ln\left|\frac{x-1}{x+3}\right|$ d $x\sin x + \cos x$

e $\frac{1}{2}\ln|x^2 + 3|$ f $x + \ln|x - 1|$

3 a $\frac{2}{1 - 2e^k}$ b $\sqrt{\frac{e^k}{1 - e^k}}$ c $\ln\left(\frac{k}{k^2 - 1}\right)$

Exercise 11.1

A and *c* are arbitrary constants

1 a $y = \frac{3}{5}x^5 + c$ b $y = \frac{1}{2}\sin 2x + c$

c $y = \frac{1}{2}\ln(x^2 + 1) + c$ d $y = x - \ln|x + 1| + c$

2 a $3y^3(4x + c) + 1 = 0$ b $e^{2y} = 2x + c$

c $y = Ae^x - 3$ d $Ae^x\cos y = 1$

3 a $y^2 - x^2 = c$ b $y = Ax$ c $y = Ae^{\frac{1}{2}x^2}$

d $y = Ax - 1$ e $y^2 = 2e^t + c$ f $y = Ate^t$

g $x = Ae^{-\frac{1}{t}} - 1$ h $At\cos x = 1$

4 a $y = \frac{1}{3}x^3 + \frac{1}{2}x^2 + x + 2$ b $\frac{1}{2}y^2 - y = x - 1$

c $3y = 2x^{\frac{3}{2}} + 10$ d $-\frac{1}{y} = \ln(1 + x) + \frac{1}{1 + x} - 2$

e $\sin y = \sin x - 1$ f $e^y = 3 - \frac{1}{e^x}$

5 a $y = \ln A(x + 1)$ b $e^{3y} = 3x + c$

c $y^2 - 4y = x^2 + 4x + c$ d $\sin y = Ae^x$

e $y^2 = x^2 + 2x + c$ f $\sin y = Ax$

g $3y = \frac{1}{2}x^2 - x + c$ h $y + \ln|y - 1| = c - \frac{1}{2}x$

i $y = \frac{Ax}{x + 1}$ j $y^2(2y + 3) = x^2(2x + 3) + c$

k $\sec y = A\sin x$ l $y(x + 1) = Ae^x$

m $y = \tan x - x + c$ n $x = y\sin y + \cos y + c$

o $c - 4e^{-x} = \ln\left|\frac{2 + y}{2 - y}\right|$

p $y = \frac{1}{2}\ln|\sec 2x| + \frac{1}{4}\ln|\sec 2x + \tan 2x| - \frac{1}{2}x + c$

6 $(y - 1)^2 = 8x + 9$

7 $k = 2$

8 $y^2 = 2(x\sin x + \cos x + 1)$

9 $x = \tan\theta$

10 $2y^3 - 3y^2 + 6y = 3x^2 + c$

11 a $\frac{dy}{dx} = x\frac{dz}{dx} + z$

Exercise 11.2

1 a $v = \frac{1}{200}t^2 + \frac{1}{10}t + 6$ b $13.5\ \mathrm{m\,s^{-1}}$

2 a 403 b 23 days

3 6.2 years

4 26.3 days

5 a $i = Ae^{-\frac{R}{L}t}$ b $\frac{L}{R}\ln 100$

6 a 2.56 m b 50 hours

7 c 10 760 years

8 b Further 12 mins c 30.1°C

9 b 2 cm

10 a $y = \frac{3(e^t - 1)}{3e^t - 1}$ b 0.906 grams c 1 gram

11 Number of shakes to reduce to 10 pins = $\frac{n \times \ln 10}{\ln 2}$

where n = number of shakes for the half-life

Review 11

1 a $y = \frac{1}{3}x^3 + 2\ln|x + c$ b $y = \ln|x^3 - 1| + c$

c $y = \ln|\sin x| + c$ d $y = \frac{1}{(c - x)}$

e $\frac{1}{3}\ln|3x - c|$ f $y = 1 - ke^{-x}$

2 a $y = \frac{2}{3}x^3 - x^2 + x - 8$ b $y^2 = e^{2x-1} + 1$

3 a $y^3 = 3(x + e^x) + c$ b $y = k\sqrt{x^2 + 1}$

c $\tan y = \sin x + c$ d $y - \ln|y + 1| = \frac{1}{3}x + c$

e $y = k\frac{x - 2}{x - 1}$ f $y\sin y + \cos y = x^2 + c$

4 a $3y^4 = 4x^3 + 44$ b $y = 3x - 1$

c $\frac{1}{2}y^2 - \ln|y| = -\cos(x - 1) + 3 - \ln 2$

5 $k = \pi$

6 $y = \frac{1}{1 - x\sin x - \cos x}$

7 a i $v = \sqrt{60t + 100}$ ii $v = \sqrt{700}$

b $v = 40$

8 a $\frac{dA}{dt} = kA$, $a = 2.5\pi 2^{0.1t}$, $10\pi\ \mathrm{m}^2$ b 29 days

9 a $k = -0.2\ln(0.8)$ b 6 minutes

10 a $\frac{dx}{dt} = kx$ b 11.6 minutes

Before you start Answers

Chapter 12

1 a i $\begin{pmatrix}1\\2\end{pmatrix}$ ii $\begin{pmatrix}1\\-5\end{pmatrix}$ iii $\begin{pmatrix}-1\\3\end{pmatrix}$

b square with vertices (9,1), (11,3), (9,3), (11,1)

2 a $\sqrt{41}$ b $\sqrt{74}, \sqrt{24}$

3 a $x=2, y=0, z=3$ b $x=1, y=-1, z=2$

Exercise 12.1

1 a 13 b 13 c $\sqrt{21}$ d $\sqrt{14}$

2 a $\sqrt{70}$ b $3\sqrt{5}$ c $\sqrt{141}$ d 0

3 a $\frac{3}{5}\mathbf{i}+\frac{4}{5}\mathbf{j}$ b $\frac{5}{13}\mathbf{i}-\frac{12}{13}\mathbf{j}$

c $\frac{1}{\sqrt{3}}\mathbf{i}+\frac{1}{\sqrt{3}}\mathbf{j}+\frac{1}{\sqrt{3}}\mathbf{k}$ d $\frac{2}{3}\mathbf{i}-\frac{1}{3}\mathbf{j}+\frac{2}{3}\mathbf{k}$

4 a $\begin{pmatrix}\frac{5}{2}\sqrt{3}\\\frac{5}{2}\end{pmatrix}$ b $\begin{pmatrix}-1\\\sqrt{3}\end{pmatrix}$

5 a $16\mathbf{i}+12\mathbf{j}$ b $\frac{10}{3}\mathbf{i}-\frac{10}{3}\mathbf{j}-\frac{5}{3}\mathbf{k}$

6 a ± 2 b $\pm\sqrt{6}$

7 a 33.7° b 34.5° c $\frac{1}{\sqrt{10}}\mathbf{i}-\frac{3}{\sqrt{10}}\mathbf{j}$

8 a $9\mathbf{i}+\mathbf{j}$ b $2\mathbf{i}+6\mathbf{j}-4\mathbf{k}$ c $-5\mathbf{i}-\mathbf{j}-2\mathbf{k}$

9 a $\lambda=6, \mu=-12$ b -4

10 a $\lambda=8, \mu=-6$ b $\lambda=-4, \mu=-3$

11 a $\begin{pmatrix}2\\4\end{pmatrix}$ b $\begin{pmatrix}3\\1\end{pmatrix}$ c $\begin{pmatrix}5\\5\end{pmatrix}$ d $\begin{pmatrix}-1\\3\end{pmatrix}$

12 a $\begin{pmatrix}5\\6\end{pmatrix}$ b $\begin{pmatrix}8\\4\end{pmatrix}$ c $\begin{pmatrix}7\\-1\end{pmatrix}$ d $\begin{pmatrix}3\\-3\end{pmatrix}$

14 $4\mathbf{i}+\mathbf{j}-3\mathbf{k}$

15 a $\mathbf{m}=\frac{\mathbf{a}+\mathbf{b}}{2}$ $\mathbf{p}=\frac{3\mathbf{a}+\mathbf{b}}{4}$ $\mathbf{q}=\frac{\mathbf{a}+3\mathbf{b}}{4}$

b Trisection $\frac{2\mathbf{a}+\mathbf{b}}{3}$ $\frac{\mathbf{a}+2\mathbf{b}}{3}$

c Quintisection $\frac{4\mathbf{a}+\mathbf{b}}{5}$ $\frac{3\mathbf{a}+2\mathbf{b}}{5}$ $\frac{2\mathbf{a}+3\mathbf{b}}{5}$ $\frac{\mathbf{a}+4\mathbf{b}}{5}$

Exercise 12.2

1 a $\sqrt{6}$ b $\sqrt{20}$

2 $2\sqrt{3}, \sqrt{3}$

3 a $\frac{1}{3}$ b $\frac{1}{5}$ c $\frac{1}{4}$ d $\frac{2}{3}$ (i.e $n=1.5$)

4 $\mathbf{n}-\mathbf{m}, 2\mathbf{n}-2\mathbf{m}$

AB is twice the length of *MN* and is parallel to *MN*

5 a $\frac{2\mathbf{a}+\mathbf{b}}{3}$ b $\frac{3\mathbf{a}+\mathbf{b}}{4}$ c $\frac{3\mathbf{a}+2\mathbf{b}}{5}$

d $\frac{5\mathbf{a}+3\mathbf{b}}{8}$ e $\frac{n\mathbf{a}+m\mathbf{b}}{m+n}$

6 $\frac{\mathbf{q}-\mathbf{p}}{1+k}, \frac{MN}{PQ}=\frac{1}{1+k}, MN \parallel PQ$

7 $\frac{3}{4}(\mathbf{a}+\mathbf{b}), \frac{RS}{OC}=\frac{3}{4}, RS \parallel OC$

8 1:3

9 $k=2$

10 $\overrightarrow{PQ}=\overrightarrow{SR}=\frac{1}{2}\mathbf{b}$

11 a $\sqrt{2}$ b $\sqrt{\frac{14}{3}}$

Exercise 12.3

1 a 24 b 2 c -2

2 a 3 b 6 c 9

d 9 e -3 f -3

3 a 62.8° b 73.6° c 39.7° d 69.5°

e 82.6° f 71.6° g 85.5° h 90°

4 a -1 b $\frac{1}{2}$

5 a -3 b ± 4

6 a parallel b neither c perpendicular

7 a -3 b $\frac{4}{3}$ c $10, -\frac{2}{5}$

8 a i 80.8° ii 76.1° iii 16.7°

b 54.7° with each axis

9 29.5°

10 47.6°

12 a 70.7°, 23.3°, 86.0° b $\sqrt{38}, \sqrt{6}, \sqrt{34}$

c 7.12 square units

13 b $\mathbf{i}+4\mathbf{j}+7\mathbf{k}$

14 a $2\mathbf{p}\cdot\mathbf{r}$ b $2\mathbf{p}\cdot\mathbf{r}$

c $|\mathbf{p}|^2-|\mathbf{r}|^2$ d $(\mathbf{p}-\mathbf{r})^2$

17 $\mathbf{i}-2\mathbf{j}+\mathbf{k}$

18 180°

19 $\frac{11}{15}$

Exercise 12.4

1 a $\mathbf{r}=(1+2\lambda)\mathbf{i}+(2-3\lambda)\mathbf{j}+(-3+\lambda)\mathbf{k}$

b $\mathbf{r}=(4-2\lambda)\mathbf{i}+(-1+\lambda)\mathbf{j}+(3+2\lambda)\mathbf{k}$

c $\mathbf{r}=(3+4\lambda)\mathbf{i}+(1+2\lambda)\mathbf{k}$

d $\mathbf{r}=(1+\lambda)\mathbf{i}+(-1+\lambda)\mathbf{j}-\lambda\mathbf{k}$

e $\mathbf{r}=(2+\lambda)\mathbf{i}+\mathbf{j}+3\lambda\mathbf{k}$

f $\mathbf{r}=2\lambda\mathbf{i}+\mathbf{k}$

2 a $\mathbf{r}=(2+\lambda)\mathbf{i}+(3-3\lambda)\mathbf{j}+(1+3\lambda)\mathbf{k}$

b $\mathbf{r}=(2+3\lambda)\mathbf{i}+(\lambda-1)\mathbf{j}+\mathbf{k}$

c $\mathbf{r}=(1+3\lambda)\mathbf{i}+5\lambda\mathbf{j}-2\lambda\mathbf{k}$

d $\mathbf{r}=\lambda\mathbf{i}+2\lambda\mathbf{j}+3\lambda\mathbf{k}$

e $\mathbf{r}=(1+\lambda)\mathbf{i}+(-2+\lambda)\mathbf{j}+(-3+\lambda)\mathbf{k}$

f $\mathbf{r}=(2+3\lambda)\mathbf{i}+(-3+3\lambda)\mathbf{j}-\mathbf{k}$

3 a $\mathbf{r}=(2+\lambda)\mathbf{i}+\mathbf{j}-2\mathbf{k}$

b $\mathbf{r}=2\mathbf{i}+\mathbf{j}+(-2+\lambda)\mathbf{k}$

4 a 2 b -5 c 0

5 a 75.3°, intersect at (4,7,4)

b 24.5°, intersect at (4,8,–6)

c 0°, parallel d 75.0°, skew e 30.7°, skew

6 a Possible equations are

$\mathbf{r}=(-1+3t)\mathbf{i}+(1+2t)\mathbf{j}+(3+t)\mathbf{k}$ for $\overrightarrow{PQ}$

and $\mathbf{r}=s\mathbf{i}+(5-s)\mathbf{j}+(2+s)\mathbf{k}$ for $\overrightarrow{RS}$

b (2,3,4) c 123.0°

7 a $\mathbf{r}=(1+t)\mathbf{i}+(2-3t)\mathbf{j}+(3-4t)\mathbf{k}$

b 2 c 3

8 $\lambda=-\frac{8}{7}, \mu=\frac{3}{7}, \frac{2}{7}\sqrt{42}$

Review 12

1 a 13 b $\frac{1}{13}(3\mathbf{i}+12\mathbf{j}+4\mathbf{k})$

c $\frac{5}{13}(3\mathbf{i}+12\mathbf{j}+4\mathbf{k})$ d 76.7°

2 $\overrightarrow{AB}=2\mathbf{i}+2\mathbf{j}+3\mathbf{k}$

3 a $\frac{1}{2}\sqrt{38}\sqrt{21}$ b $K(4,8,-3)$

4 b 72.5°

5 a $\overrightarrow{AB} = 3\mathbf{i} + 4\mathbf{j} - \mathbf{k}$ $\overrightarrow{BC} = -2\mathbf{i} + \mathbf{j} - 4\mathbf{k}$; 94.9°
b $k = -14$
6 $\frac{1}{3}(-\mathbf{i} + 2\mathbf{j} + 2\mathbf{k})$
7 a $-\frac{4}{9}$ b $\frac{\sqrt{65}}{2}$ d 2 : 1
8 a $\mathbf{i} - 2\mathbf{j} + 4\mathbf{k} + \lambda(2\mathbf{i} + 3\mathbf{j} - \mathbf{k})$
b $\mathbf{i} - 2\mathbf{j} + 4\mathbf{k} + \mu(-\mathbf{i} + 3\mathbf{j} - \mathbf{k})$
c 49.9°, 2.54
9 b 8.3° c 1.95
10 b (5, 0, 1) c 80.4°
11 a 78.9° b skew
12 a $t = 3, s = -1$ c $\sqrt{11}$ d 33.6°

Revision 4

1 a $\frac{3}{x} - \frac{3}{x+2}$ b $-\frac{1}{2(x+1)} + \frac{1}{(x+1)^2} + \frac{1}{2(x-3)}$
c $x + \frac{2}{x-2} + \frac{2}{x+2}$
2 a $A = 1, B = 0, C = 2$
b $\frac{11}{9} + \frac{50}{27}x + \frac{110}{27}x^2 + \frac{1936}{243}x^3$
3 a $A = -1, B = \frac{1}{2}, C = \frac{5}{2}$ b $p = -2, q = 3, r = 3$
4 a $\frac{\cos t}{2\tan t \sec^2 t}$ b $y = \frac{1}{4\sqrt{2}}(x+3)$
c $y^2 = \frac{x}{1+x}$
5 a $-\frac{2\cos 2t}{\sin t}$ b $t = \frac{\pi}{4}, \frac{3\pi}{4}, \frac{5\pi}{4}, \frac{7\pi}{4}$
c $\left(\frac{1}{\sqrt{2}}, \pm 1\right), \left(\frac{-1}{\sqrt{2}}, \pm 1\right)$
e $y = -2x\sqrt{1-x^2}$
6 a $45\pi^2$ b $240(\pi - 1)$ c 13.6%
7 a 11.75 b $57\frac{3}{8}$
8 a $6y = 3a - \sqrt{3}x$ b $\frac{a^2\sqrt{3}}{16}$
9 a $1 + 3x + 9x^2 + 27x^3$, $|x| < \frac{1}{3}$
b $1 + \frac{1}{6}x - \frac{1}{36}x^2 + \frac{5}{648}x^3$, $|x| < 2$
c $1 - \frac{1}{2}x - \frac{5}{8}x^2 + \frac{3}{16}x^3$, $|x| < 1$
d $1 + \frac{5}{8}x + \frac{11}{128}x^2 + \frac{17}{1024}x^3$, $|x| < 4$
e $3 + \frac{3}{2}x + \frac{9}{4}x^2 + \frac{15}{8}x^3$, $|x| < 1$
10 a $1 - \frac{1}{3}nx + \frac{1}{18}n(n-1)x^2 - \frac{1}{162}n(x-1)(n-2)x^3$, $|x| < 3$
b i −34 ii $\frac{22\,015}{27}$
11 b i $1 - \frac{3}{2}x - \frac{9}{8}x^2 - \frac{27}{16}x^3$ ii 0.866 210 937
c 0.018%
12 a $\frac{y}{2y - x}$ b $\frac{x^2 - 2xy}{x^2 - y^2}$
c $-\frac{y^2(2y + x)}{x^2(y + 2x)}$ d $\frac{y(y - x\ln y)}{x(x - y\ln x)}$

13 a $\frac{\cos x}{\sin y}$ b $\left(\frac{\pi}{2}, -\frac{2\pi}{3}\right)$ and $\left(\frac{\pi}{2}, \frac{2\pi}{3}\right)$
14 b $64\ln 2$
15 a $4\pi r^2$ b $\frac{dr}{dt} = \frac{250}{\pi r^2(2t+1)^2}$
c $V = \frac{1000t}{2t+1}$ d i 4.77 cm
16 a 1, 1.202 69, 1.414 21 b 0.8859 d 0.51%
17 a $\frac{2}{3}(e^x - 2)^{\frac{3}{2}}$ b $\frac{1}{72}(x+2)^8(8x - 11)$
c $x + \frac{1}{2}\sin 2x$ d $\frac{1}{5}\sin^5 x$
e $\ln|x - 1| + 8\ln|x + 2| + \frac{12}{x+2}$
f $e^x(x + 2)$ g $\frac{2}{5}e^x \sin 2x + \frac{1}{5}e^x \cos 2x$
h $\frac{1}{3}\ln|x^3 + 3|$ i $\frac{1}{2}\left(x - \frac{1}{2}\sin 2x\right)$
18 $\frac{1}{6\ln 2}$
19 a $\lambda = 7, \mu = 3$ b $-\frac{1}{14}\cos 7x + \frac{1}{6}\cos 3x$
c $\frac{5\sqrt{2}}{21}$
20 c $y = -x + 2 - \ln 2$
21 b $p = 45, q = 7$
22 c $\frac{1}{e^{\frac{\pi}{6}} - 1}$
23 b 20π d $\alpha = 0.345$
24 a 6.75 b overestimate c $\frac{376}{15}\pi$
25 $\pi\left(\frac{b - a}{(2b + 1)(2a + 1)}\right)$
26 a i $t = 1, 0.5$ ii $V = \frac{21}{4}\pi$
27 a i $y = \ln\left|\frac{1}{\sqrt{k - 2x}}\right|$, k is a constant ii $y = k(x + 1)$
iii $y = ke^{\frac{1}{6}x^2(2x-3)}$ iv $y = \sin^{-1}(ke^{\tan x})$
b i $y = \left(\frac{x^2}{4} + 1\right)^2$ ii $y = 10\sqrt{\frac{x^2 + 1}{2}}$
28 a $\frac{dM}{dt} = -kM$ c 7.84 grams
29 c 17.1 min d 0.25 m
30 a $\overrightarrow{PR} = -4\mathbf{j} + 4\mathbf{k}$ $\overrightarrow{PQ} = 2\mathbf{i} - 4\mathbf{j} + 5\mathbf{k}$, 96.1°
b $PR = 4\sqrt{2}, PQ = 3\sqrt{5}$ c 4.72
31 a $\overrightarrow{AB} = 3\mathbf{i} + 6\mathbf{j} + 6\mathbf{k}$ b $\frac{2}{3}$
d 2 e (2, 4, −5)
32 a $9\mathbf{i} - 2\mathbf{j} + \mathbf{k} + \lambda(-3\mathbf{i} + 4\mathbf{j} + 5\mathbf{k})$ b $p = 6, q = 11$
c 39.8° d $\frac{36}{5}\mathbf{i} + \frac{2}{5}\mathbf{j} + 4\mathbf{k}$
33 a $\overrightarrow{AB} = \mathbf{i} - 2\mathbf{j} + 2\mathbf{k}$ b $2\mathbf{i} + 6\mathbf{j} - \mathbf{k} + \lambda(\mathbf{i} - 2\mathbf{j} + 2\mathbf{k})$
c 45° d $5\mathbf{i} + 5\mathbf{k}$
34 a (5, 8, −4) d (7, 12, −5)

Index